Indeterminate Structural Analysis

Indeterminate Structural Analysis

DIPTI RANJAN SAHOO

Associate Professor
Department of Civil Engineering
Indian Institute of Technology Delhi

PHI Learning Private Limited

Delhi-110092
2025

Published by Pushpita Ghosh, PHI Learning Private Limited, Rimjhim House, 111, Patparganj Industrial Estate, Delhi-110092 and Printed by Syndicate Binders, A-20, Hosiery Complex, Noida, Phase-II Extension, Noida-201305 (N.C.R. Delhi).

₹895.00

INDETERMINATE STRUCTURAL ANALYSIS
Dipti Ranjan Sahoo

ISBN-978-93-5443-295-8 (Print Book)
ISBN-978-93-5443-294-1 (e-Book)

Contents

CHAPTER 3:   STRUCTURAL STATICS 35–73

CHAPTER 4:   DEFORMATIONS 74–129

CHAPTER 5:   METHOD OF CONSISTENT DEFORMATION 130–157

Preface

This book is intended to serve as a textbook for undergraduate students of Civil Engineering. It provides a clear and thorough understanding of fundamental concepts and structural analysis application as it applies to beams, trusses, and frames. The main goal is to develop one's ability to develop the analysis models of the realistic structures and analyze them for all applicable loadings as encountered in the practice.

In practical applications, the analyses of structures are mostly carried out through computers using matrix methods. Though this is much faster and more efficient in analyzing the complex structures as compared to the analysis by the classical methods, the accuracy and reliability of the results so obtained are largely dependent on the knowledge of modelling of realistic structures. Besides the advantages of the matrix method, it is very important and necessary for students to be well versed in the classical methods as well. These classical methods also provide a deeper understanding of basic knowledge of structural statics and mechanics of material. In addition, classical methods also serve as a means of checking the accuracy of computer results.

The book is broadly divided into four parts. Chapters 1 to 4 provide the structural classifications, the fundamental concepts of analysis of determinate structures and a review of structural statics and mechanics. These chapters provide a foundation and background for the methods of analysis of indeterminate structures. The second part (Chapters 5 to 7) presents the classical methods of analysis of indeterminate structures. These methods are suitable for hand calculation in which the deformations and internal forces of structural elements are computed using the basic structural analysis concepts. These methods form the basis of the matrix methods which are covered in the third part of the book in Chapters 8 to 10. The matrix methods of analysis based on the force and displacement unknowns along with the direct stiffness matrix method are covered in these chapters. The fourth part (Chapters 11 to 12) covers the influence lines for moving loads and the approximate methods of analysis.

Every topic is illustrated with numerous worked examples to help the students follow a step by step procedure of analyzing a structure. At the end of the chapters, a wide range of exercise problems is provided to help the students practice various analysis techniques. These problems are expected to develop critical thinking ability of students. The answers to selected problems are provided at the end of this book to help the students check their solutions to the exercise problems. It is believed that the book will prove very useful to both students and instructors of the Civil Engineering discipline.

DIPTI RANJAN SAHOO

Acknowledgements

I am thankful to my colleagues and students who provided valuable assistance in the preparation of the textbook. In particular, the help received by my Doctoral and Master's students, in particular, Abhishek Jain, Deepak Yadav, Pratik Patra, Ram Avtar Liler, Rahul Bhartiya, A. Gautham, and Saurabh Sharma, in the preparation of solution manual is much appreciated. I would like to thank my Master's students who helped in preparing the solutions to the exercise problems.

I would like to acknowledge the tremendous help and support received from my wife Awani and my daughter Tejasvi in writing this book. My wife has been the source of inspiration for writing this book. Her assistance in the preparation and formatting of figures is also much accredited. Without her continuous pursuance, motivation, and encouragement, the timely completion of this book would not have been possible.

I am thankful to the Department of Civil Engineering, IIT Delhi for providing an excellent platform for the timely completion of this book.

I also extend my thanks and appreciations to the reviewers and editors for their critical review and comments on the book. I would really appreciate if the readers could send me their comments and suggestions regarding the contents of this edition of the book.

Dipti Ranjan Sahoo
drsahoo@iitd.ac.in

CHAPTER 1
Introduction

1.1 GENERAL

A structure is defined as a system of connected elements designed to serve a specified function. Engineering structure means that is built or constructed for its intended use. Some of the examples of important civil engineering structures are buildings, bridges, towers, dams, walls, shells, and cable structures. These structures are composed of one or more elements arranged in such a way that the whole structures and their components can resist the applied loads without appreciable change in their geometries.

The design of a structure should satisfy four primary objectives, namely, (a) It should be aesthetically good; (b) It must meet the intended performance requirement; (c) It must support the applied loads safely; (d) It should be economical in terms of material consumption as well as construction. In addition to these parameters, i.e., aesthetic, serviceability, safety, and economy, the environmental constraints related to the construction of a structure must be taken into consideration.

The design process requires a fundamental knowledge of material properties and the basic laws of mechanics. To begin with, a preliminary design of a structure is proposed based on the experience. The proposed structure must be analyzed under the applicable loading conditions to ascertain its strength and deformation, which represent the safety and serviceability objectives, respectively. Analysis results provide the detailed information on the acceptability of the proposed design. If the strength (or capacity) of a structure or its components are either too high or too low in comparison to the corresponding demand obtained from the analysis for the applicable loads, the preliminary design should be revised to achieve the economy in the design of the structure. The revised design again should be analyzed to compute the demand on the structure as well as its components. Thus, the economical design of a structure is achieved by following a series of successive approximations. In each iteration, both analysis and design are carried out to compare the demand *vs.* capacity of a structure. In this book, various analysis methods applicable to the civil engineering structures are described. However, some of these methods can also be applied to the structures related to other fields of engineering.

1.2   TYPES OF STRUCTURAL ELEMENTS

A civil engineering structure is composed of one or more elements. It is, therefore, necessary to recognize the types of elements, which helps in the classification of structures depending on their form and function.

Some of the common types of elements used to form various types of civil engineering structures are presented in the following sections:

1. Ties and struts: Structural elements designed to carry tensile forces predominantly are referred to as *tie members*. These members are usually very slender, i.e., the length of these members is significantly higher than the size of the cross-section. A bar carrying only axial compression force is called a *strut*. The axial members are often referred to as *bracing members*. Rods, wires, bars, angles, channels, etc. as shown in Figure 1.1 are often used as tie members.

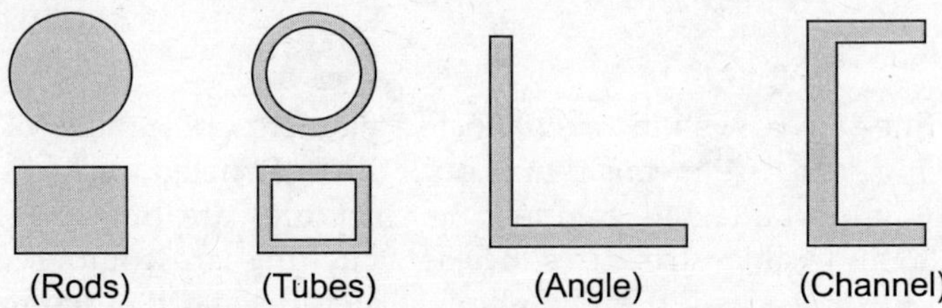

FIGURE 1.1   Examples of tie members.

2. Beams: Structural members carrying primarily lateral loads as called beams. These members are usually horizontal in buildings or bridges. Thus, loads applied to these members are vertical. These elements are designed to resist bending moments and shear forces. In steel structures, individual standard rolled I-shaped sections (i.e., mono sections) or a combination of two or more standard rolled sections (i.e., built-up sections) are used as beams. In concrete structures, beams of rectangular sections are commonly used. However, other shapes are also adopted as beams depending on the architectural requirements. These beams may be supported by the other structural elements (mostly, columns) or may rest directly on external supports, such as, *fixed* (*encastere*), *hinged*, and *pined*. Figure 1.2 shows the different types of beams defined based on the support conditions.

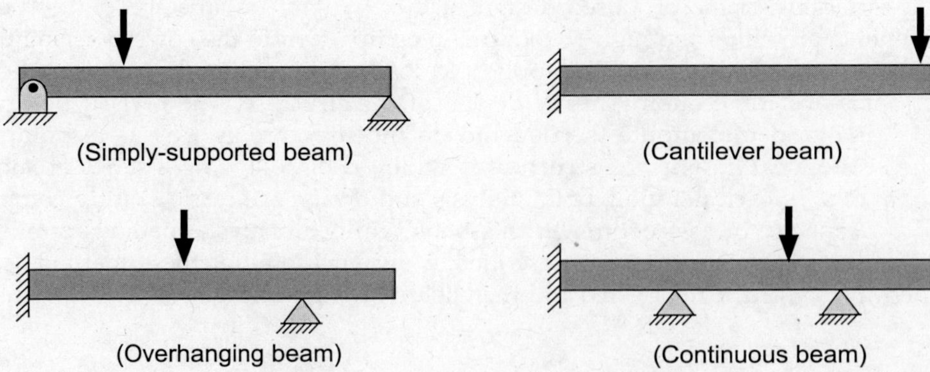

FIGURE 1.2   Different types of beams.

3. Columns: Structural elements designed primarily to resist axial compressive forces are called *columns*. These compression members may also carry shear forces, bending moments, and torsion. Similar to beams, standard structural steel sections or built-up steel sections are used as columns in the steel structures. Hollow square or circular sections are also popular as the columns in bridges and buildings. Rectangular, square, and circular sections of columns are preferred in the concrete structures. Composite columns composed of concrete and structural steel with or without reinforcement bars are also adopted in buildings and bridges.

4. Beam-columns: Structural members subjected to axial load and bending moment are referred to as beam-column. Figure 1.3 shows the comparison of columns and beam-columns.

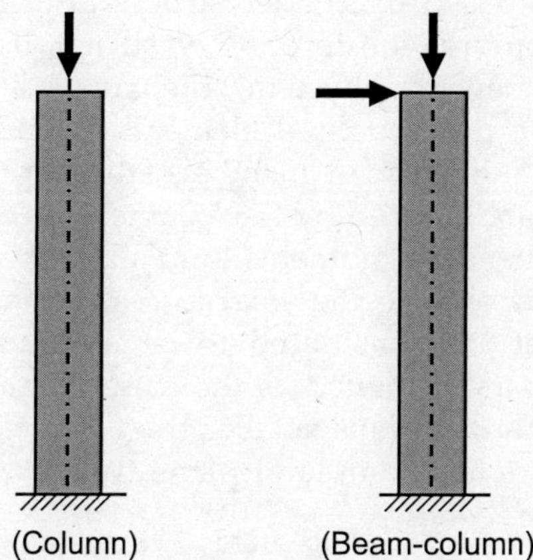

FIGURE 1.3 Example of column and beam-column.

5. Cables: Cables are the flexible structural elements/systems carrying the tensile forces. The difference between the ties and cables is that the axial forces are applied along the axes of tie members, whereas the loads are applied transversely at any point along the length of cables. Thus, cables have some defined sags along their lengths.

6. Arches: Arches are the structural elements/systems that are rigid and carry compressive forces. The curvature of these systems is exactly opposite to that of cables. Arches are widely used in bridges, roofs, and openings in buildings. Both cables are arches are used in the long-span structures.

7. Plates: Plates are the flat-shaped planner structural elements having a very small thickness as compared to their other dimensions. Plates can be sub-divided into *slabs* and *walls*. Slabs are usually in a horizontal plane and resist transverse forces by a combination of axial forces, bending moment, shear forces, and torsion. Walls consist of planner elements, usually in a vertical plane, and withstand both in-plane and transverse forces. Slabs are commonly used as the floor systems of buildings as well as the decks of bridges, whereas walls form the part of most buildings and are often used to retain water and soil.

8. Shells: Shells are the curve-shaped planner structural elements having a very small thickness as compared to their other dimensions. Like plates, shells are also capable of resisting axial forces, bending moment, torsion, and shear.

9. **Membranes:** Membranes are the planer elements that resist the applied loads by developing bi-directional tension forces. Plates, shells, and membranes can be of any mathematical shapes, such as cylindrical, parabolic, hyperbolic, etc. these structures support loads primarily in tension or compression similar to cables or arches.

1.3 TYPES OF STRUCTURAL SYSTEMS

A structural system is a combination of structural elements. These systems are classified according to the type of elements and their connections. Various types of structural system commonly used in the practice are discussed below:

1. Trusses: Trusses are composed of slender structural elements design to carry either tension forces (i.e., ties) or compression forces (i.e., struts). If all these structural members lie in the same plane (two-dimensional system), the trusses are called *plane trusses*. On the other hand, *space trusses* have the structural elements extending in all three dimensions. Because of the geometric arrangement of members, the applied lateral loads causing the bending effect on the trusses are induced by the axial tension and compression forces in the members. The magnitude of bending moments (and shear forces) in the members between the joints is negligible as compared to the axial forces in them. Thus, forces are assumed to act only at the joints that are considered to be non-moment-resisting (i.e., pinned) connections and all members are designed as the truss members (i.e., ties or struts). As a result, the material required for a truss is smaller than that required by a beam to support a given load. Figure 1.4 shows an example of plane truss.

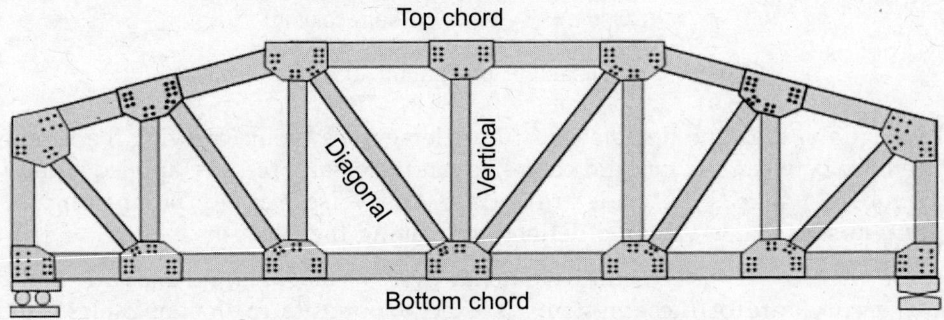

FIGURE 1.4 Example of a plane truss.

2. Frames: Frames are composed of a series of beams and columns connected to each other by means of rigid, pinned, or semi-rigid connections. In the case of rigid connections, no relative rotation can occur between the members connected at the joint. On the other hand, these members are allowed to rotate freely at the joint in the pinned connections. In the case of the semi-rigid connections, limited rotation of the members is allowed at the joint. Similar to trusses, frames can be of *plane frames* or *space frames* depending on whether the members are extended in two-dimensions or three-dimensions. Depending on the nature of loading applied on a frame, these beams and columns are considered as the generalized *beam-columns* if they carry the axial forces as well as bending moments. Figure 1.5 shows an example of plane frame.

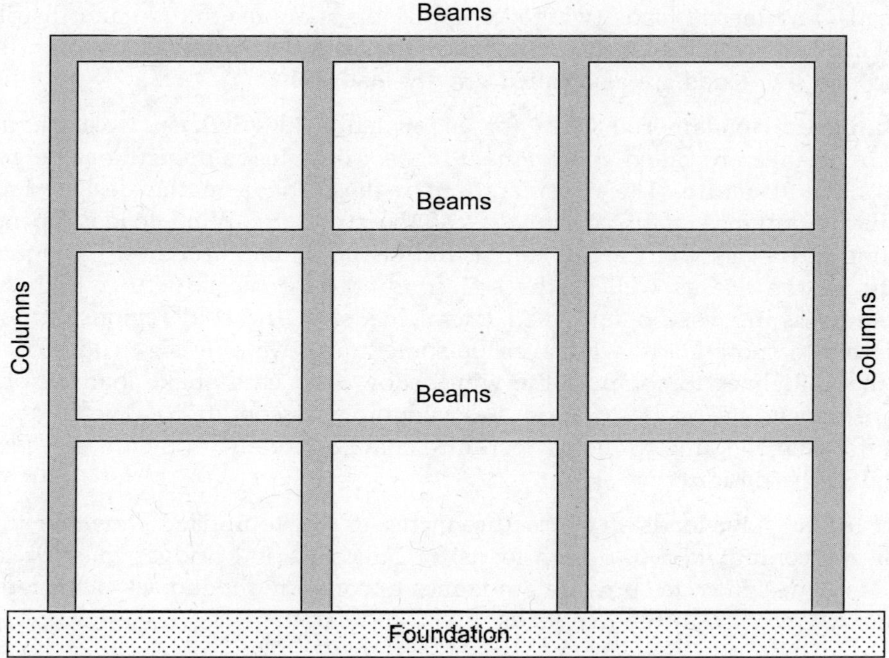

FIGURE 1.5   Example of a plane frame.

3. Cable structures: In these structures, flexible cables are used as at least one of the primary load-resisting elements. These cables can resist only axial tension. Cable structures are extensively used in many long-span bridges, roofs of stadiums, and airports.

4. Surface structures: These structures are built using planner elements, such as plates and shells. These elements can be arranged to produce a wide range of structures, such as flat slabs, tanks and silos, cooling towers, tunnels, bridge decks, and domes.

1.4   TYPES OF LOADS

The loads to be considered in the design of a structure depend on the types of structure, its location, and its intended use. Some of the loads are imposed on the structure, whereas loads may develop owing to the restraint of deformation due to several actions, such as temperature change, support settlement, and shrinkage of members. Various types of loads used in the design are discussed in the following sections.

1. Dead loads: These loads are imposed by the structural and non-structural components of a structure. These loads are also referred to as the permanent actions and include the self-weight of structural elements and any permanent constructions. Dead loads are fixed in position and can be estimated to a reasonable accuracy from the density of the relevant materials of the elements.

2. Live loads: These loads, also called imposed actions, are attributed to the intended use of the structure. Live loads are specified by the design codes of a country and depend on the expected occupancy of the structures. These loads include allowances for possible overload, impact loads, and inertia loads (wherever applicable). The magnitude and distribution of the actual live loads are not constant and hence, vary during the service life of the structure. Both dead and live loads are also called gravity loads.

3. Environmental loads: The loads (or actions) due to wind, snow, earthquake, and temperature change are called environment loads. These loads depend on the geographic location of the structure. The consideration of these loads in the design depends on the relative importance of the design life of the structure. Wind loads depend on the surrounding terrain as well as the height and shape of the structure. Earthquake loads dependent on the site as well as the soil conditions of the structure. Both wind and earthquake loads are very often called lateral loads. In the cold regions, structures are also subjected to snow loads, which can be sometimes severe. Design codes of a country provide the guidelines to compute the wind, snow, and earthquake loads applicable for a structure depending on its location and geographical conditions. The forces imposed on structural elements due to the temperature changes are also sometimes required to be considered in the design.

4. Impact loads: Live loads on a structure increases due to impact. Thus, the influence of impact on the computation of design loads for buildings and bridges must be taken into account. Many design codes provide guidelines to compute the impact factor representing the percentage increase in the live loads.

5. Hydrostatic and earth pressure: Structures used to retain the soil and water are subjected to the pressure developed by these materials. These loads are required to be considered in the design. Some of the examples of these structures are tanks, dams, retaining walls, and tunnels.

1.5 STRUCTURAL IDEALIZATION

It is practically impossible to carry out an exact analysis of a structure. The real structures have complex geometric properties and variations in material properties, which are never known exactly at the time of analysis. Structural elements designed to be perfectly straight are never exactly straight after construction. The cross-sectional dimensions also vary along the length of the members. Actual material properties are different from those considered in the analysis and also vary from point to point. The magnitude and distribution of the loads imposed on the structures also vary in the real structures and are not known precisely during the analysis and design. Further, the support conditions considered in the analysis are different from the actual ones. For example, it is not possible to construct real fixed supports or connections in the actual structures. Therefore, structural analysis is undertaken on the *idealized structures* in which assumptions are made concerning the geometry, support conditions, material properties, and loading conditions. Hence, the aim of the idealization to simplify the analysis process so that the calculated loads, internal actions, reactions, stresses, and deformations are not too different from those in a real structure and adequately represent the behaviour of the structure.

Accordingly, the applied loads on a structure are often idealized as concentrated loads, distributed loads, line loads, or surface loads. Similarly, the structural elements are also idealized as *one-dimensional* or *two-dimensional* elements. Truss elements (i.e., ties and struts) carrying the only axial forces are idealized as the one-dimensional line elements.

Beams and columns carrying bending moments, shear forces, and axial forces with or without torsion are also idealized one-dimensional line elements. Similarly, plates and shells including membrane members are idealized the *two-dimensional (planer)* elements.

1.6 THEORY OF STRUCTURES

Structural theories may be classified into different categories depending on the various aspects. These are discussed below:

(a) Statics *vs.* **dynamics:**   *Static analysis* is carried out on structures subjected to dead loads, live loads, snow loads, etc., without explicitly considering the effects of dynamic loadings. The influence of the dynamic nature of loadings is treated as a fraction of the moving loads to simplify the design and analysis. *Structural dynamics* deals with the dynamic effects of loadings due to earthquakes, wind gusts, blasts, and moving loads.

(b) Plane *vs.* **space:**   No structure in practice is planer (*two-dimensional*). However, for simplicity, analysis of various structures (i.e., beams, rigid frame buildings, and truss bridges) is conducted assuming them as *plane problems*. If the members of a structure are interrelated in such a way that analysis cannot be simplified into planner structures, such structures should be dealt with in the *space framework* under non-coplanar force systems.

(c) Linear *vs.* **nonlinear:**   Linear structure means that the relation between the applied loads and resulting displacements is assumed to be linear. In such cases, the material of the structure is linearly elastic and the changes in the geometry of the structure are so small that their influences are neglected in the stress calculation. Therefore, the principle of superposition is applicable in the linear analysis. A nonlinear relationship between the applied loads and the resulting displacements may exist in either of the two possible scenarios. There are (i) the material of the structure is not elastic (*material nonlinearity*), and (ii) the material is elastic, but the changes in the geometry are significant (*geometrical nonlinearity*). In these cases, *plastic analyses* and *buckling analyses* of structures are conducted.

(d) Determinate *vs.* **indeterminate:**   If the analysis of a structural system can be carried out using the principles of statics only, the systems are called *determinate structures*. However, in the *indeterminate structures*, the effects of geometric properties as well as material properties are necessary for the analysis in addition to the principle of statics.

(e) Force *vs.* **displacement:**   The objective of structural analysis to compute the forces and displacements of a structure and to analyze their relationships pertaining to the geometric and material properties of the structural elements. In a broader sense, structural analysis can be divided into two categories: the *force method* and the *displacement method*. In the *force method*, the forces are considered as the basic unknowns and the displacements are expressed in terms of forces, and vice-versa. In the matrix method of analysis, the *force method* is referred to as the *flexibility method*, and the *displacement method* is called the *stiffness method*.

1.7 SCOPE OF THIS BOOK

This book is confined exclusively to the planer structures, namely, beams, trusses, and frames. The analysis of three-dimensional structures has not been covered in this book.

Stability and Determinacy

2.1 GENERAL

This chapter presents an overview of the fundamental concepts based on which the principles of statics have been derived. The types of supports used to restrain the movement of the structures are described. Internal forces in various types of structural elements are presented. The use of free-body diagrams to identify all actions acting on a structure have been discussed. Based on the equilibrium equations, the procedures to compute the support reactions and internal forces in the elements of a structure are presented. This chapter also deals with the identification of structural systems based on the number of unknown support reactions and internal forces in excess of the available equilibrium equations. Accordingly, the methods to classify a structure as *statically determinate* or *statically indeterminate* are presented.

2.2 EQUATIONS OF STATIC EQUILIBRIUM

Three fundamental laws (i.e., laws of motion) of Sir Isaac Newton (1642–1727) form the basis of engineering mechanics and the principles of statics or dynamics. Newton's first law states that a body initially at rest remains at rest if the resultant force acting on the body is zero. This condition is known as *static equilibrium*. Newton's second law pertains to the unbalanced forces on a body. This means that if all the sum of all active and reactive forces is not zero, then the body is subjected to a resultant force (F) given by the product of mass (m) of the body and the acceleration (a) in the direction of the force, $F = ma$. This forms the basis of the principle of dynamics. The third law represents that the forces of actions and reactions between the interacting bodies are equal in magnitude but opposite in direction. This means that the sum of active forces and reactions along a particular direction is zero. The first and third laws of motion are used to establish the *equations of static equilibrium*.

A structure is in *equilibrium* only if the force acting on it has *zero resultant force* and *zero resultant moment*. If all active and reactive forces on a structure are expressed by their components in a rectangular Cartesian coordinate system (x, y and z) as shown in Figure 2.1, the sum of all components of the forces acting in any direction must be zero.

Accordingly, the *equations of static equilibrium* (or simply, *equation of equilibrium*) are derived for a structural system.

For the space system shown in Figure 2.1(a), there are six *equations of equilibrium* as follows:

Three force equilibrium equations:

$$\sum F_x = 0; \quad \sum F_y = 0; \quad \sum F_z = 0 \tag{2.1}$$

Three moment equilibrium equations:

$$\sum M_x = 0; \quad \sum M_y = 0; \quad \sum M_z = 0 \tag{2.2}$$

In the case of the planer system shown in Figure 2.1(b), two forces and one moment only act on a structure. Accordingly, there are three *static equilibrium equations* for such a system as follows:

Two force-equilibrium and one moment-equilibrium equations are:

$$\sum F_x = 0; \quad \sum F_y = 0; \quad \sum M_z = 0 \tag{2.3}$$

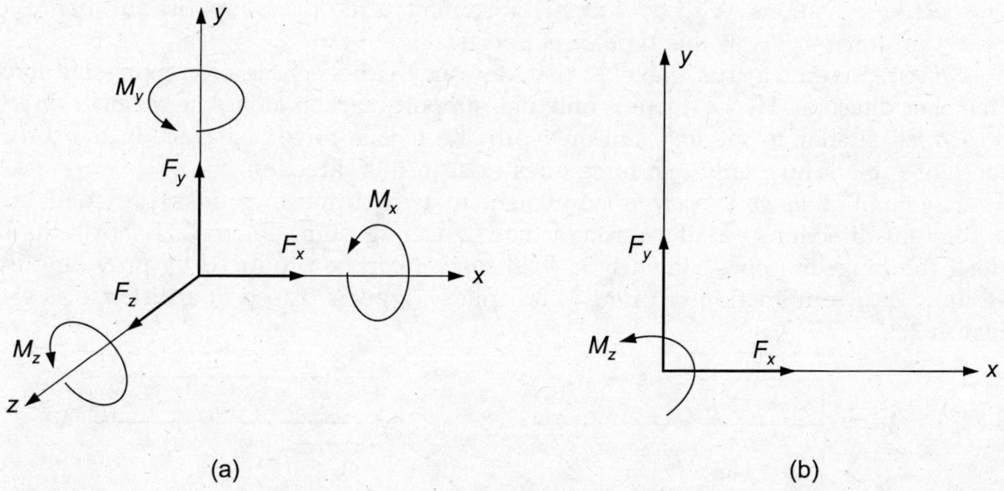

FIGURE 2.1 Coordinate systems: (a) Three-dimensional (space), (b) Two-dimensional (planer).

2.3 TYPES OF SUPPORTS

The external loads applied to a structure are transmitted to other parts of the structure or the foundations by means of supports. These supports prevent the movement of the structure in different directions when subjected to various forces. As a result, restraining forces at developed at the supports, which are called *reaction forces*, or *support reactions*, or *simply reactions*.

Support reactions depend on the type of support and hence, the type of movement (*translation* or *rotation*) restrained by the support. The *translation* movement is produced when the body is subjected to a *force* action, whereas a *moment* action induces the *rotational* movement of the body.

2.3.1   Two-dimensional (Coplanar) Coordinate System

Table 2.1 summarizes various types of supports for two-dimensional (co-planer) structures along with their respective reaction components. The two-dimensional representation is very useful for practical applications as many structural systems, such as beams, columns, and frames are often idealized and analyzed as coplanar (two-dimensional) structures. These structural elements are modelled as the "line" elements, which usually represent the centrelines of the elements.

A *fixed* (or *built-in*) support has three reactions, i.e., two reaction forces in the horizontal and vertical directions, and one reactive moment. These supports do not permit either translation or rotation in any direction.

A *pinned* (or *hinged*) support has two support reactions, i.e., two reaction forces in the horizontal and vertical directions. Thus, the translations along these reactions are zero. However, this support is free to rotate as there is no moment reaction, i.e., zero moments.

A *roller* support provides only one reaction force perpendicular to the member and prevents the translation along the force direction. The horizontal translation and rotation are unrestrained (i.e., non-zero) as there are no reaction forces and moments in the respective directions. A *rocker* support is similar to the *roller* support and hence, only one reaction force exists at this type of supports.

A *link* represents a truss member that can carry either tensile or compressive force in the member direction. Hence, there is only one support reaction along the member direction.

A *cable*, similar to the *link*, can only provide tensile force resistance in the direction of the taut cable. Thus, only one force does exist in this direction.

In general, a *hinge support* is equivalent to two supporting links provided in any two different directions passing through the connecting point. Figure 2.2 shows the links forming the hinge support. Similarly, a *fixed support* can be produced by providing either three links or a combination of pinned and roller supports placed at a distance as shown in Figure 2.3.

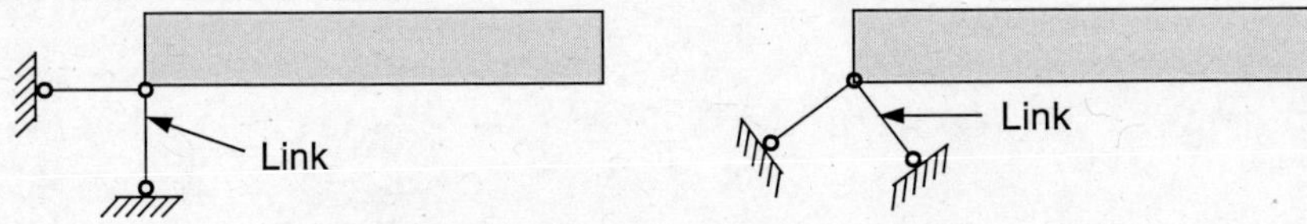

FIGURE 2.2   Equivalent hinge supports.

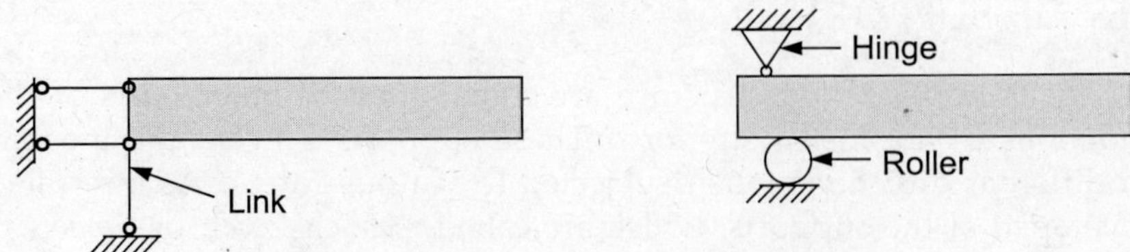

FIGURE 2.3   Equivalent fixed supports.

The pinned and the roller supports are often known as *simple supports*. A single-span beam with a pinned support at one end and the roller support at the other is known as the *simply supported* beam.

A beam with the fixed support at one end and free at the other is called the *cantilever* beam.

A beam with the fixed support at one end and the pinned (or roller) support at the other is known as the *propped cantilever* beam.

A beam with the fixed supports at both ends is called the *fixed-fixed* beam (or simply, *fixed* beam).

2.3.2   Three-dimensional (Space) Coordinate System

A *hinged (pinned) support* in the space system has three *force reactions* along three orthogonal directions. The translation movements of this support are zero along these directions. Since the *moment reactions* are zero in these directions, the rotational movements are allowed about each coordinate axis. Thus, the number of *unknown reactions* is three and the number of *unknown displacements* is also three (all rotations).

Similarly, in the case of *roller support* in the space system, only one force reaction exists which acts along the direction that is normal (perpendicular) to the surface of the support. All other *force* and *moment* reactions are zero. Accordingly, the number of *unknown reactions* is one and the number of *unknown displacements* is five (i.e., 3 rotations and 2 translations) in a roller support in the space system.

For the *fixed support* in the three-dimensional system, there are three *force reactions* and three *moment reactions*. This leads to the number of *unknown reactions* equal to six in the fixed support system. All three translation and three rotational movements are restrained and hence, all displacement and rotation values are zero for this support system.

TABLE 2.1   Types of supports for two-dimensional structures

Support type	*Symbol*	*Possible reactions*	*Support movements*
Fixed	Fixed	Reactions: 3(H, V, M)	No movements are restrained, i.e., all translations and rotation are zero.
Hinged (or) Pinned	Pin / Hinge	Reactions: 2(H, V)	Both translations are restrained (i.e., zero). The rotation is unrestrained (non-zero).
Roller	Roller / Roller / Roller	Reaction: 1(V)	Vertical translation is restrained (i.e., zero). Horizontal translation and the rotation are unrestrained (i.e., non-zero).

Support type	Symbol	Possible reactions	Support movements
Guided roller	Guided roller	H M Reactions: 2(H, M)	Horizontal translation and the rotation are restrained (i.e., zero). Vertical translation is unrestrained (i.e., non-zero).
Hinge	Hinge	H H V V Reactions: 2(H, V)	Relative translations between the members are restrained. Relative rotation between the members is not restrained.
Link	Link	F Reaction: 1 (F)	Translation in the direction along the length of link is zero.
Cable	Cable	F Reaction: 1 (F)	—

2.4   FREE-BODY DIAGRAMS

A free-body diagram is a very useful sketch of the structure showing all applied external forces and all supports are replaced by their corresponding possible reactions, as summarized in Table 2.1. A free-body diagram can be drawn for the entire structure or a part of it. In the latter case, all possible internal reactions (discussed below) are also considered in the free-body diagram in addition to the applicable external forces and the support reactions. All free-body diagrams of the complete or partial structure must satisfy the equations of equilibrium.

Figure 2.4(a) shows a portal frame AB' with one fixed support and one hinged support. The frame is subjected to a uniformly distributed load and a concentrated load on the member AB and a horizontal concentrated load on the member BC. As an example, Figure 2.4(b) shows the free-body diagram of the member AB in which the fixed support at A is replaced by two force reactions and a moment reaction. The force transferred from the member BC has been represented by two forces and one bending moment (as it is frame member) at the joint 'B'. It is worth to note that all applicable external loads acting on the member AB are also considered in the free-body diagram.

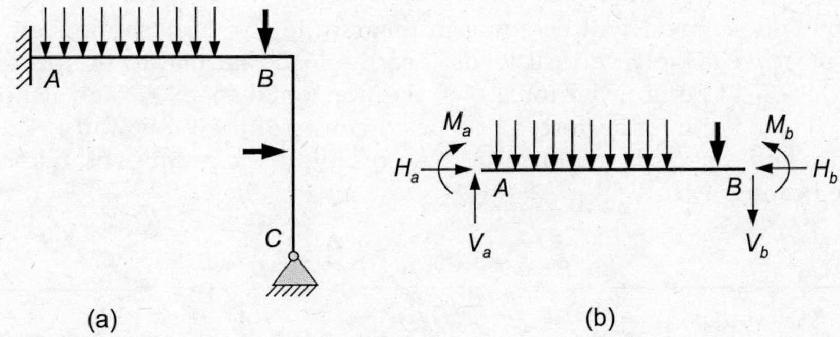

FIGURE 2.4 (a) A portal frame with external loads, (b) Free-body diagram of member AB.

2.5 INTERNAL FORCES

Internal forces are induced in a structural element under the applied loading. These forces are represented through a cut section of the member. Truss members, beams, and frame members are considered as the structural elements in the following sections.

1. Truss member: Figure 2.5(a) shows a two-dimensional (planer) truss. In a truss, all external forces are applied only at the joints. A member *a*, as shown in the figure, has been considered here for the discussion. At joints '*i*' and '*j*', four members are connected in this truss. As shown in Figure 2.5(b), these joints are subjected to the resisting forces in the member directions and the applied loadings. The resultant (F) of these forces along the direction of member '*a*' will act at both joints, i.e., only one force each is present at joint '*i*' as well as joint '*j*'. Hence, the truss members are also called the *two-force* members.

Since a truss member carries only either a tensile or compressive axial force, only one force is expected at the cut section shown in the figure. Further, since there are no additional forces applied at any intermediate points on the member between these joints, the resultant forces (F) at both these joints must be equal in magnitude and opposite sign. Therefore, the number of unknown internal forces in the truss member is one.

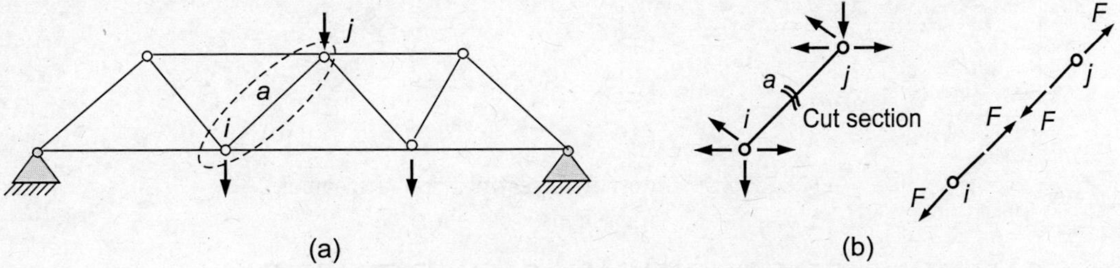

FIGURE 2.5 Internal forces of a truss member.

2. Beam: Figure 2.6(a) shows a *simply supported* beam *AB* subjected to arbitrary loading. If a section is cut at any point of the beam, two forces (H and V) and one bending moment (M) will be obtained as shown in Figure 2.6(b). Considering the three equilibrium of equations applicable for the planer members, these three forces and bending moment at the cut section for the left segment will be equal in magnitude and opposite in sign. In practice, beams are subjected to only vertical loads (i.e., the loads normal to the beam length). In such cases, there will be no axial force (i.e., the horizontal force, H) and hence, the value of H is zero. Thus, the internal forces of a beam consist of only one force, V representing the shear force and a bending moment (M). Accordingly, the number of unknown internal forces in the beam is two.

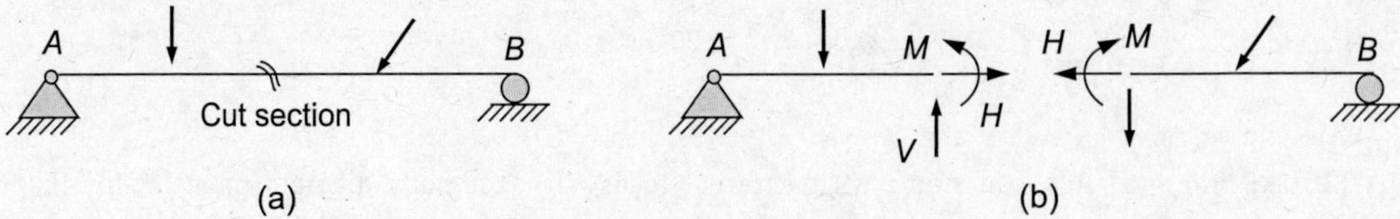

FIGURE 2.6   Internal forces of a beam.

3. Frame member: Figure 2.7(a) shows a two-bay and three-story planer frame under arbitrary loading. Typically, frames are subjected to the external forces/moments at the joints as well as on the members. If a member is cut at any point, two forces and bending moment are expected in the frame member, similar to those of a beam. Horizontal force, H represents the axial force in the members, whereas the vertical force, V represents the shear force acting on the beam as shown in Figure 2.7(b). Unlike a beam, the axial force in a frame member is not zero. Hence, the number of unknown internal forces in a frame member is three.

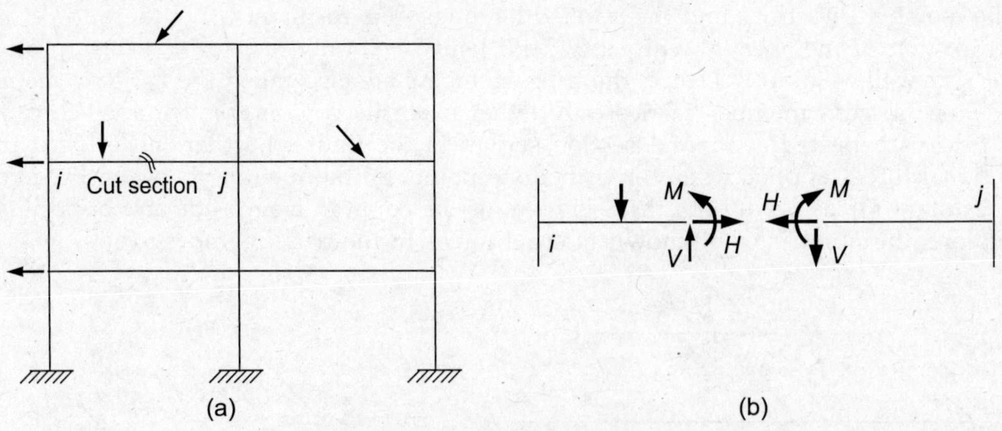

FIGURE 2.7   Internal forces in a frame member.

2.6   EQUATIONS OF CONDITION OR CONSTRUCTION

For simplifications, structural systems are often idealized as a rigid body supported by a number of supports. A simple or compound form of the rigid body is developed by interconnecting several members by means of hinges, links, or rollers. In any case, the idealized structural system is subjected to the same external forces. The support reactions must satisfy the equations of equilibrium if the structure is to remain at rest. However, the compound for the idealized structure connected by the devices (such as hinges, links, etc.) imposes further restrictions on the force system acting on it. This provides the additional equations of statics to supplement the equation of equilibrium. The equations obtained by using the method of special construction (other than external supports) are called *the equilibrium of condition or construction*.

2.7   STRUCTURAL STABILITY AND STATIC DETERMINACY

A structure at rest is considered to be *geometrically stable* if it remains at rest after the load application and the configuration and orientation of all its members do not change virtually. The change in configuration can be visualized by drawing the deflected shape of the structure qualitatively. If the excessive displacement (or rotation) of the structure is expected in any direction, the structure is said to be *unstable*.

The magnitude of deflection (or rotation) of a structure depends on the support conditions and the arrangement of structural elements. The excessive displacement contributing to the structural stability may be induced either by the arrangement of supports or the geometry of the structure. Accordingly, structural stability can be classified into two types, namely, (a) *external stability*, which is related to the support arrangements, and (b) *internal stability*, which is primarily dependent on the geometry of the structure. For a structure to be statically stable, both external and internal stability criteria must be satisfied.

External stability

For a structure at rest, three equilibrium of equations must be satisfied in a coplanar system. Therefore, *at least three elements of support reactions are necessary to restrain a structure in stable equilibrium*. However, *this condition, though necessary, is not always sufficient*. If there are more than three elements of reactions, the structure is necessarily more stable because of the additional restraints. When the external stability due to support arrangements is determined, the internal conditions are not considered and hence, the structure can be assumed as a rigid body.

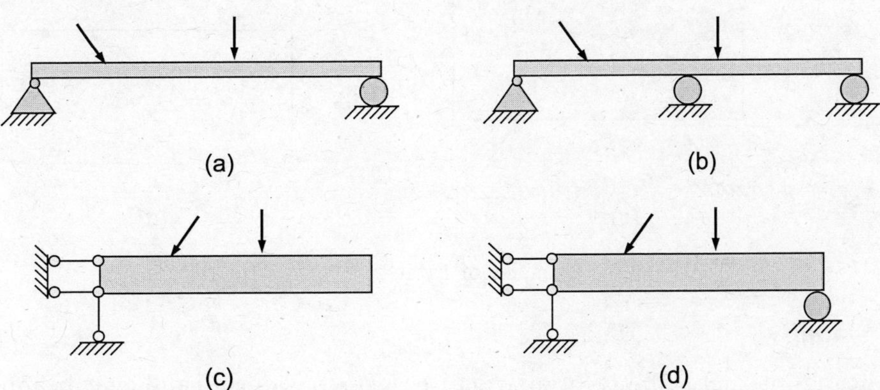

FIGURE 2.8   Examples of externally stable structures.

Figure 2.8 shows some examples of structures that are *statically stable externally*. In Figure 2.8(a), the total number of reactions is three (i.e., two at the hinged support and one at the roller support), which is equal to the available three numbers of equilibrium equations. The beam, shown in Figure 2.8(b), has three supports having four numbers of support reactions, which is higher than the minimum number of three required for equilibrium. Hence, the beam is externally stable. Similarly, beams shown in Figures 2.8(c) and (d) have three and four support reactions against the minimum required number of 3. Hence, they are also externally stable.

There are many cases where the structural systems are not externally stable even though the number of support reaction components is three or more. If the reaction forces are either *colinear* as shown in Figure 2.9(a), or *parallel* as shown in Figure 2.9(b), or *concurrent* as shown in Figure 2.9(c), the structural system is said to be externally unstable irrespective of the total number of reaction components. If the reactions are parallel, they cannot prevent the sliding of the body because there is no resistance in the sliding direction. For the structure shown in Figure 2.9(b), the equilibrium equation $\sum F_x = 0$ is not satisfied. In the case of colinear or concurrent reactions, they cannot resist the bending moment about the concurrent point. Structure supported only on one support as shown in Figure 2.9(d) is obviously externally unstable as there is no resistance to the bending action due to the applied external forces.

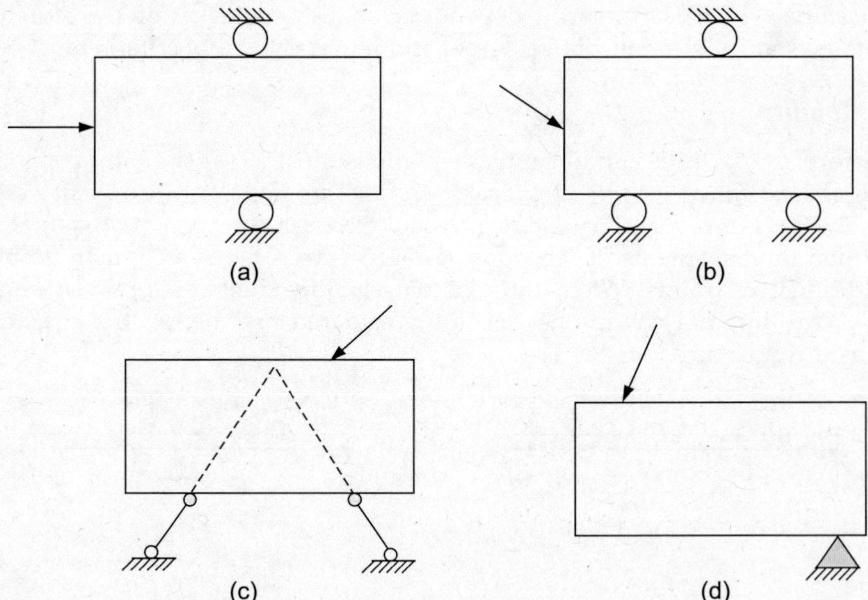

FIGURE 2.9   Examples of externally unstable structures.

Internal stability

To explain the internal stability of a structure, let's consider a cantilever beam as shown in Figure 2.10(a). Under the application of any transverse loading, the deflection of the beam is represented by a continuous curve as shown by the dashed line. At any point on the deflection curve, the rotations (or slopes) of the beam segments to the left and right of this point must be the same to maintain continuity. The magnitudes of the translation and rotation at any point are small such that the geometry of the beam does not change significantly. Hence, this cantilever beam can be said to be *geometrically stable*.

Now, let's consider the same beam with an internal hinge as shown in Figure 2.10(b). It is to be remembered that the value of the bending moment at the hinge is zero. In other words, two segments of the beam at the hinge location are free to rotate independently. It can be visualized that any load applied in the beam segment to the left of the hinge would be safely resisted by the beam without significant deflection. However, even a small load on the beam segment on the right side of the hinge would result in the significant (uncontrolled) rotation and displacement. Hence, this cantilever beam with an internal hinge is said to be *geometrically unstable*.

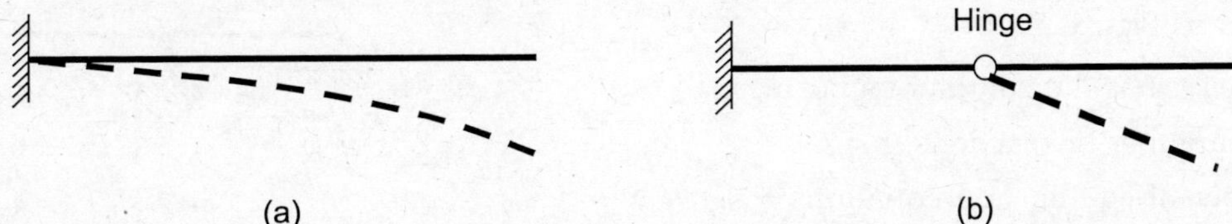

(a)   (b)

FIGURE 2.10   Structural stability: (a) Cantilever beam, (b) Cantilever beam with an internal hinge.

A structure is said to be *statically determinate* if the structure is geometrically stable and the number of unknown reaction components can be determined using the available equations of static equilibrium. Thus, *an indeterminate structure may be defined as the one for which the reaction components and internal forces can't be completely determined by using the equilibrium of equations.* Accordingly, the degree of static indeterminacy (DSI) for a given structure is the number of unknowns over and above the number of equilibrium equations available for the solution. Here, the equilibrium equations include the equations of condition as well.

Similar to the structural stability, the *static indeterminacy* is dependent on the support arrangement as well as the member configurations. If the DSI is computed with respect to the components of support reactions only, it is called *external determinacy*. On the other hand, if the DSI is determined with respect to the member forces, it is said to be *internal determinacy*. A combination of external and internal determinacy represents the *total determinacy (or indeterminacy)* of a structure. These are discussed in the following sections in detail.

2.7.1   External Stability and Determinacy

The degree of external static indeterminacy (D_{se}) is the difference between the number of support reactions (r) and the number of available equilibrium equations. For the two-dimensional (planer) structure, the number of equilibrium equations available is three, whereas this value is six for the three-dimensional (space) system. Mathematically,

$$\text{Plane (two-dimension) system, } D_{se} = r - 3 \qquad (2.4)$$

$$\text{Space (three-dimension) system, } D_{se} = r - 6 \qquad (2.5)$$

EXAMPLE 2.1   Determine the degree of *external static determinacy* of the propped cantilever beam shown in Figure E2.1.

Solution: The propped beam is in the plane system. Total number of reactions, $r = 4$.

Number of equilibrium equations = 3.

$$D_{se} = r - 3$$

Hence, the value of $D_{se} = 4 - 3 = 1$.

    Thus, the beam has *one degree of external static indeterminacy.*

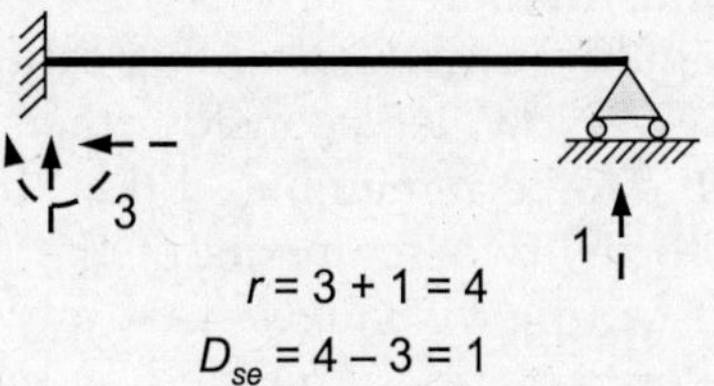

FIGURE E2.1

EXAMPLE 2.2   Determine the degree of *external static determinacy* of the continuous beam shown in Figure E2.2.

Solution: For the continuous beam,

Total number of reactions, $r = 4$.

Number of equilibrium equations = 3.

Degree of external static determinacy, $D_{se} = r - 3$

Hence, the value of $D_{se} = 4 - 3 = 1$

FIGURE E2.2

Thus, the beam has a *one degree of external static indeterminacy.*

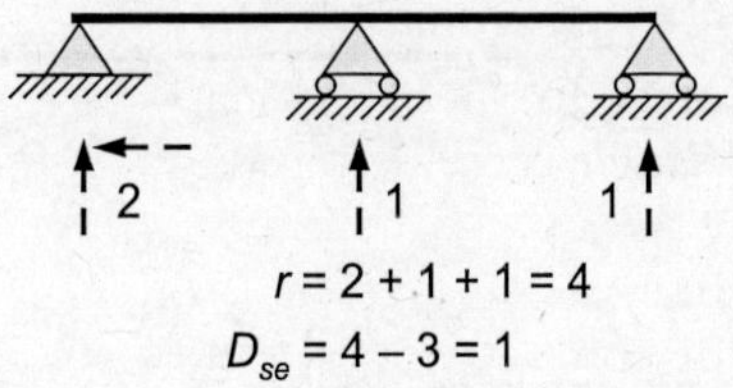

EXAMPLE 2.3   Determine the degree of *external static determinacy* of the plane truss shown in Figure E2.3.

Solution: For the given plane truss,
Total number of reactions, $r = 5$.
Number of equilibrium equations = 3.
Degree of external static determinacy,

$$D_{se} = r - 3$$

Hence, the value of $D_{se} = 5 - 3 = 2$.

Thus, the beam has *two degrees of external static indeterminacy.*

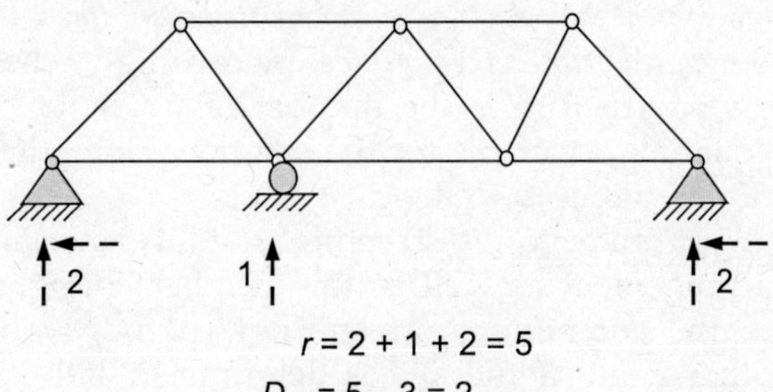

FIGURE E2.3

EXAMPLE 2.4   Determine the degree of *external static determinacy* of the space truss shown in Figure E2.4.

Solution: For the given space truss,
Total number of reactions, $r = 5$.
Number of equilibrium equations = 3.
Degree of external static determinacy, $D_{se} = r - 6 = 9 - 6 = 3$

Thus, the beam has *three degrees of external static indeterminacy.*

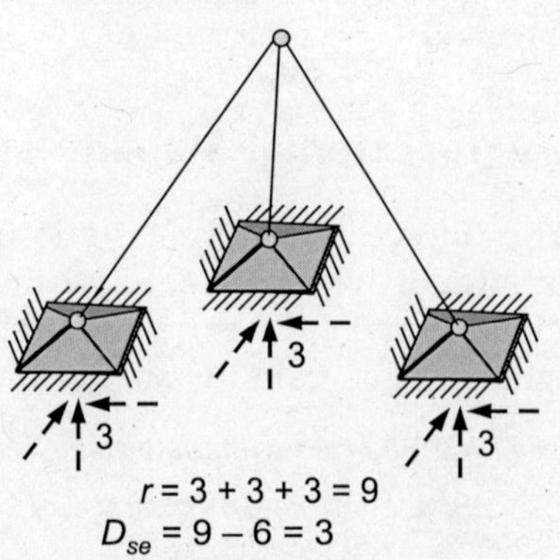

FIGURE E2.4

2.7.2 Internal Stability and Determinacy

The degree of internal static indeterminacy is the difference between the number of member forces and the number of available equilibrium equations. Since the number of member forces depends on the type of structural element, the expressions to determine the degree of internal static determinacy are different for different structural systems. Three types of structural systems are discussed below:

1. Beams: Beams are considered as the best example of internally statically determinate structures as the number of internal member forces is equal to the number of available equilibrium equations. Hence, the internal instability does not arise in a beam unless special conditions are present. It is worth mentioning that the beam is always considered as a single continuous member without any joints at the intermediate supports if any.

2. Trusses: All joints of a truss are pin-jointed, and each member carries only one axial force. In the case of a plane truss, the geometry of the structural system must form a *triangle* configuration to ensure stability. A minimum of three members and three joints are required to make a simple and stable truss. If there are additional members and joints, they must a series of triangles for stability. This means that two more members are required for connecting each additional joint if at least one triangle has already been formed using three members and three joints.

   If there are j number of joints in a truss, the first three members are required for three joints to make a triangle. For remaining $(j - 3)$ joints, the number of members is required for stability is $2(j - 3)$. Thus, the total number of members, m required for the stability of a plane truss is given by the following equation:

$$m = 2(j - 3) + 3 = (2j - 3) \tag{2.6}$$

If $m < (2j - 3)$, then a plane truss is said to be unstable. If $m \geq (2j - 3)$, the plane truss is stable.

   To determine the degree of internal static indeterminacy of a truss, we have to assume that the same structure is externally stable and determinate. This means that number of support reactions (r) should be equal to the number of equilibrium equations, which is three for the plane truss. Thus, the total number of unknowns is the sum of the number of member forces (m) and support reactions (r). It is worth mentioning that the member forces are equal to the number of members as each member carries only one force. Accordingly, total unknowns in a plane truss is $(m + r)$. At each pin-joint of the plane truss, two equilibrium equations are available, i.e., $\sum F_x = 0;\ \ \sum F_y = 0$. Thus, the total available equation of equilibrium is 2 times the number of joints $(= 2j)$. Degree internal static indeterminacy of a plane is given by the following equation:

$$D_{si} = (m + 3) - 2j \tag{2.7}$$

   In the case of a space truss, the most elementary form of structure required for stability is a *tetrahedron* having four joints and six members. These members and joints basically form four surfaces of triangular configuration. Three members are required for connecting each additional joint to create further tetrahedrons. Thus, for a space truss having j number of joints, the number of members required for the remaining $(j - 4)$ joints (i.e., excluding the first four joints) would be $3(j - 4)$ in order to ensure the stability of the structure. Thus, the total number of members, m required for the stability of a space truss is given by the following equation:

$$m = 3(j - 4) + 6 = (3j - 6) \tag{2.8}$$

Similar to the plane truss, if $m < (3j - 6)$, then a space truss is unstable and otherwise stable. The number of unknowns is given by $(m + 6)$ as at least 6 nos. of reactions are required to make the structure statically stable and determinate externally. The available equations of equilibrium are equal to $3j$ in a space truss. Degree internal static indeterminacy of a space truss is given by the following equation:

$$D_{si} = (m + 6) - 3j \tag{2.9}$$

Table 2.2 summarizes the conditions of internal stability and determinacy of trusses. *It is reiterated that though these conditions are necessary, they may not be sufficient in many instances.*

TABLE 2.2   Internal static determinacy of trusses

Condition	Plane truss	Space truss
Unstable internally	$m < (2j - 3)$	$m < (3j - 6)$
Stable and statically determinate internally	$m = (2j - 3)$	$m = (3j - 6)$
Stable and statically indeterminate internally	$m > (2j - 3)$	$m > (3j - 6)$
Degree of internal static indeterminacy (D_{se})	$(m + 3) - 2j$	$(m + 6) - 3j$

EXAMPLE 2.5   Determine the degree of *internal static determinacy* of the plane truss shown in Figure E2.5.

Solution: For the given plane truss,

Total number of members, $m = 22$.

Total number of joints, $j = 12$.

$$2j - 3 = 2 \times 11 - 3 = 19$$

Since $m > (2j - 3)$, the truss is *internally stable and indeterminate.*

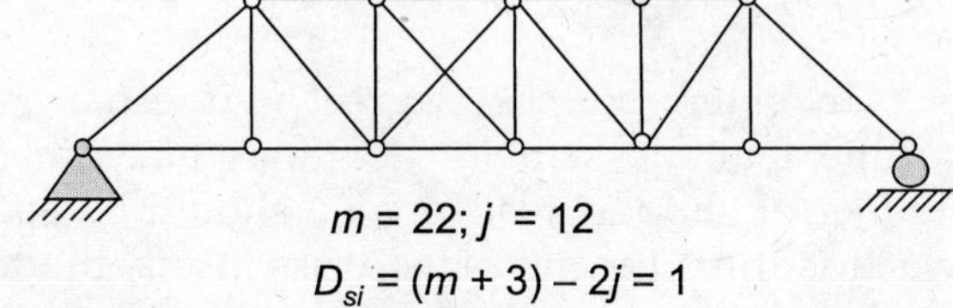

FIGURE E2.5

Degree of internal indeterminacy can be determined as follows:

$$D_{si} = (m + 3) - 2j$$
$$D_{si} = (22 + 3) - 2 \times 12 = 1$$

Thus, the truss is *internally indeterminate to first degree.*

EXAMPLE 2.6   Determine the degree of *internal static determinacy* of the space truss shown in Figure E2.5.

Solution: For the given space truss,

Total number of members, $m = 3$

Total number of joints, $j = 3$

$$3j - 6 = 3 \times 3 - 6 = 3$$

Since, $m = (3j - 6)$, the truss is *internally stable and determinate.*

It can be verified that the degree of internal indeterminacy is $D_{si} = (m+6) - 3j = 0$.

3. Frames: The degree of internal static indeterminacy of plane frames can be derived using the same procedure as discussed above. Usually, all joints of a frame are considered rigid, unless specified otherwise. In the case of a plane frame, each member has three internal forces (i.e., two forces and one bending moment). Three equations of equilibrium are also available at each joint due to rigid connections. Further, a minimum of three support reactions is required to make a plane frame externally stable and determinate. Thus, the total number of unknowns in a plane frame is $(3m + 3)$ and the total number of available equations is $3j$. The degree of internal static determinacy of a plane frame is given as follows:

Plane frame: $D_{si} = (3m + 3) - 3j$ (2.10)

In the case of a space frame, each member has six internal forces, and six number of support reactions are required for external stability. There are six equilibrium equations are available at each joint of the space frame. Thus, the degree of internal static determinacy of a space frame is given as follows:

Space frame: $D_{si} = (6m + 6) - 6j$ (2.11)

EXAMPLE 2.7 Determine the degree of *internal static determinacy* of the frame shown in Figure E2.7.

Solution: The structure shown is a *plane frame.*

Total number of members, $m = 15$.

Total number of joints, $j = 12$.

Degree of internal indeterminacy can be determined as follows:

$$D_{si} = (3m + 3) - 3j$$

$$D_{si} = (3 \times 15 + 3) - 3 \times 12 = 12$$

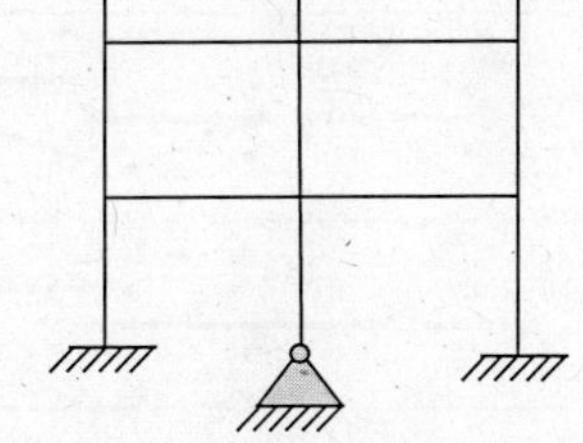

FIGURE E2.7

Thus, the frame is *internally stable and indeterminate to twelfth degree.*

It is worth noting that exact number of support reactions are not considered in the computation of internal stability and determinacy.

2.7.3 Overall Stability and Determinacy

The total degree of static indeterminacy (DSI) is the sum of the external static indeterminacy and the internal static indeterminacy. Mathematically,

$$DSI = D_{si} + D_{se}$$ (2.12)

If $DSI < 0$, the structure is unstable. If $DSI = 1$, structure is stable and statically determinate. If $DSI > 1$, the structure is stable and statically indeterminate.

The expressions to compute the DSI of beams, trusses, and frames are given below:

1. Beams: The degree of static indeterminacy (DSI) for a beam may be given by:

$$DSI = \begin{cases} r-(c+3) & \text{(Two-dimension system)} \\ r-(c+6) & \text{(Three-dimension system)} \end{cases} \qquad (2.13)$$

where, r = Total number of reactions, c = Number of equations of condition/construction. For internal hinge in a two-dimension beam, the bending moment is zero. Hence, the value of $c = 1$ for an internal hinge. Similarly, the value of $c = 2$ for a roller. If a beam has no internal connection, the value of $c = 0$.

For illustration, some examples of beams are discussed in Table 2.3.

TABLE 2.3   Degree of static indeterminacy of beams

Beam	DSI	Classification
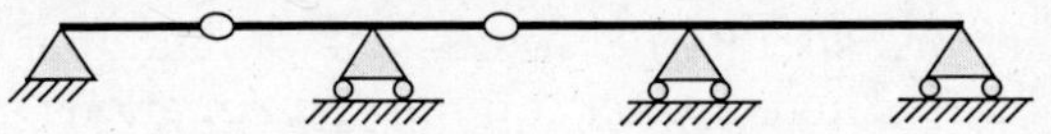	$r = 5; c = 2$ $DSI = 5 - (2+3) = 0$	Stable and statically determinate
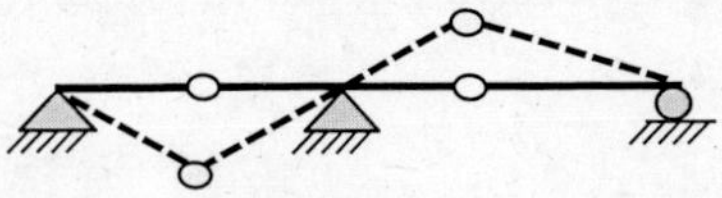	$r = 5; c = 2$ $DSI = 5 - (2+3) = 0$	Unstable[*]
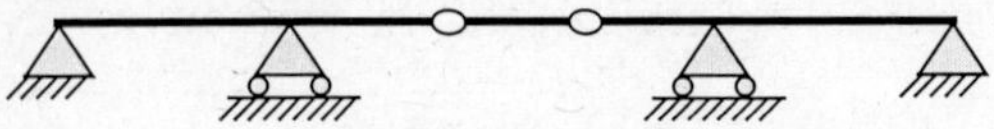	$r = 6; c = 2$ $DSI = 6 - (2+3) = 1$	Stable and statically indeterminate to first degree
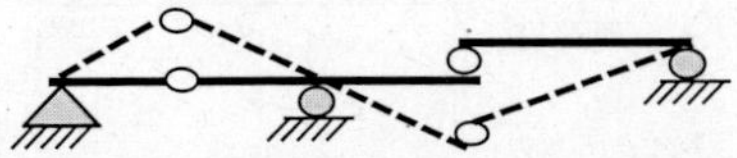	$r = 4; c = 1+2 = 3$ $DSI = 4 - (3+3) < 1$	Unstable[*]
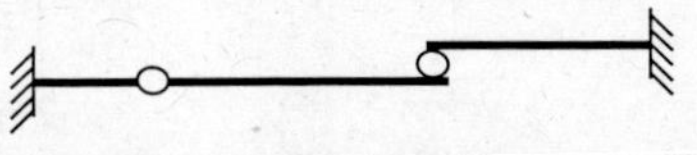	$r = 6; c = 1+2 = 3$ $DSI = 6 - (3+3) = 0$	Stable and statically determinate
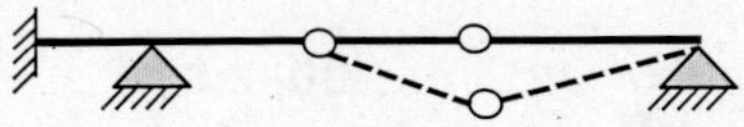	$r = 7; c = 2+1 = 3$ $DSI = 7 - (3+3) = 1$	Unstable[*]
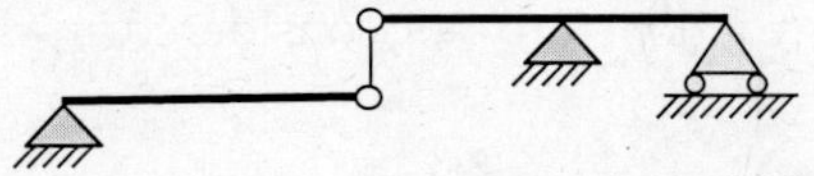	$r = 5; c = 1+1 = 2$ $DSI = 5 - (2+3) = 0$	Stable and statically determinate[**]

Internal geometric instability; Dashed line indicates the possible deformed shape of the structure.
***Bending moment about any pin-joint of link and the force normal to link is zero (i.e., c = 2)*

Even though the equation of stability and determinacy is satisfied for the beam shown in the second row of Table 2.3, the structure is found to be unstable. The equations given in the preceding sections to determine the stability of a structure is not sufficient in many cases. It is, therefore, important to visualize the stress path and the deflected shape for determining the stability of a structure.

2. Trusses: The degree of static indeterminacy (DSI) of a truss is the sum of external (D_{se}) and internal (D_{si}) static indeterminacies. Using the Eqns. (2.4) and (2.7), the expression for *DSI* of a plane truss can be obtained from Eq. (2.12). Similarly, the equation for *DSI* of a space truss can be derived using Eqns. (2.5) and (2.9) as follows:

$$DSI = \begin{cases} (m+r)-2j & \text{(Plane truss)} \\ (m+r)-3j & \text{(Space truss)} \end{cases}$$

(2.14)

Table 2.4 illustrates the stability and determinacy of example plane trusses.

TABLE 2.4   Degree of static indeterminacy of trusses

Truss	*Degree of static indeterminacy*	*Classification*
	$m = 7;\ j = 5;\ r = 3$ $DSI = (7 + 3) - 2 \times 5 = 0$ $D_{si} = (7 + 3) - 2 \times 5 = 0$ $D_{se} = 3 - 3 = 0$	Unstable[*]
	$m = 7;\ j = 5;\ r = 3$ $DSI = (7 + 3) - 2 \times 5 = 0$ $D_{si} = (7 + 3) - 2 \times 5 = 0$ $D_{se} = 3 - 3 = 0$	Stable and statically determinate
	$m = 7;\ j = 5;\ r = 3$ $DSI = (7 + 3) - 2 \times 5 = 0$ $D_{si} = (7 + 3) - 2 \times 5 = 0$ $D_{se} = 3 - 3 = 0$	Unstable[**]
	$m = 9;\ j = 6;\ r = 6$ $DSI = (9 + 6) - 2 \times 6 = 3$ $D_{si} = (9 + 3) - 2 \times 6 = 0$ $D_{se} = 6 - 3 = 3$	Stable and statically indeterminate to *third* degree
	$m = 22;\ j = 12;\ r = 3$ $DSI = (22 + 3) - 2 \times 12 = 1$ $D_{si} = (22 + 3) - 2 \times 12 = 1$ $D_{se} = 3 - 3 = 0$	Stable and statically indeterminate to *first* degree
	$m = 10;\ j = 6;\ r = 4$ $DSI = (10 + 4) - 2 \times 6 = 2$ $D_{si} = (10 + 3) - 2 \times 6 = 1$ $D_{se} = 4 - 3 = 1$	Stable and statically indeterminate to *second* degree

Truss	Degree of static indeterminacy	Classification
	$m = 9;\ j = 6;\ r = 3$ $DSI = (9 + 3) - 2 \times 6 = 0$ $D_{si} = (9 + 3) - 2 \times 6 = 0$ $D_{se} = 3 - 3 = 0$	Unstable internally[**] (*All three forces carried by member connecting inner triangle to outer triangle are concurrent*)

*External geometric instability because of parallel lines of reaction
**Internal geometric instability; Dashed line indicates the possible deformed shape of the structure.

3. Frames: Using the same procedure described above, the DSI of plane and space frames can be determined using the following expressions:

$$DSI = \begin{cases} (3m + r) - (3j + c) & \text{(Plane frame)} \\ (6m + r) - (6j + c) & \text{(Space frame)} \end{cases} \tag{2.15}$$

Table 2.5 illustrates the stability and determinacy of example plane frames.

TABLE 2.5 Degree of static indeterminacy of frames

Frame	Degree of static indeterminacy	Classification
	$m = 11;\ j = 9;\ r = 8$ $DSI = (3 \times 11 + 8) - 3 \times 9 = 17$ $D_{si} = (3 \times 11 + 3) - 3 \times 9 = 9$ $D_{se} = 8 - 3 = 5$	Stable and statically indeterminate to fourteenth degree
	$m = 10;\ j = 9;\ r = 7$ $DSI = (3 \times 10 + 7) - 3 \times 9 = 10$ $D_{si} = (3 \times 10 + 3) - 3 \times 9 = 6$ $D_{se} = 7 - 3 = 4$	Stable and statically indeterminate to tenth degree
	$m = 19;\ j = 16;\ r = 10$ $DSI = (3 \times 19 + 10) - 3 \times 16 = 19$ $D_{si} = (3 \times 19 + 3) - 3 \times 16 = 12$ $D_{se} = 10 - 3 = 7$	Stable and statically indeterminate to nineteenth degree

Frame	Degree of static indeterminacy	Classification
	$m = 4;\; j = 5;\; r = 9$ $DSI = (3 \times 4 + 9) - 3 \times 5 = 6$ $D_{si} = (3 \times 4 + 3) - 3 \times 5 = 0$ $D_{se} = 9 - 3 = 6$	Stable and statically indeterminate to sixth degree
	$m = 4;\; j = 5;\; r = 9;\; c = 3 - 1 = 2^*$ $DSI = (3 \times 4 + 9) - (3 \times 5 + 2) = 4$	Stable and statically indeterminate to fourth degree
	$m = 4;\; j = 5;\; r = 9;\; c = 2 - 1 = 1^*$ $DSI = (3 \times 4 + 9) - (3 \times 5 + 1) = 5$	Stable and statically indeterminate to fifth degree

**Number of equations of conditions at the hinge joint = Number of members at connected at the joint −1. This is because one equation of equilibrium (i.e., bending moment) has already been considered in the number of available equations. Since the number of equations of conditions for a hinge joint is equal to the number of members connected at this joint, the number of additional equations to be considered in the computation of DSI is as follows: c = Number of members connected at the joint −1.*

2.7.4 Simplified Methods

Sometimes, it is tedious and time-consuming to determine the degree of static determinacy (DSI) of trusses and frames using the expressions presented in the previous sections. This is particularly true for long-span trusses as well as high-rise frames having a large number of members and joints. Simplified methods provide an effective and alternative way to compute the DSI of trusses and frames. These methods applied to trusses and frames are discussed in the following sections:

1. Truss: It is known that the plane trusses of triangular configurations are stable and determinate. Further, it is much easier to determine the DSI of a beam, which is internally stable and determinate. Thus, DSI of a beam is dependent on its external indeterminacy only, which is smaller in number as compared to those of members of a truss. Any triangular arrangement of members in a truss can be regarded as the equivalent beams. The DSI of a plane truss would then be computed based on the number of support reactions and special conditions.

The same procedure can be utilized for a space truss, in which the tetrahedral shaped configurations are considered as stable and determinate in a three-dimension system. All tetrahedral shaped configurations can be replaced by equivalent beams in the space system. The DSI of a space truss would then be computed based on the number of support reactions and special conditions.

This has been explained in the following examples.

EXAMPLE 2.8   Determine the *degree of static determinacy* of the truss shown in Figure E2.8.

Solution: For this plane truss, all joints and members form a series of triangles, which are stable and determinate. Hence, the entire geometry of the truss can be replaced by an equivalent beam with the same support conditions as before as shown in the figure. Now, one can easily find that the equivalent beam has three support reactions against three available equilibrium equations. Hence, the truss is *statically determinate*. The same value of DSI can be obtained using the equations as well.

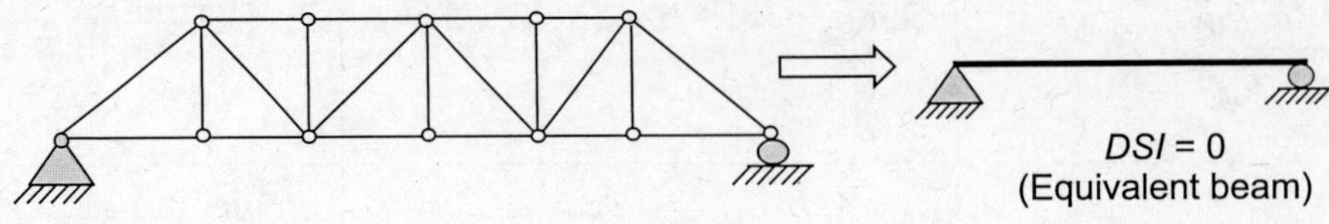

FIGURE E2.8

EXAMPLE 2.9   Determine the *degree of static determinacy* of the truss shown in Figure E2.5.

Solution: In this truss, two diagonals are present in the third panel from the left. If one of them is removed, the panel will have the configuration so two triangles. Thus, one diagonal is the extra member (or redundant). Since there is no redundant support reaction, the DSI of this truss is *one*.

EXAMPLE 2.10   Determine the *degree of static determinacy* of the truss shown in Figure E2.10.

Solution: Consider another truss as shown in Figure 2.18. In this truss, any one member can be safely removed to make the structure internally stable and determinate. In addition, there are four support reactions against the requirement of three. Thus, there are two redundant, i.e., one internal and one external. Thus, DSI of this truss is computed to be *two*.

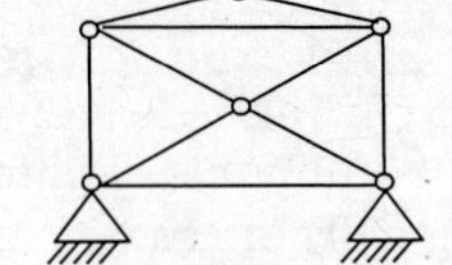

FIGURE E2.10

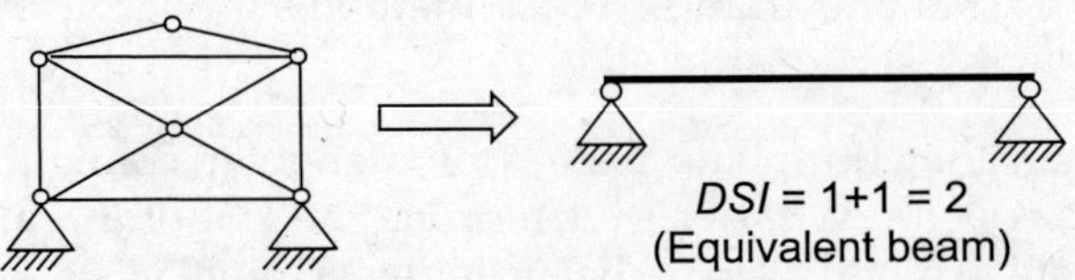

EXAMPLE 2.11   Determine the *degree of static determinacy* of the bridge trusses shown in Figure E2.11.

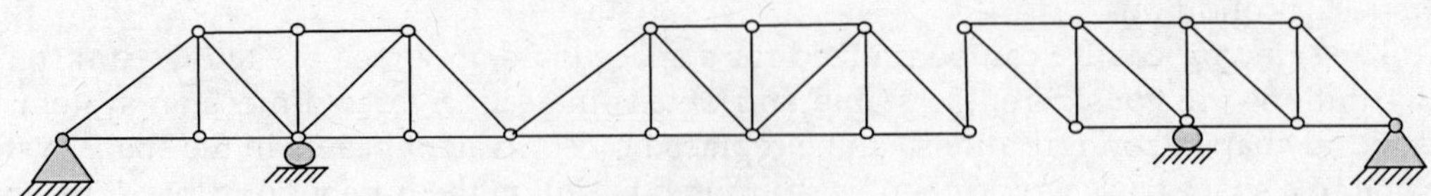

FIGURE E2.11

Solution: There are three trusses, namely, Truss-1, Truss-2, and Truss-3. All these trusses consist of all triangles and hence, can be replaced by three equivalent beams connected by an internal hinge and a link beam. There are six support reactions and three equations of equilibrium. However, the presence of an internal hinge provides one equation of condition. Similarly, the vertical link provides two additional equations of conditions. Thus, there are six unknowns against six available equations of equilibrium and conditions. Hence, the given bridge truss is *statically determinate.*

One can find the same value of DSI using the equations. In this case, $m = 40$, $j = 23$, and $r = 6$. Thus, DSI = $40 + 6 - 2 \times 23 = 0$ (*statically determinate*).

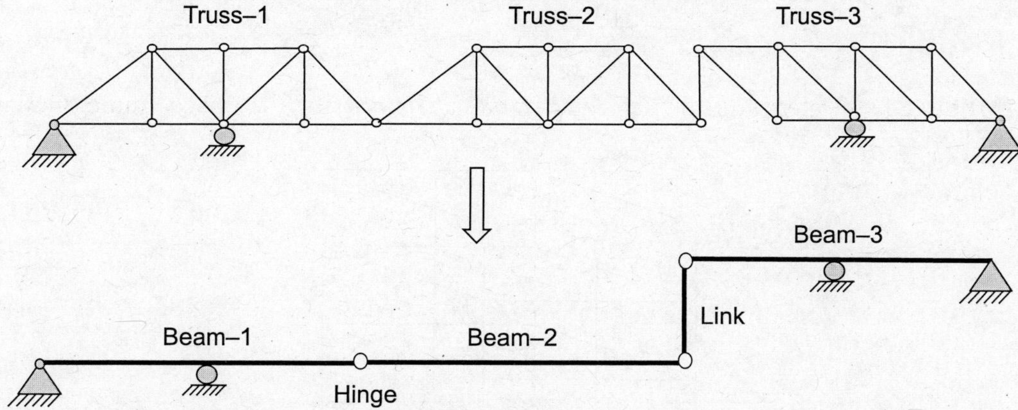

2. Frames: The computation of DSI for a frame system is much simpler and quicker than a truss system. The simplest form of a stable and determinate structure is a *"tree"*. A tree is fixed at the base and free at other ends, and hence, it is often called a *"cantilever tree"*. A cantilever beam with a fixed end is stable and externally determinate. The beam itself is also statically determinate internally. Hence, a *cantilever tree* is statically stable and determinate. The entire frame is cut at several places in order to make the maximum number of independent *cantilever trees*. If any frame member is cut at any section along its length, either three or six numbers of internal forces are expected in a plane or space frame, respectively. The sum of all unknown internal forces at all the cut sections is equal to the DSI of the frame. Mathematically,

$$DSI = \begin{cases} 3 \times \text{No. of cut sections} & \text{(Plane frame)} \\ 6 \times \text{No. of cut sections} & \text{(Space frame)} \end{cases} \tag{2.16}$$

EXAMPLE 2.12   Determine the *degree of static determinacy* of the plane frame shown in Figure E2.12.

Solution: For the frame shown in the figure, if three independent cantilever trees can be obtained by cutting all beams. Three internal forces are obtained at each size cut sections. This makes the total number of unknowns equal to 18 (= 6×3). However, the tree corresponding to the middle column is not a cantilever tree since the base is hinged, not fixed. Thus, one moment is required

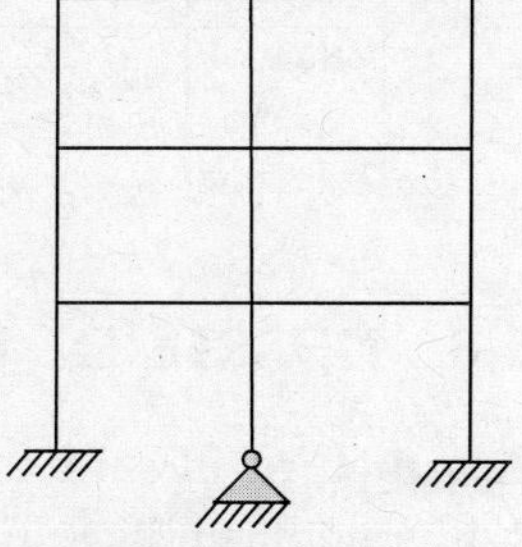

FIGURE E2.12

to be added to the support of the middle column to make it a cantilever tree. Thus, DSI of the frame = 18 − 1 = 17.

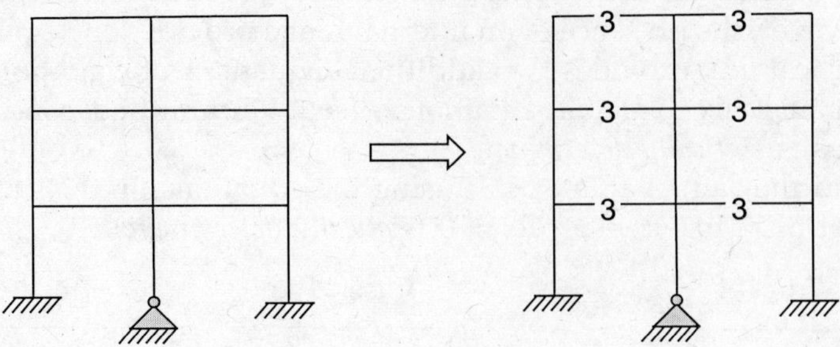

EXAMPLE 2.13 Determine the *degree of static determinacy* of the space frame shown in Figure E2.13.

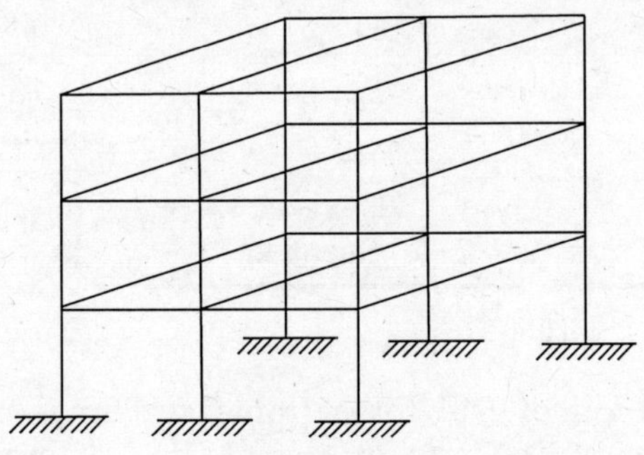

FIGURE E2.13

Solution: For this space frame, 21 number of cuts are required to make six number of cantilever trees. Each cut results in 6 unknowns. Hence, total number of unknowns = 21 × 6 = 126. Since all bases are fixed, the value of DSI is 126.

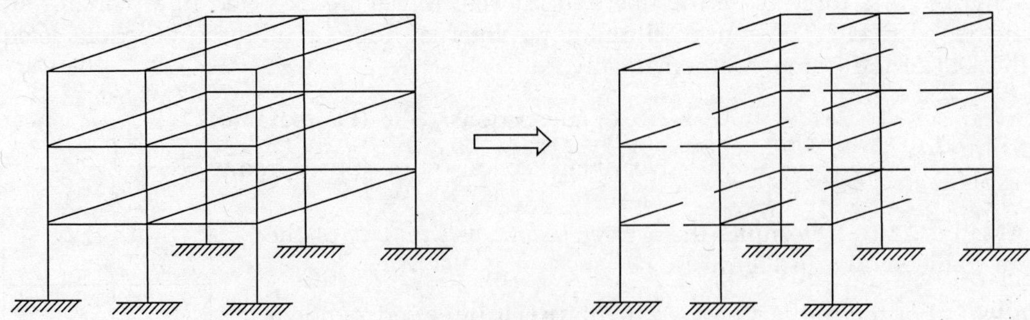

2.8 DEGREE OF KINEMATIC INDETERMINACY

A structure is said to be kinematically determinate if the displacement (and rotation) components of the joints can be determined by compatibility equations. Thus, a kinematically indeterminate structure has the number of unknown displacement components higher than the number of compatibility equations. The number of independent unknown displacement components at the joints of a structure is called the *degree of kinematic indeterminacy* (DKI) or *degree of freedom.*

If any joint of a structure is fixed, all compatibility equations related to the displacement components are known (i.e., all displacement components are zero). Since the number of unknown displacements (i.e., DKI) is zero, the fixed support is *kinematically determinate*. A hinged support in a plane system has two compatibility equations, i.e., the displacement components along both axes are zero. However, the rotation at this support is not known. Hence, the value of DKI for a hinged support is *one*. Similarly, a roller support inplane system has two unknowns (i.e., one displacement and one rotation) and one zero (known) displacement which is normal to the plane of roller sliding surface. Hence, the DKI of a roller support is *two*.

In the case of a plane truss, all joints are free to displacement in both orthogonal directions. Hence, there are two unknowns at each joint of a plane truss and three unknowns at each joint of a space truss. All these unknowns are displacements as the rotation of these joints are not considered due to the absence of bending moment demand at these pined joints. Thus, the DKI of a truss is the sum of all possible independent joint displacements. Figure 2.11 shows the commutation of DKI for a plane truss.

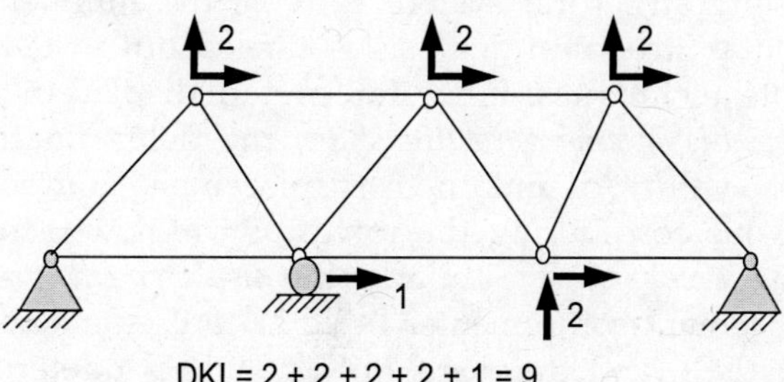

FIGURE 2.11　Degree of kinematic indeterminacy of a plane truss.

In the case of a plane frame, each joint has three displacement components. These are: (i) axial displacement along the length of the member, (ii) shear displacement normal to the length of the member, and (iii) rotation about the axis normal to the plane. Similarly, each joint of a space frame has six displacement components, namely, three displacements and three rotations. Depending on the number of joints, the value DKI can be determined for a plane or a space frame.

In many practical framed structures, the magnitude of axial (longitudinal) deformation of a frame member is much smaller than its shear (normal) deformation. Therefore, the effect of axial deformation is often neglected in the structural analysis of a frame for simplicity. Such members are considered to be *axially inextensible*. This results in two displacement components (i.e., one normal displacement and one rotation) at each joint of a frame member. This approach is generally used in the civil engineering structures and hence, has been adopted in the analysis of frame structures in the subsequent chapters of this book. Figure 2.12 shows the computation of DKI of a plane frame with or without considering the axial inextensibility of members.

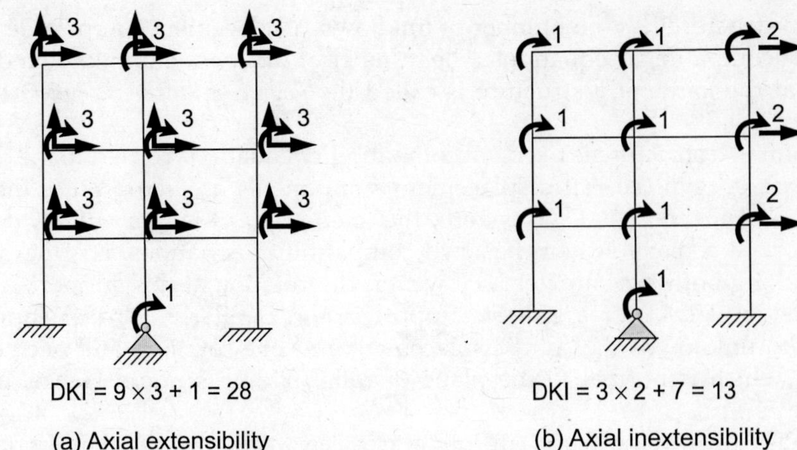

DKI = 9 × 3 + 1 = 28   DKI = 3 × 2 + 7 = 13

(a) Axial extensibility   (b) Axial inextensibility

FIGURE 2.12 Degree of kinematic indeterminacy of a plane frame.

2.9 SUMMARY

Both DSI and DKI of any structural system provide the number of unknown forces or displacement components. Structural analysis is carried out to determine these unknown quantities using suitable techniques. Since forces and displacements are related to each other by means of *force-displacement relationships*, the determination of unknown forces would also provide the values of unknown displacement components and *vice-versa*. If the analysis is carried out considering the forces as unknown, the method of structural analysis is known as the *force method*. Similarly, the *displacement method* of analysis is based on the unknown displacement components. Table 2.6 presents a comparison of the degree of static indeterminacy (DSI) and the degree of kinematic indeterminacy (DKI) of beams, plane trusses, and plane frames (with axial inextensibility). It can be noted that the number of unknown forces (DSI) is smaller than unknown displacements (DKI) for a plane truss. Similarly, the value of DKI is smaller than that of DSI for a plane frame. Though both the *force method* and the *displacement method* can be adopted for a particular structural system, the complexity and time required to solve a structure are greatly reduced if the analysis method involves the least number of unknowns. Hence, it is preferred that the *displacement method* may be selected for the analysis of *frames*, whereas the *force method* may be considered for the analysis of *trusses*.

TABLE 2.6 Comparison of degree of static and kinematic indeterminacy

Structural system	*DSI*	*DKI*
(Beam)	DSI = 5	DKI =7

Structural system	DSI	DKI
(Truss)	DSI = 1 (One small diagonal in the central panel is redundant)	DKI =37
(Truss)	DSI = 3 (Two longest diagonals and one central diagonal are the redundant)	DKI = 4x2+1 = 9
(Frame)	DSI = 15-2 = 13	DKI = 11
(Frame)	DSI = 9	DKI = 9 (Note: 3 rotations at the hinge)

2.10 PROBLEMS

Determine the degree static indeterminacy (DSI) and degree of kinematic indeterminacy (DKI) of the following structures.

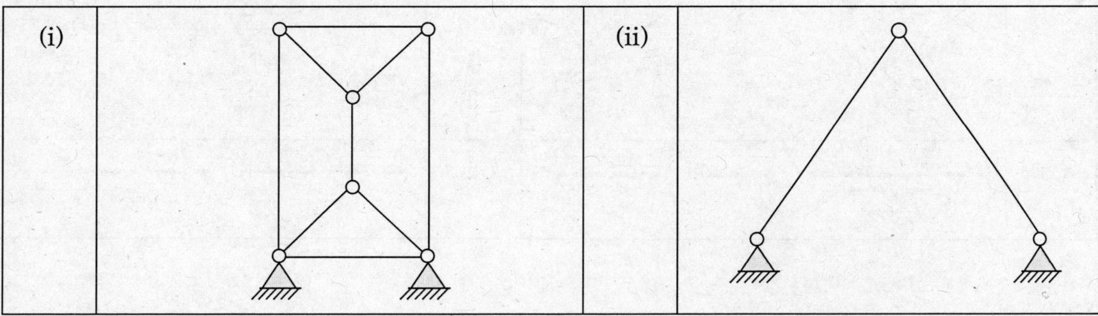

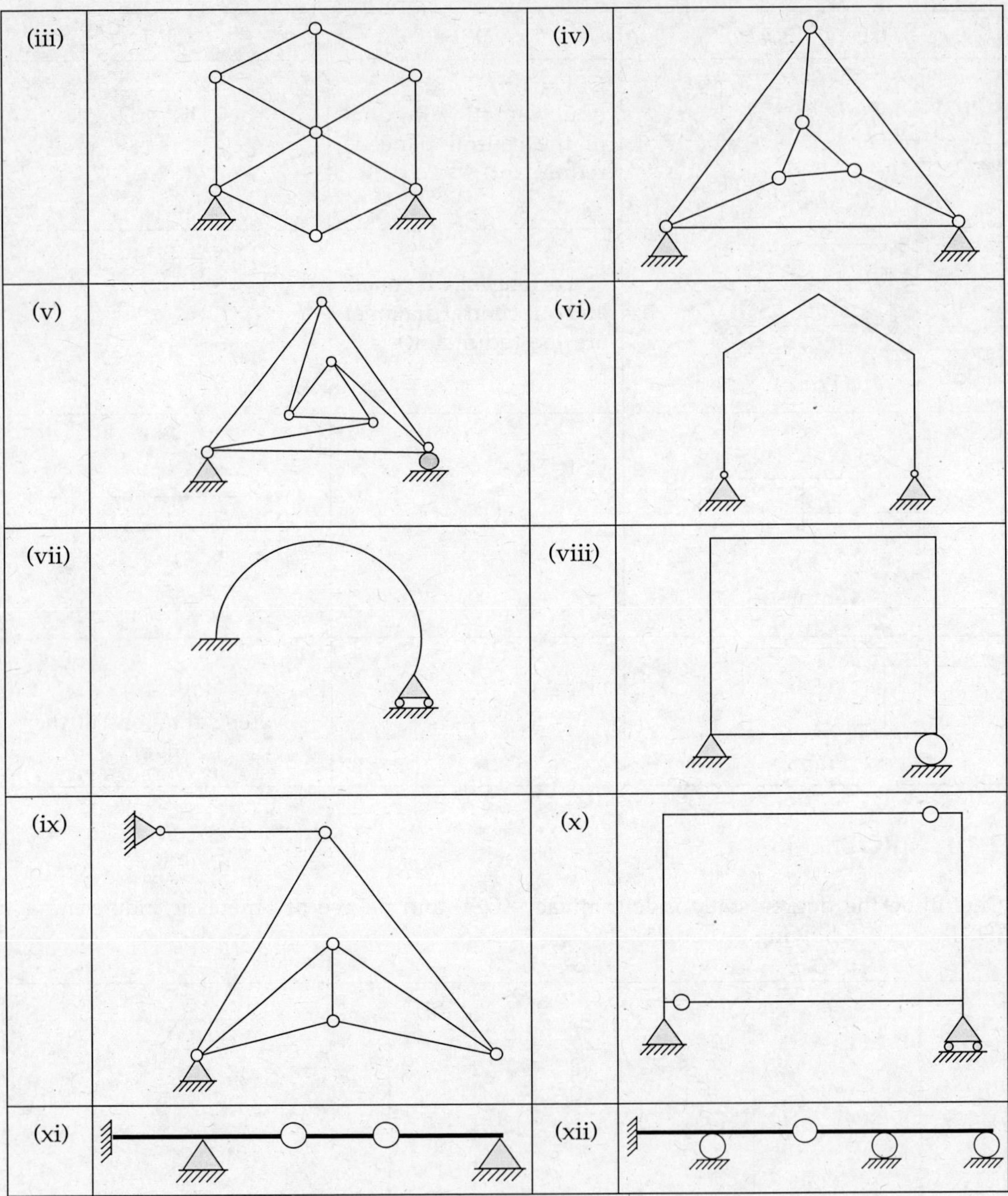

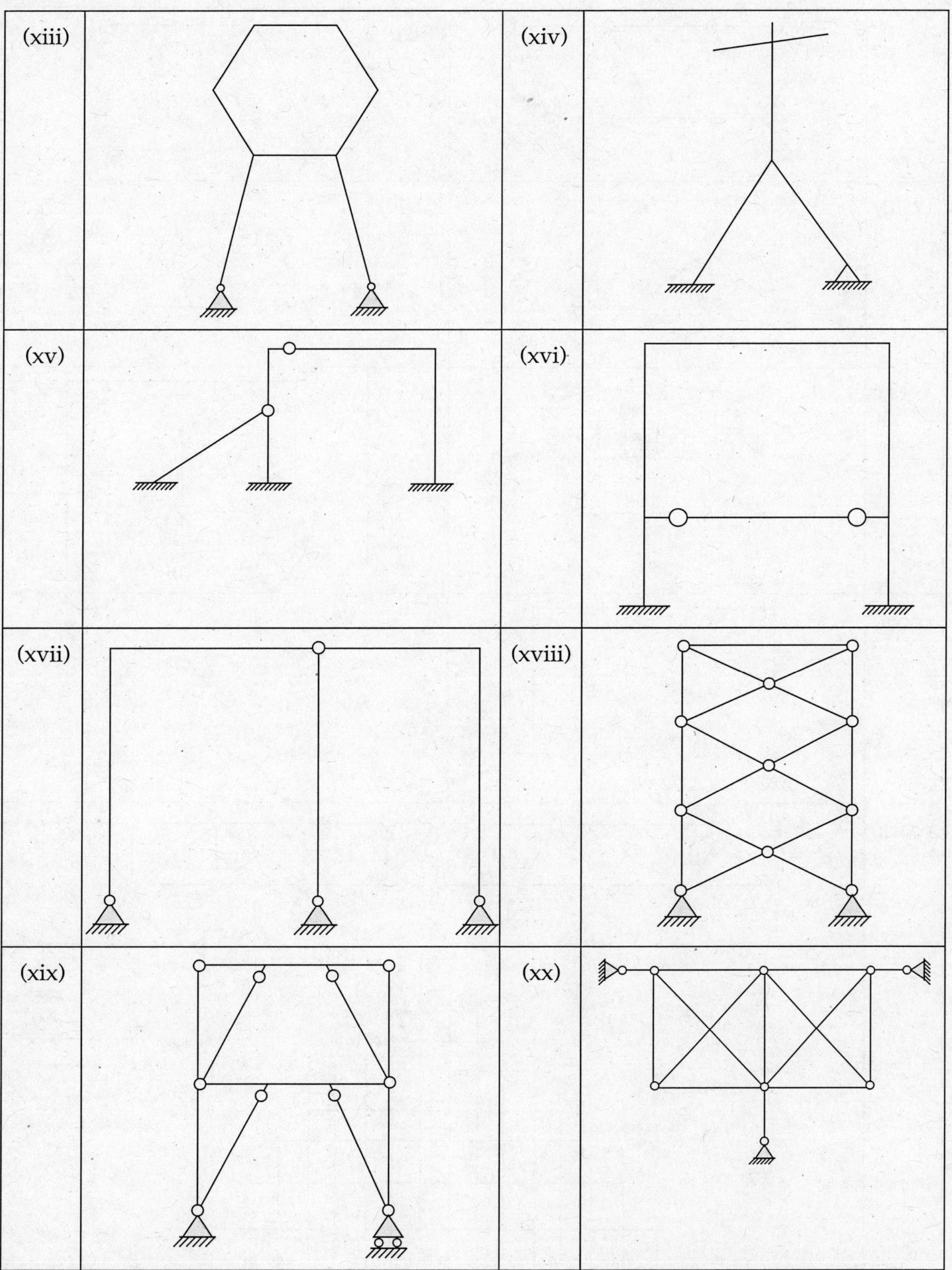

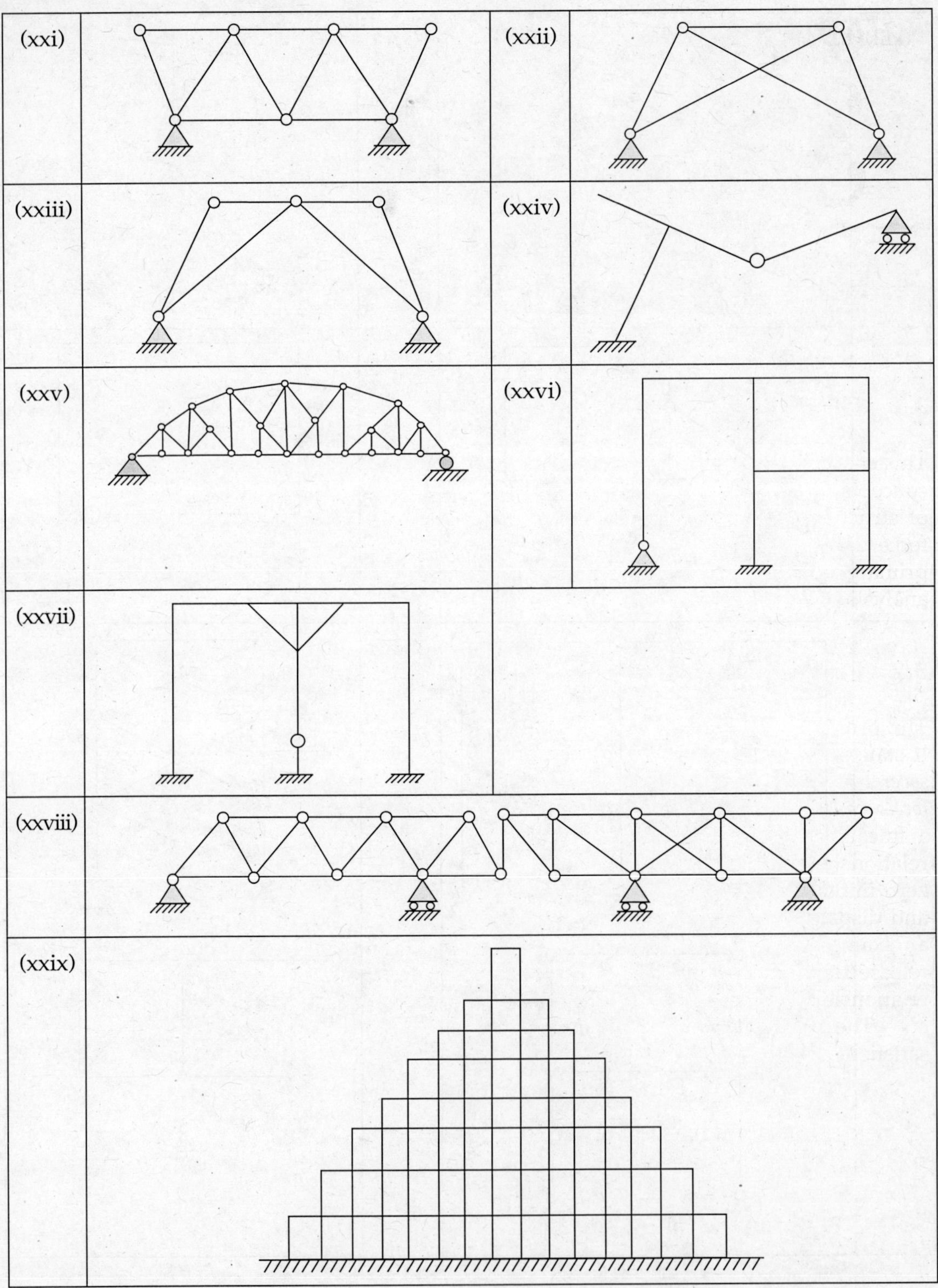
(xxi)
(xxii)
(xxiii)
(xxiv)
(xxv)
(xxvi)
(xxvii)
(xxviii)
(xxix)

Structural Statics

3.1 GENERAL

This chapter presents the analysis of the statically determinate structure. The fundamental concepts of analysis and the relationship between the bending moment and the shear force of structural components are discussed. The bending moment diagram as well as shear force diagram for the structures under various loading conditions are presented. Three primary structural systems, namely, beams, trusses, and frames are considered for the analysis in this chapter.

3.2 PRINCIPLES OF SUPERPOSITION

The principle of superposition is one of the fundamental concepts of structural analysis. It states that *"the joint displacements and the member internal forces of a structure subjected to several external loadings is the algebraic sum of the joint displacements and the member internal forces caused by each of the external loads acting separately"*. This principle is only valid if a linear relationship exists between the loads, stresses, and displacements. This linear relationship means that if the internal forces and deflections of a structural element are zero initially and the same member is subjected to the external loads, the internal forces and displacements will be zero upon the removal of external loads. Consider a spring for an example. When we press the spring axially, its length is reduced. But, as soon as, we release the load, the spring comes back to its original position. This shows that a linear relationship exists between load and displacement.

The principle of superposition is applicable if the following two requirements are satisfied:

1. The material is assumed to be linearly elastic so that Hooke's law is valid. Thus, the displacements are proportional to the applied external loads.
2. The geometry of the structure should not change significantly under the application of external loading. In other words, the magnitude of the displacement components is small. Thus, the small displacement theory is valid.

Figure 3.1 shows the application of the principle of superposition to a simply supported beam subjected to a uniformly distributed load (UDL) and a concentrated load. This beam can be split into two beams, namely, one with the UDL only and the other with the concentrated load only. The vertical displacement (δ) at any point of the beam will be the sum of the displacements due to the UDL (δ_1) and the point load (δ_2) as shown in the figure. All member forces or displacement components can be derived using the same procedure.

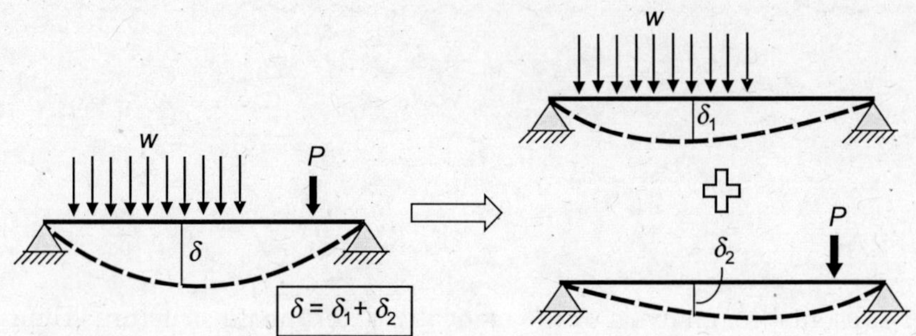

FIGURE 3.1   Principle of superposition.

3.3   BENDING MOMENT AND SHEAR FORCE

The flexural members under the transverse loading induce internal forces, such as *shear force* and *bending moment*. These internal forces usually vary along the length of the member.

Consider a simply supported beam shown in Figure 3.2(a). The beam subjected to an arbitrary load varying with the distance x along the length. Consider an elementary length dx of the bema at a distance of x from the support as shown in Figure 3.2(a). The free-body diagram of the element is shown in Figure 3.2(b). The shear force and bending moment at the left section of the element are assumed as V and M, respectively. These values are assumed as $(V + dV)$ and $(M + dM)$ at the right section.

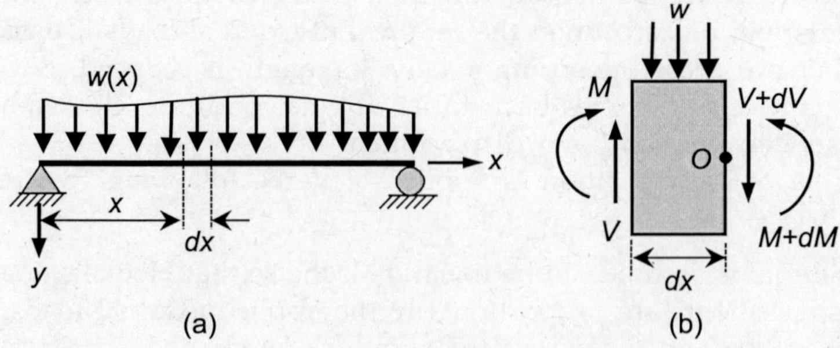

FIGURE 3.2   (a) Simply supported beam with loading, (b) Free-body diagram of the element.

Since the element must be in equilibrium and there are no horizontal forces, the equations of equilibrium are $\sum F_y = 0$ and $\sum M_z = 0$.

$$\sum F_y = 0 \Rightarrow V + dV - V + w.dx = 0$$

or
$$w = -\frac{dV}{dx} \tag{3.1}$$

Taking moment about point 'O' on the right section,

$$\sum M_o = 0 \Rightarrow M + dM - M - Vdx + w.dx.\frac{dx}{2} = 0$$

Neglecting the square of dx, as dx is small and the square of a small term would be further smaller, one would get

$$V = \frac{dM}{dx} \tag{3.2}$$

Remember that Eqn. (3.1) is not valid for the concentrated load. Similarly, Eqn. (3.2) is not valid for a couple applied on a beam. These equations are valid between the concentrated loads and couples, respectively.

From the above equations, one would get $V = -\int w.dx$, and $M = \int V.dx$. If the variation of load, w with the distance can be expressed as an n-the degree curve, then the variation of shear force V will be $(n + 1)$ degree and the variation of bending moment, M would $(n + 2)$ degree along the length. The variation in bending moments, shear forces, and axial forces along the length of the beam is represented by the bending moment diagram (BMD), shear force diagram (SFD), and axial force diagram (AFD), respectively.

Sign Convention

It is very important to adopt a consistent sign convention for drawing the BMD, SFD, and AFD of a flexural member. Figure 3.3 shows the sign convention assumed for axial force, shear force, and bending moment. Tensile axial forces are considered as *positive*, whereas the compressive axial forces are considered as *negative*. If the shear forces acting on both ends of a member induces *a clockwise angle of rotation* (or moment), it is considered as *positive*, else assumed to be *negative* as shown in the figure. If the bending moment acting on both sides of an element results in the compression at the top and the tension at the bottom, this bending moment is a *sagging moment* and hence, is considered as *positive*. Similarly, the end moment results in the tension at the top and compression at the bottom, the bending moment is said to be the *hogging moment* and considered as *negative*. It is worth mentioning that these sign conventions do not change while drawing the BMD, SFD, and AFD of the structural system irrespective of the analysis method adopted.

3.4   TRUSSES

As mentioned earlier, a truss is a structure in which the members are tied together in such a way that they form a series of stable triangular units. All external forces are applied

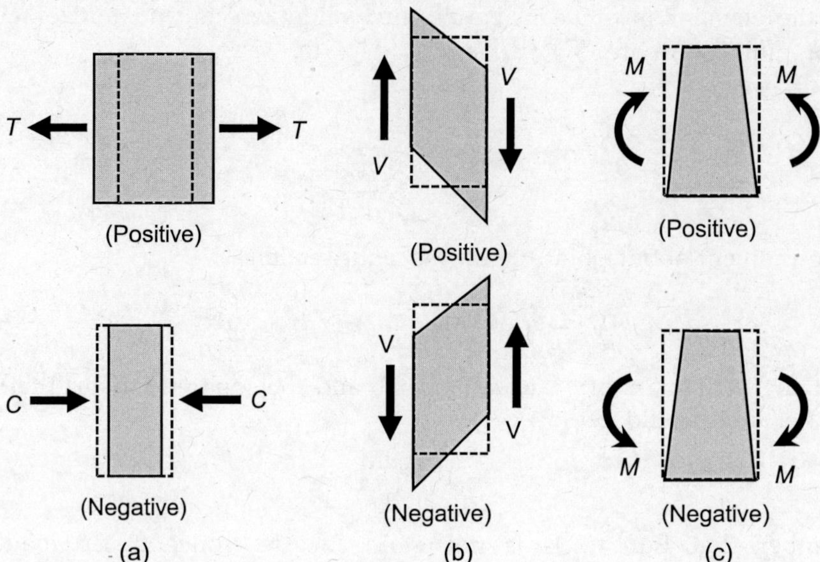

FIGURE 3.3 Sign conventions for (a) axial force, (b) shear force, and (c) bending moment.

at the joints, which are considered as the pined connections for the purpose of analysis. In reality, these joints are not exactly pinned connections as they are generally made by either welding or bolting and some of the members may even be continuous through the joints. However, members of a truss are often relatively slender, and they carry axial forces predominantly. As a result, the joint fixity has a minor effect on the member internal forces. Further, there is a little tendency for the member ends to rotate under the applied external loads. Therefore, the assumption of pinned connections not only simplifies the analysis of a truss, but also results in the fairly accurate estimation of member forces.

Trusses are widely used in the bridges or roof systems of buildings depending on the requirements. The primary components of a truss system are the *top* and *bottom* chords. These chords are held together by means of *diagonals* and *verticals*. The end diagonal is called as *end post*, whereas the side (or end) verticals are known as *hip verticals*.

Qualitatively, trusses can be classified into three types, namely, *simple*, *compound*, and *complex*. Figure 3.4 shows the various configurations of *simple trusses* used in bridges and roof systems. The principal types of truss used in the bridges are the *Pratt*, *Howe*, and *Warren* trusses. The basic difference between these trusses is the arrangement of members and the nature of forces they carry under a specified loading condition. In the *Pratt truss*, all *diagonals* except the end posts carry *tension forces*, and all *verticals* except the hip verticals are subjected to the *compression* forces when the truss is subjected to the dead (gravity) loads only. In the case of the *Howe* truss, all *diagonals* are subjected to the *compression* forces and all *verticals* carry *tension* forces. In the Warren trusses, the diagonals may carry tension or compression forces. The examples of the compound and complex trusses are shown in Figures 3.5 and 3.6, respectively.

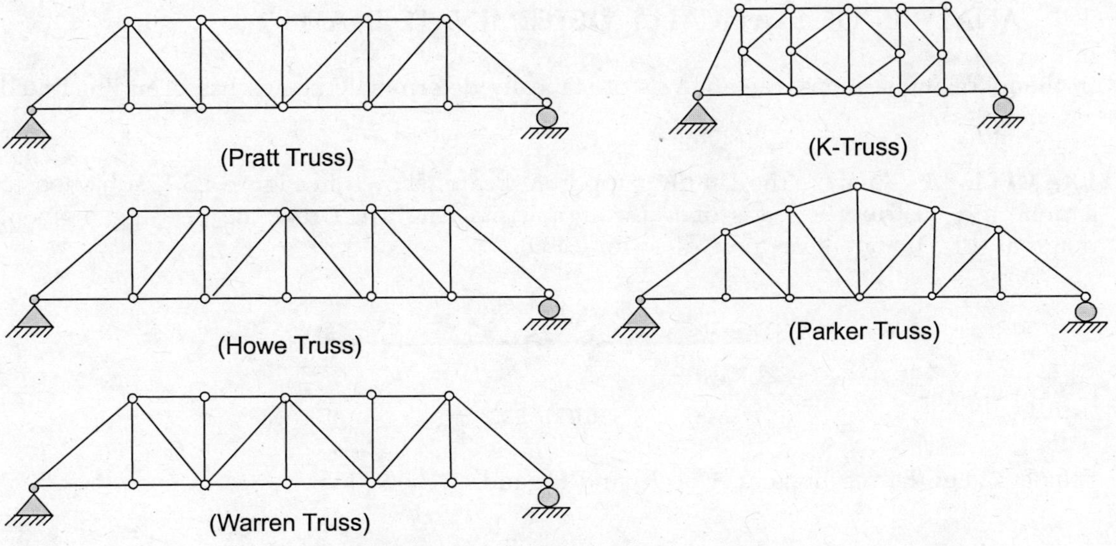

FIGURE 3.4 Examples of simple trusses.

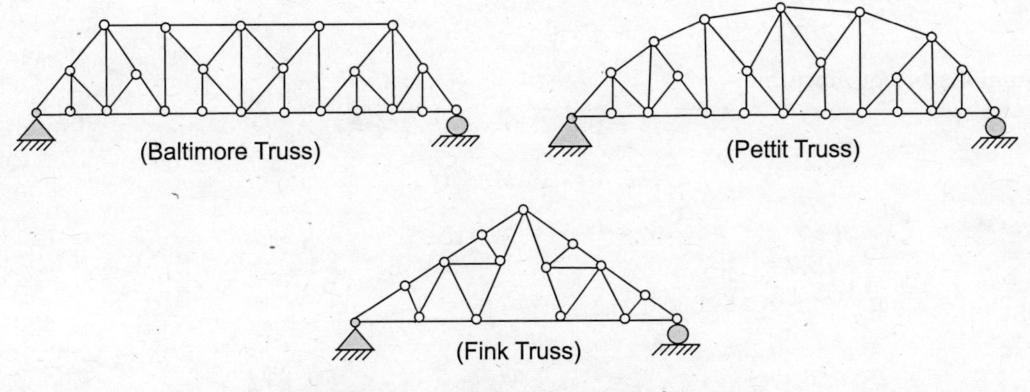

FIGURE 3.5 Examples of compound trusses.

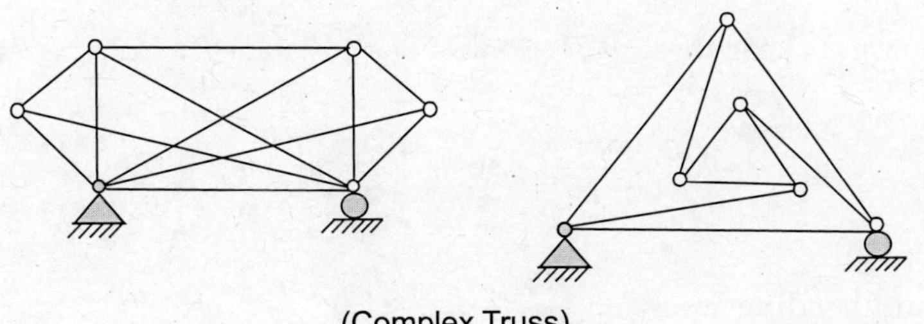

FIGURE 3.6 Examples of complex trusses.

3.5   ANALYSIS OF STATICALLY DETERMINATE BEAMS

In the following sections, the analysis of statically determinate beams has been illustrated as examples.

EXAMPLE 3.1   Analyze the simply supported beam, shown in Figure E3.1, subjected to a uniformly distributed loads of w throughout its length l. Draw the bending moment diagram (BMD) and shear force diagram (SFD).

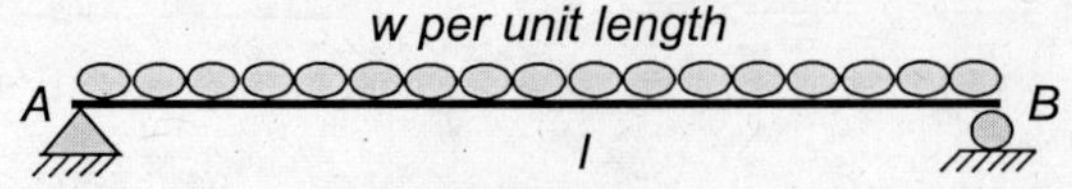

FIGURE E3.1

Solution: Let the reactions at A be R_a and H_a, and at B be R_b.

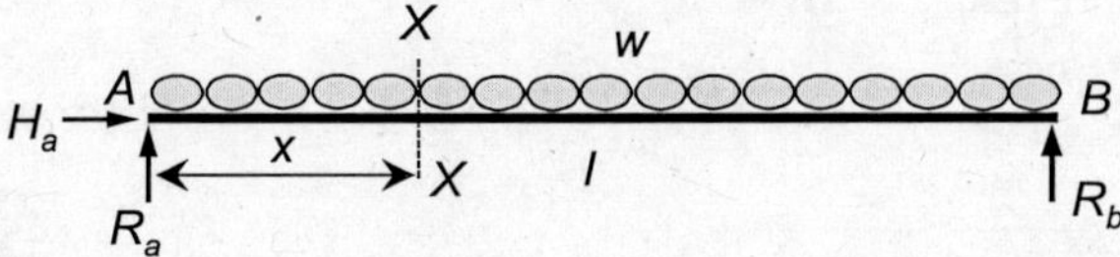

Equations of equilibrium:

$$\Sigma H = 0; \quad \Sigma V = 0; \quad \Sigma M = 0$$

$$\Sigma H = 0 \quad \Rightarrow H_a = 0$$

$$\Sigma V = 0 \quad \Rightarrow R_a + R_b = w.l$$

Taking bending moment at A, $M_a = R_b l - w.l.\dfrac{l}{2}$

$$\sum M_a = 0 \quad \Rightarrow R_b = \frac{wl}{2} \ (\uparrow) \therefore R_a = \frac{wl}{2} \ (\uparrow)$$

Bending moment and shear force at section X-X at a distance x from A, $M_x = R_a x - w.x.\dfrac{x}{2}$; $V_x = R_a - w.x$

$$\therefore \qquad M_x = \frac{wx}{2}(l-x); \qquad V_x = w\left(\frac{l}{2} - x\right)$$

For the maximum bending moment, $\dfrac{dM_{xx}}{dx} = 0 \quad \Rightarrow \quad \dfrac{wl}{2} - wx = 0 \quad \therefore \ x = \dfrac{l}{2}.$

Maximum bending moment, Bending moment and shear force diagrams are shown below:

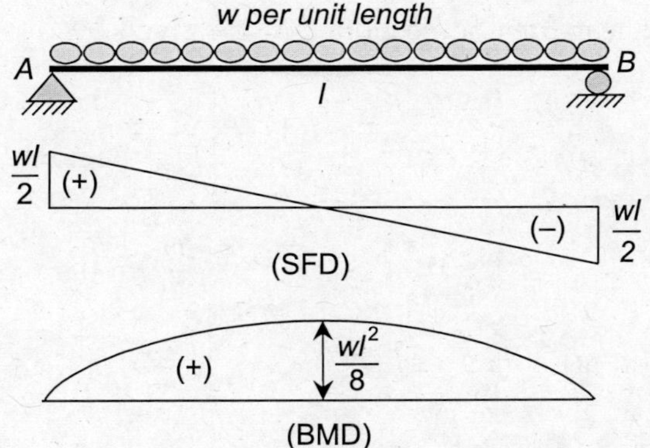

EXAMPLE 3.2 Analyze the simply supported beam subjected to a concentrated load, P as shown in Figure E3.2. Draw the bending moment diagram (BMD) and shear force diagram (SFD).

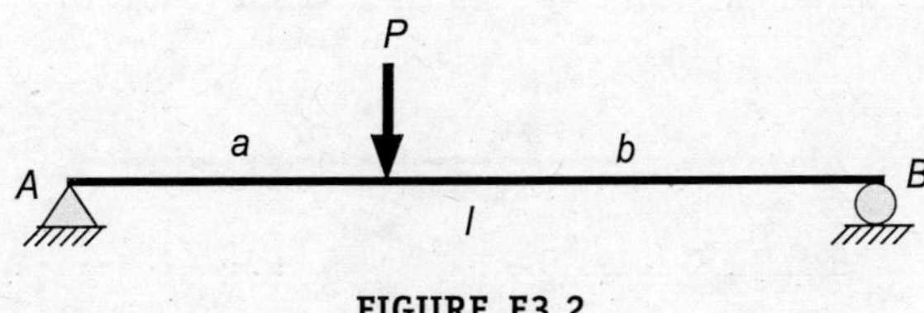

FIGURE E3.2

Solution: Let the reactions at A be R_a and H_a, and at B be R_b.

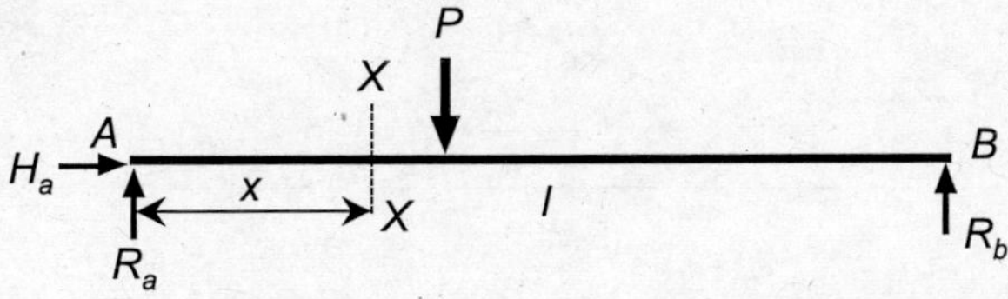

Equations of equilibrium:

$$\Sigma H = 0; \quad \Sigma V = 0; \quad \Sigma M = 0$$

$$\Sigma H = 0 \quad \Rightarrow H_a = 0$$

$$\Sigma V = 0 \quad \Rightarrow R_a + R_b = P$$

Taking bending moment at A, $M_a = R_b l - Pa$

$$\sum M_a = 0 \quad \Rightarrow R_b = \frac{Pa}{l} \ (\uparrow)$$

$$\therefore \quad R_a = P - \frac{Pa}{l} = P\left(\frac{l-a}{l}\right) = \frac{Pb}{l} \ (\uparrow)$$

Bending moment and shear force at section X-X are given by

$$M_x = \begin{cases} R_a x & 0 \le x \le a \\ R_a x - P(x-a) & a \le x \le l \end{cases} \qquad ; \qquad V_x = \begin{cases} R_a & 0 \le x \le a \\ R_b & a \le x \le l \end{cases}$$

$$\therefore \quad M_x = \begin{cases} \dfrac{Pbx}{l} & 0 \le x \le a \\ \dfrac{Pbx}{l} - P(x-a) & a \le x \le l \end{cases} \qquad ; \qquad V_x = \begin{cases} \dfrac{Pb}{l} & 0 \le x \le a \\ \dfrac{Pa}{l} & a \le x \le l \end{cases}$$

For the maximum bending moment, $x = a$

Maximum bending moment, $M_{max} = \dfrac{Pab}{l}$

Bending moment and shear force diagrams are shown below:

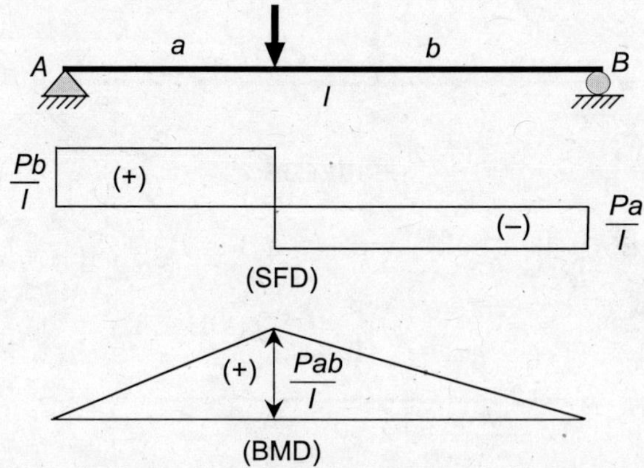

EXAMPLE 3.3   Analyze the simply supported beam subjected to a concentrated bending moment, M as shown in Figure E3.3. Draw the bending moment diagram (BMD) and shear force diagram (SFD).

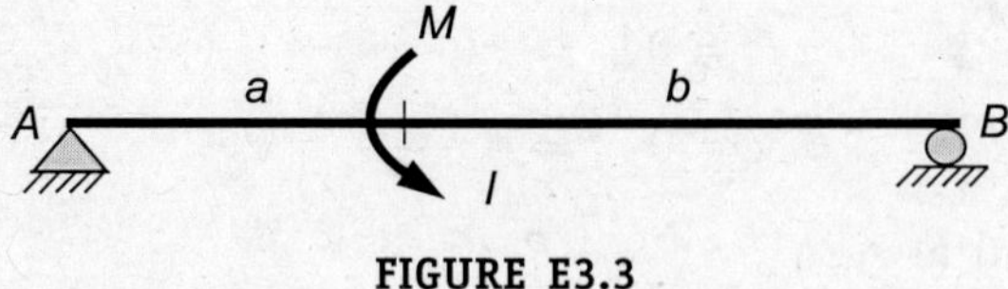

FIGURE E3.3

Solution: Let the reactions at A be R_a and H_a, and at B be R_b.

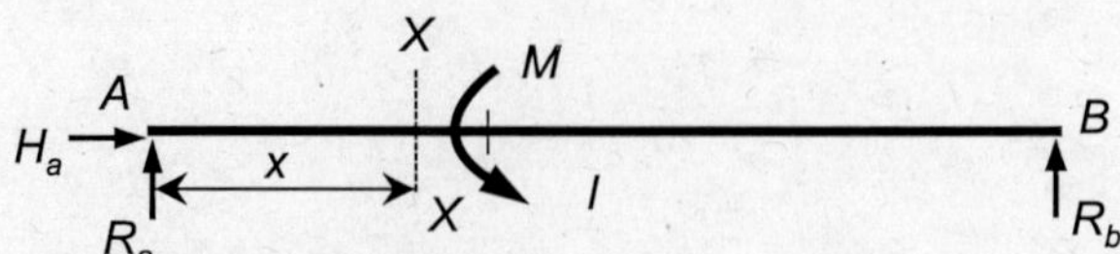

Equations of equilibrium:

$$\Sigma H = 0 \quad \Rightarrow H_a = 0$$
$$\Sigma V = 0 \quad \Rightarrow R_a + R_b = 0$$

Taking bending moment at A, $M_a = R_b l + M$

$$\sum M_a = 0 \quad \Rightarrow R_b = -\frac{M}{l} \ (\downarrow) \qquad \therefore R_a = \frac{M}{l} \ (\uparrow)$$

Bending moment and shear force at section X-X are given by

$$M_x = \begin{cases} R_a x & 0 \leq x \leq a \\ R_a x - M & a \leq x \leq l \end{cases} \quad ; \qquad V_x = \begin{cases} R_a & 0 \leq x \leq a \\ R_b & a \leq x \leq l \end{cases}$$

$$\therefore \quad M_x = \begin{cases} \dfrac{Mx}{l} & 0 \leq x \leq a \\ -M\left(\dfrac{l-x}{l}\right) & a \leq x \leq l \end{cases} \quad ; \qquad V_x = \begin{cases} \dfrac{M}{l} & 0 \leq x \leq a \\ -\dfrac{M}{l} & a \leq x \leq l \end{cases}$$

Bending moment and shear force diagrams are shown below:

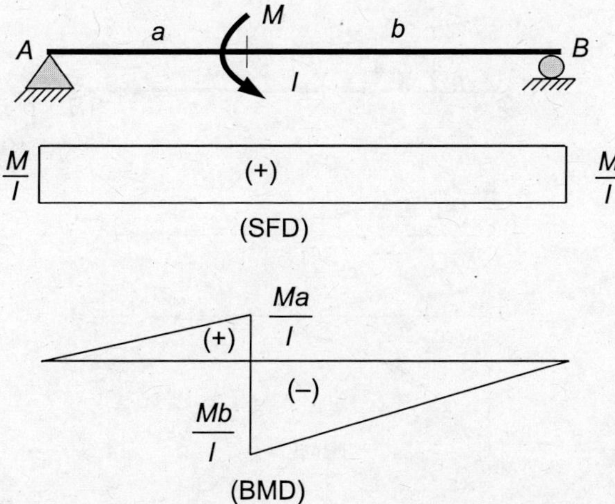

EXAMPLE 3.4　Analyze the cantilever beam subjected to a triangular loading of maximum intensity, w as shown in Figure E3.4. Draw the bending moment diagram (BMD) and shear force diagram (SFD).

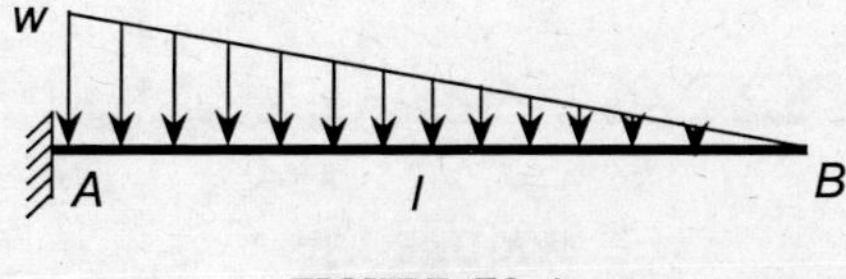

FIGURE E3.4

Solution: Let the reactions and bending moment at A be R_a, H_a and M_a.

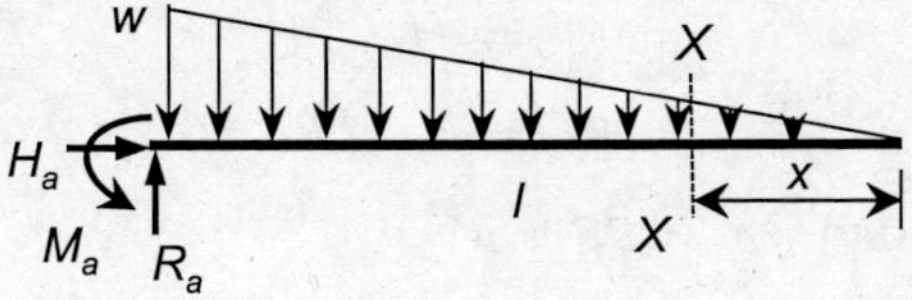

Equations of equilibrium:

$$\sum H = 0 \quad \Rightarrow H_a = 0; \qquad \sum V = 0 \quad \Rightarrow R_a = \frac{wl}{2}$$

Let's take a section X-X at a distance of x from the free end.

Bending moment at section X-X, $M_x = \frac{1}{2}\left(\frac{wx}{l}\right)(x)\left(\frac{x}{3}\right) = \frac{wx^3}{6l}$

Bending moment at the support A, i.e., $x = l$, $M_a - \frac{wl^2}{6} = 0 \quad \Rightarrow M_a = \frac{wl^2}{6}$

Shear force at section X-X , $V_x = \frac{1}{2}\left(\frac{wx}{l}\right)(x) = \frac{wx^2}{2l}$

Shear force at the support A, i.e., $x = l$, $V_a = R_a \quad \Rightarrow V_a = \frac{wl}{2}$

The bending moment and shear force diagrams are shown below:

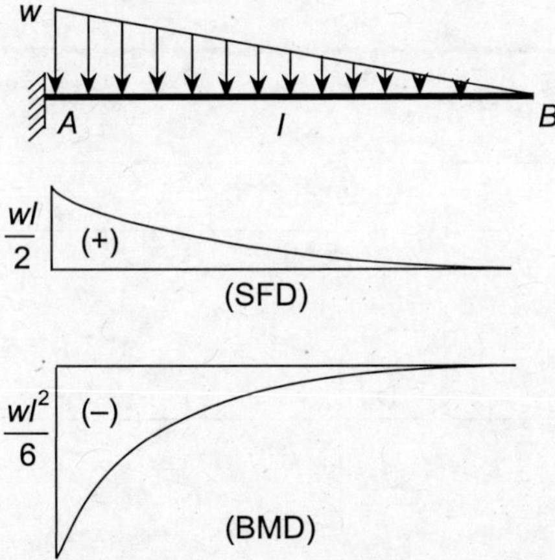

EXAMPLE 3.5   Analyze the beam shown in Figure E3.5 subjected to a concentrated load on one span and a uniformly distributed load on the other span. Two spans are connected using a hinge connection. Draw the bending moment diagram (BMD) and shear force diagram (SFD).

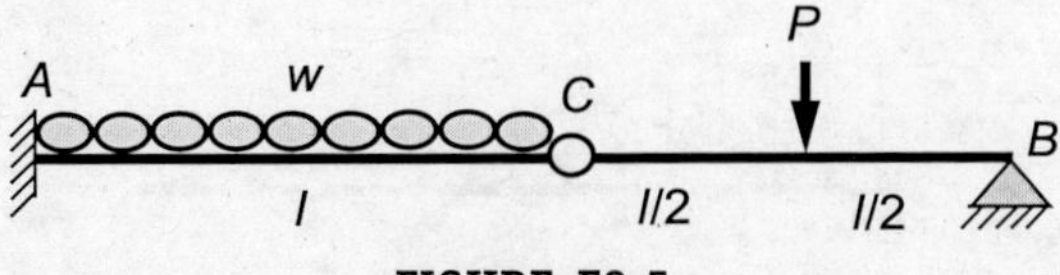

FIGURE E3.5

Solution: Since the beam is not subjected to any external load along the length of the beam, the horizontal component of reactions at the fixed support as well as the hinged support is zero.
Thus, total number of unknown reactions = 2 + 1 = 3.

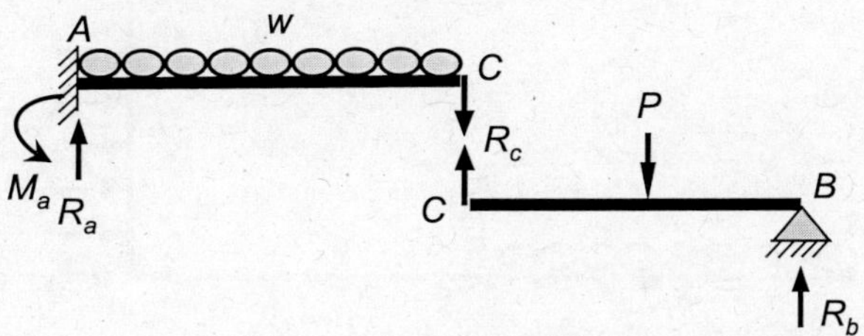

The presence of an internal hinge at 'C' results in one equation of condition, i.e., the bending moment at 'C' is zero.

Accordingly, number of available equilibrium equations = 2 + 1 = 3.

Hence, the beam is *statically determinate*.

In order to solve this determinate beam, the beam must be split into two segments about the hinge 'C'. The free-body diagram of each span is shown in the figure. At point 'C', the vertical reaction, R_c for the spans AC and CB would be equal in magnitude but opposite in sign in order to maintain the shear equilibrium between these spans at this point.

For span CB:

$$\Sigma V = 0 \quad \Rightarrow R_b + R_c = P$$

Taking the moment about C, $M_{cb} = R_b l - P\dfrac{l}{2}$

Since, $M_{cb} = 0;$ $R_b = R_c = \dfrac{P}{2}$

For span AC:

$$R_a = \frac{P}{2} + wl\,; \qquad\qquad M_a = \frac{Pl}{2} + \frac{wl^2}{2}$$

Bending moment and shear force diagrams are shown below:

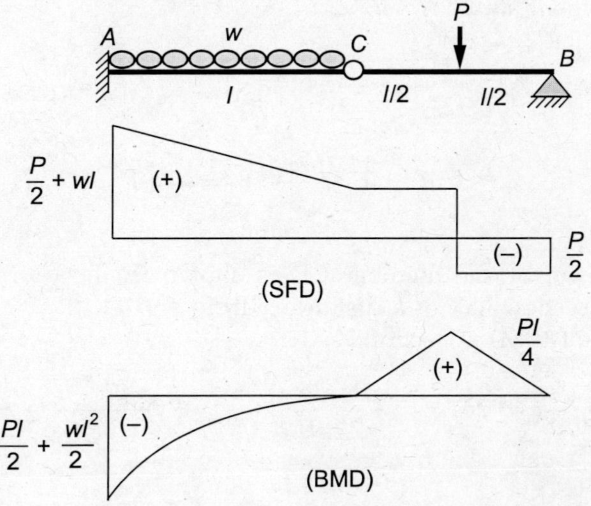

EXAMPLE 3.6 Analyze the beam shown in Figure E3.6 and draw the bending moment diagram (BMD) and shear force diagram (SFD).

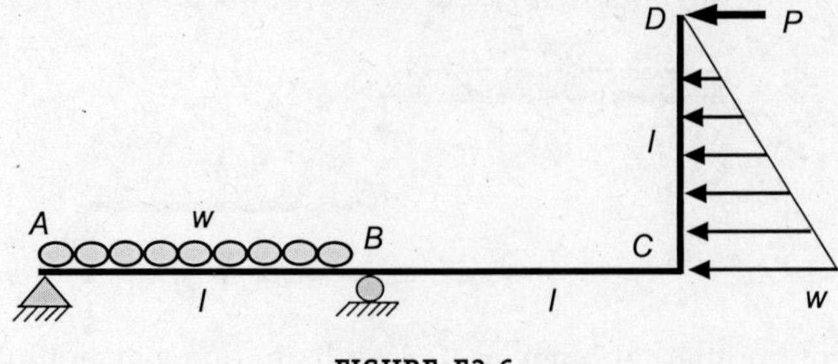

FIGURE E3.6

Solution: Let the reactions at A be R_a and H_a, and at B be R_b.

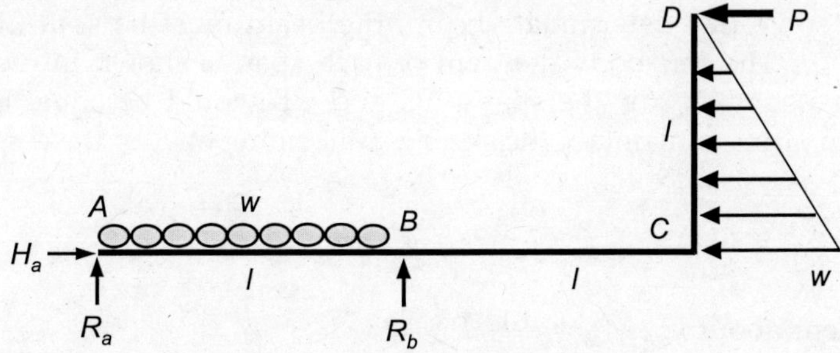

The support reactions can be determined using the equations of equilibrium.

$$\sum H = 0 \quad \Rightarrow H_a = P + \frac{wl}{2}$$

$$\sum V = 0 \quad \Rightarrow R_a + R_b = \frac{wl}{2}$$

Sum of bending moment about A, i.e., $\sum M_a = 0$

$$\Rightarrow R_b l - \frac{wl^2}{2} + Pl + \frac{1}{2} wl \left(\frac{l}{3}\right) = 0 \qquad \therefore R_b = \frac{wl}{3} - P$$

Thus,
$$R_a = \frac{wl}{2} - \left(\frac{wl}{3} - P\right) = \frac{wl}{6} + P$$

Member DC
The free-body diagram of the member CD is shown in the Figure. Let's take a section X-X at a distance x from end D.
    Bending moment (BM) at section X-X,

$$M_x = Px + \frac{1}{2}\left(\frac{wx}{l}\right)x\left(\frac{x}{3}\right) = Px + \frac{wx^3}{6l}$$

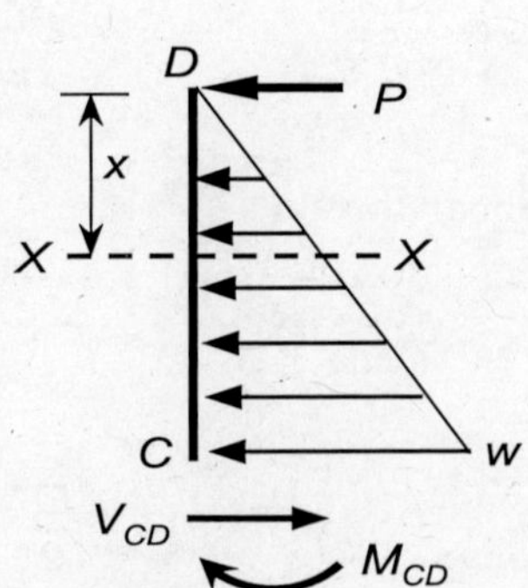

Thus, BM at point D ($x = 0$) is $M_D = 0$

BM at point C ($x = l$) is $M_{CD} = Pl + \dfrac{wl^2}{6}$

Shear force (SF) at section X-X, $V_x = -P - \dfrac{1}{2}\left(\dfrac{wx}{l}\right)x = -\left(P + \dfrac{wx^2}{2l}\right)$

At point D ($x = 0$), SF is $V_D = -P$;   At point C ($x = l$), SF is $V_{CD} = -\left(P + \dfrac{wl}{2}\right)$

Span BC

The free-body diagram of the member AC is shown in the figure. The shear force and bending moment at joint D for member CD will act as axial force and bending moment at the same joint for member AC, respectively.

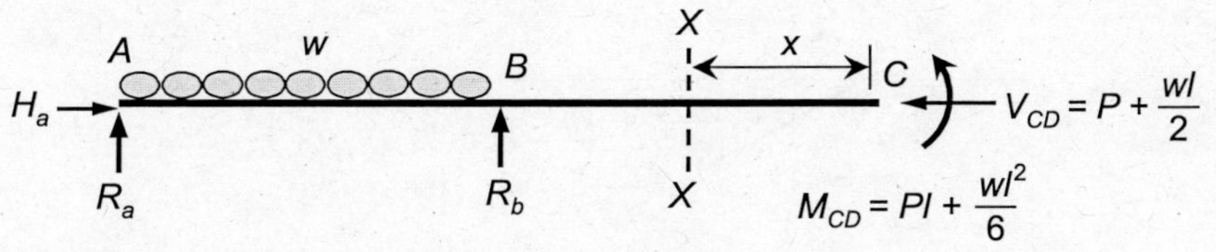

Let's take a section X-X at a distance x from end C.

Bending moment (BM) at section X-X, $M_x = Pl + \dfrac{wl^2}{6}$

Thus, BM at point C ($x = 0$) and B ($x = l$) is $M_{CB} = M_{CD} = Pl + \dfrac{wl^2}{6}$

Shear force on span BC is zero as there is no transverse force acting on it.

Span AB

Let's take a section X-X at a distance x from support A.

Bending moment (BM) at section X-X, $M_x = R_a x - \dfrac{wx^2}{2} = \left(\dfrac{wl}{6} + P\right)x - \dfrac{wx^2}{2}$

At joint A ($x = 0$), $M_a = 0$

At joint B ($x = l$), $M_b = \left(\dfrac{wl}{6} + P\right)l - \dfrac{wl^2}{2} = Pl - \dfrac{wl^2}{3}$

Shear force (SF) at section X-X, $V_x = R_a - wx = \left(\dfrac{wl}{6} + P\right) - wx$

At joint A ($x = 0$), $V_a = R_a = \dfrac{wl}{6} + P$

At joint B ($x = l$), $V_b = \left(\dfrac{wl}{6} + P\right) - wl = P - \dfrac{5wl}{6}$

Bending moment and shear force diagrams are shown below:

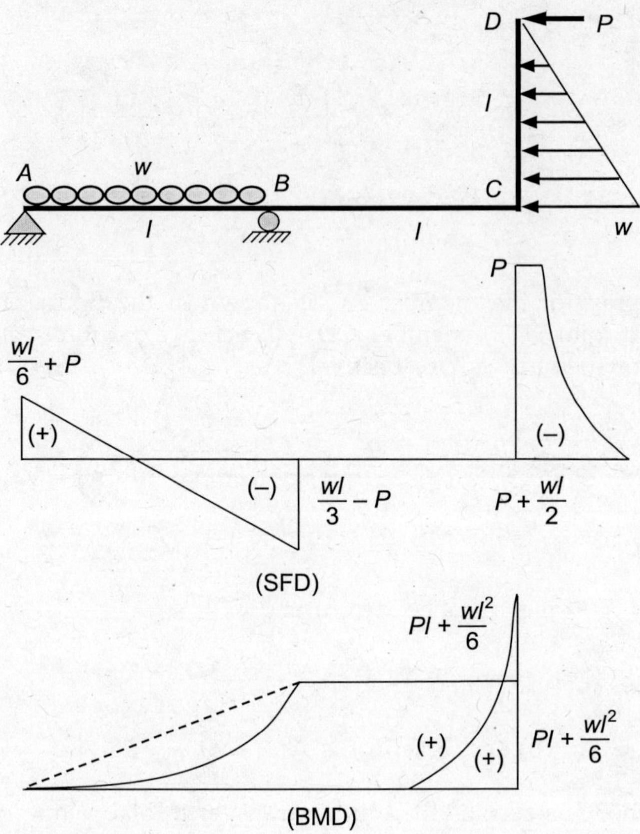

In the above example, the bending moment for the span AB is obtained by superimposing two BMDs, namely a triangular shape due to end moment and a parabolic shape due to UDL.

If a beam is subjected to several external loads, the bending moment diagram as well as the shear force diagram can be obtained by combining the individual respective diagrams for each load using the *principle of superposition*. Table 3.1 shows the application of the principle of superposition to derive the bending moment diagrams of beams subjected to more than one loading.

TABLE 3.1   Bending moment diagrams for beams with different loading and end conditions

Beam	BMD for each load	Combined BMD
M_1 M_2 l	M_1 (+) (+) M_2	M_1 (+) M_2

Beam	*BMD for each load*	*Combined BMD*
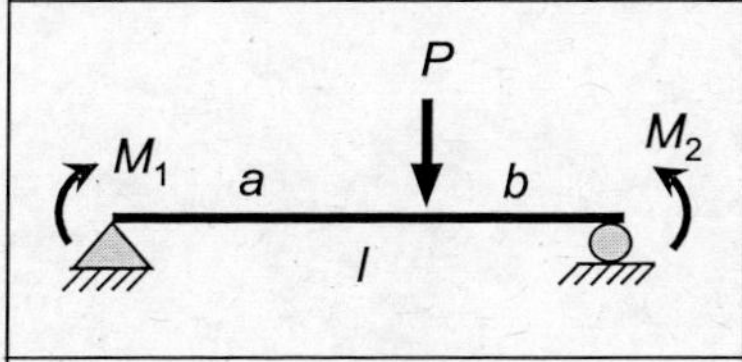 P applied at distance a, b along span l; end moments M_1, M_2 (pinned–roller)	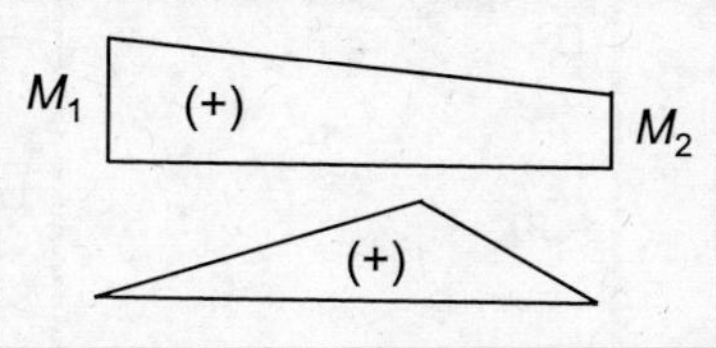 M_1, $(+)$, M_2; $(+)$	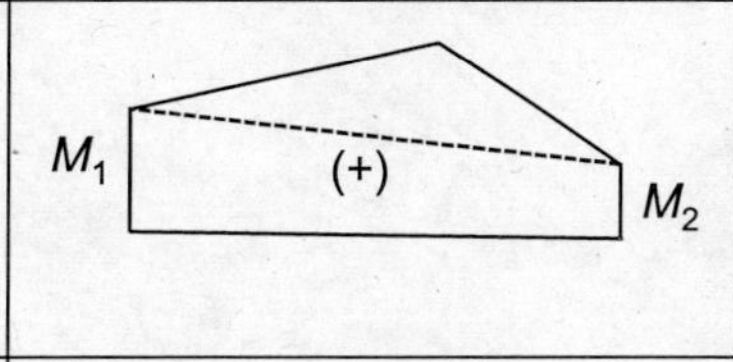 M_1, $(+)$, M_2
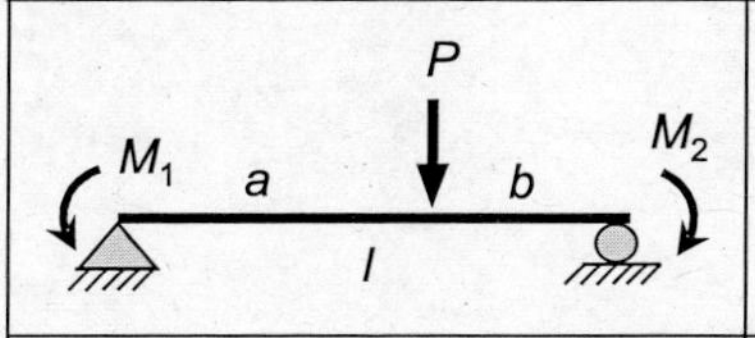 P applied at distance a, b along span l; end moments M_1, M_2	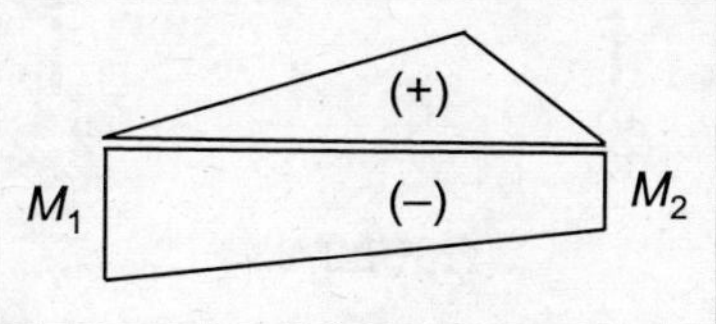 $(+)$; M_1, $(-)$, M_2	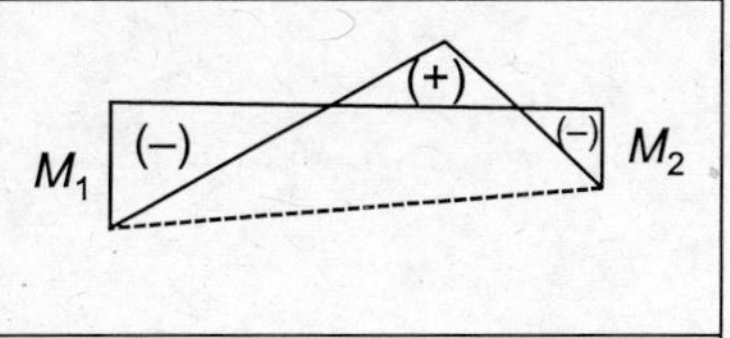 $(+)$; M_1, $(-)$, $(-)$, M_2
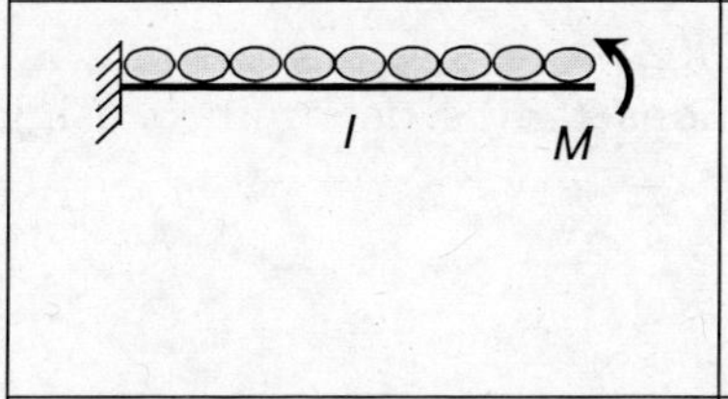 Uniformly distributed load over l; end moment M (cantilever)	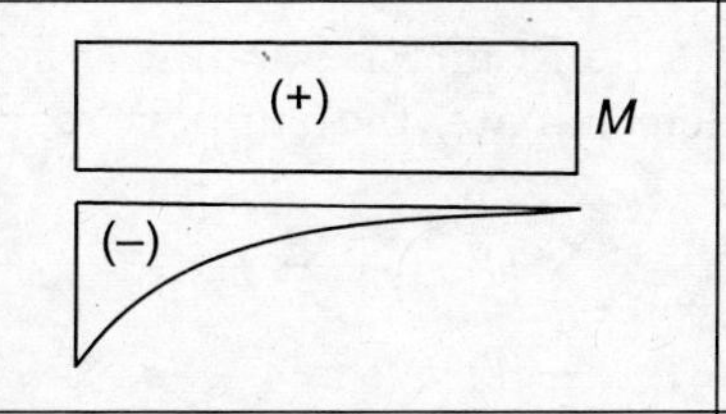 $(+)$, M; $(-)$	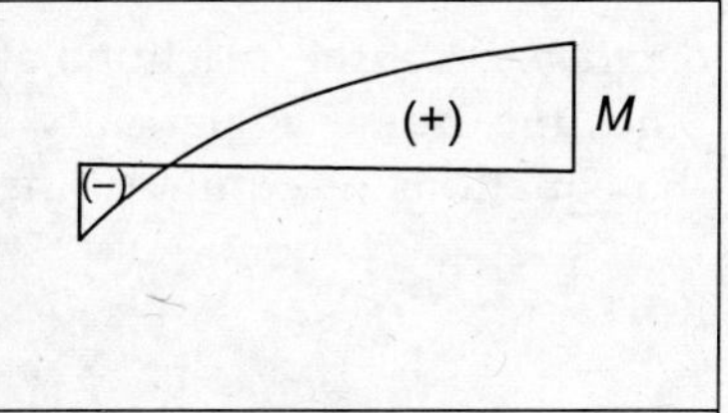 $(-)$, $(+)$, M
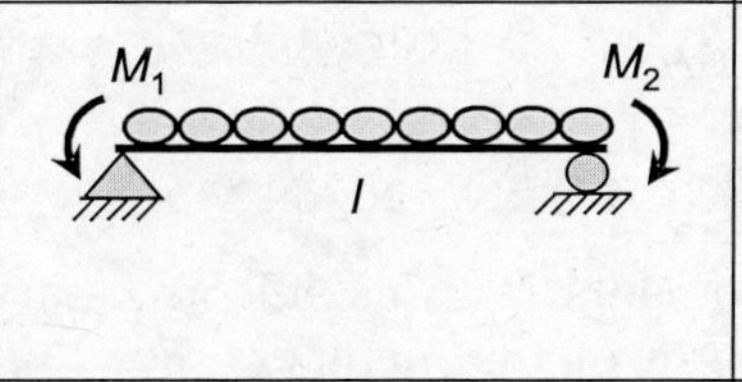 Uniformly distributed load over l; end moments M_1, M_2	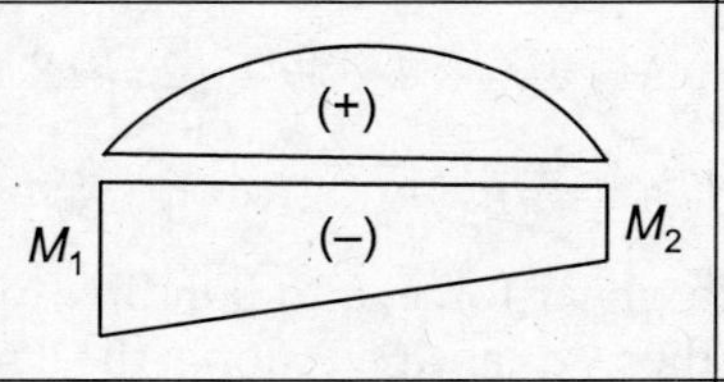 $(+)$; M_1, $(-)$, M_2	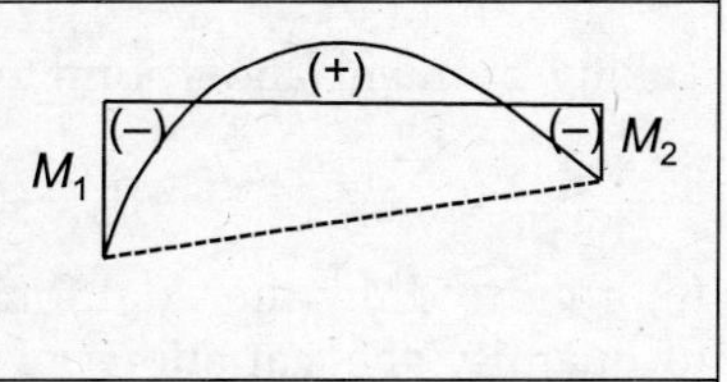 $(+)$; M_1, $(-)$, $(-)$, M_2
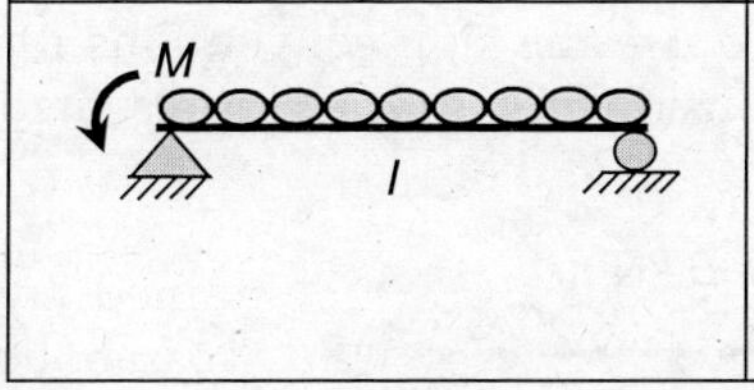 Uniformly distributed load over l; end moment M	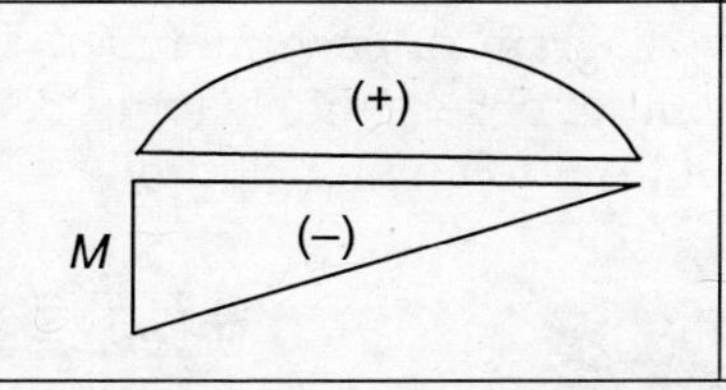 $(+)$; M, $(-)$	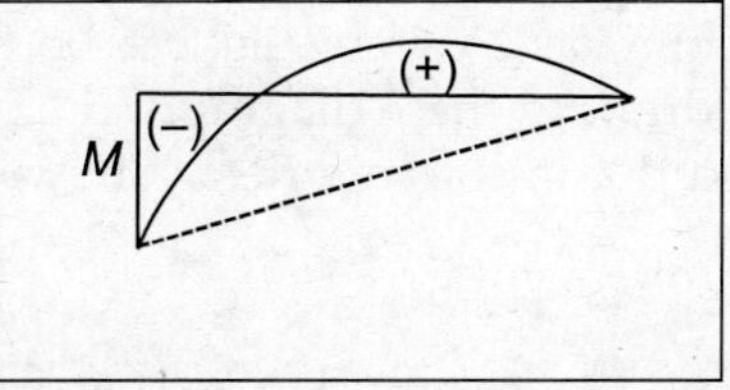 $(+)$; M, $(-)$

3.6   ANALYSIS OF STATICALLY DETERMINATE FRAMES

The analysis of statically determinate frames is similar to that of the beams. The internal forces in the frame members consist of axial force, shear force, and bending moment. These forces are determined by drawing the free-body diagram at any section of the frame. Accordingly, the axial force, the shear force, and the bending moment diagrams can be drawn. While selecting the sign convention for a frame member, it is assumed that the point of observation is within the frame, not outside the frame. The following examples illustrate the procedure to analyze the determinate rigid frames.

EXAMPLE 3.7   Analyze the rigid frame shown in Figure E3.7 and draw the axial force, bending moment, and shear force diagrams.

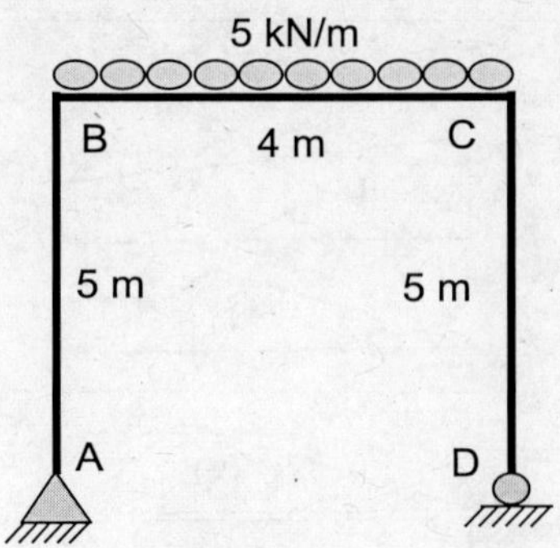

FIGURE E3.7

Solution: Let the reactions at A be R_a and H_a, and at D be R_d.

Since the frame is statically determinate, the support reactions can be determined using the equations of equilibrium.

$$\sum H = 0 \quad \Rightarrow H_a = 0$$

$$\sum V = 0 \quad \Rightarrow R_a + R_b = 4 \times 5 \text{ kN} = 20 \text{ kN}$$

Taking moment about joint A, $M_a = R_d \times 4 - 5 \times 4 \times 2 = 4R_d - 40$

Since $\sum M_a = 0$; $4R_d - 40 = 0 \Rightarrow R_d = 10$ kN $\quad \therefore R_a = (20 - 10)$ kN $= 10$ kN

In order to draw the axial force, shear force, and bending moment diagrams, it is necessary to quantify the variations of these quantities along the length of the members. For this, one has to draw the free-body diagram of each member. To use the sign conventions for drawing these diagrams, the point of observation should be inside the frame as represented a dot in the middle as shown in the following figure.

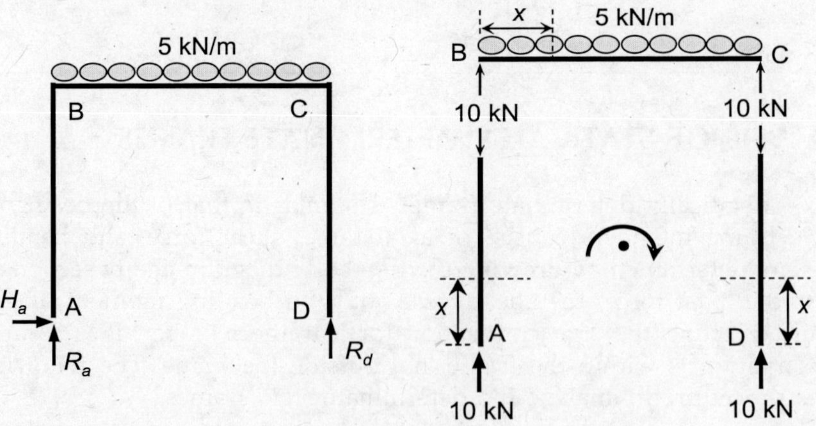

Member AB

Let's take a section at a distance of x from end A.

Axial force, $P_x = -10 \quad (0 \le x \le 5)$

Shear force, $V_x = 0$ $(0 \leq x \leq 5)$

Bending moment, $M_x = 0$ $(0 \leq x \leq 5)$

Member BC

Let's take a section at a distance of x from end B.

Axial force, $P_x = 0$ $(0 \leq x \leq 4)$

Shear force, $V_x = (10 - 5x)$ $(0 \leq x \leq 4)$

Bending moment, $M_x = (10x - 2.5x^2)$ $(0 \leq x \leq 4)$

Member CD

Let's take a section at a distance of x from end D.

Axial force, $P_x = -10$ $(0 \leq x \leq 5)$

Shear force, $V_x = 0$ $(0 \leq x \leq 5)$

Bending moment, $M_x = 0$ $(0 \leq x \leq 5)$

Axial force, shear force, and bending moment diagrams are shown in the following figure.

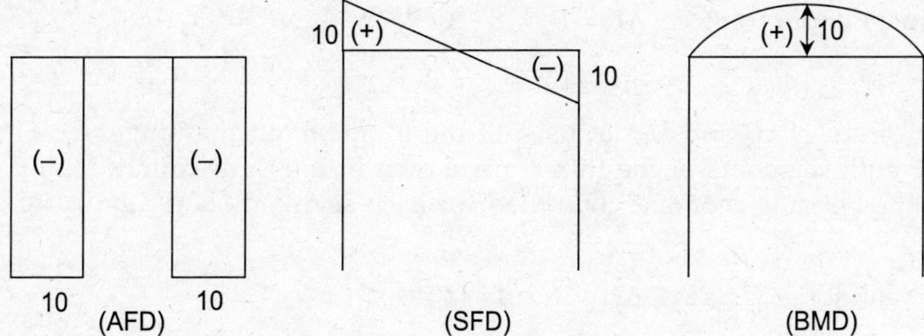

EXAMPLE 3.8 Analyze the rigid frame shown in Figure E3.8 and draw the axial force, bending moment, and shear force diagrams.

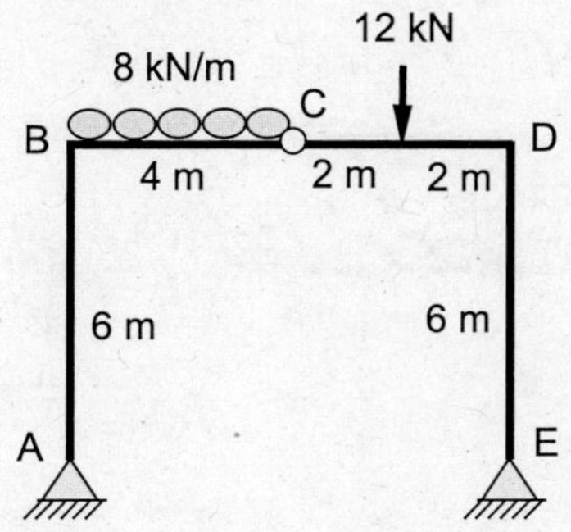

FIGURE E3.8

Solution: Let the reactions at support A be R_a and H_a, and at support, E be R_e and H_e.

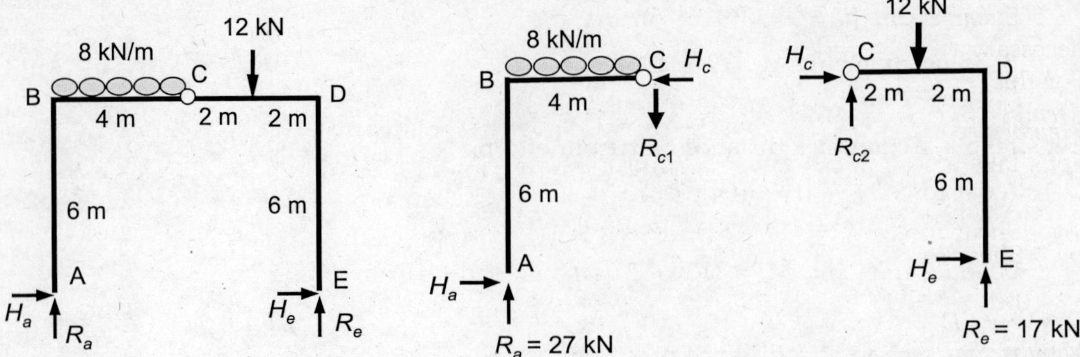

Since the frame is statically determinate, the support reactions can be determined using the equations of equilibrium.

$$\sum H = 0 \quad \Rightarrow H_a + H_e = 0$$

$$\sum V = 0 \quad \Rightarrow R_a + R_e = (4 \times 8 + 12)\ \text{kN} = 44\ \text{kN}$$

Taking moment about joint A, $M_a = R_e \times 8 - 8 \times 4 \times 2 - 12 \times 6 = 8R_e - 136$

$$\sum M_a = 0 \quad \Rightarrow 8R_e - 136 = 0 \quad \therefore R_e = 17\ \text{kN}\,;\ \ R_a = 27\ \text{kN}$$

To find the values of H_a and H_e, let us split the frame about the hinge 'C'. The free-body diagrams of both segments of the frame are shown in the figure. Since joint C is a hinge, the sum of the bending moment computed for each segment should be zero.

Segment ABC
Taking moment about C, $4R_a - 6H_a - 8 \times 4 \times 2 = 0$

$$\Rightarrow 4 \times 27 - 6H_a - 8 \times 4 \times 2 = 0 \qquad \therefore H_a = 7.33\ \text{kN}\ (\rightarrow)$$

Segment CDE
Taking moment about C, $4R_e + 6H_e - 12 \times 2 = 0$

$$\Rightarrow 4 \times 17 + 6H_e - 24 = 0 \qquad \therefore H_e = -7.33\ \text{kN}\ (\leftarrow)$$

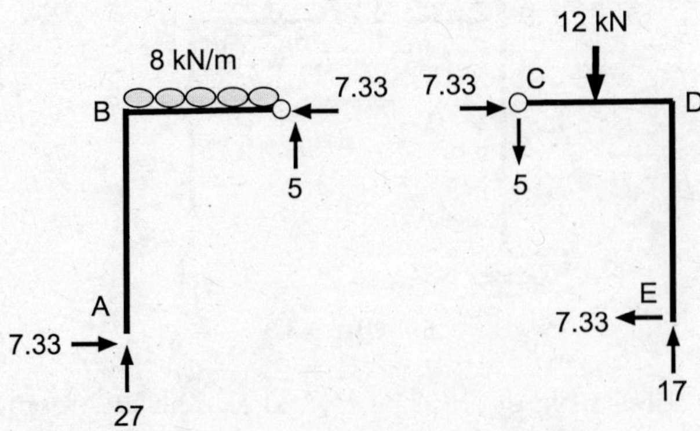

Check that $H_a + H_e = 0$!

In order to draw the axial force, shear force and bending moment diagrams, it is necessary to quantify the variations of these quantities along the length of the members. For this, one has to draw the free-body diagram of each member.

Member AB

Let's take a section at a distance of x from end A.

Axial force, $P_x = -27$ kN $(0 \le x \le 6)$

Shear force, $V_x = -7.33$ kN $(0 \le x \le 6)$

Bending moment, $M_x = -7.33x$ kNm $(0 \le x \le 6)$

Member BC

Let's take a section at a distance of x from end C.

Axial force, $P_x = -7.33$ kN $(0 \le x \le 4)$

Shear force, $V_x = (8x - 5)$ $(0 \le x \le 4)$

Bending moment, $M_x = (5x - 4x^2)$ $(0 \le x \le 4)$

Member CD

Let's take a section at a distance of x from end C.

Axial force, $P_x = -7.33$ kN $(0 \le x \le 4)$

Shear force, $V_x = \begin{cases} -5.0 \text{ kN} & (0 \le x \le 2) \\ -17.0 \text{ kN} & (2 \le x \le 4) \end{cases}$

Bending moment, $M_x = \begin{cases} -5x \text{ kNm} & (0 \le x \le 2) \\ -5x - 12(x-2) \text{ kNm} & (2 \le x \le 4) \end{cases}$

Member DE

Let's take a section at a distance of x from end E.

Axial force, $P_x = -17$ kN $(0 \le x \le 6)$

Shear force, $V_x = 7.33$ kN $(0 \le x \le 6)$

Bending moment, $M_x = -7.33x$ kNm $(0 \le x \le 6)$

Axial force, shear force, and bending moment diagrams are shown in the following figures.

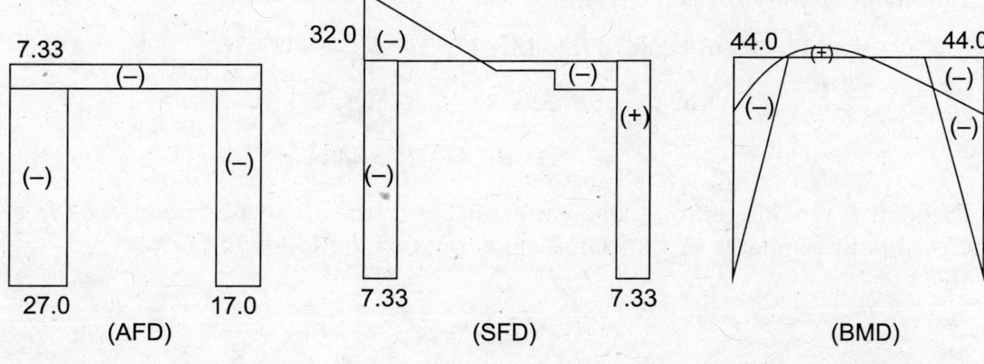

EXAMPLE 3.9   Analyze the rigid frame shown in Figure E3.9 and draw the axial force, bending moment, and shear force diagrams.

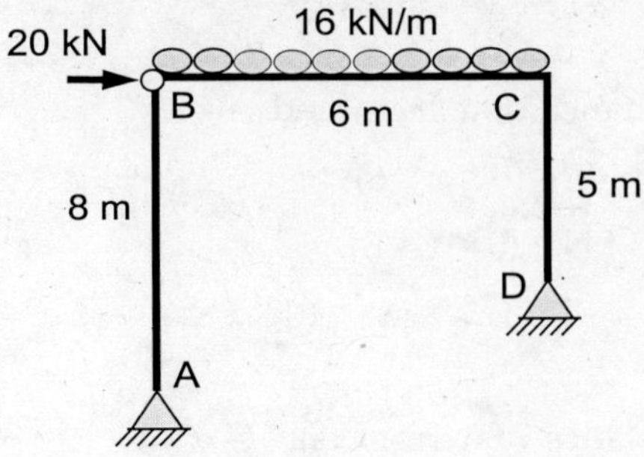

FIGURE E3.9

Solution: Let the reactions at support A be R_a and H_a, and at support, D be R_d and H_d.

Since the frame is statically determinate, the support reactions can be determined using the equations of equilibrium.

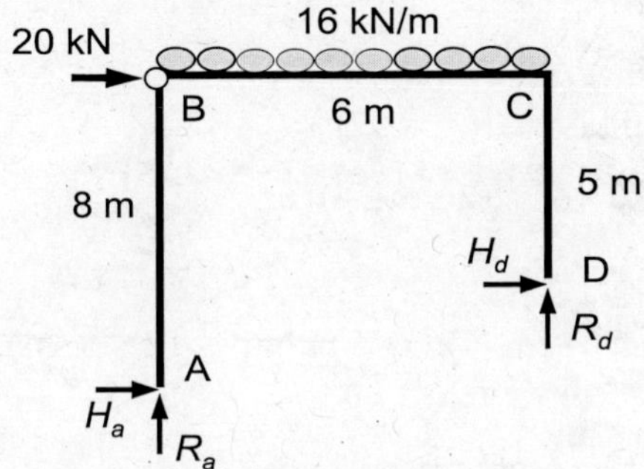

Equilibrium Equations:

$$\sum H = 0 \quad \Rightarrow H_a + H_d + 20 = 0 \tag{i}$$

$$\sum V = 0 \quad \Rightarrow R_a + R_d = (16)(6) \text{ kN} = 96 \text{ kN} \tag{ii}$$

Taking moment about joint A,

$$M_a = R_d(6) - 20(8) - 16(6)(3) - H_d(3) = 6R_d - 448 - 3H_d$$

$$\sum M_a = 0 \quad \Rightarrow 6R_d - 448 - 3H_d = 0$$

$$\Rightarrow 6R_d - 3H_d = 448 \tag{iii}$$

Since the joint B is a hinged one, one can split the frame about this point. The free-body diagrams of both segments of the frame are shown in the following figure.

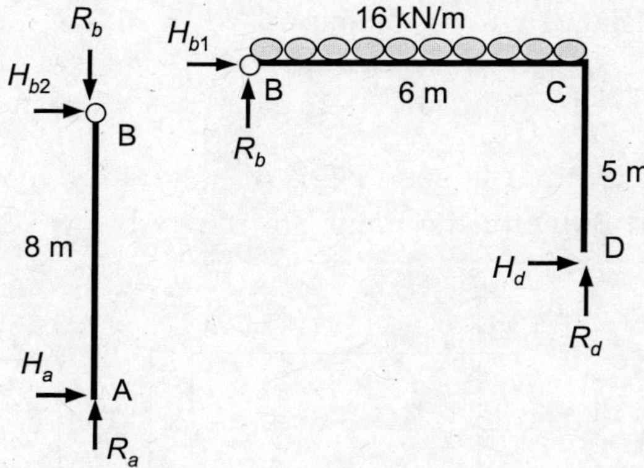

Since joint B is a hinge, the sum of bending moments computed for each segment should be zero.

Segment BCD

Taking moment about B, $6R_d + 5H_d - 16(6)(3) = 0$

$$\Rightarrow 6R_d + 5H_d = 288 \tag{iv}$$

Solving Eqns. (iii) and (iv), one can get $H_d = -20$ kN ($\leftarrow$) $\therefore H_a = 0$

The same value of the horizontal component of support reactions can be obtained if the equilibrium of segment AB is considered. Taking moment about A, one would get $6H_a = 0$ $\Rightarrow H_a = 0$ $\therefore H_d = -20$ kN ($\leftarrow$).

Putting the value of H_d in Eqn. (iv) or (iii) and using Eqn. (ii),

$$R_d = 64.67 \text{ kN}(\uparrow) \quad \therefore R_a = (96 - 64.67) \text{ kN} = 31.33 \text{ kN}(\uparrow)$$

In order to draw the axial force, shear force, and bending moment diagrams, it is required to draw the free-body diagram of the frame with external loading and support reactions as shown below.

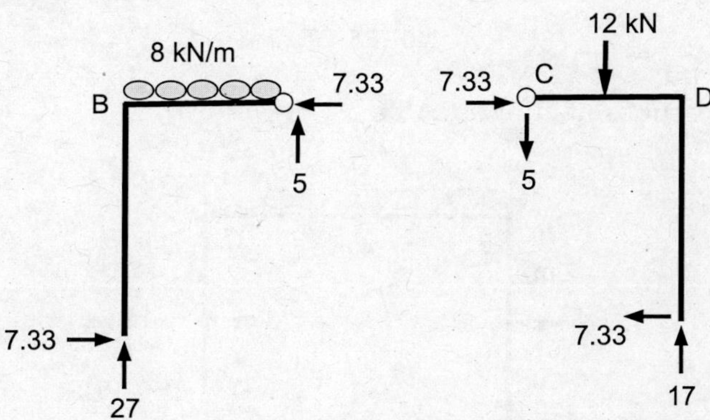

Member BC

For the maximum positive bending moment in span BC, $\dfrac{dM_x}{dx} = 0$

Bending moment at a distance x from the joint B, $M_x = 31.33x - 8x^2$

$$\because \qquad \frac{dM_x}{dx} = 0 \quad \Rightarrow 31.33 - 16x = 0 \quad \therefore\ x = 1.96 \text{ m}$$

Thus, the maximum positive BM in span BC is $M_{max} = 31.33x - 8x^2 = 30.67$ kNm

The AFD, SFD, and BMD for the frame are shown below.

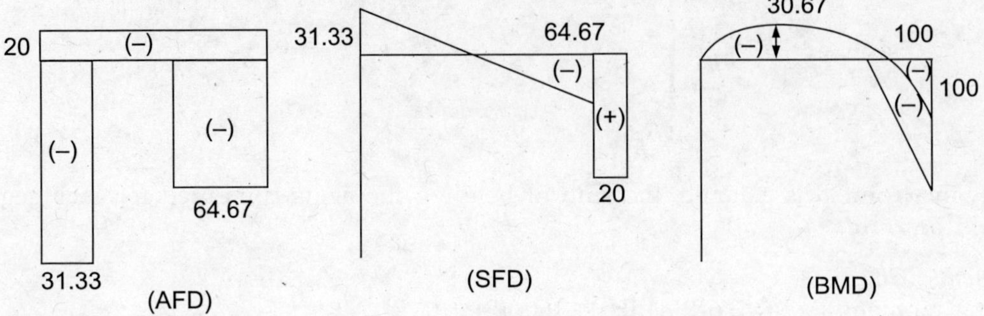

EXAMPLE 3.10   Analyze the rigid frame shown in Figure E3.10 and draw the axial force, bending moment, and shear force diagrams.

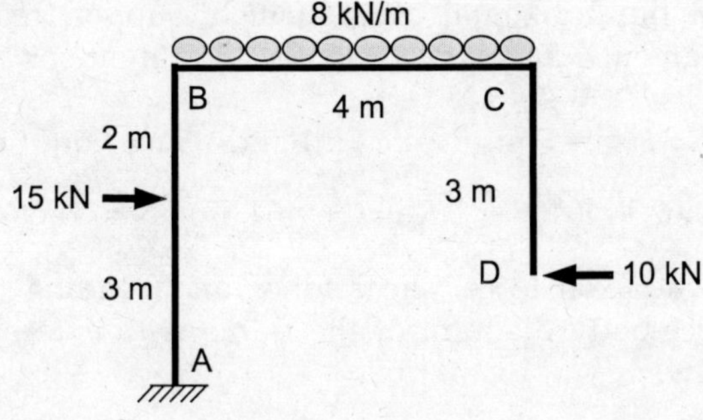

FIGURE E3.10

Solution: Let the reactions at support A be R_a, H_a, and M_a.

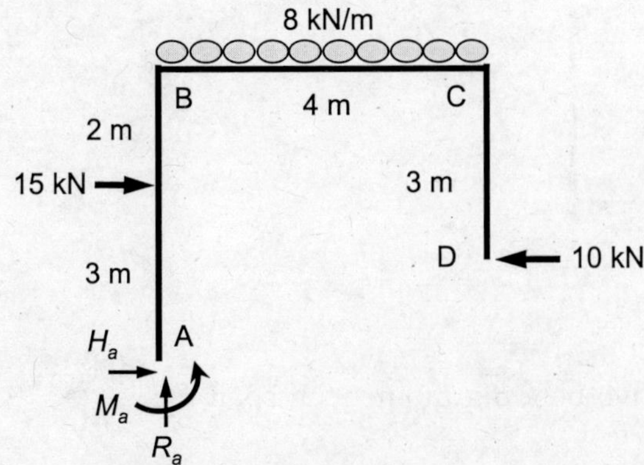

Equilibrium Equations:

$$\sum H = 0 \quad \Rightarrow H_a + 15 - 10 = 0 \qquad \therefore H_a = -5 \text{ kN}(\leftarrow)$$

$$\sum V = 0 \quad \Rightarrow R_a = (8)(4) \text{ kN} = 32 \text{ kN } (\uparrow)$$

Taking moment about joint A,

$$M_a + 10(2) - 15(3) - 8(4)(2) = 0 \qquad \left(\because \sum M = 0 \right)$$

$$\therefore \qquad\qquad M_a = 89 \text{ kNm}$$

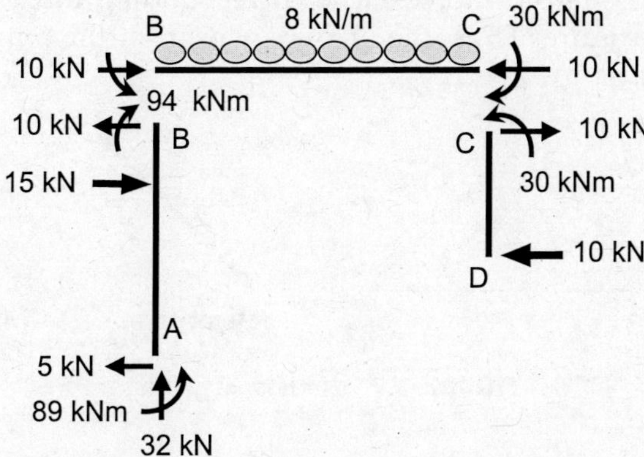

The free-body diagram showing the support reactions and member forces are shown in the following figure.

The AFD, SFD, and BMD for the frame are shown in the following figure.

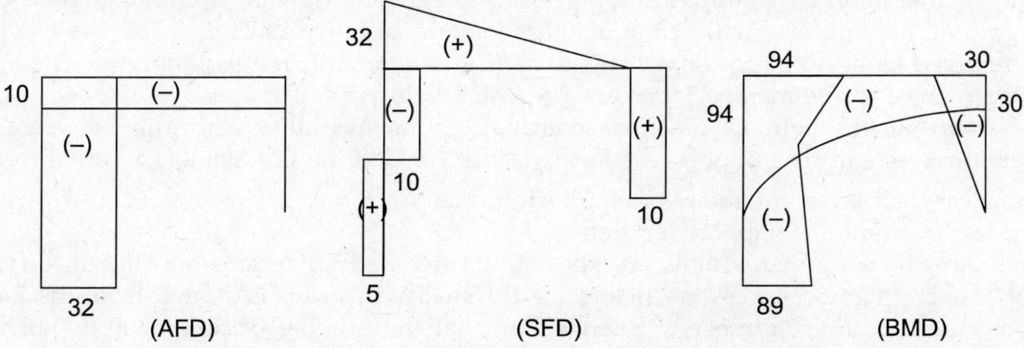

3.7 ANALYSIS OF STATICALLY DETERMINATE TRUSSES

A truss in static equilibrium represents that all joints and members of the truss are also in equilibrium. This means that the equations of equilibrium are applicable to each member as well as to each joint. Accordingly, two methods are primarily adopted to analyze the statically determinate trusses. These are (i) *method of joints* and (ii) *method of sections*. These methods are discussed in the following sections by means of examples.

3.7.1 Methods of Joints

The method of joints relies on the determination of internal member forces by applying the equations of equilibrium at the joints. Since all joints of a truss are pinned connections and all external forces are applied at the joints only, there are two equations of equilibrium at each joint.

Consider a truss as shown in Figure 3.7. Since the member forces are not known at the beginning, it is necessary to assume the nature of member forces as either *tension* or *compression*. If the *tensile* member forces can be assumed as *positive*, the *compression* forces should be considered as negative. In Figure 3.7, the direction of internal forces in the member '*m*' represent the tension forces. The corresponding force component at the joint '*p*' is also shown in the figure. Thus, the arrow away from the joint represents the tensile internal forces in the truss member.

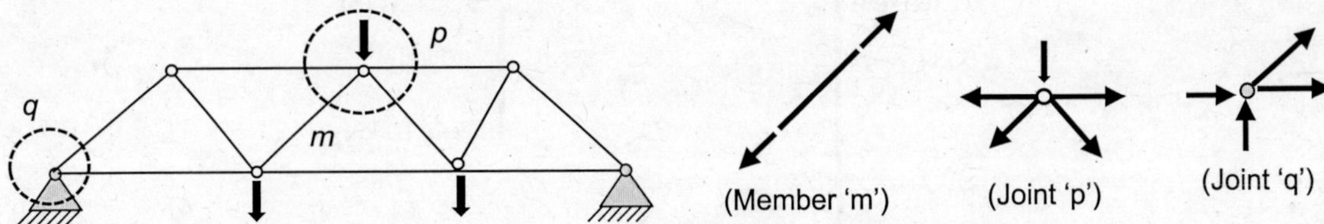

FIGURE 3.7 Method of joints.

In the *method of joint*, the equilibrium equations are required to be satisfied at each joint of the truss. Usually, two equilibrium equations (i.e., $\Sigma H = 0; \Sigma V = 0$) are applied written along two coordinate axes in a plane truss system. Therefore, free-body diagrams are drawn at each joint considering all the applicable internal forces and the external forces acting at that joint. As examples, the free-body diagrams of joints '*p*' and '*q*' showing all forces (internal and external) acting at these joints are shown in Figure 3.7. These forces are resolved along two coordinate axes to satisfy two equations of equilibrium. The basic steps involved in the method joints are presented below:

Compute the support reactions considering the overall equilibrium between the external forces and the support reactions without considering the member internal forces. In this case, all three equations of equilibrium (i.e., $\Sigma H = 0; \Sigma V = 0; \Sigma M = 0$) may be used to determine the support reactions.

Once the support reactions are known, the free-body diagrams are drawn for each joint and the member forces are determined. Usually, this step is started from the joints having the minimum number of members so that the number of unknowns to solve is minimized.

In this context, it is important to visualize the *zero-force members*, if present, in a truss to simply the analysis process. A member connected at a joint, where no external force is acting and two other connected members are colinear, is called a *zero-force* member as the magnitude of internal force in this member is zero.

This process is continued till all member forces are determined for the given truss. As a common practice, a schematic diagram showing all member forces are drawn at the end of the analysis.

The method of joints applied to plane trusses has been explained in the following examples.

EXAMPLE 3.11 Determine the internal force in each member of the truss shown in Figure E3.11 using the method of joint.

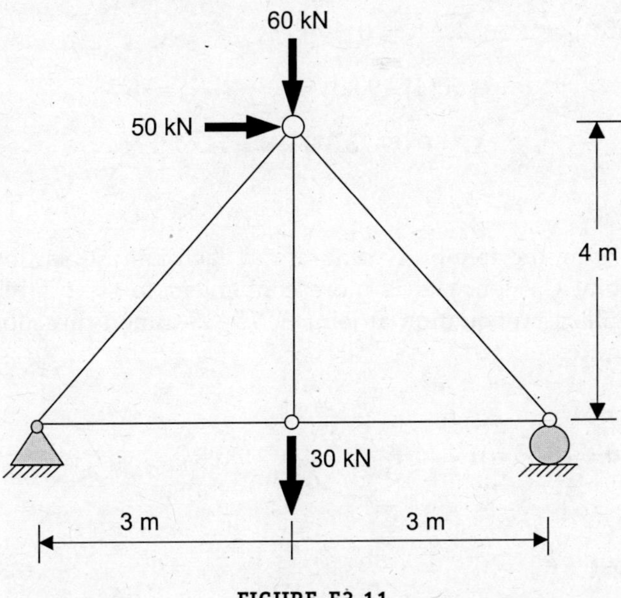

FIGURE E3.11

Solution: There two reactions at support A and one reaction at support C.

Support reactions

Support reactions can be obtained considering the overall equilibrium of the whole truss system.

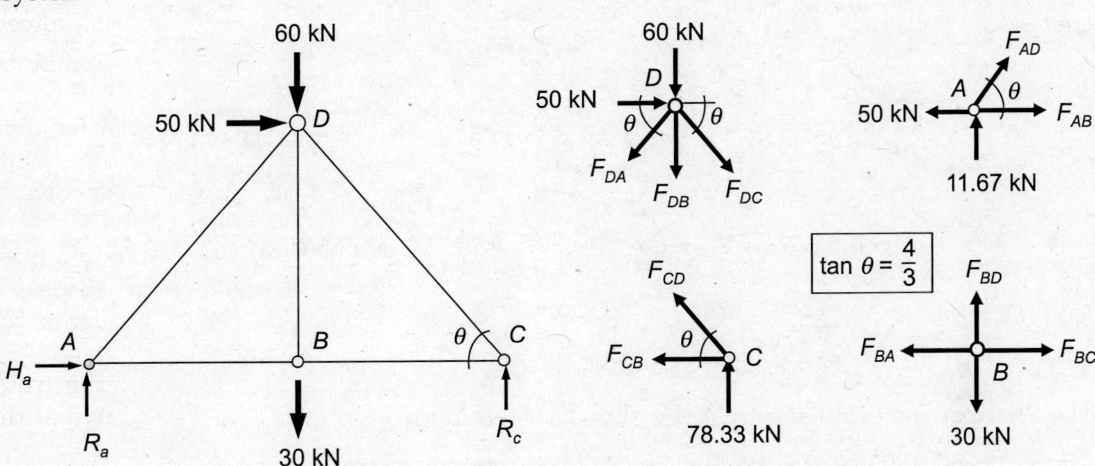

Equilibrium Equations:

$$\sum H = 0 \quad \Rightarrow H_a + 50 = 0 \quad \therefore H_a = -50 \text{ kN} (\leftarrow)$$

$$\sum V = 0 \quad \Rightarrow R_a + R_c = (60 + 30) \text{ kN} = 90 \text{ kN}$$

Taking moment about joint A, $\sum M_a = 0$

$$\Rightarrow 6R_c - 50(4) - 90(3) = 0 \quad \therefore R_c = 78.33 \text{ kN}$$

$$\therefore \qquad R_a = (90 - 78.33) \text{ kN} = 11.67 \text{ kN}$$

Joint equilibrium

The next step is to draw the free-body diagram at each joint. It can be seen that there are two unknowns at joint C. Hence, it is more convenient to start solving these unknowns by applying the equilibrium equation at joint C. The assumed directions of member forces are shown in the figure.

Joint 'C':

$$F_{CD} \sin\theta + 78.33 = 0 \quad \Rightarrow F_{CD}\left(\frac{4}{5}\right) + 78.33 = 0 \quad \therefore F_{CD} = -97.91 \text{ kN}$$

$$F_{CD} \cos\theta + F_{CB} = 0 \quad \Rightarrow (-97.91)\left(\frac{3}{5}\right) + F_{CB} = 0 \quad \therefore F_{CB} = 58.75 \text{ kN}$$

Joint 'B':

$$F_{BC} = 58.75 \text{ kN}$$

$$F_{BA} - F_{BC} = 0 \quad \Rightarrow F_{BA} - 58.75 = 0 \quad \therefore F_{BA} = 58.75 \text{ kN}$$

$$F_{BD} - 30 = 0 \quad \therefore F_{BD} = 30 \text{ kN}$$

Joint 'A':

$$F_{AB} = 58.75 \text{ kN}$$

$$F_{AD} \sin\theta + 11.67 = 0 \quad \Rightarrow F_{AD}\left(\frac{4}{5}\right) + 11.67 = 0 \quad \therefore F_{AD} = -14.59 \text{ kN}$$

$$F_{AD} \cos\theta + F_{AB} = (-14.59)\left(\frac{3}{5}\right) + 58.75 = 50 \text{ kN } (Check!)$$

Joint 'D':

$$F_{DA} = -14.59 \text{ kN}; \quad F_{DC} = -97.91 \text{ kN}; \quad F_{DB} = 30 \text{ kN}$$

The final bar forces for the truss are shown in the following figure.

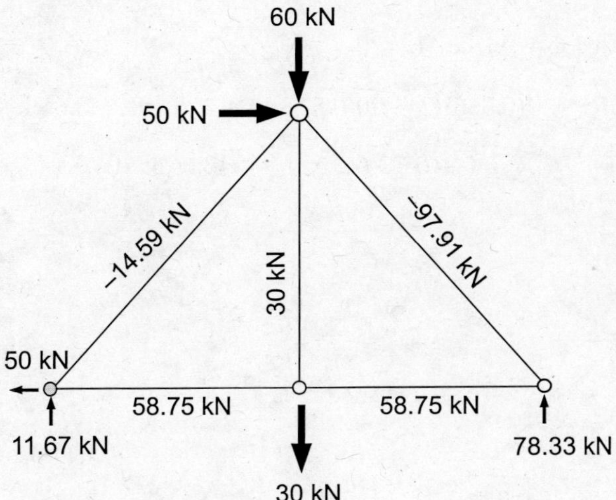

EXAMPLE 3.12 Determine the internal force in each member of the truss shown in Figure E3.12 using the method of joint.

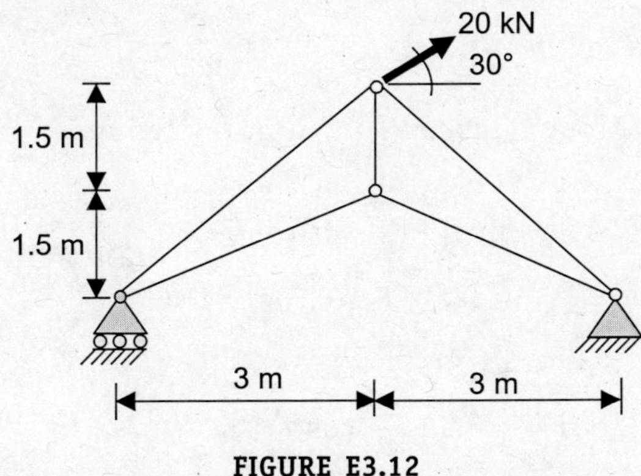

FIGURE E3.12

Solution: Let's reaction at support A be R_a and the reactions at support C be R_c and H_c.

Support reactions

Support reactions can be obtained considering the overall equilibrium of the whole truss system.

Equilibrium Equations

$$\sum H = 0 \quad \Rightarrow H_c - 20 \cos 30 = 0$$

$$\Rightarrow H_c - 20(0.866) = 0 \quad \therefore H_c = 17.32 \text{ kN } (\leftarrow)$$

$$\sum V = 0 \quad \Rightarrow R_a + R_c + 20 \sin 30 = 0 \quad \Rightarrow R_a + R_c = -10 \text{ kN}$$

Taking moment about joint A, $\Sigma M_a = 0$

$$\Rightarrow 6R_c - 20(0.866)(3) + 20(0.5)(3) = 0 \qquad \therefore R_c = 3.66 \text{ kN}(\uparrow)$$

$$\therefore \qquad R_a = (-10 - 3.66) \text{ kN} = -13.66 \text{ kN}(\downarrow)$$

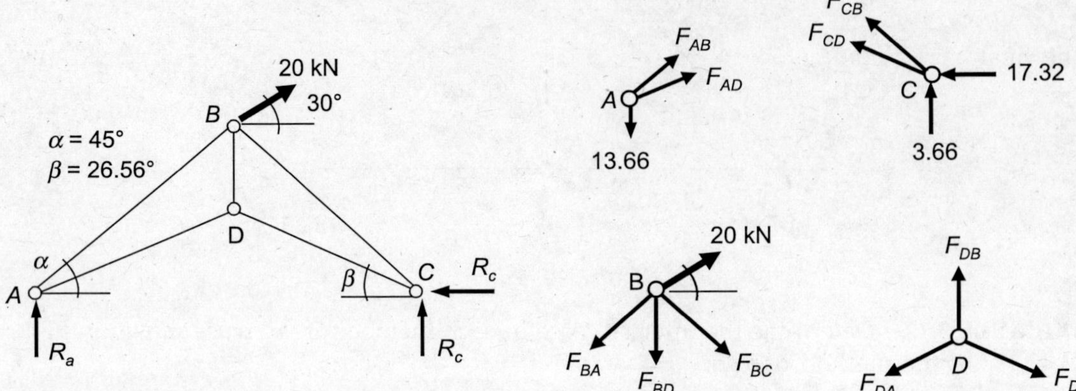

Joint equilibrium

The free-body diagrams of all joints are shown in the figure.

Joint 'A':

$$F_{AD} \sin \beta + F_{AB} \sin \alpha = 13.66$$

$$\Rightarrow F_{AD}(0.447) + F_{AB}(0.707) = 13.66$$

$$\Rightarrow F_{AD} + 1.58 F_{AB} = 30.56 \tag{i}$$

$$F_{AD} \cos \beta + F_{AB} \cos \alpha = 0$$

$$\Rightarrow F_{AD}(0.894) + F_{AB}(0.707) = 0$$

$$\Rightarrow F_{AD} + 0.791 F_{AB} = 0 \tag{ii}$$

Solving Eqs. (i) and (ii), one can get $F_{AB} = 38.73$ kN $\quad \therefore F_{AD} = -30.64$ kN

Joint 'C':

$$F_{CD} \cos \beta + F_{CB} \cos \alpha + 17.32 = 0$$

$$\Rightarrow F_{CD}(0.894) + F_{CB}(0.707) + 17.32 = 0$$

$$\Rightarrow F_{CD} + 0.791 F_{CB} = -19.37 \tag{iii}$$

$$F_{CD} \sin \beta + F_{CB} \sin \alpha + 3.66 = 0$$

$$\Rightarrow F_{CD}(0.447) + F_{CB}(0.707) = -3.66$$

$$\Rightarrow F_{CD} + 1.581 F_{CB} = -8.19 \tag{iv}$$

Solving Eqs. (i) and (ii), one can get $F_{CB} = 14.15$ kN $\quad \therefore F_{CD} = -30.56$ kN

Joint 'D':

$$F_{DA} = -30.64 \text{ kN}; \quad F_{DC} = -30.56 \text{ kN}$$

$$(F_{DA} + F_{DC}) \sin \beta - F_{DB} = 0$$

$$(-30.64 - 30.56)(0.447) - F_{DB} = 0$$

$$F_{DB} = -27.35 \text{ kN}$$

The final bar forces for the truss are shown in the following figure.

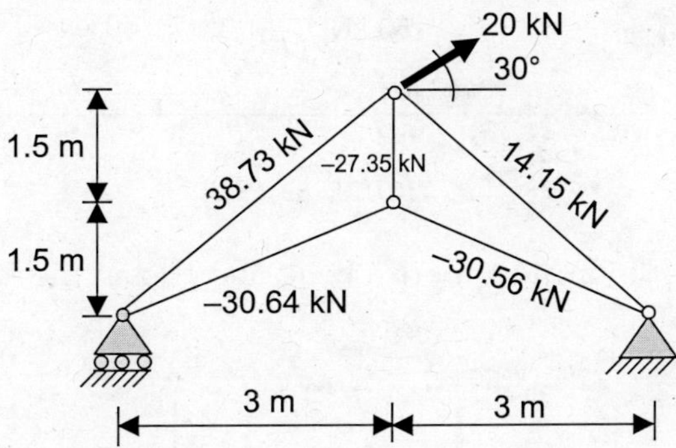

3.7.2 Methods of Sections

In many truss configurations, the method of joints may not be adequate to determine all internal member forces. In such cases, the method of sections can be very useful. In this method, the truss is sub-divided into two free-body diagrams. The truss is cut in such a way that no more than three members connecting at a joint so that all member forces can be determined using the available equilibrium equations. Consider a truss shown in Figure 3.8. The truss is cut into two sub-divisions as shown in the figure. At the cut sections, member internal forces are produced. These unknown forces are determined using the three equations of equilibrium.

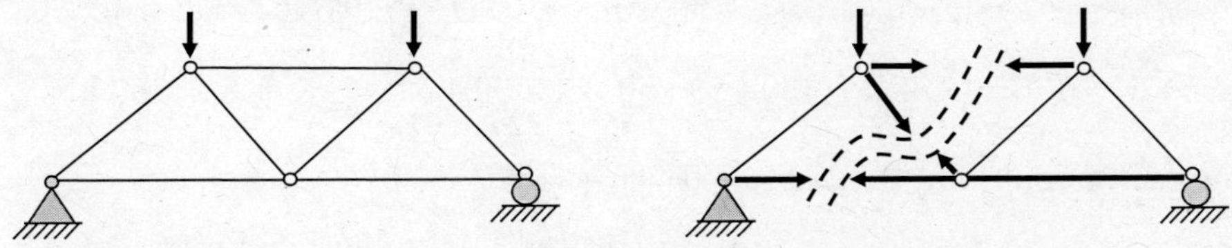

FIGURE 3.8 Method of sections.

The method of sections applied to the plane trusses has been discussed in the following examples.

EXAMPLE 3.13 Determine the internal forces in members (1)–(6) of the truss shown in Figure E3.13 using the method of section.

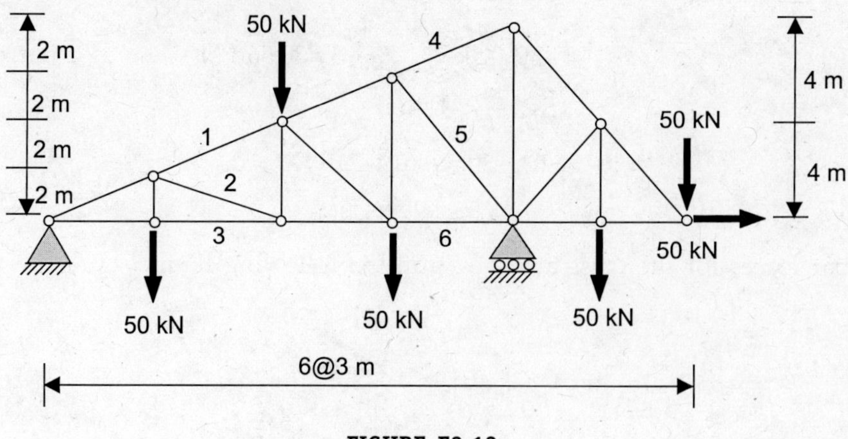

FIGURE E3.13

Solution: Assume the reaction at support A be R_a and H_a and the reaction at support J be R_j.

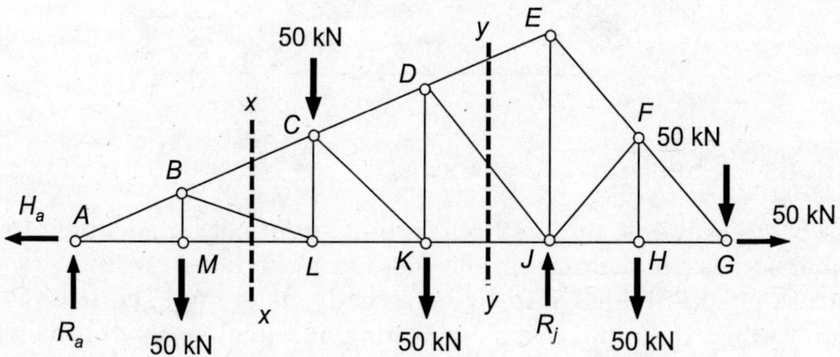

Support reactions

Equilibrium Equations:

$$\sum H = 0 \quad \Rightarrow H_c - 50 = 0 \qquad \therefore H_c = 50 \text{ kN}(\leftarrow)$$

$$\sum V = 0 \quad \Rightarrow R_a + R_j = (50 + 50 + 50 + 50 + 50)\,\text{kN} \quad \therefore R_a + R_j = 250 \text{ kN}$$

Taking moment about joint A, $\sum M_a = 0$

$$\Rightarrow 50(3 + 6 + 9 + 15 + 18) - R_j(12) = 0 \qquad \therefore R_j = 212.5 \text{ kN}(\uparrow)$$

$$\therefore \qquad R_a = (250 - 212.5) \text{ kN} = 37.5 \text{ kN}(\uparrow)$$

Member forces

In order to determine the member forces, two cut sections x-x and y-y shown by the dashed lines are considered. The free-body diagrams at these two sections are shown below.

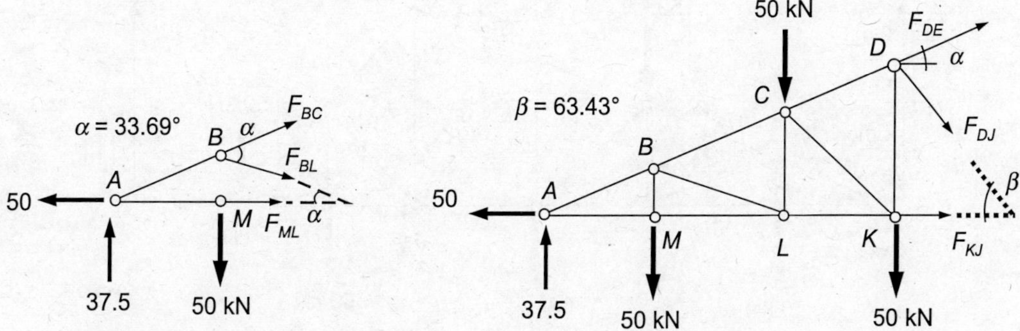

Let's consider the free-body diagram of the left portion of the truss at cut-section x-x. Three member forces, namely, F_{BC}, F_{BL}, and F_{ML}, will be obtained at the cut section x-x.

Taking moment about joint 'A', $\sum M_a = 0$

$$(50)(3) + F_{BL}(6)(\sin \alpha) = 0$$

$$\Rightarrow 300 + F_{BL}(6)(0.555) = 0 \quad \therefore \ F_{BL} = -45.05 \text{ kN}$$

Taking moment about joint 'B', $\sum M_b = 0$

$$(F_{ML} - 50)(2) - (37.5)(3) = 0 \quad \therefore \quad F_{ML} = 106.25 \text{ kN}$$

Taking vertical equilibrium,

$$37.5 - 50 - (F_{BL} - F_{BC})(\sin \alpha) = 0$$

$$-12.5 - (-45.05 - F_{BC})(0.555) = 0 \quad \therefore \ F_{BC} = -22.53 \text{ kN}$$

Now, let's consider the free-body diagram of the left portion of the truss at cut-section y-y. Three member forces, namely, F_{DE}, F_{DJ}, and F_{KJ}, will be obtained at the cut section y-y.

Taking moment about joint 'A', $\sum M_a = 0$

$$(50)(3 + 6 + 9) + F_{DJ}(12)(\sin \beta) = 0$$

$$\Rightarrow 900 + F_{DJ}(12)(0.894) = 0 \quad \therefore \ F_{DJ} = -83.87 \text{ kN}$$

Taking moment about joint 'D', $\sum M_d = 0$

$$(F_{KJ} - 50)(6) - (37.5)(9) + (50)(3 + 6) = 0 \quad \therefore \ F_{KJ} = 31.25 \text{ kN}$$

Taking vertical equilibrium,

$$37.5 - 150 + (F_{DE})(\sin \alpha) - (F_{DJ})(\sin \beta) = 0$$

$$(F_{DE})(0.555) - (-83.87)(0.894) = 112.5 \quad \therefore \ F_{DE} = 67.60 \text{ kN}$$

The final bar forces for the truss are shown in the following figure.

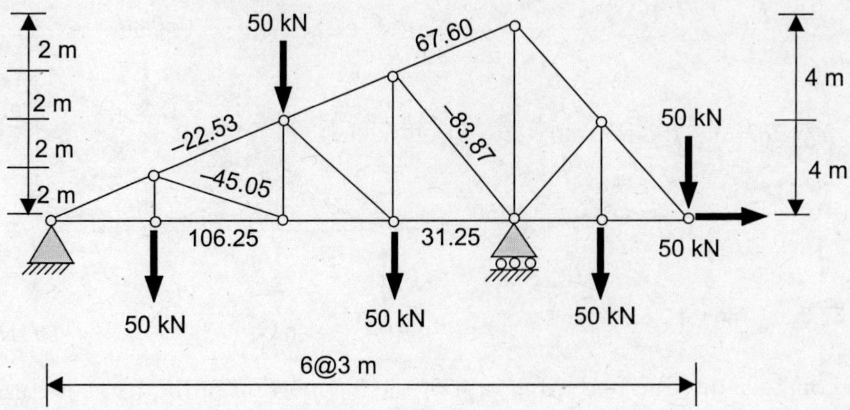

EXAMPLE 3.14 Determine the internal forces in all members of the truss shown in Figure E3.14 using the method of section.

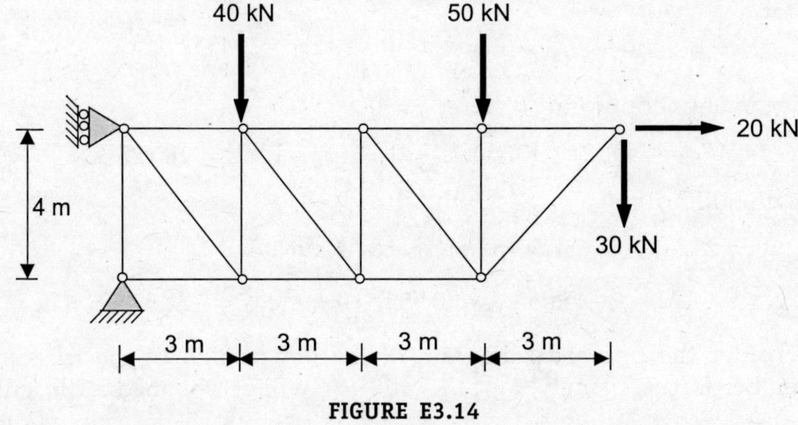

FIGURE E3.14

Solution: Assume the reaction at support K be R_k and H_k and the reaction at support A be H_a.

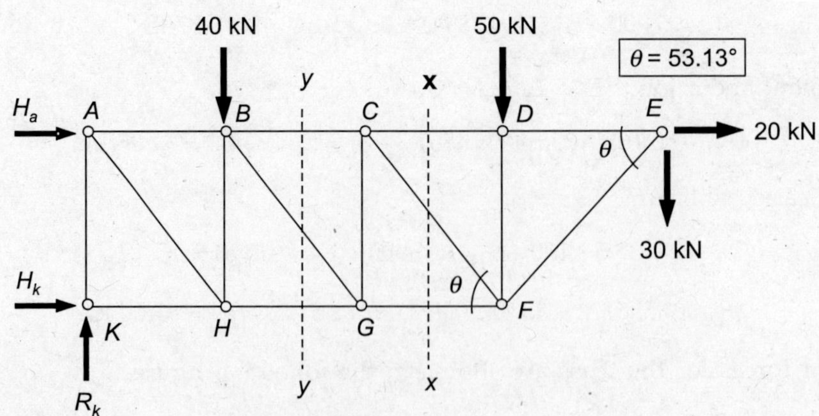

Support reactions

Equilibrium Equations:

$$\sum H = 0 \quad \Rightarrow H_a + H_k + 20 = 0$$

$$\sum V = 0 \quad \Rightarrow R_k = (40 + 50 + 30)\ \text{kN} \quad \therefore R_k = 120\ \text{kN}$$

Taking moment about joint A, $\sum M_a = 0$

$$\Rightarrow (40)(3) + (50)(9) + 30(12) - H_k(4) = 0 \quad \therefore H_k = 232.5\ \text{kN}(\rightarrow)$$

$$\therefore \qquad H_a = -(232.5 + 20)\ \text{kN} = -252.5\ \text{kN}(\leftarrow)$$

Member forces

In order to determine the member forces, two cut sections as shown by the dashed lines are considered. Let's consider the free-body diagram of the left portion of the truss about the cut section x-x. Three member forces, namely, F_{CD}, F_{CF}, and F_{GF}, will be obtained at this cut section.

Vertical shear at the cut section, $V = (120 - 40)\ \text{kN} = 80\ \text{kN}$

Since, $V = F_{CF} \sin \theta = F_{CF}(0.8);\ \therefore F_{CF} = 100\ \text{kN}$

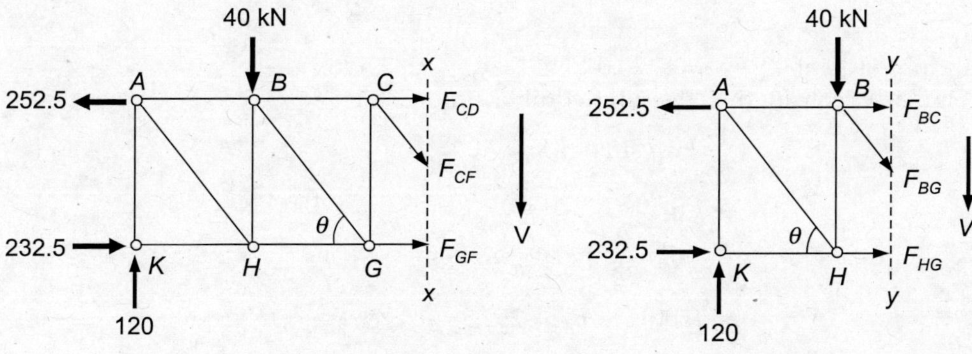

Taking moment about joint 'K', $\sum M_k = 0$

$$(252.5 - F_{CD})(4) - (40)(3) - F_{CF}(6 \sin \theta + 4 \cos \theta) = 0$$

$$4F_{CD} = 890 - 720 \quad \therefore F_{CD} = 42.5\ \text{kN}$$

Considering the equilibrium of forces in horizontal directions,

$$(252.5 - 232.5) = F_{CD} + F_{GF} + F_{CF} \cos \theta$$

$$\Rightarrow F_{GF} + 42.5 + (100)\left(\frac{3}{5}\right) = 20 \quad \therefore F_{GF} = -82.5\ \text{kN}$$

Thus, $F_{DE} = 82.5\ \text{kN};\ F_{DF} = -50.0\ \text{kN}$.

Considering the force equilibrium in the vertical direction at joint 'E'

$$F_{EF}\sin\theta + 30 = 0 \quad \therefore\ F_{EF} = -37.5\ \text{kN}$$

Considering the force equilibrium in the vertical direction at joint 'C',

$$F_{CG} + F_{CF}\sin\theta = 0 \quad \therefore\ F_{CG} = -80.0\ \text{kN}$$

Similarly, three member forces, i.e., F_{BC}, F_{BG}, and F_{HG}, are obtained at the cut section y-y.

Vertical shear at the cut section, $V = 80$ kN $\quad \therefore\ F_{BG} = 100$ kN

Taking moment about joint 'K', $\sum M_k = 0$

$$(252.5 - F_{BC})(4) - (40)(3) - F_{BG}(3\sin\theta + 4\cos\theta) = 0$$

$$\therefore \qquad F_{BC} = 102.5\ \text{kN}$$

Force equilibrium in the vertical direction at joint B would provide the following expression:

$$40 + F_{BG}(\sin\theta) + F_{BH} = 0 ; \qquad \therefore\ F_{BH} = -120\ \text{kN}$$

Considering the equilibrium of forces in the horizontal direction,

$$(252.5 - 232.5) = F_{BC} + F_{HG} + F_{BG}\cos\theta$$

$$\Rightarrow F_{HG} + 102.5 + (100)\left(\frac{3}{5}\right) = 20 \quad \therefore\ F_{HG} = -142.5\ \text{kN}$$

Equilibrium at joint K, $F_{AK} = -120.0$ kN; $F_{KH} = -232.5$ kN

Similarly, considering the joint equilibrium at 'A',

$$F_{AH}(\sin\theta) + F_{AK} = 0$$

$$\Rightarrow F_{AH}(0.8) - 120 = 0 \qquad \therefore\ F_{AH} = 150\ \text{kN}$$

$$F_{AH}(\cos\theta) + F_{AB} - 252.5 = 0$$

$$\Rightarrow 150(0.6) + F_{AB} - 252.5 = 0 \quad \therefore\ F_{AB} = 162.5\ \text{kN}$$

The final member forces in the truss are shown in the following figure.

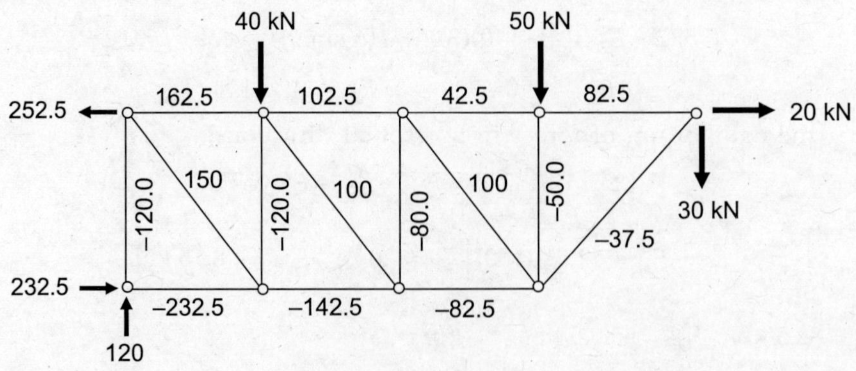

3.8 PROBLEMS

3.1 Determine the support reactions and draw the bending moment and shear force diagrams of the beam shown below.

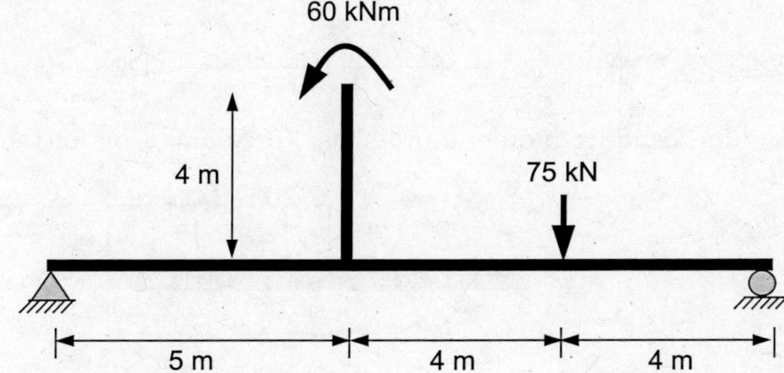

3.2 Determine the support reactions and draw the bending moment and shear force diagrams of the beam shown below.

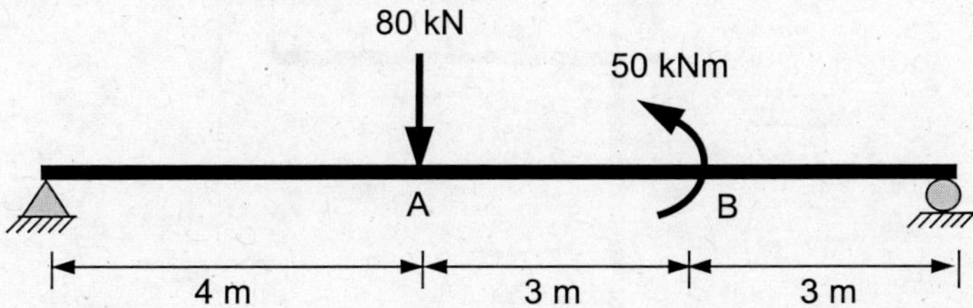

3.3 Draw the bending moment and shear force diagrams of the beam shown below.

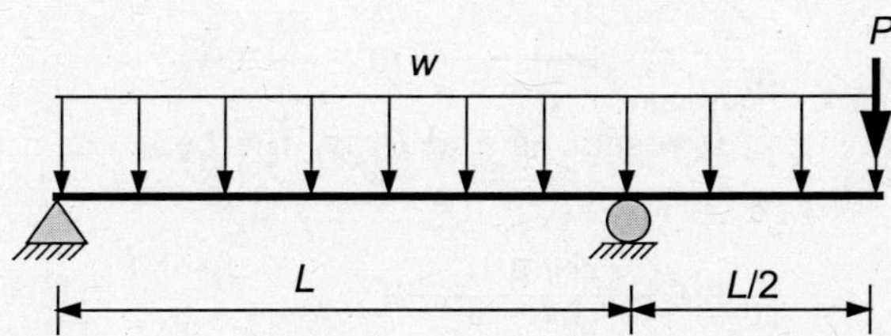

3.4 Draw the bending moment and shear force diagrams of the beam shown below.

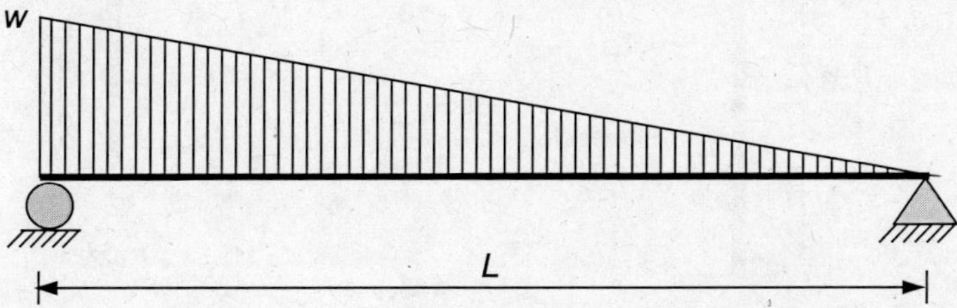

3.5 Draw the bending moment and shear force diagrams of the beam shown below.

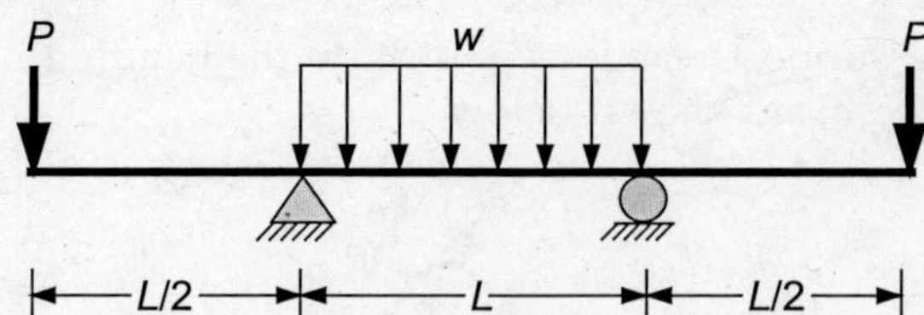

3.6 Draw the bending moment and shear force diagrams of the beam shown below.

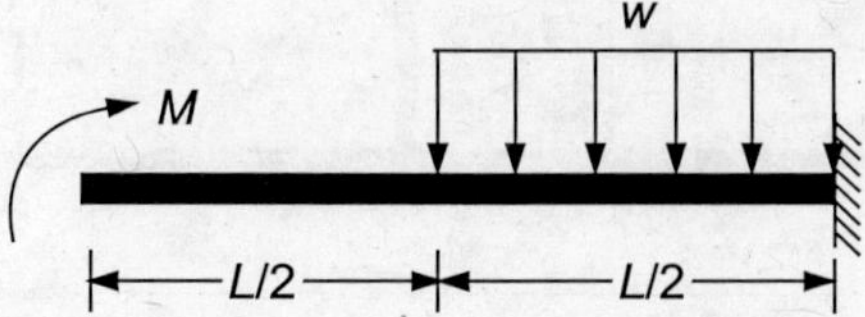

3.7 Determine the support reactions and draw the bending moment and shear force diagrams of the portal frame shown in the following figure.

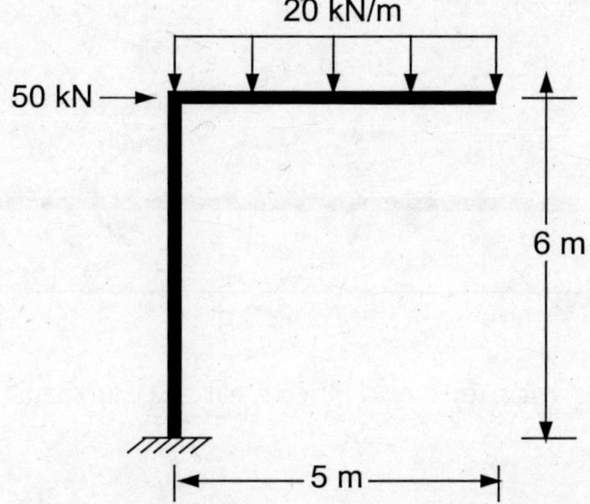

3.8 Determine the support reactions and draw the bending moment and shear force diagrams of the portal frame shown in the following figure.

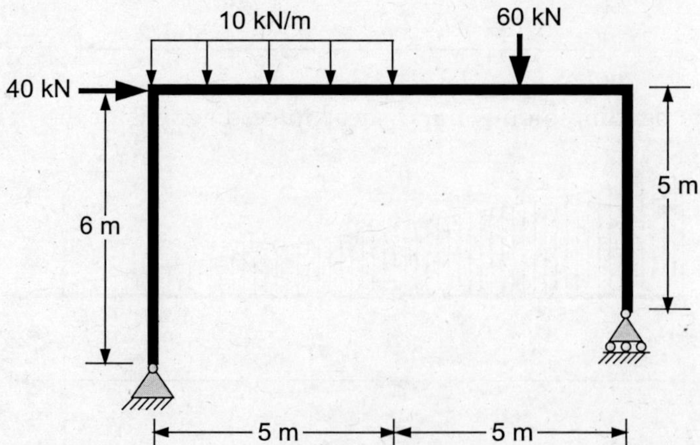

3.9 Determine the support reactions and draw the bending moment and shear force diagrams of the portal frame shown in the following figure.

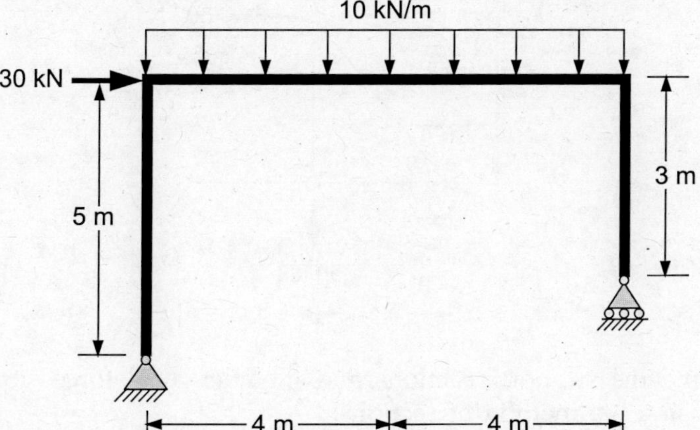

3.10 Determine the support reactions and member axial forces for the truss shown below using the method of sections.

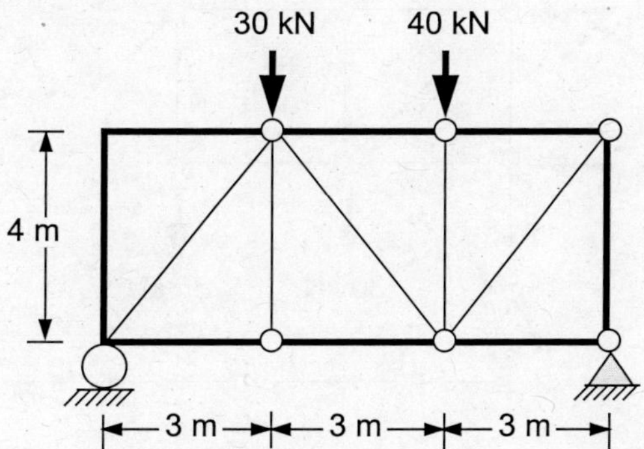

3.11 Determine the support reactions and member axial forces for the truss shown below using the method of joints.

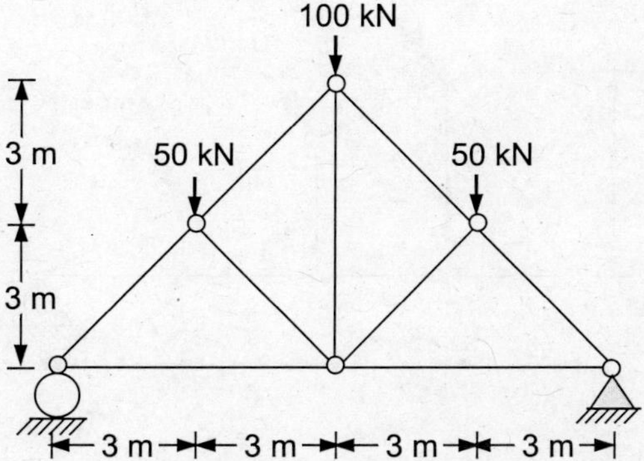

3.12 Determine the support reactions and member axial forces for the truss shown below using the method of joints.

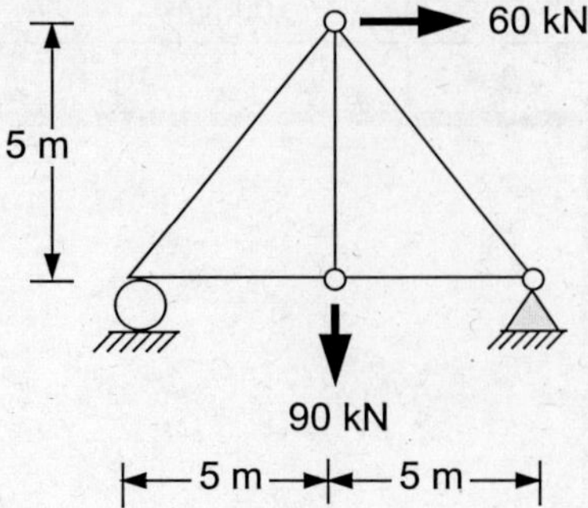

3.13 Determine the support reactions and member axial forces for the truss shown below using the method of sections.

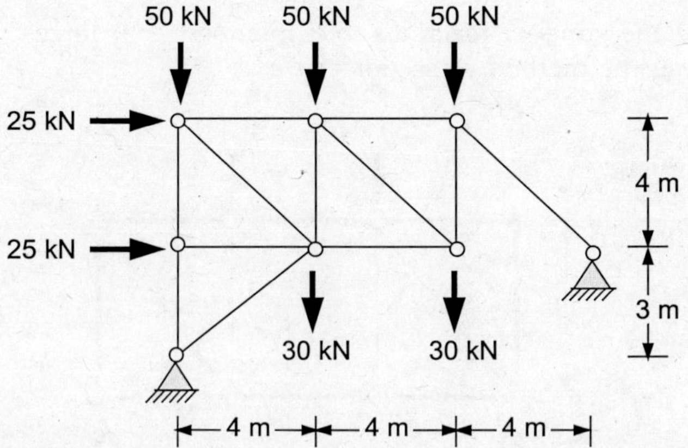

3.14 Determine the support reactions and member axial forces for the truss shown below using the method of joints.

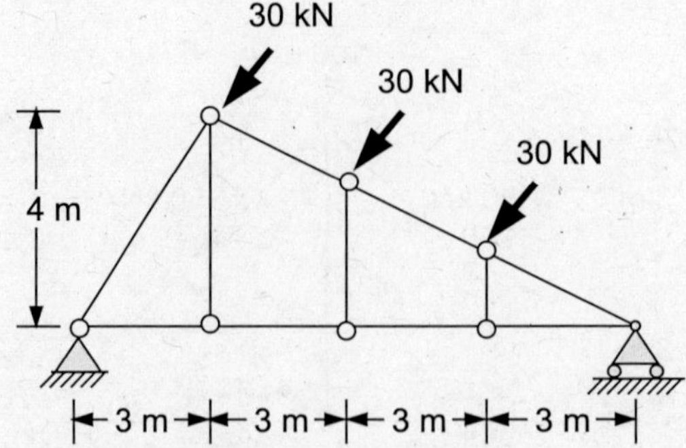

3.15 Determine the support reactions and member axial forces for the truss shown below using the method of sections.

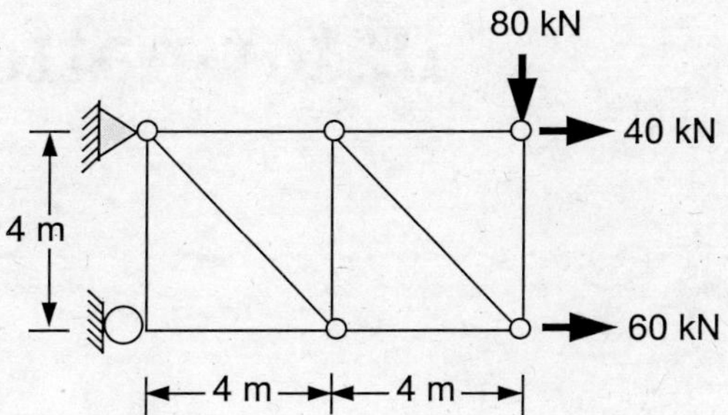

Deformations

4.1 GENERAL

A structural member when subjected to external loads undergoes strain deformations. This results in the deflection of the overall structural system. Excessive deflection may cause structural failure, inhibit serviceability, or impede the aesthetics of a structure. Therefore, the accurate estimation of the deflection of members as well as structures is an important step in the design process.

In the previous chapter, the concept of *equilibrium condition* has been utilized to determine the internal member forces of structures. This chapter is primarily focused on the computation of the deflection of the structural systems under various loading conditions. The basic principles related to the elastic deformation of a structure have been presented. Though numerous methods are available to compute the deflection of structures, the most common methods are discussed in this chapter. The underlying principles of these methods and their applications to various structural systems, such as beams, trusses, and frames, have been illustrated by means of several examples.

4.2 BASIC CONCEPTS

The term *deflection* discussed in this book represents *elastic deformation*. This means that the behaviour of the structure is linearly elastic such that the deformed structure would return back to its original configuration upon the removal of the external loads. Stresses in a member are directly proportional to their strains. Further, it is assumed that the deformations of members are very small (*Small deflection theory*) such that the stresses and deformations are computed on the basis of the original dimensions and the undeformed shapes of structural systems. In addition, the consequences of any change in the geometry of structures due to applied loads are negligible.

Figure 4.1 shows a two-span beam subjected to various external loads. The qualitative deflected shape of the beam is shown in the figure. The change in length of the beam due to axial deformation is generally neglected. Hence, the displacement of the beam in the transverse direction only has been considered. The displacement and rotation values change along the length of the beam. However, the beam is free to rotate at these supports (i.e., θ_A, θ_C). At any intermediate point (for example, D, E, ...), one can determine the displacement and rotation of the beam at that point as shown in the figure. At support B, the vertical displacement is zero being roller support. However, as the beam is continuous, the rotation at B(i.e., θ_B) of the beam for the left segment will be the same as that of the right segment. This is termed as *displacement condition equation* or *compatibility equation*. The zero displacement values at the supports A, B, and C are also other examples of *compatibility equations*. Similar to the *equilibrium equations*, the *compatibility conditions* are the fundamental requirements for the analysis of any structural systems.

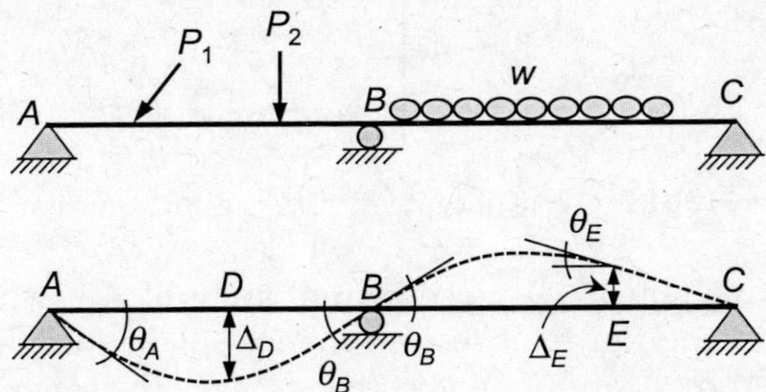

FIGURE 4.1 Assumed deflected shape of a two-span continuous beam.

The basic principles used to determine the deflection of a structural system are discussed in the following sections.

4.2.1 Principle of Virtual Displacements

The *method of virtual work* is one of the most versatile methods used to compute the defections of structures. This principle was originated by Johann Bernoulli (1977). A *virtual displacement* represents a hypothetical finite or infinite displacement of a point or system of points on a rigid body such that the equations of equilibrium of the body are violated. Virtual displacements are caused by some actions or actions other than loads that hold the rigid body in equilibrium.

Figure 4.2 illustrates the principle of virtual displacements. The rigid body shown in the figure is subjected to a system of external forces (i.e., P_1, P_2, P_3, P_4) and bending moments (i.e., M_1, M_2). The body is in equilibrium under the action of these forces and moments. All forces can be resolved along both coordinate axes, x and y, as shown for P_1.

Since the body is in equilibrium, three equations of equilibrium must be satisfied. Mathematically,

$$\sum H = 0 \quad \Rightarrow \sum P_x = 0 \tag{4.1}$$

$$\sum V = 0 \quad \Rightarrow \sum P_y = 0 \tag{4.2}$$

$$\sum M + \sum P_x \cdot y + \sum P_y \cdot x = 0 \tag{4.3}$$

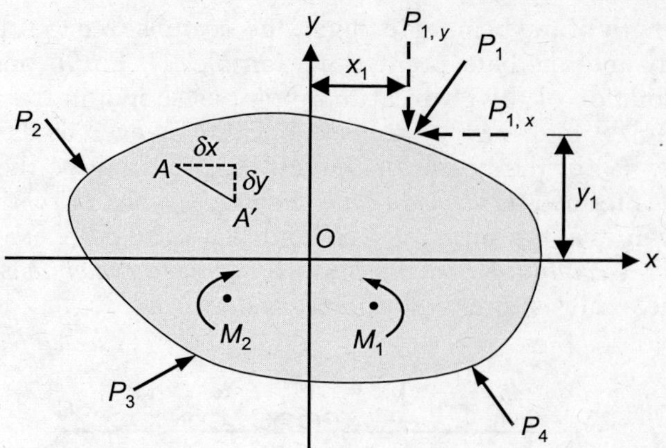

FIGURE 4.2   Concept of virtual displacements.

Assume that a virtual (small) displacement of amount AA' without rotation is applied to the body as shown in the figure. The components of this displacement parallel to the x- and y-axes are denoted by δ_x and δ_y, respectively. Since the magnitude of displacement is very small, it is assumed that the applied forces or moments do not change their direction and magnitude under the influence of virtual displacement.

The work done by the forces during the translation is given by

$$WD = \sum P_x . \delta_x + \sum P_y . \delta_y \tag{4.4}$$

Since the values of δ_x and δ_y are constant at all points, the above equation can be written as

$$WD = \delta_x \sum P_x + \delta_y \sum P_y \tag{4.5}$$

Using the Eqns. (4.1) and (4.2), the work done is zero, i.e., $WD = 0$.

Similarly, let's assume that the body is applied a rotational displacement such that it is rotated by an angle α about origin O. The translation components along x- and y-axes are $y\alpha$ and $x\alpha$, respectively. Total work done by the forces and bending moments applied on the body is given by

$$WD = \sum M . \alpha + \sum P_x . y\alpha + \sum P_y . x\alpha = \alpha\left(\sum M + \sum P_x . y + \sum P_y . x\right) \tag{4.6}$$

Using Eqn. (4.3), the work done given by the above equation is zero.

Since any small displacement can be expressed as the sum of a translation and a rotation about some point, total work done by the forces and bending moments would be zero.

Accordingly, Johan Bernoulli's *principle of virtual displacement* states that *if a rigid body is held in equilibrium by a system of forces and/or couples (moments), the total virtual work done by this system of forces and/or couples (moments) during a virtual displacement is zero.*

4.2.2 Principle of Virtual Work

The principle of virtual displacements is utilized to develop the principle of virtual work. To explain this principle, let's consider a deformable body as shown in Figure 4.3 subjected to a system of forces. The body as a whole as well as any small elements within the body must be in equilibrium under the application of applied loading.

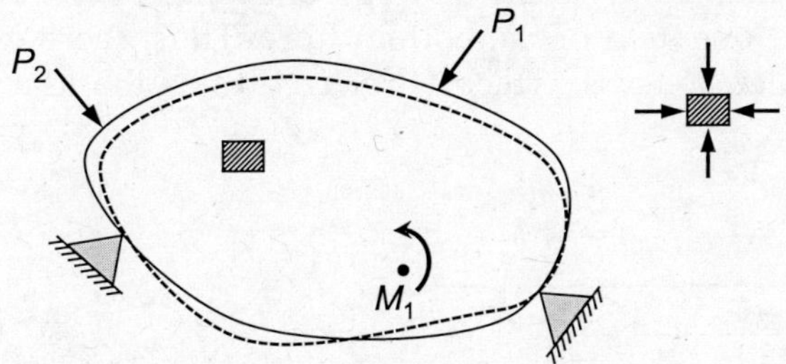

FIGURE 4.3 Deformation of a body under a system of forces.

Now assume the body is subjected to a virtual force (or moment) action. This will result in the virtual deformations of the whole body as well as all its segments. The virtual displacements along the directions for the applied forces would result in the virtual work, which is the product of force and virtual displacement. Let this external virtual work due to the external loads be dw_e. The external work done should be equal to the sum of the energy dissipated (i) by the rigid body movement and (ii) due to the internal deformation (namely, translation, rotation, or strain) of the system. Let the energy dissipated by the rigid body movement and internal deformation is represented by dw_{rt} and dw_i, respectively. Considering the principle of energy conservation, the external work done must be equal to the internal work done.

$$dw_e = dw_{rt} + dw_i \tag{4.7}$$

From the principle of virtual displacement, the work done by rigid body movement, $dw_{rt} = 0$. Therefore,

$$dw_e = dw_i \tag{4.8}$$

For the whole body, the integration of the above equation would result in

$$W_e = W_i \tag{4.9}$$

where, W_e is the total virtual work done by the external forces and W_i is the internal virtual strain energy of the body. It is worth mentioning that the work done by the intersegmental stresses would be zero as they are equal in magnitude and opposite in direction for maintaining the equilibrium condition.

Thus, the *principle of virtual work* can be stated as follows: *If a virtual displacement is applied to a deformable structure in equilibrium carrying a system of forces, the external virtual work of the given system of forces is equal to the internal virtual work of the stresses and deformation caused by the applied loading.*

4.2.3 Moment-curvature Relationship

Mathematically, a *curvature* can be defined as *the rate at which a curve changes its direction*. To derive the relationship between the curvature and bending moment, let's consider a beam of length 'L' subjected to constant end moment 'M' as shown in Figure 4.4(a). Under the action of applied bending moments, the beam would bend into a concave shape as shown in Figure 4.4(b). The top portion of the beam would be subjected to compression, whereas the bottom portion of the beam would undergo tension. This results in the shortening of the top fibres and the elongation of the bottom fibre. Thus, there exists a plane where the fibres do not change their lengths, which is referred to as the *neutral axis*.

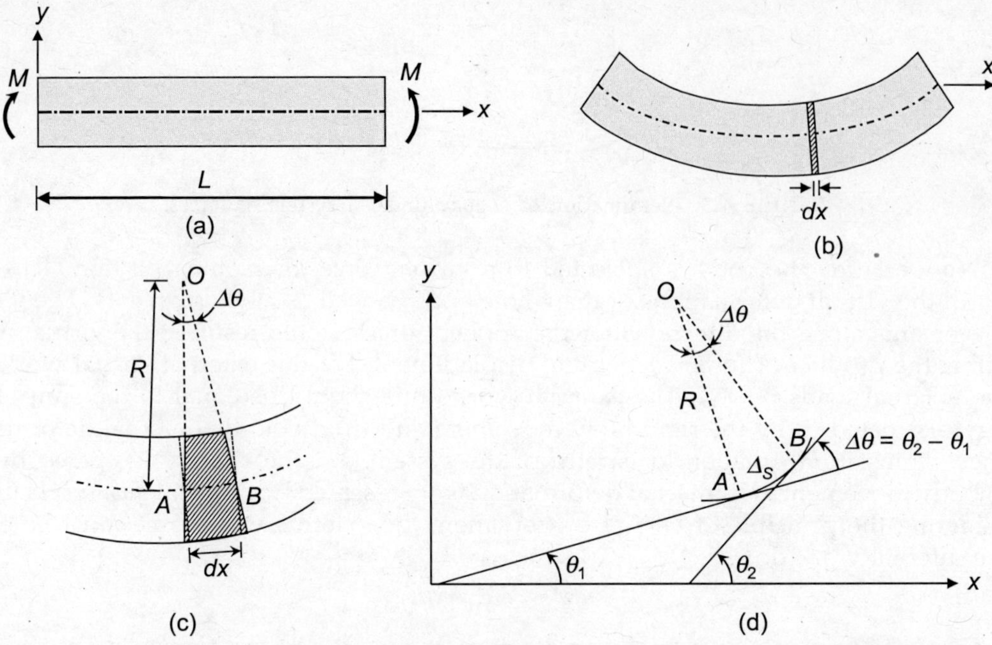

FIGURE 4.4   Moment-curvature relationship.

Let's consider an infinitesimal segment of length dx as shown in Figure 4.4(b). Both sides of the segment which were initially parallel to each other before bending would rotate with respect to each other by an angle $\Delta\theta$ after bending as shown in Figure 4.4(c). This rotation produces the strain that varies linearly with the depth of the cross-section. At the deflected configuration, let the length of the segment at the level of neutral axis (represented by elastic curve) shown by AB is Δs. The average rate of change of direction between points A and B is $\Delta\theta / \Delta s$. The limiting value of this rate of change as the value of Δs approaches zero is called *curvature*. If the radius of curvature is R, then $(\Delta\theta)R = \Delta s$. Thus, the *radius of curvature* is the reciprocal of the curvature (κ). Mathematically, the curvature (κ) can be expressed as follows:

$$\kappa = \frac{1}{R} = \lim_{\Delta s \to 0} \frac{\Delta\theta}{\Delta s} = \frac{d\theta}{ds} \tag{4.10}$$

One can find that $\tan\theta = \dfrac{dy}{dx}, \ \dfrac{d}{dx}(\tan\theta) = \dfrac{d^2 y}{dx^2}$

$$\Rightarrow \quad (1+\tan^2\theta)\frac{d\theta}{dx} = \frac{d^2 y}{dx^2}$$

$$\Rightarrow \frac{d^2y}{dx^2} = \left[1 + \left(\frac{dy}{dx}\right)^2\right]\frac{d\theta}{dx}$$

$$\therefore \quad \frac{d\theta}{dx} = \frac{d^2y/dx^2}{[1 + (dy/dx)^2]} \tag{4.11}$$

Since, $ds = \sqrt{dx^2 + dy^2}$;

$$\therefore \quad \frac{ds}{dx} = \sqrt{1 + \left(\frac{dy}{dx}\right)^2} = \left[1 + \left(\frac{dy}{dx}\right)^2\right]^{1/2} \tag{4.12}$$

Eqn. (4.10) can be rewritten as $\kappa = \dfrac{d\theta}{ds} = \left(\dfrac{d\theta}{dx}\right)\left(\dfrac{dx}{ds}\right)$

Using Eqns. (4.11) and (4.12),

$$\kappa = \frac{d^2y/dx^2}{[1 + (dy/dx)^2]^{3/2}} \tag{4.13}$$

For small deflection theory, the value of dy/dx representing the slope (or rotation) of a member is very small. The square of this term would be further smaller and hence, can be considered to be zero. Accordingly, the curvature can be expressed as follows:

$$\kappa = \frac{d\theta}{ds} \cong \frac{d^2y}{dx^2} \tag{4.14}$$

In order to derive the relationship between the bending moment and the curvature of a beam, let's consider a segment of a beam shown in Figure 4.5. The change in angle between PQ and $P'Q'$ is $d\theta$. Considering the distance of top/bottom fibres from the neutral axis is c, the maximum shortening/elongation of these fibres is $(c.d\theta)$. The original length of the segment is ds. Thus, the strain (ε) at the level of these fibres is $(c.d\theta/ds)$. If the modulus of elasticity of the material is E, the maximum bending stress (f) can be given by $f = E\varepsilon$.

Accordingly,
$$\frac{c.d\theta}{ds} = \frac{f}{E} \quad \text{or} \quad \frac{d\theta}{ds} = \frac{f}{Ec}$$

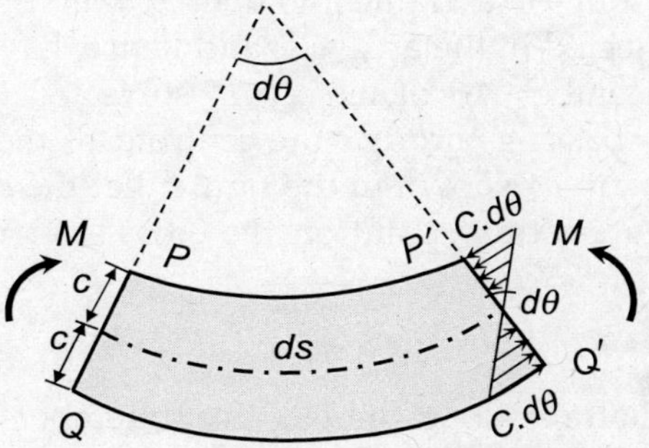

FIGURE 4.5 Curvature of a segment of beam under bending.

Since,
$$\frac{M}{I} = \frac{f}{c}; \quad \frac{d\theta}{ds} = \frac{M}{EI}$$

Using Eqn. (4.14), the approximate curvature of a loaded beam can be related to the bending moment as follows:

$$\frac{d^2y}{dx^2} = \frac{M}{EI} \quad \text{or} \quad M = EI\frac{d^2y}{dx^2} \tag{4.15}$$

Generally, the value of curvature (κ) reduces as the magnitude of the bending moment (M) is increased along the length of a beam. Hence, the generalized form of the moment-curvature relationship is expressed by the following expression:

$$M = -EI\frac{d^2y}{dx^2} \tag{4.16}$$

The main assumptions made in the derivation of the above expressions are as follows:

1. The deflection of the beam is small (*Small deflection theory*)
2. The material is linearly elastic (*Hooke's law* is applicable)
3. The beam is predominantly subjected to bending moment only
4. The plane sections before being remains plane after bending (*Linear strain distribution*)

The moment-curvature relationship (or stress-strain relationships) represents the relationship between the generalized force and displacements of structural members. The generalized force-displacement relationships are also referred to as the *constitutive equations* in structural analysis.

4.2.4   Strain-displacement Relationship

Strain-displacement relationships represent the correlation between the strain deformations at a particular cross-section of a member and its displacement under the applied external loading. These relationships are also referred to as *kinematic equations*.

In general, a two-dimensional frame member carries an axial force, a shear force, and a bending moment. All these actions would result in deformations in the respective directions. Out of three possible deformations, axial and flexural (or bending) deformations usually govern the behaviour of a frame member. Other deformations, such as shear deformations are often neglected in the analysis, and hence, have not been considered here.

Figure 4.6 shows a member *AB* of length '*L*' subjected to an axial tensile force '*P*' along the longitudinal axis passing through the centroid of the cross-section. Consider an infinitesimal segment of length dx shown in the figure. Let the deformation of this segment be du. Thus, strain (ε) at an arbitrary point on the cross-section can be defined as

$$\varepsilon = \frac{du}{dx} \tag{4.17}$$

The variation of axial strain along the length of members depends on the geometry, loading, and boundary conditions. Accordingly, the displacement of the member can be given by

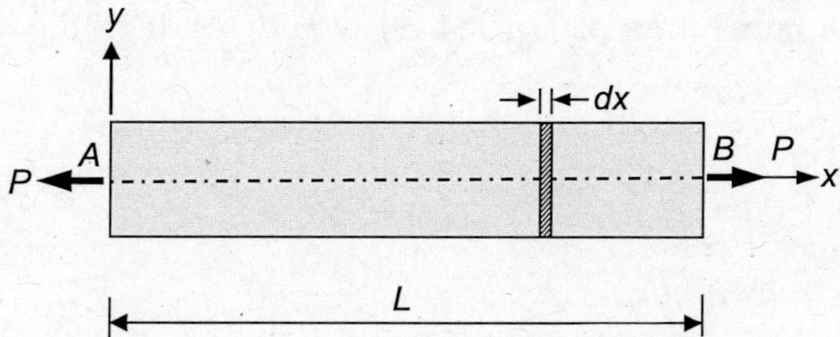

FIGURE 4.6 Strain-displacement relationship.

$$u = \int \varepsilon.dx + C \qquad (4.18)$$

For the member AB shown in the figure, total displacement can be computed as follows:

$$u = \int_{x_A}^{x_B} \varepsilon.dx = \int_{x_A}^{x_B} \frac{f}{E}.dx \qquad (4.19)$$

where, f = axial stress, E = Modulus of elasticity.

A similar strain-deflection relationship can be established for the flexural deformation. The deflection of a beam at a point depends on the support conditions and the curvature at each point along its length.

4.3 METHODS OF COMPUTING DEFLECTIONS OF BEAMS

Several methods are available to determine the slope and deflection of the beam under various loading conditions. Depending on the geometrical complexity and loading conditions, one can choose any of these methods to solve a deflection problem. Further, the ability to solve a deflection problem by several methods provide the means to perform an independent check as well. Some of the common methods adopted to compute beam deflections are discussed in the following sections.

4.3.1 Double Integration Method

The double integration method used for computing slope and deflection at any point along the length of the beam is based on the simple flexural theory. As discussed earlier, the moment-curvature relationship (or constitutive equations) of a beam under pure bending can be given by Eqn. (4.16). Rearranging this equation, one would get and reproduced below:

$$\frac{d^2 y}{dx^2} = -\frac{M}{EI} \qquad (4.20)$$

Since the curvature of a beam is the rate of change of slope along the length, the change in slope at a point can be obtained by performing a single integration Eqn. (4.19) as follows:

$$\theta = \frac{dy}{dx} = -\int \frac{M}{EI}.dx + C_1 \qquad (4.21)$$

Similarly, the double integration of Eqn. (4.19) would result in the deflection of a beam. Mathematically,

$$y = \iint \frac{d^2 y}{dx^2} dx.dx + C_1.x + C_2 = -\iint \frac{M}{EI} dx.dx + C_1.x + C_2$$

or $\qquad\qquad y = \int \theta.dx + C_2$ $\qquad\qquad\qquad\qquad\qquad\qquad\qquad\qquad$ (4.22)

The integration constants (C_1 and C_2) can be obtained from the end and support conditions of the beam. In some cases, the number of integration constants is higher than the displacement condition (or compatibility) equations available at the supports. In such cases, the additional *compatibility equations* must be written at the intermediate points along the length of the beam depending on the loading conditions. Since the deflection of a member is obtained by integrating the moment-curvature relationship twice, the method is known as the *Double integration method*.

The method of Double Integration to determine the deflection and slope of beams has been illustrated in the following examples.

EXAMPLE 4.1   A simply supported beam, shown in Figure E4.1, is subjected to a uniformly distributed loads of w throughout its length L. Determine the slope at the support and the deflection at the mid-span of the beam using the *Double Integration method*. Assume constant flexural rigidity, EI.

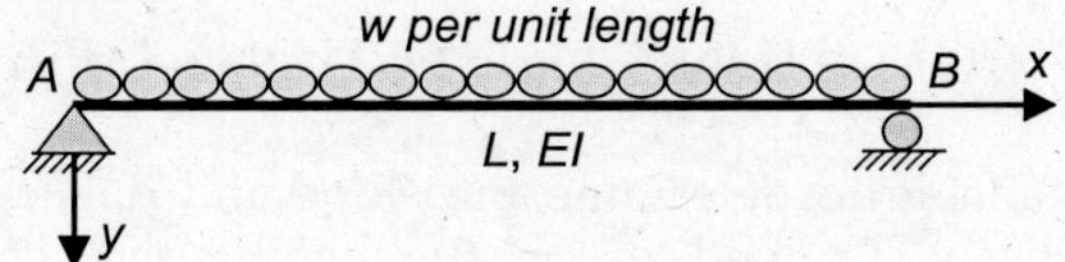

FIGURE E4.1

Solution: Using the equations of equilibrium, support reactions can be determined as follows:

$$R_a = R_b = \frac{wl}{2} (\uparrow)$$

Bending moment at section X-X at a distance x from A, $M_x = \dfrac{wL}{2}x - \dfrac{w.x^2}{2}$

We know, $\dfrac{d^2 y}{dx^2} = -\dfrac{M}{EI}$

$$EI\frac{d^2 y}{dx^2} = -\left(\frac{wL}{2}x - \frac{wx^2}{2}\right) \qquad\qquad\qquad (1)$$

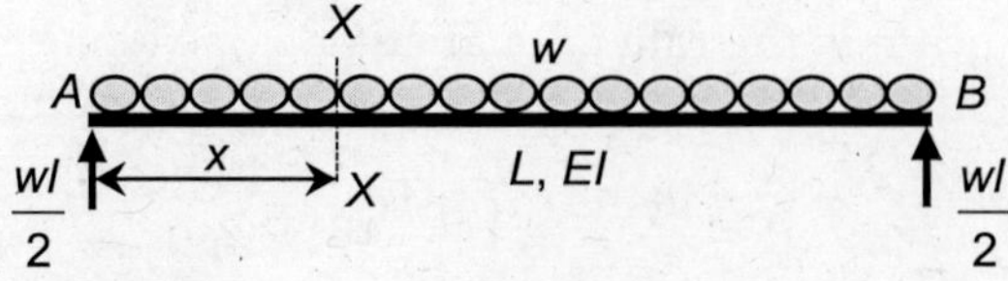

Expression of slope can be obtained by integrating the above equation,

$$EI \int \frac{d^2y}{dx^2} dx = - \int \left(\frac{wL}{2} x - \frac{wx^2}{2} \right) dx + C_1$$

$$EI \frac{dy}{dx} = - \left(\frac{wLx^2}{4} - \frac{wx^3}{6} \right) + C_1 \qquad (2)$$

Expression of deflection can be obtained by integrating Eqn. (2) as follows:

$$EI \int \frac{dy}{dx} dx = - \int \left(\frac{wLx^2}{4} - \frac{wx^3}{6} \right) dx + \int C_1 . dx + C_2$$

$$EIy = - \left(\frac{wLx^3}{12} - \frac{wx^4}{24} \right) + C_1 . x + C_2 \qquad (3)$$

Both integration constants, C_1, and C_2 can be determined by using the boundary conditions. For the given simply supported beam, the deflection at the supports is zero, i.e., At $x = 0$; $y = 0$ and at $x = L$; $y = 0$.

Using these boundary conditions in Eqn. (3), $C_2 = 0$, and $C_1 = wL^3/24$.

Thus, the final expression to determine the slope at any point along the length of the beam is given by

$$EI \frac{dy}{dx} = \frac{wL^3}{24} - \left(\frac{wLx^2}{4} - \frac{wx^3}{6} \right)$$

Slope at support A (i.e., $x = 0$), $\theta_A = \left. \frac{dy}{dx} \right|_{x=0} = \frac{wL^3}{24EI}$ (Clockwise)

Slope at support B (i.e., $x = L$), $\theta_B = \left. \frac{dy}{dx} \right|_{x=L} = - \frac{wL^3}{24EI}$ (Anti-clockwise)

One can also check that the slope at the mid-span (i.e., $x = L/2$) is zero.

The expression to determine the deflection of the beam can be written as follows:

$$EIy = \frac{wL^3}{24} x - \left(\frac{wLx^3}{12} - \frac{wx^4}{24} \right)$$

Deflection at the mid-span (i.e., $x = L/2$) of the beam, $\Delta = y \big|_{x=L/2} = \frac{wL^4}{384 \, EI}$ (Downward)

It is worth mentioning that the maximum deflection of the beam is noted at the point where the value of the slope is zero in this example. In general, the maximum deflection of a beam is observed where the slope changes sign from positive to negative or vice-versa.

EXAMPLE 4.2   A cantilever beam of flexural rigidity EI and length L is subjected to a linearly varying loads of intensity w as shown in Figure E4.2. Determine the slope and deflection at the free end of the beam using the *Double Integration method*.

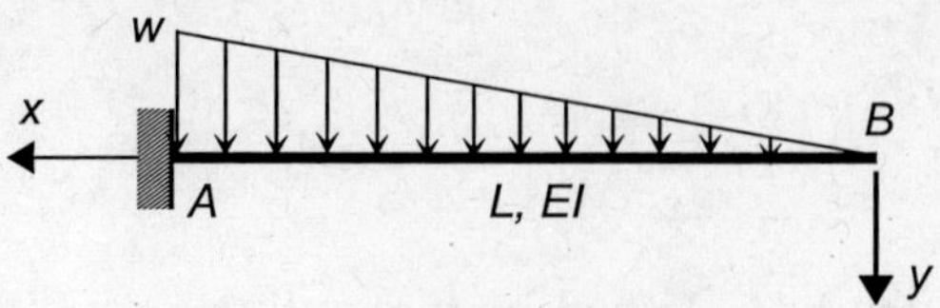

FIGURE E4.2

Solution: Let's consider a section at a distance x from the free end B.
Bending moment at this section,

$$M_x = -\left(\frac{1}{2} \cdot \frac{wx}{L} \cdot x\right)\left(\frac{x}{3}\right) = -\frac{wx^3}{6L}$$

Since,

$$EI\frac{d^2y}{dx^2} = -M_x, \quad EI\frac{d^2y}{dx^2} = \frac{wx^3}{6L}$$

Integrating the above equation once, $EI\displaystyle\int \frac{d^2y}{dx^2}\,dx = \int \left(\frac{wx^3}{6L}\right)dx + C_1$

$$EI\frac{dy}{dx} = \left(\frac{wx^4}{24L}\right) + C_1 \tag{1}$$

Deflection can be obtained by integrating Eqn. (1) as follows:

$$EI\int \frac{dy}{dx}\,dx = \int \left(\frac{wx^4}{24L}\right)dx + \int C_1\,.dx + C_2$$

$$EIy = \frac{wx^5}{120L} + C_1\,.x + C_2 \tag{2}$$

At the fixed end, both slope and deflections are zero.

Boundary conditions:

1.   At $x = L$; $y = 0$ and
2.   At $x = L$; $dy/dx = 0$.

Putting these conditions in Eqns. (1) and (2), one would get

$$C_1 = -\frac{wL^3}{24} \quad \text{and} \quad C_2 = \frac{wL^4}{30}$$

Thus, the final expression to determine the slope at any point along the length of the beam is given by

$$EI \frac{dy}{dx} = \frac{wx^4}{24L} - \frac{wL^3}{24} = \frac{w(x^4 - L^4)}{24L}$$

Slope at support B (i.e., $x = 0$), $\theta_B = \left. \dfrac{dy}{dx} \right|_{x=0} = -\dfrac{wL^3}{24EI}$

Similarly, the final expression to determine the deflection of the beam can be written as follows:

$$EIy = \frac{wx^5}{120L} - \frac{wL^3 x}{24} + \frac{wL^4}{30}$$

Deflection at the free end (i.e., $x = 0$) of the beam, $\Delta = y\big|_{x=0} = \dfrac{wL^4}{30EI}$.

EXAMPLE 4.3   A simply supported beam of flexural rigidity EI and length L is subjected to a triangular loading of intensity w as shown in Figure E4.3. Determine the slope at the supports and the deflection at the mid-span of the beam using the *Double Integration method*.

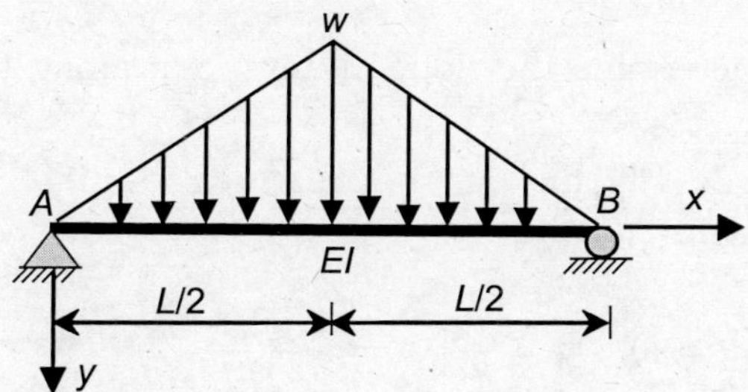

FIGURE E4.3

Solution: Total applied load = $wL/2$.

Since the loading is symmetric, reaction at each simple support = $wL/4$.

Let's consider a section at a distance x from the support A.

Bending moment at this section,

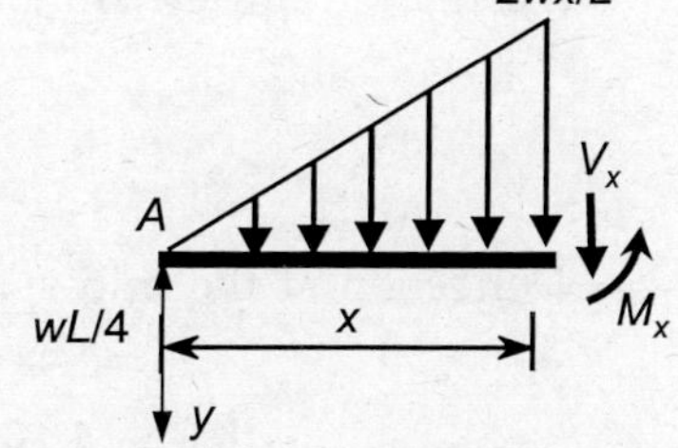

$$M_x = \frac{wL}{4}x - \left(\frac{1}{2} \cdot \frac{2wx}{L} \cdot x \right)\left(\frac{x}{3} \right) = \frac{wL}{4}x - \frac{wx^3}{3L} \quad (0 \le x \le L/2)$$

Since, $EI\dfrac{d^2y}{dx^2} = -M_x$, $EI\dfrac{d^2y}{dx^2} = \dfrac{wx^3}{3L} - \dfrac{wLx}{4} \quad (0 \le x \le L/2)$ (1)

Integrating the above equation once,

$$EI \int \frac{d^2y}{dx^2}\, dx = \int \left(\frac{wx^3}{3L} - \frac{wLx}{4} \right) dx + C_1$$

$$EI\frac{dy}{dx} = \left(\frac{wx^4}{12L} - \frac{wLx^2}{8}\right) + C_1 \quad (0 \le x \le L/2) \tag{2}$$

Expression of deflection can be obtained by integrating Eqn. (2) is

$$EI\int \frac{dy}{dx}\,dx = \int \left(\frac{wx^4}{12L} - \frac{wLx^2}{8}\right)dx + \int C_1\,.dx + C_2$$

$$EIy = \frac{wx^5}{60L} - \frac{wLx^3}{24} + C_1\,.x + C_2 \quad (0 \le x \le L/2) \tag{3}$$

Boundary conditions ($0 \le x \le L/2$):

1. At $x = 0$; $y = 0$ and
2. At $x = L/2$; $dy/dx = 0$.

Putting these conditions in Eqns. (2) and (3), one would get

$$C_2 = 0 \quad \text{and} \quad C_1 = \frac{5wL^3}{192}$$

The expression to determine the slope at any point along the length of the beam is given by

$$EI\frac{dy}{dx} = \frac{wx^4}{12L} - \frac{wLx^2}{8} + \frac{5wL^3}{192}$$

Slope at support A (i.e., $x = 0$), $\theta_A = \left.\frac{dy}{dx}\right|_{x=0} = \frac{5wL^3}{192\,EI}$.

Expression to determine the deflection of the beam can be written as follows:

$$EIy = \frac{wx^5}{60L} - \frac{wLx^3}{24} + \frac{5wL^3x}{192}$$

Deflection at the mid-span (i.e., $x = L/2$) of the beam, $\Delta = \left.y\right|_{x=L/2} = \frac{wL^4}{120\,EI}$

Since the loading is symmetric, the slope at support B is $\theta_B = -\frac{5wL^3}{192\,EI}$.

EXAMPLE 4.4 A simply supported beam AB of flexural rigidity EI and length L is subjected to a concentrated load of P at a distance of $2L/3$ from support A as shown in Figure E4.4. Determine the slope at the supports and the deflection at point C of the beam using the *Double Integration method*.

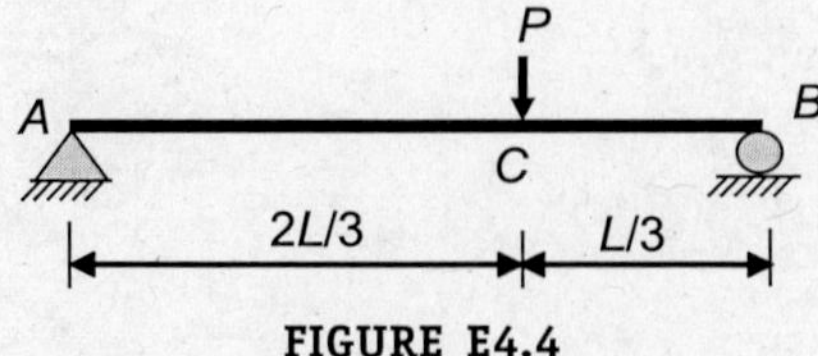

FIGURE E4.4

Solution: *Calculation of support reactions:*
Let the reactions at A and B are R_a and R_b, respectively.
   Taking moment about A,

$$R_b . L = P . \frac{2L}{3} \qquad \therefore R_b = \frac{2P}{3}$$

$$R_a = P - \frac{2P}{3} \qquad \therefore R_a = \frac{P}{3}$$

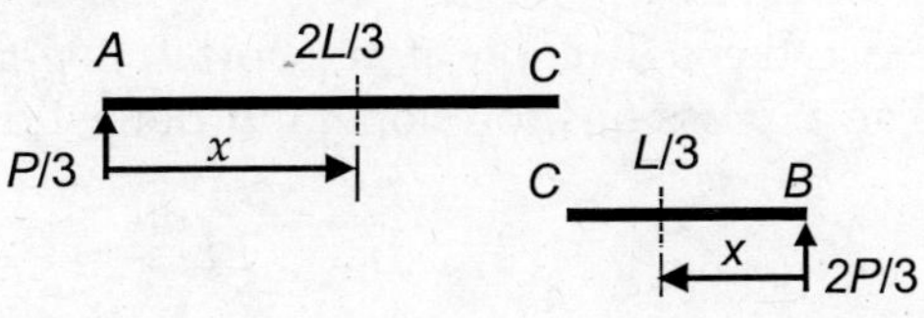

Since it is not possible a single equation of bending moment for the entire beam length, it is better to write the bending moment equations for segments AC and BC independently. Considering a section at a distance x from the support A within segment AC, the bending moment at this section is

$$M_x = \frac{P}{3} x \quad (0 \le x \le 2L/3)$$

Similarly, taking a section at a distance x from the support B within segment AC, the bending moment at this section can be written as follows:

$$M_x = \frac{2P}{3} x \quad (0 \le x \le L/3)$$

Segment AC $(0 \le x \le 2L/3)$:

$$EI \frac{d^2 y}{dx^2} = -M_x = -\frac{Px}{3}; \quad EI \frac{d^2 y}{dx^2} = -\frac{Px}{3}$$

$$EI \frac{dy}{dx} = -\int \frac{Px}{3} dx + C_1; \quad EI \frac{dy}{dx} = -\frac{Px^2}{6} + C_1$$

$$EIy = -\int \frac{Px^2}{6} dx + C_1 x + C_2; \quad EIy = -\frac{Px^3}{18} + C_1 x + C_2$$

Since the deflection at support A is zero, putting the value of $x = 0$, one can get $C_2 = 0$. The expressions for slope and deflection of the beam within the segment AC are given by

$$EI \frac{dy}{dx} = -\frac{Px^2}{6} + C_1 \quad (0 \le x \le 2L/3) \tag{1}$$

$$EIy = -\frac{Px^3}{18} + C_1 x \quad (0 \le x \le 2L/3) \tag{2}$$

Segment BC $(0 \le x \le L/3)$:

$$EI \frac{d^2 y}{dx^2} = -M_x = -\frac{2Px}{3}; \quad EI \frac{d^2 y}{dx^2} = -\frac{2Px}{3}$$

$$EI \frac{dy}{dx} = -\int \frac{2Px}{3} dx + C_1; \quad EI \frac{dy}{dx} = -\frac{Px^2}{3} + C_3$$

$$EIy = -\int \frac{Px^2}{3}\, dx + C_3 x + C_4 \; ; \qquad EIy = -\frac{Px^3}{9} + C_3 x + C_4$$

Since the deflection at support B is zero, putting the value of $x = 0$, one can get $C_4 = 0$. The expressions for slope and deflection of the beam within the segment BC are given by

$$EI\frac{dy}{dx} = -\frac{Px^2}{3} + C_3 \quad (0 \le x \le L/3) \tag{3}$$

$$EIy = -\frac{Px^3}{9} + C_3 x \quad (0 \le x \le L/3) \tag{4}$$

The integration constants C_1 and C_3 can be determined using the compatibility equations. Two equations are required to solve two unknown constants.

Compatibility equations

Deflection at point C for the segment AC must be equal to that of the segment BC, i.e., $y|_{c,AC} = y|_{c,BC}$

Slope at point C for the segment AC must be equal in magnitude and opposite in direction that of the segment BC, i.e., $\left.\dfrac{dy}{dx}\right|_{c,AC} = \left.\dfrac{dy}{dx}\right|_{c,BC}$

Putting $x = 2L/3$ in Eqn. (2) and $x = L/3$ in Eqn. (4) and using first the compatibility equation, one can get

$$-\frac{4PL^3}{243} + \left(\frac{2L}{3}\right) C_1 = -\frac{PL^3}{243} + \left(\frac{L}{3}\right) C_3$$

On simplification, the above equation can be written as

$$2C_1 - C_3 = \frac{PL^2}{27} \tag{5}$$

Similarly, putting $x = 2L/3$ in Eqn. (1) and $x = L/3$ in Eqn. (3) and using the second compatibility equation, one can get

$$-\frac{4PL^2}{54} + C_1 = \frac{PL^2}{27} - C_3$$

$$C_1 + C_3 = \frac{3PL^2}{27} \tag{6}$$

Solving Eqns. (5) and (6), both unknown constants can be determined as follows:

$$C_1 = \frac{4PL^2}{81} \; ; \; C_3 = \frac{5PL^2}{81}$$

The final expressions for slope and deflections of the beam can be written as follows:

$$\frac{dy}{dx} = \frac{1}{EI}\begin{cases} -\dfrac{Px^2}{6} + \dfrac{4PL^2}{81} & \left(0 \le x \le \dfrac{2L}{3}\right) \\[3mm] -\dfrac{Px^2}{3} + \dfrac{5PL^2}{81} & \left(0 \le x \le \dfrac{L}{3}\right) \end{cases}$$

$$y = \frac{1}{EI}\begin{cases} -\dfrac{Px^3}{18} + \dfrac{4PL^2 x}{81} & \left(0 \le x \le \dfrac{2L}{3}\right) \\[3mm] -\dfrac{Px^3}{9} + \dfrac{5PL^2 x}{81} & \left(0 \le x \le \dfrac{L}{3}\right) \end{cases}$$

To compute the slope at support A, put $x = 0$ in the first equation as applicable for the segment AC. The slope at support B can be determined by putting $x = 0$ in the second equation as applicable for the segment BC. Thus,

$$\theta_A = \left.\frac{dy}{dx}\right|_{x=0} = \frac{4\,PL^2}{81\,EI} \quad \text{and} \quad \theta_B = \left.\frac{dy}{dx}\right|_{x=0} = \frac{5PL^2}{81\,EI}$$

The deflection at the point C of the beam can be obtained by putting $x = 2L/3$ in the first equation as applicable for the segment AC (or $x = L/3$ in the second equation as applicable for the segment BC):

$$y_c = \frac{1}{EI}\left(-\frac{8PL^3}{486} + \frac{8PL^3}{243}\right) = \frac{4\,PL^3}{243\,EI}$$

4.3.2 Moment-area Method

Moment-area theorems are developed Charles E. Greene in 1873 to determine the deflection of beams. This deflection method is based on the linear relationships between the shape of the elastic (bending) curve, the bending moment, and the flexural rigidity of the beam.

Referring to Section 4.2.3, we know that $d^2y/dx^2 = -M/EI$, where y = the transverse displacement of beam and $\theta = dy/dx$ = the change in slope at any point on the elastic curve. Accordingly, the moment-curvature expression can be rewritten as follows:

$$\frac{d^2y}{dx^2} = \frac{d\theta}{dx} = -\frac{M}{EI} \tag{4.23}$$

$$d\theta = -\frac{M.dx}{EI} \tag{4.24}$$

In order to understand the usefulness of the above expression, let's consider a prismatic simply supported beam AB of length 'L' subjected to an arbitrary loading as shown in Figure 4.7(a). The bending moment diagram of the beam under the applied loading is shown in Figure 4.7(b) in which the value of bending moment (M) is divided by the flexural rigidity (EI). In other words, Figure 4.7(b) shows the *M/EI diagram* of the beam. Consider an infinitesimal segment pq of length dx at a distance of x from the support A.

Figure 4.7(c) shows the assumed deflected shape of the beam *AB*. The beam is perfectly straight on the unloaded stage. Thus, if the tangents are drawn at p and q, they will be concurrent with the beam axis represented by the dashed line shown in the figure. In the deflected position, the change in slope between the tangents drawn at these points is represented by $d\theta$. Using Eqn. (4.24), this change in slope ($d\theta$) is equal to the area of the *M/EI diagram* between the points p and q, which is represented by the shaded area. Thus, the change in slope between the tangents at any two points on the elastic curve (for example, A and B) can be given by the following expression:

$$\theta = -\int_{A}^{B} \frac{M.dx}{EI} = A_{M/EI} \tag{4.25}$$

where, $A_{M/EI}$ is the area of the *M/EI diagram* between A and B. This equation is valid if the elastic curve is a continuous function between A and B. If a discontinuity exists in the elastic curve (e.g., an internal hinge) of the beam, then the above equation can only be applied to compute the change in slope between two points lying on a continuous segment on either side of the point of discontinuity. In other words, this equation is not applicable if both points are falling on two different continuous curves that are lying on both sides of the point of discontinuity.

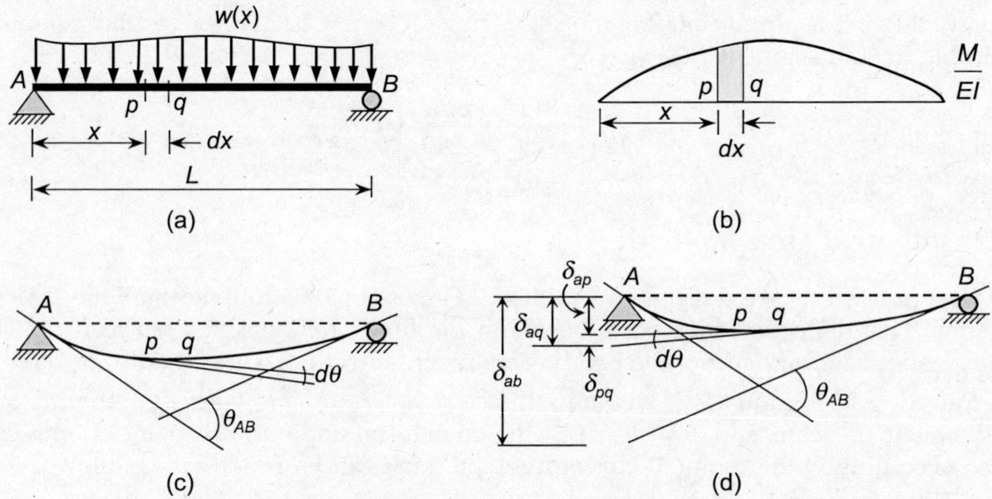

FIGURE 4.7   Moment-area method.

The deformed elastic curve of the beam *AB* is again shown in Figure 4.7(d). Two tangents are drawn at both ends of the infinitesimal segment *pq*. The angle between these tangents is dq. The segment is at a distance of x from the support A. The circular arc length for a radius x sweeping through an angle $d\theta$ is equal to $x.d\theta$. Considering small deformation theory, this arc length can be safely assumed to be equal to δ_{pq}. Using the Eqn. (4.24), the relative vertical displacement, between the tangents drawn at points p and q, just below the support *A* can be expressed as follows:

$$\delta_{pq} = x.d\theta = -x.\left(\frac{M.dx}{EI}\right) \tag{4.26}$$

The right side of the above expression represents the product of the distance from the support A and the ordinate of the M/EI diagram for the segment pq. If the entire length of the elastic curve can be assumed to be a combination of n number of infinitesimal segments, the vertical distance between point A and the tangent drawn at B is equal to the sum of arc lengths between the all adjacent segments, i.e., $x_1.dq_1 + x_2.dq_2 + \ldots + x_n.dq_n$. In the integral form, one can get

$$\delta_{ab} = \int_A^B x.d\theta = -\int_A^B x.\left(\frac{M.dx}{EI}\right) = \bar{x}.A_{M/EI} \tag{4.27}$$

where, $\bar{x}$ is the distance measured from support A from the centroid of the area of M/EI *diagram* between A and B. Similar to Eqn. (4.25), Eqn. (4.27) is valid when the two points lie on a continuous elastic curve. These two equations represent the two *moment-area theorems* as stated below:

Theorem 1: *The change in slope between two points on a continuous elastic curve is equal to the area of the M/EI diagram between these points.*

Theorem 2: *The vertical intercept measured between point A and a tangent drawn at point B on the elastic curve is equal to the product of the area of the M/EI diagram between these points and the distance of the centroid of the M/EI diagram about point A.*

The moment-area method is particularly useful when the geometric properties (i.e., area and centroidal distance) of the M/EI diagram is known. Table 4.1 shows the properties of different types of geometric shapes. If a beam is subjected to several types of loads, the slope and deflections at any point can be computed as the sum of these quantities for all loads considered independently using the principle of superpositions. The following examples illustrate the application of the moment-area method in different structures.

TABLE 4.1 Properties of moment-area diagrams

Shape	Properties	Shape	Properties
(Rectangle)	$A = bh$ $\bar{x} = \dfrac{b}{2}$	(Triangle)	$A = \dfrac{bh}{2}$ $\bar{x} = \dfrac{b}{3}$
(Trapezoid)	$A = \dfrac{b(h_1 + h_2)}{2}$ $\bar{x} = \dfrac{b}{3}\dfrac{(2h_2 + h_1)}{(h_2 + h_1)}$	(Semi-parabola)	$A = \dfrac{2bh}{3}$ $\bar{x} = \dfrac{3b}{8}$
(nth degree spandrel curve)	$A = \dfrac{bh}{(n+1)}$ $\bar{x} = \dfrac{b}{(n+2)}$	(nth degree semisegment curve)	$A = \dfrac{bhn}{(n+1)}$ $\bar{x} = \dfrac{b}{2}\dfrac{n+1}{n+2}$

EXAMPLE 4.5 A cantilever beam flexural rigidity *EI* and length *L* is subjected to a uniformly distributed load, *w* throughout its length as shown in Figure E4.5. Determine the slopes and deflections at points *B* and *C* of the beam using the *Moment-Area method*.

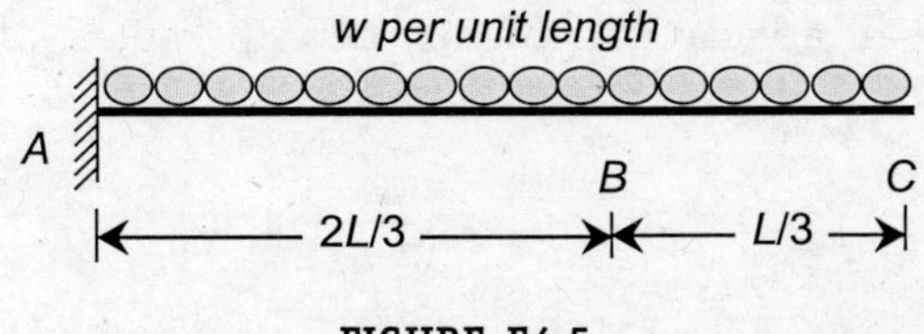

FIGURE E4.5

Solution: Let's draw the qualitative deflected shape of the beam. Since the *A* is fixed, the tangent drawn at *A* would coincide with the undeformed shape since the slope and deflection at the fixed support are zero.

Slope at any point on the elastic curve of the beam would be equal to the angle between the tangent drawn at the same point with the horizontal (i.e., tangent drawn at *A*). Thus, the slope at any point on the beam is the area of the *M/EI* diagram between point *A* and the point of consideration.

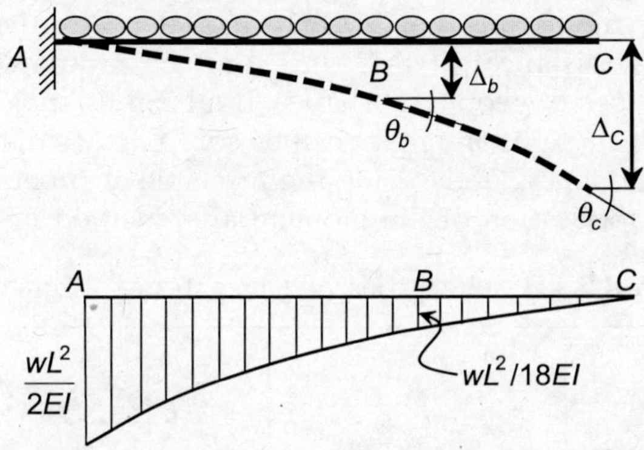

The *M/EI* diagram can be obtained by dividing all ordinates of the bending moment diagram by *EI* as shown in the figure. The ordinate of *M/EI* diagram at A is $\dfrac{wL^2}{2EI}$ and that of *B* is $\dfrac{wL^2}{18EI}$.

Slope at free end *C* of the beam = Area of *M/EI* diagram between *A* and *C*

i.e.,
$$\theta_c = \frac{1}{3} \cdot \left(\frac{wL^2}{2EI} \right) L = \frac{wL^3}{6EI}$$

Slope at point *B* of the beam = Area of *M/EI* diagram between *A* and *B*

i.e.,
$$\theta_b = \left(\frac{wL^2}{18EI} \right) \frac{2L}{3} + \frac{1}{3} \cdot \left(\frac{wL^2}{2EI} - \frac{wL^2}{18EI} \right) \frac{2L}{3} = \frac{11\,wL^3}{81\,EI}$$

Deflection at any point of the beam can be obtained by taking the moment of the area of M/EI diagram between this point and point A about the point of consideration. Thus, deflection at C = Area of M/EI diagram between points A and C times the centroidal distance measured from point C as given below:

$$\Delta_c = \frac{1}{3}\left(\frac{wL^2}{2EI}\right)L\left(\frac{3L}{4}\right) = \frac{wL^4}{8EI} \qquad \therefore \ \Delta_c = \frac{wL^4}{8EI}$$

Similarly, the deflection at point B can be obtained as follows:

$$\Delta_b = \left(\frac{wL^2}{18EI}\right)\frac{2L}{3}\left(\frac{1}{2}\cdot\frac{2L}{3}\right) + \frac{1}{3}\left(\frac{wL^2}{2EI} - \frac{wL^2}{18EI}\right)\frac{2L}{3}\left(\frac{3}{4}\cdot\frac{2L}{3}\right)$$

$$\therefore \qquad \Delta_b = \frac{5wL^4}{81EI}$$

EXAMPLE 4.6 A cantilever stepped beam of 9 m in length is subjected to a concentrated load of 40 kN at the free end as shown in Figure E4.6. Determine the slope and deflection at the free end of the beam using the *Moment-Area method*. Assume $E = 200$ GPa, $I_1 = 40{,}000$ cm^4, and $I_2 = 20{,}000$ cm^4.

Solution: The deflected shape of the beam is shown in the figure. Being a fixed end, the slope and deflection at point A are zero.

Slope at any point on the elastic curve of the beam would be equal to the angle between the tangent drawn at the same point with the horizontal (i.e., tangent drawn at A). Thus, the slope at any point on the beam is the area of the M/EI diagram between point A and the point of consideration.

As shown in the figure, the bending moment diagram (BMD) of the beam is a triangle with the maximum BM value of 360 kNm at the fixed end A.

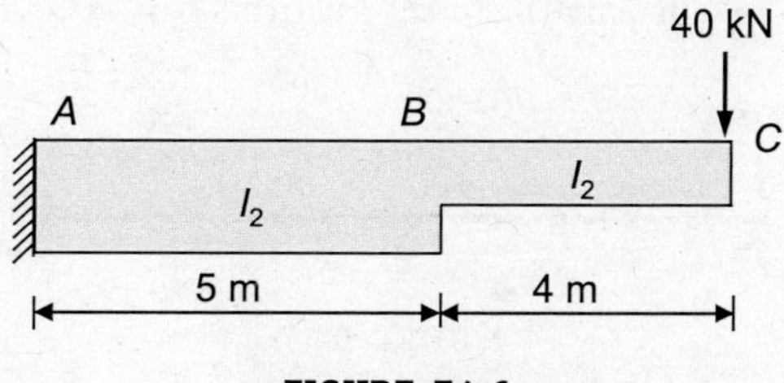

FIGURE E4.6

The M/EI diagram is obtained by dividing all ordinates of the bending moment diagram by the respective values of EI as shown in the Figure.

Slope at free end C of the beam = Area of M/EI diagram between A and C

$$\text{Thus, } \theta_c = \frac{1}{2}\cdot\left(\frac{160}{EI_2}\right)4 + \left(\frac{160}{EI_1}\right)5 + \frac{1}{2}\cdot\left(\frac{200}{EI_1}\right)5 = \left(\frac{320}{EI_2}\right) + \left(\frac{1300}{EI_1}\right)$$

Putting, $I_1 = 2I_2$, $\theta_c = \dfrac{970}{EI_2} = 0.02425$ radian

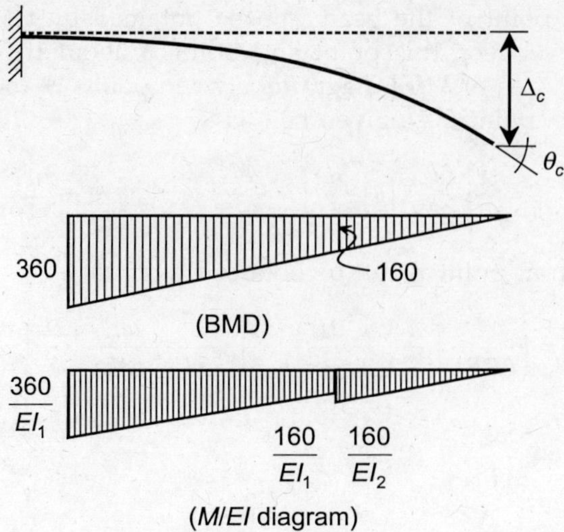

Deflection at point C can be obtained by taking the moment of the area of M/EI diagram between A and C about point C.

$$\Delta_c = \frac{1}{2}\cdot\left(\frac{160}{EI_2}\right)4\cdot\left(\frac{8}{3}\right)+\left(\frac{160}{EI_1}\right)5\left(4+\frac{5}{2}\right)+\frac{1}{2}\cdot\left(\frac{200}{EI_1}\right)5\left(4+\frac{10}{3}\right)$$

$$\Delta_c = \frac{8866.67}{EI_1}+\frac{853.3}{EI_2}=\frac{5286.67}{EI_2} \qquad \therefore\ \Delta_c = 0.1322 \text{ m}$$

EXAMPLE 4.7   A simply supported beam of 7 m in length and constant flexural rigidity (*EI*) is subjected to a concentrated load of 70 kN at a distance of 3 m from the left support as shown in Figure E4.7. Determine the slopes at the support and the deflection at the loading point using the *Moment-Area method*. Assume $E = 200$ GPa and $I = 20,000$ cm^4.

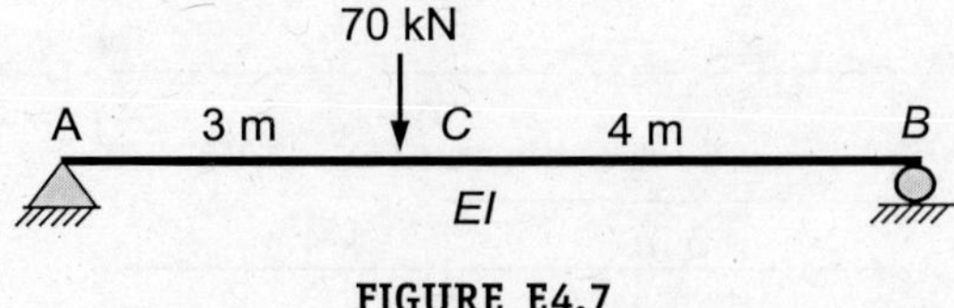

FIGURE E4.7

Solution: Let the reactions at A and B are R_a and R_b, respectively.

Taking moment about A,

$$R_b.(7) = (70).3 \qquad \therefore\ R_b = 30 \text{ kN}$$

$$R_a + R_b = 70 \qquad \therefore\ R_a = 40 \text{ kN}$$

The distribution of the bending moment is linear along the length of the beam. The bending moment is the maximum at point C and is given by

$$M_c = (40)(3) = 120 \text{ kNm}$$

The *M/EI* diagram of the beam is a triangle with a maximum ordinate of 120/*EI* as shown in the figure.

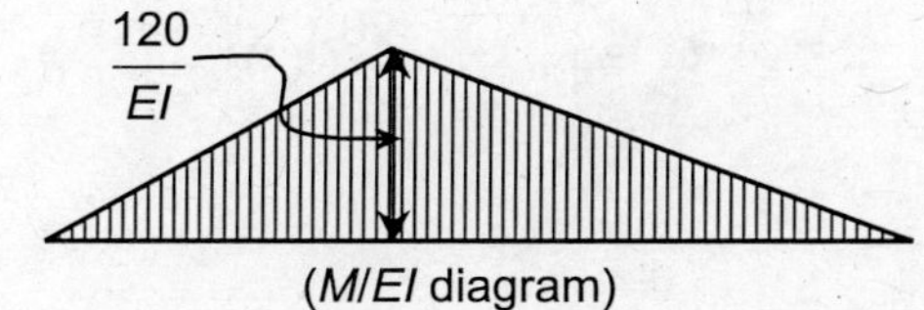

(*M/EI* diagram)

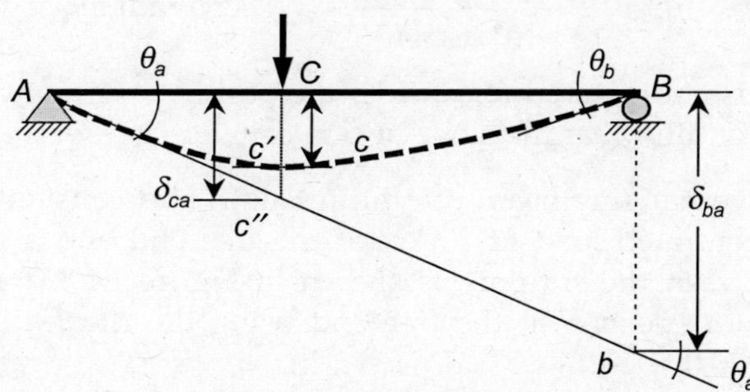

The vertical intercept *Bb* at *B* with respect to the tangent drawn to the elastic curve at *A* can be determined by taking the moment of the *M/EI* diagram about B as follows:

$$\delta_{ba} = \frac{1}{2} \cdot \left(\frac{120}{EI}\right)(4)\left(\frac{8}{3}\right) + \frac{1}{2} \cdot \left(\frac{120}{EI}\right)(3)\left(4 + \frac{3}{3}\right) \qquad \therefore \ \delta_{ba} = \frac{1540}{EI}$$

From the similar triangles *ABb* and *Acc''*, one can determine the intercept *Cc'* as follows:

$$\frac{Bb}{7} = \frac{Cc''}{3} \Rightarrow \delta_{ca} = \frac{3\delta_{ba}}{7} = \frac{660}{EI}$$

Similarly, the vertical intercept between the tangents drawn to the elastic curve at *A* and *C* is given by:

$$c'c'' = \frac{1}{2} \cdot \left(\frac{120}{EI}\right)(3)\left(\frac{3}{3}\right) \qquad \therefore \ c'c'' = \frac{180}{EI}$$

Thus, the deflection at point C of the beam is

$$\Delta_c = \delta_{ca} - c'c'' = \frac{660}{EI} - \frac{180}{EI} = \frac{480}{EI}$$

$$\therefore \qquad \Delta_c = \frac{480}{(2 \times 10^8)(2 \times 10^{-4})} = 0.012 \text{ m}$$

Slope at support *A*, $\theta_a = \dfrac{\delta_{ba}}{L} = \dfrac{220}{EI}$

$$\theta_a = \frac{220}{(2 \times 10^8)(2 \times 10^{-4})} = 0.0055 \text{ radian}$$

The vertical intercept at *A* with respect to the tangent drawn to the elastic curve at *B* can be determined by taking the moment of the *M/EI* diagram about *B* as follows:

$$\delta_{ab} = \frac{1}{2} \cdot \left(\frac{120}{EI}\right)(4)\left(3 + \frac{4}{3}\right) + \frac{1}{2} \cdot \left(\frac{120}{EI}\right)(3)\left(\frac{6}{3}\right) \quad \therefore \quad \delta_{ab} = \frac{1400}{EI}$$

Slope at support B, $\theta_b = \dfrac{\delta_{ab}}{L} = \dfrac{200}{EI}$

$$\therefore \qquad \theta_b = \frac{200}{(2 \times 10^8)(2 \times 10^{-4})} = 0.0050 \text{ radian}$$

For simplicity, it is not required to consider the positive or negative signs in the computation of slopes. One can find that the slope at *B* is counter-clockwise and at *A* is clockwise.

EXAMPLE 4.8   An overhanging beam of 6 m in length and constant flexural rigidity (*EI*) is subjected to a concentrated load of 40 kN at the free end and a uniformly distributed load of 10 kN/m between the support as shown in Figure E4.8. Determine the slope at the support *A* and the deflection at the free end using the *Moment-Area method*. Assume *E* = 200 GPa and *I* = 20,000 cm^4.

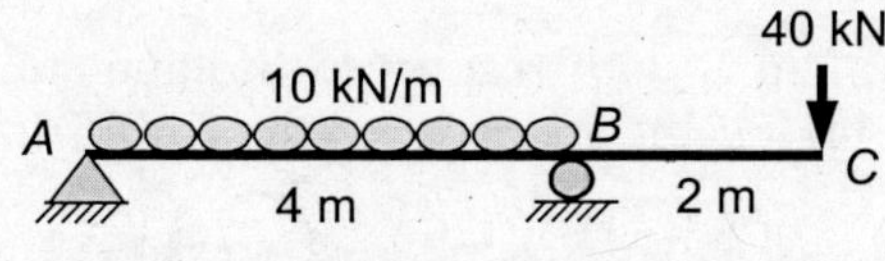

FIGURE E4.8

Solution: Bending moment diagram of the beam can be drawn by superimposing the bending moment diagrams due to point load and uniformly distributed load.

Due to a point load of 40 kN at point *C*, the bending moment diagram would be a triangle with a maximum ordinate of −80 kNm.

Similarly, one would get a parabolic distribution of the bending moment diagram due to uniformly distributed load of 10 kN/m. The maximum BM at the midspan of the span *AB* would be $wL^2/8$ = 20 kNm.

The combined *M/EI* diagram is shown in the figures.

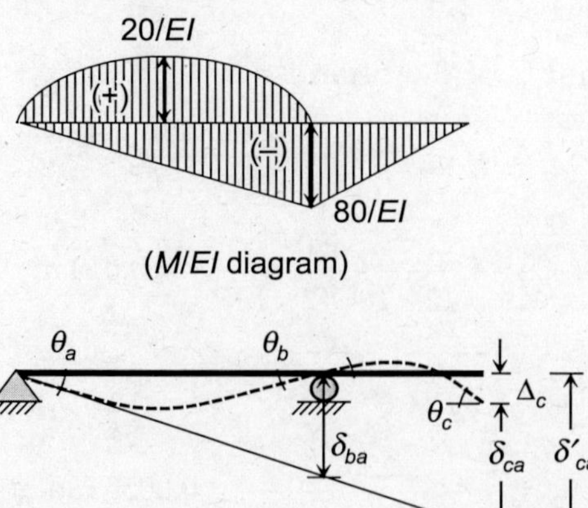

The vertical intercept at B with respect to the tangent drawn to the elastic curve at A can be determined by taking the moment of the M/EI diagram about B as follows:

$$\delta_{ba} = \frac{2}{3}\cdot\left(\frac{20}{EI}\right)(4)\left(\frac{4}{2}\right) - \frac{1}{2}\cdot\left(\frac{80}{EI}\right)(4)\left(\frac{4}{3}\right) \qquad \therefore\ \delta_{ba} = -\frac{320}{3EI}$$

By extrapolation, one can get

$$\frac{\delta_{ba}}{4} = \frac{\delta'_{ca}}{6} \quad\Rightarrow \delta'_{ca} = \frac{6\delta_{ba}}{4} = -\frac{160}{EI}$$

The vertical intercept at C to the tangent drawn to the elastic curve at A can be given by:

$$\delta_{ca} = \frac{2}{3}\cdot\left(\frac{20}{EI}\right)(4)\left(2+\frac{4}{2}\right) - \frac{1}{2}\cdot\left(\frac{80}{EI}\right)(4)\left(2+\frac{4}{3}\right) - \frac{1}{2}\cdot\left(\frac{80}{EI}\right)(2)\left(\frac{4}{3}\right)$$

$$\therefore \qquad\qquad \delta_{ca} = -\frac{1280}{3EI}$$

The deflection at point C of the beam is

$$\Delta_c = \delta'_{ca} - \delta_{ca} = -\frac{160}{EI} + \frac{1280}{3EI} = \frac{266.67}{EI}$$

$$\Delta_c = \frac{266.67}{(2\times10^8)(2\times10^{-4})} = 0.0067\text{ m}$$

The slope at support A, $\theta_a = \dfrac{\delta_{ba}}{L_{ab}} = -\dfrac{26.67}{EI}$

$$\therefore \qquad\qquad \theta_a = -\frac{26.67}{(2\times10^8)(2\times10^{-4})} = -0.00067\text{ radian}$$

4.3.3 Conjugate-Beam Method

The conjugate-beam method was originally introduced as the elastic-weight method by Otto Mohr in 1860. This method is similar to the Method-area method. The underlying principle of this method is to determine the slopes and deflections of a structure resulting from the applied loads (actual loads) by solving the shear forces and bending moments of a conjugate beam due to the elastic loads derived from the angle changes of structural elements.

As discussed previously, the slope and deflection at any point of the beam can be written as follows:

$$\theta = -\int \frac{M.dx}{EI} \qquad\qquad (4.28)$$

$$y = -\iint \frac{M.dx}{EI}dx \qquad\qquad (4.29)$$

For a beam subjected to a distributed load of $w(x)$, we know that $dV/dx = -w$ and $dM/dx = V$. Thus, $V = -\int w.dx$ and $M = \int V.dx = -\int\int w.dx.dx$.

Suppose that we have a beam, called *conjugate beam*, whose length is equal to that of the actual beam. Let this beam be subjected to an *elastic load* intensity of M/EI. The *elastic load* is also referred to as the *angle load* as the change in slope, $d\theta$ is related to the M/EI, i.e., $d\theta = M \cdot (dx/EI)$.

On comparing the expressions for the slope and shear force, one can get

$$V = -\int \frac{M.dx}{EI} \tag{4.30}$$

Similarly,

$$M = -\int\int \frac{M.dx}{EI} dx \tag{4.31}$$

Accordingly, the conjugate-beam theorems can be defined as follows:

Theorem 1: *The slope at any point on the real beam is equal to the shear force at the same point of the conjugate beam loaded with the M/EI diagram of the real beam.*

Theorem 2: *The deflection at any point on the real beam is equal to the bending moment at the same point of the conjugate beam loaded with the M/EI diagram of the real beam.*

In order to make the above theorems true, the boundary conditions and connections of the conjugate beam must be consistent with that of the real beam based on the loading conditions and displacement compatibility. In addition to maintaining the same length in the conjugate beam as the real beam, the boundary (or interior) conditions must be suitably modified. Table 4.2 summarizes the requirements of support conditions of the conjugate beam for the given support conditions of the real beam.

TABLE 4.2   Comparison of support conditions of the real beam and the conjugate beam

Real beam (Applied load)		*Conjugate beam (Elastic load)*
Fixed end $\begin{cases} \theta = 0 \\ y = 0 \end{cases}$	$\Longrightarrow$	$\left.\begin{array}{l} \bar{V} = 0 \\ \bar{M} = 0 \end{array}\right\}$ Free end
Free end $\begin{cases} \theta \neq 0 \\ y \neq 0 \end{cases}$	$\Longrightarrow$	$\left.\begin{array}{l} \bar{V} \neq 0 \\ \bar{M} \neq 0 \end{array}\right\}$ Fixed end
Simple end (Hinged/roller) $\begin{cases} \theta \neq 0 \\ y = 0 \end{cases}$	$\Longrightarrow$	$\left.\begin{array}{l} \bar{V} \neq 0 \\ \bar{M} = 0 \end{array}\right\}$ Simple end (Hinged/roller)
Interior support (Hinged/roller) $\begin{cases} \theta \neq 0 \\ y = 0 \end{cases}$	$\Longrightarrow$	$\left.\begin{array}{l} \bar{V} \neq 0 \\ \bar{M} = 0 \end{array}\right\}$ Interior connection (Hinged/roller)
Interior connection (Hinged/roller) $\begin{cases} \theta \neq 0 \\ y = 0 \end{cases}$	$\Longrightarrow$	$\left.\begin{array}{l} \bar{V} \neq 0 \\ \bar{M} = 0 \end{array}\right\}$ Interior support (Hinged/roller)

The examples of conjugate beams are shown in Figure 4.8. It is important to note that the conjugate beams may be unstable in themselves. However, they must maintain the stable equilibrium under the action of elastic loads.

It is also necessary to follow a consistent sign convention in the conjugate-beam method as mentioned below:

1. The *positive* segments of the M/EI diagram of the real beam are considered as the M/EI loading in the *downward* direction on the conjugate beam.
2. The *positive shear* on the conjugate beam corresponds to the *clockwise slope* (rotation) of the real beam and the *negative shear* corresponds to the *counterclockwise slope* (rotation) of the real beam.
3. The *positive bending moment* on the conjugate beam corresponds to the *downward displacement* of the real beam and the *negative bending moment* corresponds to the *upward displacement* of the real beam.

TABLE 4.3 Examples of conjugate beam

Real beam		Conjugate beam

EXAMPLE 4.9 Using the *Conjugate-beam method*, determine the slope at the supports and the mid-span deflection of the simply supported beam subjected to a uniformly distributed load of w as shown in Figure E4.9. Assume the beam of constant flexural rigidity EI.

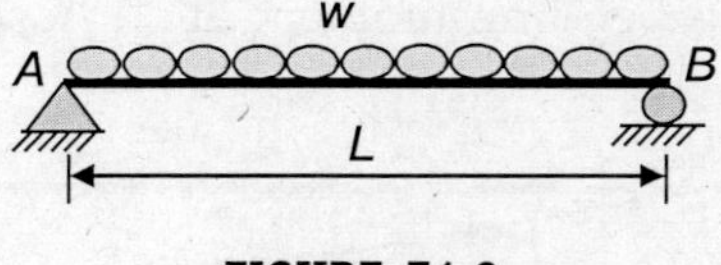

FIGURE E4.9

Solution: Bending moment diagram (BMD) of the beam can be drawn by computing the reactions. For the given loading, the BMD is shown in the figure with the maximum ordinate of $wL^2/8$.

Conjugate beam is represented by a simply supported beam with M/EI loading. The shear force and bending moment at any point of the conjugate beam represent the change in slope and deflection of the real beam.

Let's assume the reactions of the conjugate beam are R_a and R_b.

Vertical force equilibrium:

$$R_a + R_b = \frac{2}{3} \cdot \left(\frac{wL^2}{8EI}\right) L = \frac{wL^3}{12EI}$$

Taking moment about end A, $R_b(L) = \frac{2}{3} \cdot \left(\frac{wL^2}{8EI}\right) L \left(\frac{L}{2}\right)$ $\therefore$ $R_b = \frac{wL^3}{24EI}$ ($\uparrow$)

$$R_a = \frac{wL^3}{12EI} - \frac{wL^3}{24EI} = \frac{wL^3}{24EI} \, (\uparrow)$$

Slope at support A of the real beam = Shear force at A of the conjugate beam

$$\theta_a = V_a = +\frac{wL^3}{24EI} \quad \text{(clockwise)}$$

Slope at support B of the real beam = Shear force at B of the conjugate beam

$$\therefore \qquad \theta_b = V_b = -\frac{wL^3}{24EI} \quad \text{(anti-clockwise)}$$

Deflection at the mid-span of the real beam = Bending moment of the conjugate beam at the same point.

$$\Delta = BM\big|_{x=L/2} = \left(\frac{wL^3}{24EI}\right)\left(\frac{L}{2}\right) - \frac{2}{3} \cdot \left(\frac{wL^2}{8EI}\right)\left(\frac{L}{2}\right)\left(\frac{3}{8} \cdot \frac{L}{2}\right)$$

$$\therefore \qquad \Delta = +\frac{5wL^4}{384EI} \quad \text{(Downward)}$$

EXAMPLE 4.10   Determine the bending moment, M required to cause a rotation θ at support B of the beam AB of length, L as shown in Figure E4.10 using the *Conjugate-beam method*. Also, find the fixed end bending moment at A. Assume the flexural rigidity as EI.

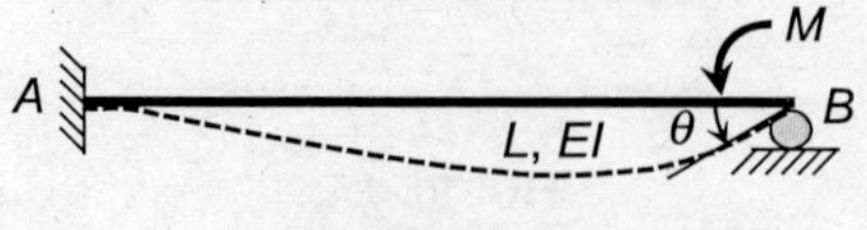

FIGURE E4.10

Solution: Let's assume the vertical reaction and the fixed end bending moment at A be R_a and M_a, respectively. The reaction at support B is R_b.

Bending moment diagram of the beam can be drawn considering the effects of the external moment, M, and fixed end moment, M_a separately as shown in the figure.

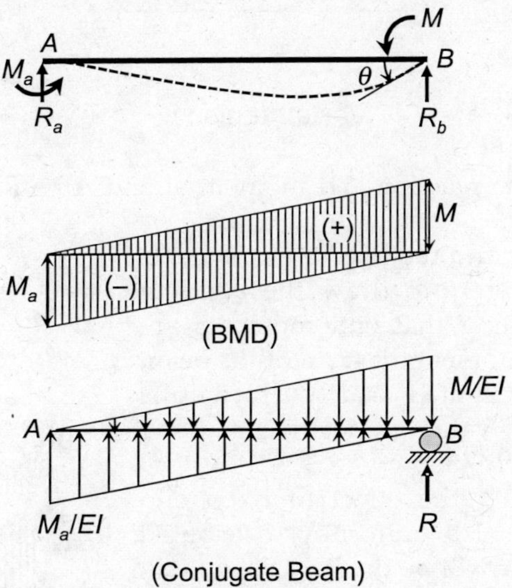

(BMD)

(Conjugate Beam)

For the assumed direction of M_a, the related bending moment diagram is negative.

The Conjugate beam is shown in the figure. Let the reaction at B of the conjugate beam is R. Since the conjugate beam must be in stable and equilibrium, all equations of equilibrium must be satisfied.

Taking moment about B of the conjugate beam,

$$\frac{1}{2} \cdot \frac{M_a}{EI} \cdot L \cdot \left(\frac{2L}{3}\right) = \frac{1}{2} \cdot \frac{M}{EI} \cdot L \cdot \left(\frac{L}{3}\right) \qquad \therefore M_a = \frac{M}{2}$$

This shows that the fixed end bending moment of a beam is equal to one-half of the bending moment applied at the other end. In addition, the direction of the fixed end bending moment is the same as that of the applied bending moment.

Now, the force equilibrium in the vertical direction of the conjugate beam would lead to the following expression:

$$R + \frac{1}{2} \cdot \frac{M_a}{EI} \cdot L = \frac{1}{2} \cdot \frac{M}{EI} \cdot L$$

Using $M_a = \dfrac{M}{2}$, $R = \dfrac{ML}{4EI}$

Slope at support B of the real beam = Shear force at B of the conjugate beam $\theta = R = \dfrac{ML}{4EI}$

$\therefore$
$$M = \frac{4EI\theta}{L} \quad \text{and} \quad M_a = \frac{2EI\theta}{L}$$

EXAMPLE 4.11   Determine the fixed end bending moments at both ends of a fixed beam subjected to a uniformly distributed load, *w* using the *Conjugate-beam method*. Assume the flexural rigidity as *EI*.

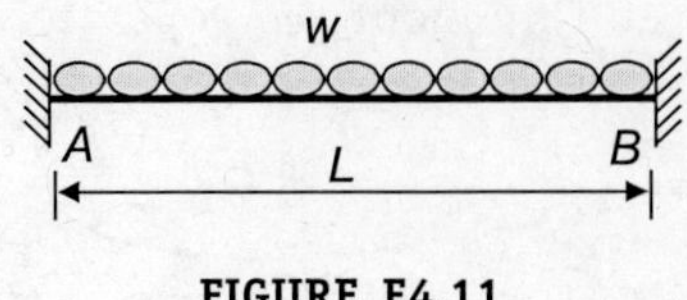

FIGURE E4.11

Solution: Let's assume the reaction and moment at end *A* be R_a and M_a, and that of *B* be R_b and M_b, respectively.

The external loading and the fixed end moments can be considered separately to draw the bending moment diagram. Assuming that end moments are not present, the bending moment diagram of the beam due to the uniformly distributed load (UDL) will be parabolic. The maximum value of positive bending moment due to UDL is $wL^2/8$.

Similarly, considering the effect of fixed end moments in the absence of a uniformly distributed load, the bending moment will be drawn as shown in the figure. The superimposition of these two moment diagrams will lead to the final bending moment diagram of the beam.

Conjugate beam for a fixed-ended beam is a free-free beam with loading equal to the *M/EI* diagram as shown in the figure. Since the conjugate beam must be stable, all equations of equilibrium must be satisfied.

In absence of any support, the equilibrium of forces in the vertical direction of the conjugate beam results in the following:

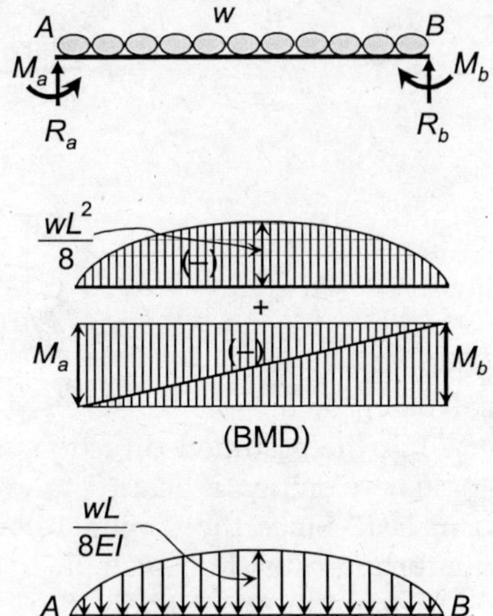

Total downward force = Total upward force

$$\frac{2}{3}\cdot\left(\frac{wL^2}{8EI}\right)L = \frac{1}{2}\cdot\frac{(M_a+M_b)}{EI}\cdot L$$

$$\therefore \qquad M_a + M_b = \frac{wL^2}{6} \qquad (1)$$

Taking moment about A,

$$\frac{2}{3}\cdot\left(\frac{wL^2}{8EI}\right)L\cdot\left(\frac{L}{2}\right) - \frac{1}{2}\cdot\frac{M_a}{EI}\cdot L\cdot\left(\frac{L}{3}\right) - \frac{1}{2}\cdot\frac{M_b}{EI}\cdot L\cdot\left(\frac{2L}{3}\right) = 0$$

$$\therefore \qquad M_a + 2M_b = \frac{wL^2}{4} \qquad (2)$$

Solving Eqns. (1) and (2), the fixed end bending moments can be obtained as follows:

$$M_a = M_b = \frac{wL^2}{12}$$

EXAMPLE 4.12   Determine the fixed end bending moments at both ends of the beam of length L if the end B is displaced vertically by Δ. Use the *Conjugate-beam method* and assume the flexural rigidity of the beam is EI.

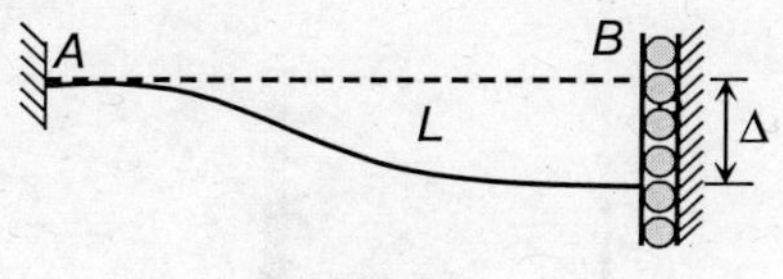

FIGURE E4.12

Solution: Let's assume the fixed end moments and reaction be M_a, M_b, and R_a.

For the assumed directions of end moments, the bending moment diagram can be drawn as shown in the figure.

The conjugate beam for the real beam is shown in the figure. Since the deflection of the real beam is equal to the bending moment in the conjugate beam at the same point, a bending moment of D is added at point B.

Since the conjugate beam is stable, all equations of equilibrium must satisfy.

Force equilibrium:

$$\frac{1}{2} \cdot \left(\frac{M_a}{EI}\right) L = \frac{1}{2} \cdot \left(\frac{M_b}{EI}\right) L \quad \therefore M_a = M_b$$

Taking moment about B,

$$\Delta + \frac{1}{2} \cdot \left(\frac{M_b}{EI}\right) L \cdot \left(\frac{L}{3}\right) - \frac{1}{2} \cdot \left(\frac{M_b}{EI}\right) L \cdot \left(\frac{2L}{3}\right) = 0$$

$$\Delta = \frac{M_a L^2}{6EI} \quad \therefore M_a = \frac{6EI\Delta}{L^2} = M_b$$

4.3.4   Energy Method

This method is based on the principle of energy conservation which states that the external work done by the applied loads is equal to the internal strain energy stored in the structure.

If a variable force F moves by a distance of ds along its direction, the word done is $F.ds$. To explain the concept of external work done, let's consider a uniform bar having one end fixed and the other end free. As shown in Figure 4.8, the bar is subjected to a

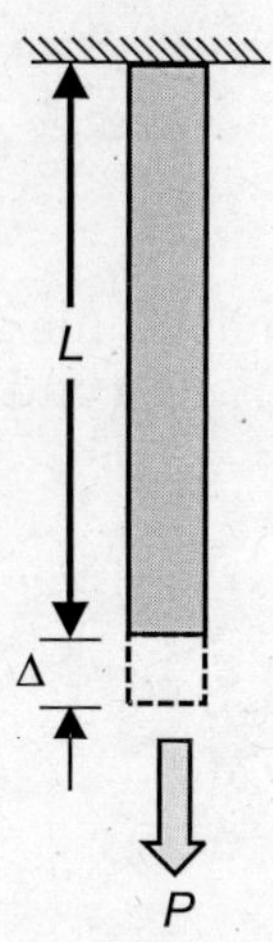
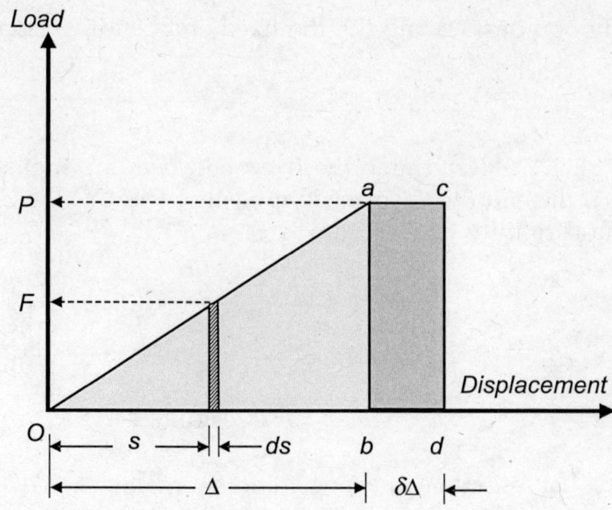

FIGURE 4.8   Force-displacement relationship.

gradually applied load at the free end. The axial displacement of the bar reaches a value of Δ as the load increases from zero to P. As shown in the figure, the load-displacement relationship is linear as represented by OA for linearly elastic material. Total work done by the load is given by

$$W_e = \int_0^\Delta F \cdot ds = \int_0^\Delta \left(\frac{P.s}{\Delta}\right) \cdot ds = \frac{1}{2}P\Delta \tag{4.32}$$

Thus, the external done by the gradually increasing load P is equal to the area of the shaded triangle *oab*. If any further load results in the additional displacement of $\delta\Delta$ in the direction P, then the additional amount of work done by the already applied load P is $P.\delta\Delta$. This additional work done is equal to the area of the shaded rectangle *abcd*.

Similarly, the external work done by a bending moment applied to a structure is equal to the product of the average bending moment and the angle of rotation, i.e.,

$$W_e = \frac{1}{2}M\theta \tag{4.33}$$

Next, let's compute the internal work done or energy stored by the structures. The amount of strain energy stored depends on the type of structure, i.e., truss or frame. In a plane truss, internal axial forces are developed in the members when subjected to static loading. These forces result in the axial deformation of the members. If a member of length, L and uniform cross-sectional area, A carries an axial force of S, then the corresponding axial deformation, δ can be obtained as follows:

Stress = Strain × Modulus of elasticity (E)

$$\frac{S}{A} = \frac{\delta}{L} \cdot E \quad \Rightarrow \delta = \frac{SL}{AE} \tag{4.34}$$

Since the internal axial force in a member gradually increases from 0 to S, the internal work (or, *strain energy*) of the member is equal to the product of the average of the internal force and the corresponding deformation. Mathematically,

$$dW_i = \frac{1}{2}S\delta = \frac{S^2 L}{2AE} \qquad (4.35)$$

Since a truss consists of more than one member, the total strain energy stored in a truss is the sum of internal work of all members as given expressed below:

$$W_i = \sum \frac{S^2 L}{2AE} \qquad (4.36)$$

Similarly, the internal work of a beam (or any flexural member) of flexural rigidity, EI, and length, L carrying a bending moment, M can be determined. Consider an infinitesimal segment of length 'dx' of beam carrying a bending moment m. The change in the angle of rotation of this segment can be determined using Eqn. (4.24). The strain energy stored in this segment is given by

$$dW_i = \frac{1}{2}M \cdot d\theta = \frac{M^2 dx}{2EI} \qquad (4.37)$$

The total strain energy stored over the entire length L of the flexural member is

$$W_i = \int_0^L \frac{M^2 dx}{2EI} \qquad (4.38)$$

Once the external work done by the loads and the strain energy stored by the members are known, the deflection at any point can be determined by using the Principle of Energy Conservation. This means that the displacement, Δ due to the applied load, P on a structure can be determined by equating the external work done to the internal strain energy stored in all members of a structure.

For plane truss (axial force): $W_e = W_i \quad \Rightarrow \frac{1}{2}P\Delta = \sum \frac{S^2 L}{2AE} \qquad (4.39)$

For beam (flexure action): $W_e = W_i \quad \Rightarrow \frac{1}{2}P\Delta = \int_0^L \frac{M^2 dx}{2EI} \qquad (4.40)$

The above equations are useful if a structure is subjected to only one concentrated loading. Further, the displacement component along the load direction can only be determined. Though its direct application is limited, this concept has been utilized to develop other methods for computing the member displacements.

4.3.5 Method of Virtual Work (Unit-load Method)

Consider a structure in equilibrium is subjected to a system of virtual (or imaginary) forces, Q. As a result, the structure would undergo small virtual displacements such that it remains in equilibrium. From the principle of energy conservation, the external virtual work, dW_e must be equal to the internal virtual work, dW_i. Therefore, the energy change in a structure must remain constant when subjected to a system of real or virtual external loads. Accordingly, *Bernoulli's principle of virtual work* can be stated as follows:

Consider a structure in equilibrium under a system of real external loads is subjected to a unit virtual load. By the principle of conservation of energy, the external work done is equal to the internal strain energy stored due to internal virtual forces undergoing real deformations.

The *virtual work method* is also referred to as the *unit-load method* or the *dummy unit-load method*. This method has been demonstrated by their applications to the truss systems as well as beams.

Plane truss

Consider a plane truss, as shown in Figure 4.9, subjected to the external loads, P_1, and P_2. Under the action of these loads, the internal axial forces (S) and axial deformations (Δ) of all members are also shown in the figure. For the known values of axial force (S), cross-sectional area (A), and length (L), the axial deformations (Δ) of all members can be determined as follows:

$$\Delta_1 = \frac{S_1 L_1}{A_1 E}; \quad \Delta_2 = \frac{S_2 L_2}{A_2 E}; \quad \text{and so on...}$$

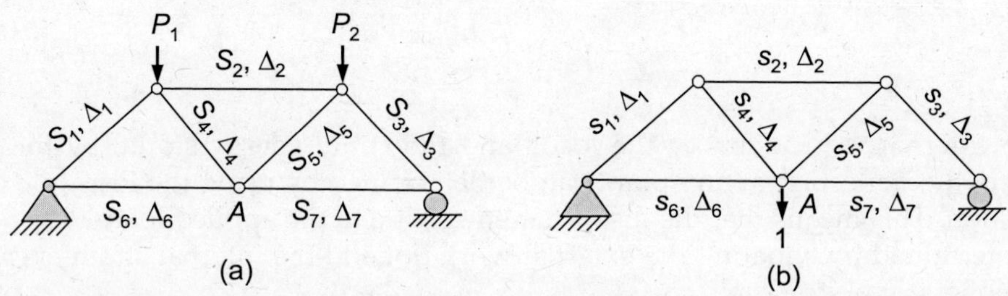

FIGURE 4.9   Member forces due to (a) applied loads, and (b) unit load at A.

Assume that it is required to determine the vertical displacement at joint A of the truss. Let this (unknown) displacement at A be δ_A. If a virtual unit load is applied at the joint A of the truss with all external loads acting, the virtual external work done (dW_e) by this unit force undergoing the displacement, δ_A can be given by

$$dW_e = 1 \cdot \delta_A \tag{4.41}$$

If the internal member forces (s) are generated due to the action of the virtual unit force at A, the total internal virtual work (or virtual strain energy) can be computed as follows.

$$dW_i = \sum_{j=1}^{n} s_j \cdot \Delta_j \tag{4.42}$$

Since,
$$dW_e = dW_i, \quad 1 \cdot \delta_A = \sum_{j=1}^{n} s_j \cdot \Delta_j \tag{4.43}$$

Thus, the unknown displacement of any joint of the plane truss subjected to a system of external loads can be determined using the following expression:

$$\delta_A = \sum \frac{SsL}{AE} \tag{4.44}$$

In the above equation, s represents the internal axial forces in the members developed due to a unit force applied at the same joint along the same direction as that of δ_A.

Beam (or frame member)

A similar procedure as followed for plane truss can be adopted to derive the deflection at any point of a beam subjected to external loading. Lets' consider a simply supported prismatic beam shown in Figure 4.10(a) subjected to a system of external loads. The length of the beam is L, and the flexural rigidity of the beam is EI. For simplicity and better understanding, it is assumed that the beam cross-section in rectangular in shape and width b and depth d as shown in Figure 4.10(b). Assume that it is required to determine the deflection (y_A) at point A of the beam under the applied loading.

Let's consider an infinitesimal segment dx at a distance of x from the left support. Assume that the bending moment at this section due to the applied loading is M. Due to the flexural action, the longitudinal deformation of a fibre at a distance y from the neutral axis (*n.a.*) on the beam cross-section of the segment dx can be determined as follows:

$$\delta = \varepsilon \cdot dx = \left(\frac{\sigma}{E}\right) \cdot dx = \left(\frac{My}{EI}\right) \cdot dx \tag{4.45}$$

Now, apply a virtual unit load at point A in the vertical direction as shown in Figure 4.10(c). Let's assume that the bending moment at location x due to this unit load be m. The internal longitudinal force on a differential strip dy at a distance of y from the neutral axis can be determined as follows:

$$dF = d\sigma \cdot dA = d\sigma \cdot b \cdot dy = \left(\frac{my}{I}\right) \cdot b \cdot dy \tag{4.46}$$

The internal virtual work at this differential strip in the beam cross-section can be found as follows:

$$dW_i = dF \cdot \delta = \left(\frac{my}{I}\right) b \cdot dy \left(\frac{My}{EI}\right) dx \tag{4.47}$$

The virtual work for the entire cross-section of the segment dx can be computed by area integration and by putting $I = \int y^2 dA = \int y^2 b \cdot dy$ as follows:

$$dW_i = \int \left(\frac{Mm}{EI^2}\right) y^2 b \cdot dy \cdot dx = \frac{M \cdot m \cdot dx}{EI} \tag{4.48}$$

The internal virtual work of the beam over the length from 0 to L is as follows:

$$W_i = \int_0^L \frac{M \cdot m \cdot dx}{EI} \tag{4.49}$$

The external virtual work, $\qquad\qquad W_e = 1 \cdot y_A$ $\qquad\qquad$ (4.50)

Equating (4.19) and (4.50), one can determine the deflection at joint A as follows:

$$y_A = \int_0^L \frac{M \cdot m \cdot dx}{EI}$$ (4.51)

Similarly, if it is required to determine the rotation at point *A* on the beam, a unit moment, instead of unit load, should be applied at that point. Accordingly, the bending moment at any section *x* would be *m'*. Eqn. (4.51) can be modified as follows to determine the rotation at point *A*.

$$\theta_A = \int_0^L \frac{M \cdot m' \cdot dx}{EI}$$ (4.52)

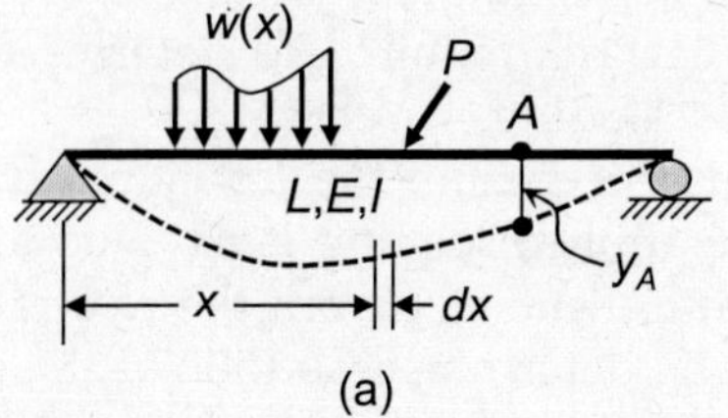
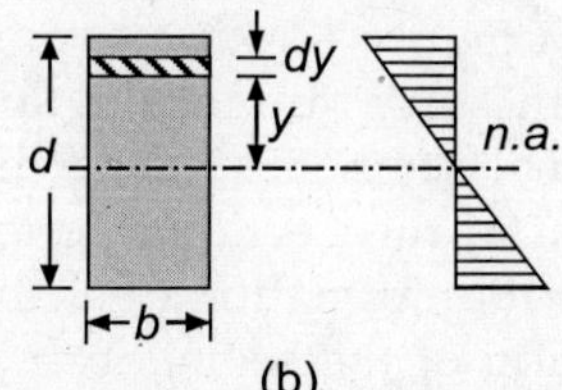
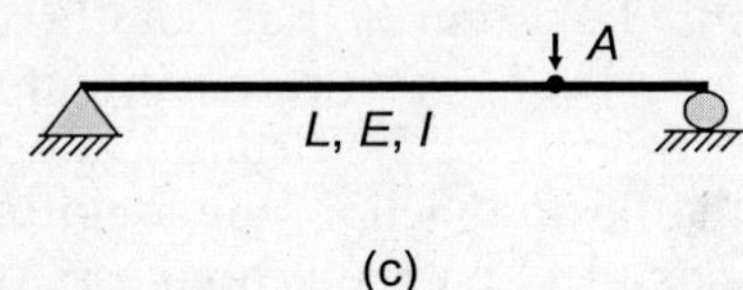

FIGURE 4.10   (a) Deformed shape of beam, (b) Variation in strain across the beam cross-section, (c) Virtual unit force applied at *A*.

EXAMPLE 4.13   Determine the vertical deflection of joint *F* of the truss shown in Figure E4.13 using the *method of virtual work*. The cross-sectional area of each bar is 4 cm^2 and the modulus of elasticity (*E*) of material is 200 GPa.

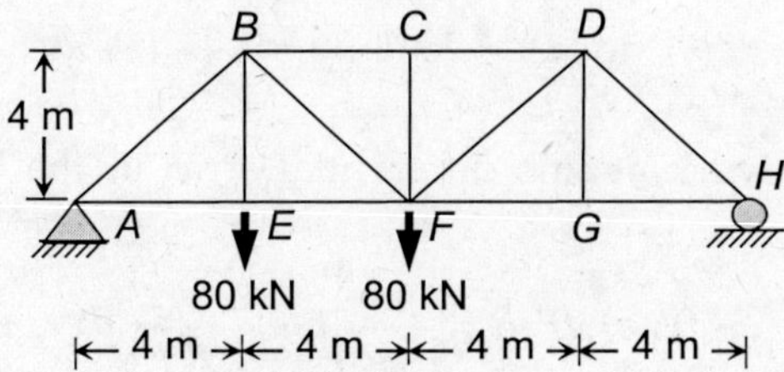

FIGURE E4.13

Solution:

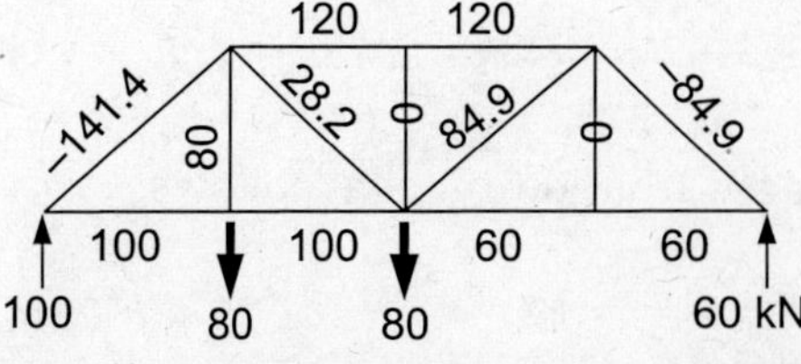
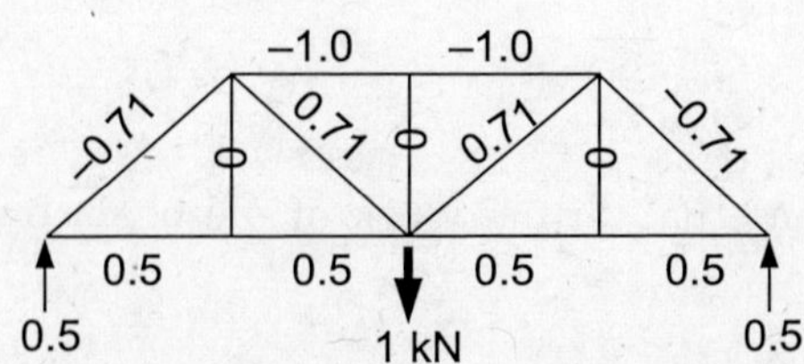

(Member forces, *S* due to external loads) $\qquad$ (Member forces, *s* due to unit load at point *F*)

Member	$L(m)$	S(kN)	s(kN)	SsL
AB	5.66	−141.4	− 0.71	568.2
BC	4.00	120.0	− 1.0	− 480.0
CD	4.00	120.0	− 1.0	− 480.0
DH	5.66	−84.9	− 0.71	336.4
AE	4.00	100.0	0.5	200.0
EF	4.00	100.0	0.5	200.0
FG	4.00	60.0	0.5	120.0
GH	4.00	60.0	0.5	100.0
BE	4.00	80.0	0	0
BF	5.66	28.2	0.71	113.3
CF	4.00	0	0	0
FD	5.66	84.9	0.71	341.2
DG	4.00	0	0	0
Total				**1039.1**

Vertical deflection at joint F, $\Delta_F = \sum \dfrac{SsL}{AE} = \dfrac{(1039.1) \times 10^3}{(4 \times 10^{-4}) \cdot (200 \times 10^6)}$ mm = 13.0 mm

EXAMPLE 4.14 Determine (i) the horizontal and vertical deflection of joint F, and (ii) the horizontal deflection at the roller support D of the truss shown in Figure E4.14 using the *method of virtual work*. The cross-sectional area of each bar is 10 cm^2 and the modulus of elasticity (E) of material is 200 GPa.

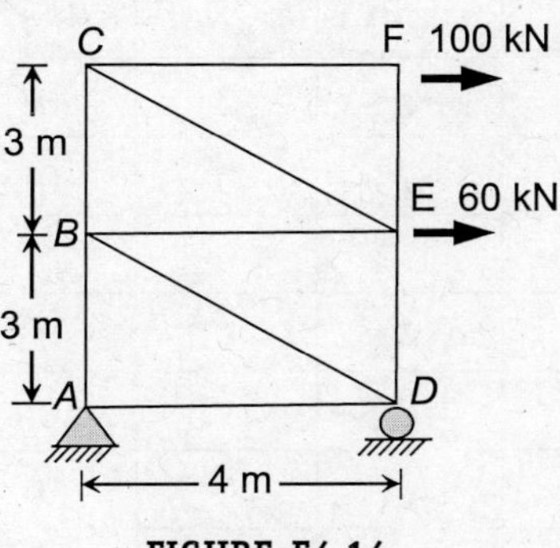

FIGURE E4.14

Solution: Let the horizontal and vertical deflections at the joint F be D_{hF} and D_{vF}, respectively. Further, let the horizontal deflection at the joint D be D_{hD}.

$$\Delta_{hF} = \sum \frac{Ss_1 L}{AE}; \quad \Delta_{vF} = \sum \frac{Ss_2 L}{AE}; \quad \Delta_{hD} = \sum \frac{Ss_3 L}{AE}$$

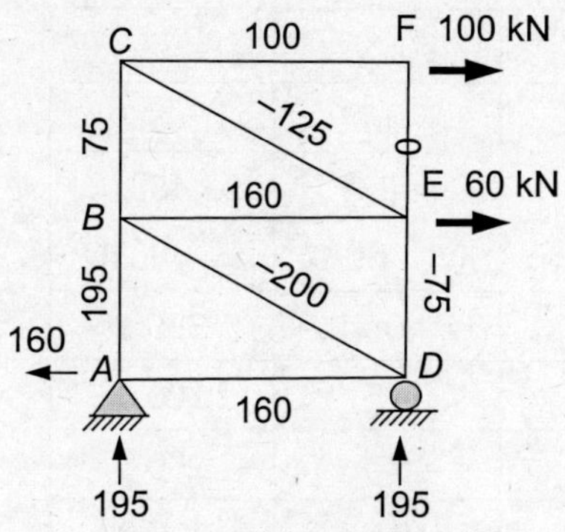

(Member forces, *S* due
to external loads)

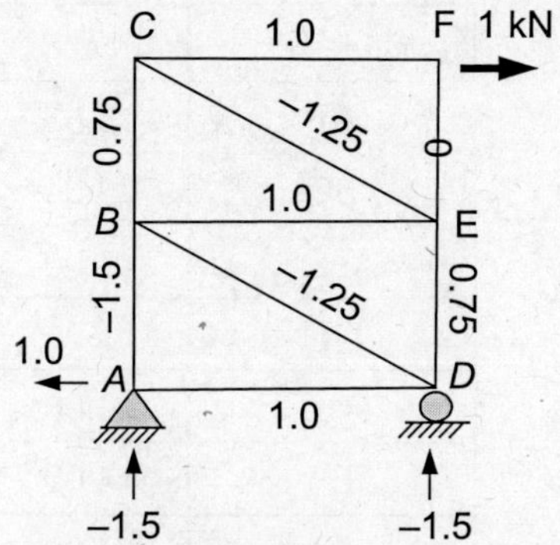

(Member forces, s_1 due to unit
horizontal load at joint *F*)

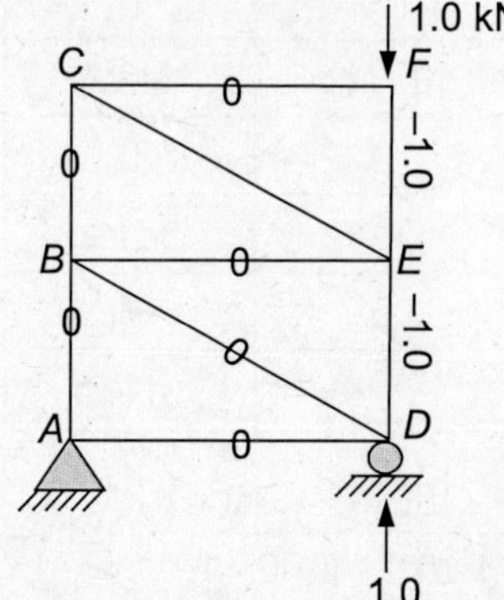

(Member forces, s_2 due to unit
vertical load at joint *F*)

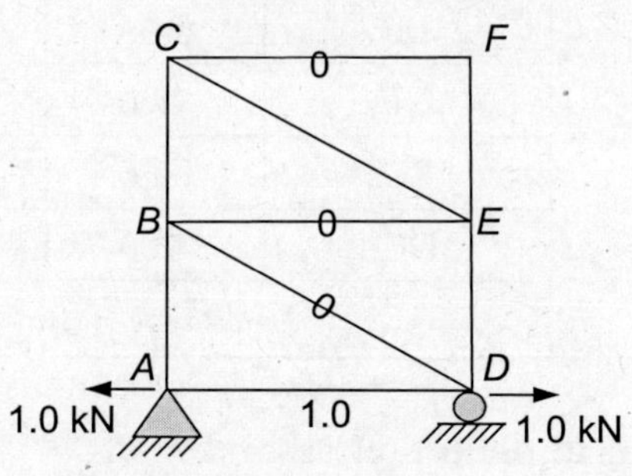

(Member forces, s_3 due to unit
horizontall load at joint *D*)

Member	L(m)	S(kN)	s_1 (kN)	s_2 (kN)	s_3 (kN)	Ss_1L	Ss_2L	Ss_3L
AB	3.0	195	−1.5	0	0	−877.5	0	0
BC	3.0	75	0.75	0	0	168.8	0	0
CF	4.0	100	1.0	0	0	400.0	0	0
EF	3.0	0	0	−1.0	0	0	0	0
DE	3.0	−75	−0.75	−1.0	0	168.8	225.0	0
AD	4.0	160	1.0	0	1.0	640.0	0	640
BD	5.0	−200	−1.25	0	0	1250.0	0	0
BE	4.0	160	1.0	0	0	640.0	0	0
CE	5.0	−125	−1.25	0	0	781.3	0	0
Total						**3171.3**	**225.0**	**640.0**

Horizontal deflection at joint F, $\Delta_{hF} = \sum \dfrac{Ss_1 L}{AE} = \dfrac{(3171.3) \times 10^3}{(10 \times 10^{-4}) \cdot (200 \times 10^6)}$ mm = 15.8 mm

Vertical deflection at joint F, $\Delta_{vF} = \sum \dfrac{Ss_2 L}{AE} = \dfrac{(225) \times 10^3}{(10 \times 10^{-4}) \cdot (200 \times 10^6)}$ mm = 1.13 mm

Horizontal deflection at joint D, $\Delta_{hD} = \sum \dfrac{Ss_3 L}{AE} = \dfrac{(640) \times 10^3}{(10 \times 10^{-4}) \cdot (200 \times 10^6)}$ mm = 3.2 mm

EXAMPLE 4.15   All upper chord members of the truss shown in Figure E4.15 are subjected to a temperature rise of 50°C, whereas the vertical and diagonal members experience a temperature rise of 40°C. All remaining members are subjected to a temperature rise of 30°C. Using the *method of virtual work*, determine (i) the vertical deflection of joint G, and (ii) the horizontal deflection at C of the truss. Assume the linear coefficient of thermal expansion of the material, $\alpha = 12 \times 10^{-6}/°C$ for all members.

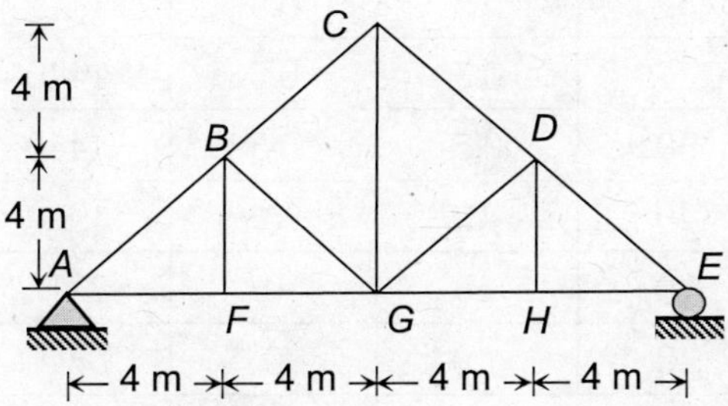

FIGURE E4.15

Solution:

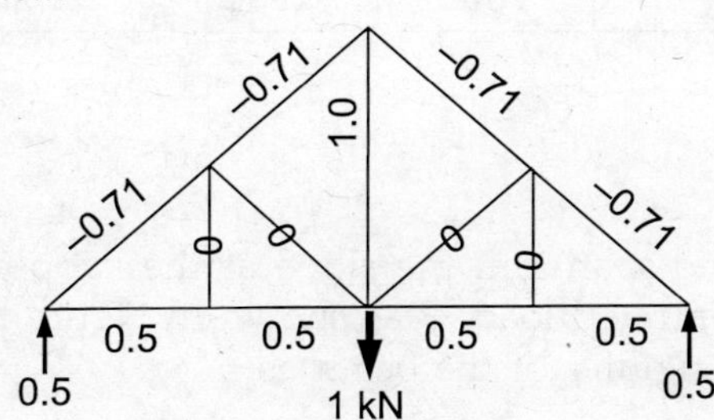

(Member forces, s_1 due to unit vertical load at joint G)

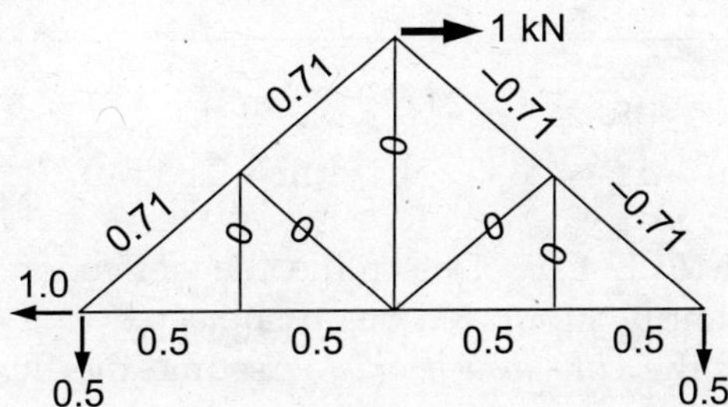

(Member forces, s_2 due to unit horizontal load at joint C)

Let the horizontal deflection at C be Δ_{hC} and the vertical deflection at the joint G be Δ_{vG}. Axial displacement of members due to the temperature change of $\Delta L = L.\alpha.(\Delta T)$

We know that $\Delta = \sum \dfrac{Ss_1 L}{AE} = \sum s_1 \left(\dfrac{SL}{AE} \right)$

Replacing the axial displacement term due to external load, S by the displacement due to thermal expansion, the following expressions can be written.

$$\Delta = \sum s_1 L . \alpha . (\Delta T)$$

Thus, $\Delta_{vG} = \sum s_1 L \cdot \alpha \cdot (\Delta T)$ and $\Delta_{hC} = \sum s_2 L \cdot \alpha \cdot (\Delta T)$

Member	L (m)	ΔT (°C)	s_1 (kN)	s_2 (kN)	$L.\alpha.\Delta T$ $(x10^{-6})$	$s_1 L.\alpha.\Delta T$ $(x10^{-6})$	$s_2 L.\alpha.\Delta T$ $(x10^{-6})$
AB	5.66	50	−0.71	0.71	3396	−2411.2	2411.2
BC	5.66	50	−0.71	0.71	3396	−2411.2	2411.2
CD	5.66	50	−0.71	−0.71	3396	−2411.2	−2411.2
DE	5.66	50	−0.71	−0.71	3396	−2411.2	−2411.2
AF	4.0	30	0.5	0.5	1440	720.0	720.0
FG	4.0	30	0.5	0.5	1440	720.0	720.0
GH	4.0	30	0.5	0.5	1440	720.0	720.0
HE	4.0	30	0.5	0.5	1440	720.0	720.0
BF	4.0	40	0	0	1920	0	0
BG	5.66	40	0	0	2717	0	0
CG	8.0	40	1.0	0	3840	3840.0	0
GD	5.66	40	0	0	2717	0	0
DH	4.0	40	0	0	1920	0	0
					Total	**−2924.6**	**2880.0**

$$\therefore \; \Delta_{vG} = -2924.6 \times 10^{-3} \, \text{mm} = -2.92 \, \text{mm}$$

$$\therefore \; \Delta_{hC} = 2880 \times 10^{-3} \, \text{mm} = 2.88 \, \text{mm}$$

EXAMPLE 4.16 Determine the deflection at mid-span and the slope at the supports of the simply supported beam subjected to a concentrated load, P as shown in Figure E4.16, using the *Unit load method*. Assume the flexural rigidity of the beam as *EI*.

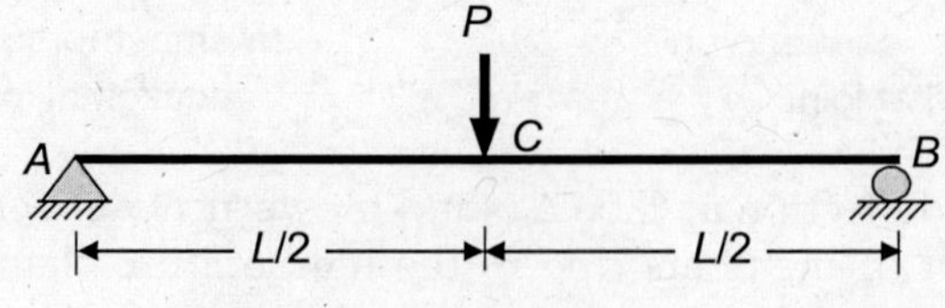

FIGURE E4.16

Solution: Deflection at any point of the beam is given by

$$\Delta = \int \frac{Mm}{EI} dx$$

where, M = Bending moment due to the external loading, m = Bending moment due to unit load applied in the desired direction at the point of consideration.

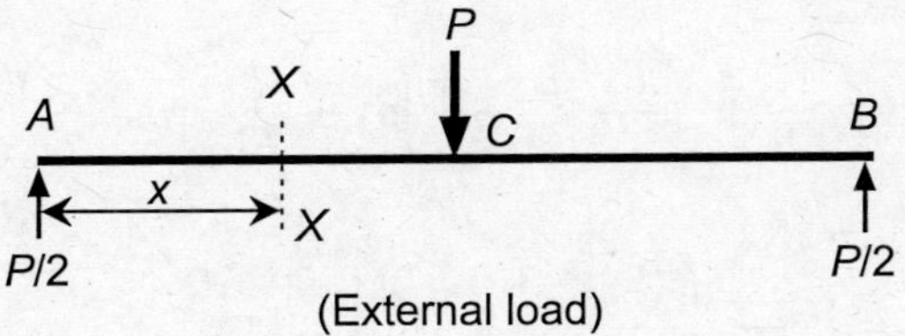

(External load)

Due to external loading, the bending moment at any section X-X is

$$M_x = \frac{P}{2}x \quad (0 \leq x \leq L/2)$$

To compute the deflection at the mid-span, a unit load is applied at the same point. The bending moment at any section X-X is

$$m_x = \frac{1}{2}x \quad (0 \leq x \leq L/2)$$

To compute the slope at support A, a unit moment needs to be applied at the same point. The bending moment at any section X-X is

$$m_x = 1 - \frac{x}{L} \quad (0 \leq x \leq L)$$

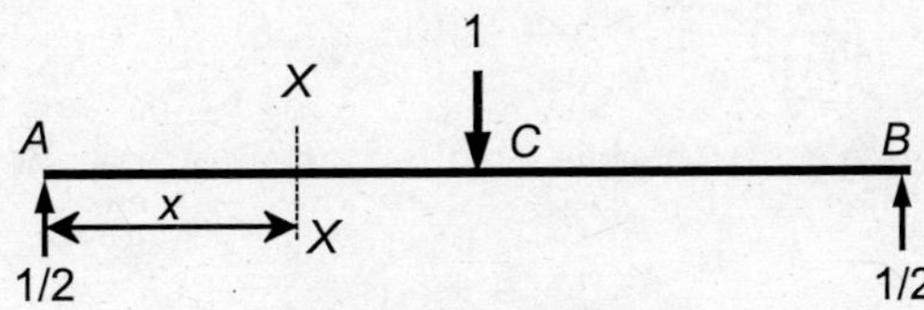

(Unit load at point C)

(Unit moment at point A)

Deflection at mid-span, $\Delta = \int \dfrac{Mm}{EI} dx$

$$\Delta = \frac{2}{EI} \int_0^{L/2} \left(\frac{Px}{2}\right)\left(\frac{x}{2}\right) dx = \left(\frac{P}{2EI}\right)\left(\frac{x^3}{3}\right)\Bigg|_0^{L/2}$$

$\therefore \qquad \Delta = \dfrac{PL^3}{48\,EI}$

Slope at the support A, $\theta_A = \int \dfrac{Mm}{EI} dx$

$$\theta_A = \frac{1}{EI} \int_0^{L/2} \left(\frac{Px}{2}\right)\left(1 - \frac{x}{L}\right) dx + \frac{1}{EI} \int_0^{L/2} \left(\frac{Px}{2}\right)\left(\frac{x}{L}\right) dx$$

$$\theta_A = \left(\frac{P}{2EI}\right)\left(\frac{x^2}{2} - \frac{x^3}{3L} + \frac{x^3}{3L}\right)\Bigg|_0^{L/2}$$

$$\therefore \qquad \theta_A = \frac{PL^2}{16EI}$$

EXAMPLE 4.17   Using the *Unit load method*, find the slope and deflection at the free end of the cantilever beam subjected to a triangular loading of maximum intensity, w as shown in Figure E4.17. Flexural rigidity of the beam is *EI*.

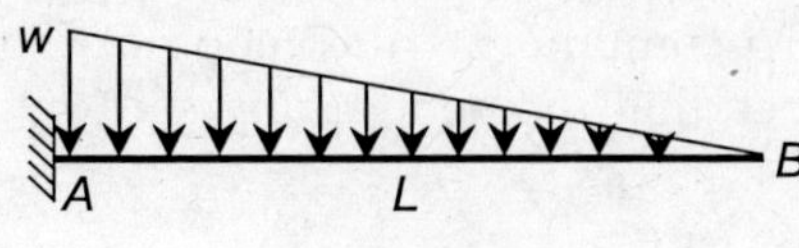

FIGURE E4.17

Solution: Let's consider a section at a distance x from the free end.

Bending moment at this section due to external loading,

(Unit load at free end)

$$M_x = \frac{1}{2} \cdot x \cdot \left(\frac{wx}{L}\right)\left(\frac{x}{3}\right) = \frac{wx^3}{6L} \qquad (0 \le x \le L)$$

To compute deflection, a unit load is applied at free end. Bending moment at a distance x from the free end,

$$m_x = x \qquad (0 \le x \le L)$$

Deflection at the free end, $\Delta = \int \dfrac{Mm}{EI} dx$

$$\Delta = \frac{1}{EI} \int_0^L \left(\frac{wx^3}{6L}\right)(x)\, dx = \left(\frac{w}{6EIL}\right)\left(\frac{x^5}{5}\right)\Bigg|_0^L \qquad \therefore \ \Delta = \frac{wL^4}{30\,EI}$$

To compute the slope at the free end, assume that a clockwise moment of 1 kNm is applied at this point. Thus,

$$\theta_B = \frac{1}{EI} \int_0^L \left(\frac{wx^3}{6L}\right)(1)\, dx = \left(\frac{w}{6EIL}\right)\left(\frac{x^4}{4}\right)\Bigg|_0^L$$

$$\therefore \qquad \theta_B = \frac{wL^3}{24\,EI}$$

EXAMPLE 4.18   Using the *Unit load method*, find the slope and deflection at the free end, B of the overhanging beam as shown in Figure E4.18. The flexural rigidity of the beam is *EI*.

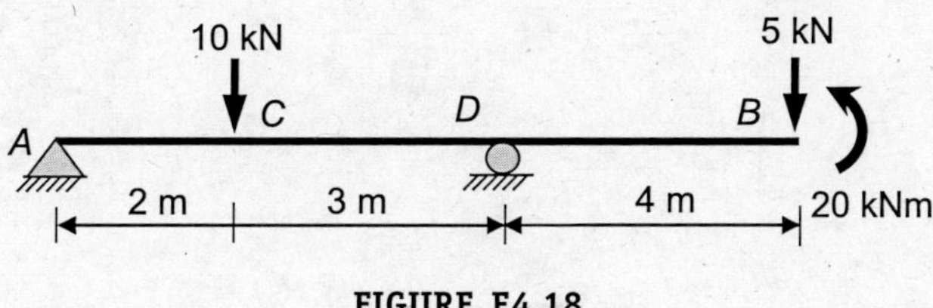

FIGURE E4.18

Solution:

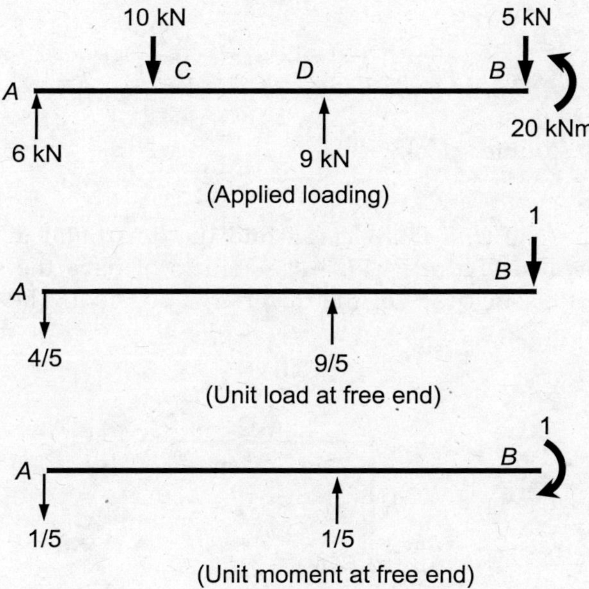

Deflection, $\Delta = \int \dfrac{Mm}{EI} dx$;   Slope, $\theta = \int \dfrac{Mm'}{EI} dx$

Member AB ($0 \leq x \leq 2$):

$$M_x = 6x; \quad m_x = -\frac{4x}{5}; \quad m'_x = -\frac{x}{5}$$

Member BC ($2 \leq x \leq 5$):

$$M_x = 6x - 10(x-2) = 20 - 4x; \quad m_x = -\frac{4x}{5}; \quad m'_x = -\frac{x}{5}$$

Member DC ($0 \leq x \leq 4$):

$$M_x = 20 - 5x; \quad m_x = -x; \quad m'_x = -1$$

Deflection at B,

$$\Delta_B = \int_0^2 6x \cdot \left(-\frac{4x}{5}\right)\left(\frac{dx}{EI}\right) + \int_2^5 (20-4x)\cdot\left(-\frac{4x}{5}\right)\left(\frac{dx}{EI}\right)$$

$$+ \int_0^4 (20-5x)\cdot(-x)\left(\frac{dx}{EI}\right)$$

$$\therefore \qquad\qquad \Delta_B = -\frac{109.33}{EI}\ (\uparrow)$$

Slope at B,

$$\theta_B = \int_0^2 6x \cdot \left(-\frac{x}{5}\right)\left(\frac{dx}{EI}\right) + \int_2^5 (20-4x)\cdot\left(-\frac{x}{5}\right)\left(\frac{dx}{EI}\right) + \int_0^4 (20-5x)\cdot\left(\frac{dx}{EI}\right)$$

$$\theta_B = -\frac{54}{EI}\ \text{(counter-clockwise)}$$

EXAMPLE 4.19   Using the *Unit load method*, find the horizontal deflection at point D of the portal frame shown in Figure E4.19. All members of have the same flexural rigidity of EI. The modulus of elasticity (E) of the material is 200 GPa. The moment of inertia (I) of members is 25000 cm^4.

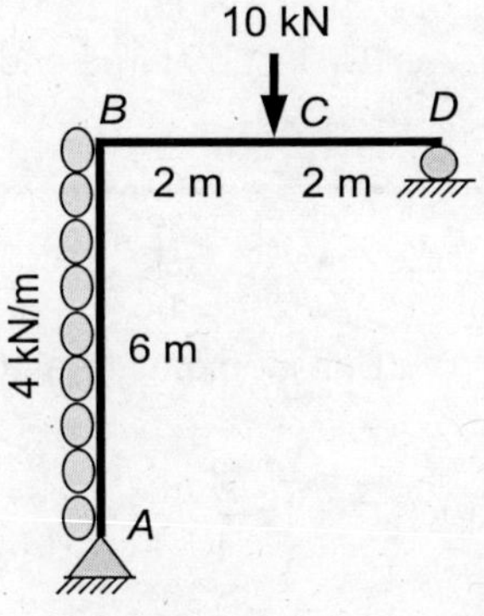

FIGURE E4.19

Solution: The reactions at the supports due to the external loading can be determined using the equations of equilibrium and are shown in the figures.

To determine the deflection of the joint D, a unit load is applied in the horizontal direction at the same joint. The reaction forces as computed using the equations of equilibrium are also shown in the figure.

$$\text{Deflection at joint } D,\ \Delta = \int \frac{Mm}{EI}dx$$

Member AB $(0 \le x \le 6)$:

$$M_x = 24x - 2x^2; \quad m_x = x$$

Member DC $(0 \le x \le 2)$:

$$M_x = 23x; \quad m_x = 1.5x$$

Member CB $(2 \le x \le 4)$:

$$M_x = 23x - 10(x-2) = 13x + 20; \quad m_x = 1.5(x-2) = 1.5x - 3$$

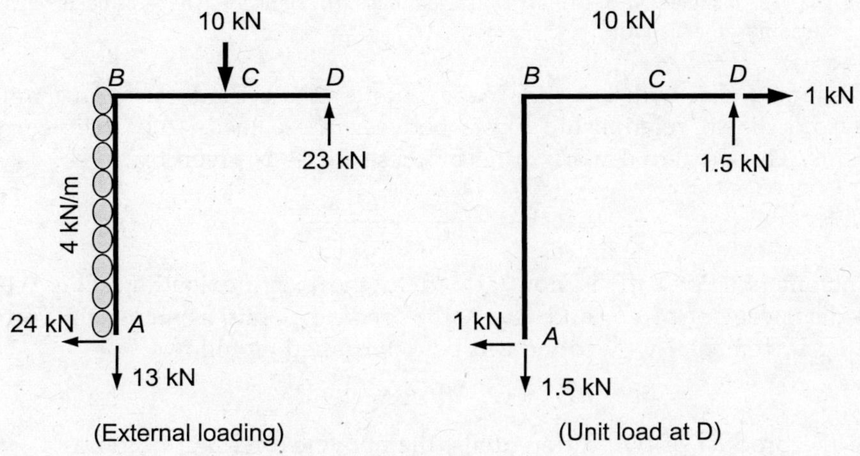

Deflection at D,

$$\Delta = \int_0^6 (24x - 2x^2) \cdot x \left(\frac{dx}{EI} \right) + \int_0^2 (23x) \cdot (1.5x) \left(\frac{dx}{EI} \right)$$

$$+ \int_2^4 (13x + 20) \cdot (1.5x - 3) \left(\frac{dx}{EI} \right)$$

$$\therefore \quad \Delta = \frac{1362}{EI} = \frac{1362 \times 10^3}{(200 \times 10^6)(25000 \times 10^{-8})} \text{mm} = 27.24 \text{ mm}$$

4.3.6   Castigliano's First Theorem

Castigliano's first theorem provides one of the most important methods for determining elastic deflections of structures. One can find any deflection component (i.e., displacement or rotation) in any direction at any point of the structure using this method. The first theorem of Castigliano can be stated as follows:

The deflection (or rotation) at the point application of a force (or couple) on a structure in the direction of the force (or couple) can be determined by taking the first partial derivative of the total internal strain energy of the structure with respect to the applied force (or couple).

To demonstrate this theorem, let us consider a simply-supported beam subjected to two concentrated forces, P and Q, as shown in Figure 4.11(a). Let the corresponding displacements in the direction of the applied forces be Δ_1 and Δ_2.

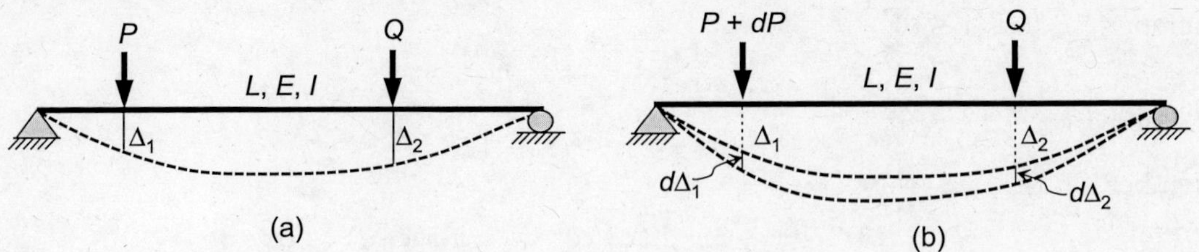

FIGURE 4.11   (a) Beam subjected to two concentrated loads, (b) Deflected shape of beam with incremental loading at one point.

It is assumed that both external loads, P and Q, are gradually and simultaneously applied. Also, a linear relationship exists between these loads and their corresponding displacements. Thus, external work done by these forces is given by

$$W_e = \frac{P \cdot \Delta_1}{2} + \frac{Q \cdot \Delta_2}{2} \tag{4.53}$$

A small incremental load dP is now added to the existing load P. This would cause additional displacement of $d\Delta_1$ and $d\Delta_2$ at the loading points as shown in Figure 4.11(b). The resulting incremental work done can be determined as follows:

$$dW_e = (P + dP) \cdot d\Delta_1 + Q \cdot d\Delta_2 \tag{4.54}$$

Neglecting the product of two differentials, the above expression becomes

$$dW_e = P \cdot d\Delta_1 + Q \cdot d\Delta_2 \tag{4.55}$$

Let's determine the above work done in a different way. For this, the loads $(P + dP)$ and Q are assumed to be applied gradually and simultaneously. Total external work of the beam due to application of these loads traversing the corresponding displacements, shown in Figure 4.11(b), can be given by

$$W_e' = \left(\frac{P + dP}{2} \right)(\Delta_1 + d\Delta_1) + \frac{Q}{2}(\Delta_2 + d\Delta_2) \tag{4.56}$$

Neglecting the product of two differentials, the above expression can be rewritten as

$$W_e' = \frac{P \cdot \Delta_1}{2} + \frac{Q \cdot \Delta_2}{2} + \frac{dP \cdot \Delta_1}{2} + \frac{P \cdot d\Delta_1}{2} + \frac{Q \cdot d\Delta_2}{2} \tag{4.57}$$

Since $dW_e = W_e' - W_e$, the incremental work done can be determined using Eqns. (4.53) and (4.57) as follows:

$$dW_e = \frac{dP \cdot \Delta_1}{2} + \frac{P \cdot d\Delta_1}{2} + \frac{Q \cdot d\Delta_2}{2} \tag{4.58}$$

Comparing Eqns. (4.55) and (4.58), one can get

$$dW_e = \frac{dP \cdot \Delta_1}{2} + \frac{dW_e}{2} \tag{4.59}$$

Therefore, $$dW_e = dP \cdot \Delta_1 \tag{4.60}$$

$$\text{Rearranging, } \Delta_1 = \frac{dW_e}{dP} \tag{4.61}$$

The above equation demonstrates the Castigliano's first theorem as stated earlier. Since more than one action (force or couple) is usually applied to a structure, the general expression for deflections by Castigliano's first method should be written in terms of partial derivative as follows:

$$\Delta = \frac{\partial W_e}{\partial P} \tag{4.62}$$

We know that the external work done is equal to the internal strain energy. Thus, Eqn. (4.61) can be rewritten for truss member or beam member using Eqns. (4.39) and (4.40), respectively.

Accordingly, the deflection at any point of a beam member can be written as

$$\Delta = \int M \left(\frac{\partial M}{\partial P} \right) \frac{dx}{EI} \tag{4.63}$$

Similarly, the deflection at any joint of a plane truss can be computed as follows:

$$\Delta = \sum S \left(\frac{\partial S}{\partial P} \right) \frac{L}{AE} \tag{4.64}$$

If there is no force (or couple) is acting at the point at which the deflection is required, an imaginary force (or couple) can be applied in the desired direction until the partial derivative of the internal strain energy has been found. The imaginary force (or couple) is then reduced to zero to compute the actual deflection of the structure. This method is demonstrated in the following examples.

EXAMPLE 4.20 Determine the vertical deflection at point C of the plane truss shown in Figure E4.20 using *Castigliano's Theorem*. Assume Modulus of elasticity (E) of the material is 200 GPa and the area (A) of all bars is 10 cm^2.

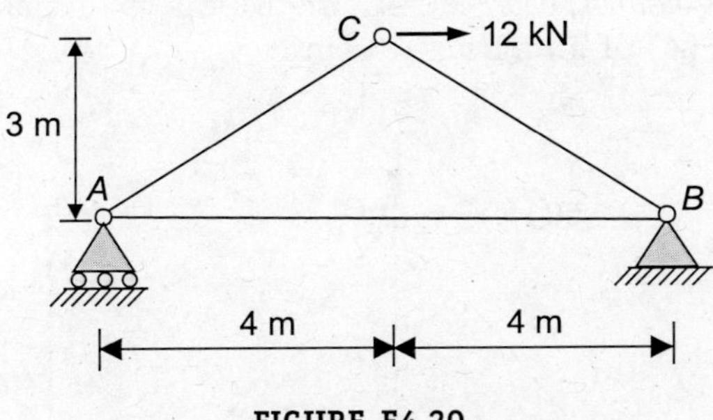

FIGURE E4.20

Solution: Let's apply a fictitious load, P in the vertical direction at the joint C of the truss. Using the equations of the equilibrium, the reaction forces as well as the internal forces in all members of the truss can be determined. All these forces are computed separately and shown in the following figures. In the computation of member forces (S), the value of P must be set to zero being a fictitious one.

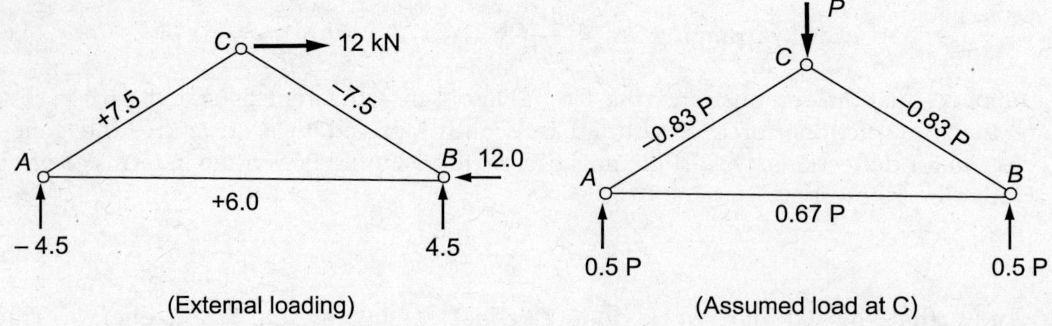

The final forces in the members are computed as the sum of member forces computed in these two cases.

Member	Length (m)	S	$\dfrac{\partial S}{\partial P}$	$S\,(P = 0)$	$S\left(\dfrac{\partial S}{\partial P}\right)L$
AB	8.0	6.0 + 0.67 *P*	0.67	6.0	32.2
BC	5.0	7.5 − 0.83 *P*	− 0.83	− 7.5	31.1
AC	5.0	7.5 − 0.83 *P*	− 0.83	7.5	− 31.1
				Total	**32.2**

Vertical deflection at joint C, $\Delta_c = \sum S\left(\dfrac{\partial S}{\partial P}\right)\dfrac{L}{AE}$

$$\Delta_c = \frac{32.2}{AE} = \frac{32.2 \times 10^3}{(200 \times 10^6)(10 \times 10^{-4})}\ \text{mm} = 0.161\ \text{mm}$$

EXAMPLE 4.21   Determine the horizontal deflection at point *C* of the plane truss shown in Figure E4.21 using *Castigliano's Theorem*. Assume Modulus of elasticity (*E*) of the material is 200 GPa and the area (*A*) of all bars is 10 cm².

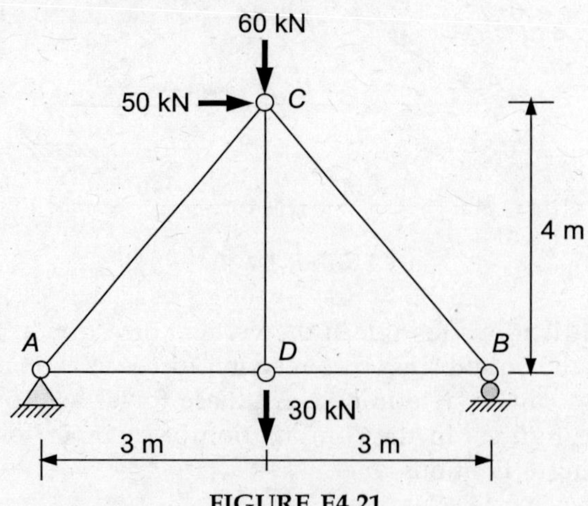

FIGURE E4.21

Solution: For the given loading conditions, the member forces can be determined as follows:

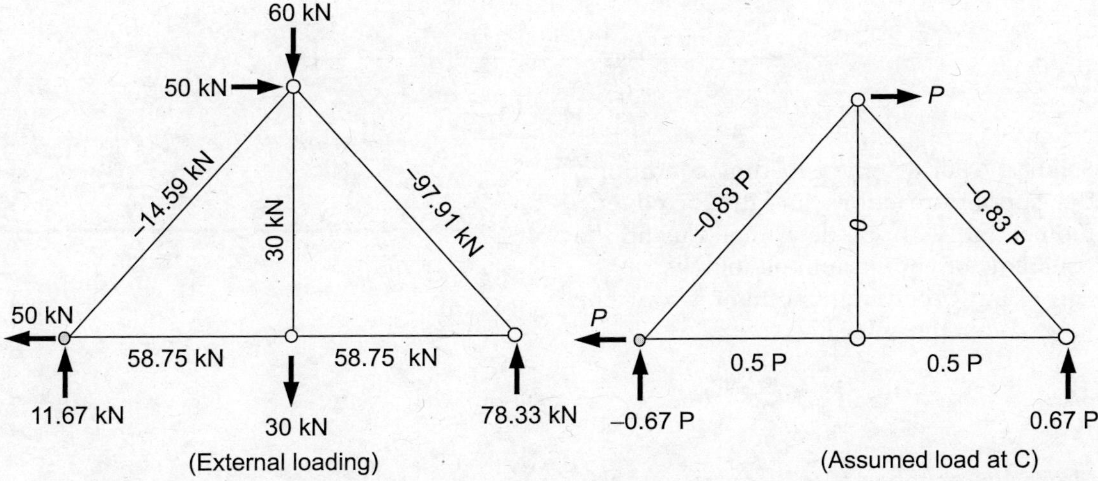

Let's apply a fictitious load P in the horizontal direction at the joint C of the truss. The internal forces in all members of the truss due to this load, P only are shown in the figure. The final forces in the members are computed as the sum of member forces computed in these two cases. In computation of member forces (S), the value of P must be set to zero being fictitious one.

Member	Length (m)	S	$\dfrac{\partial S}{\partial P}$	$S\,(P = 0)$	$S\left(\dfrac{\partial S}{\partial P}\right)L$
AD	3.0	$58.75 + 0.5P$	0.5	58.75	88.12
DB	3.0	$58.75 + 0.5P$	0.5	58.75	88.12
BC	5.0	$-97.91 - 0.83P$	-0.83	-97.91	406.33
AC	5.0	$-14.59 + 0.83P$	0.83	-14.59	-60.55
CD	4.0	30.0	0	30.0	0
				Total	**522.03**

Horizontal deflection at joint C, $\Delta_c = \sum S\left(\dfrac{\partial S}{\partial P}\right)\dfrac{L}{AE}$

$$\Delta_c = \frac{522.03}{AE} = \frac{522.03 \times 10^3}{(200 \times 10^6)(10 \times 10^{-4})}\, \text{mm} = 2.61 \text{ mm}$$

EXAMPLE 4.22: Determine the vertical deflection and the rotation at the hinge B of the beam shown in Figure E4.22 using *Castigliano's Theorem*. Assume the flexural rigidity of beam is EI.

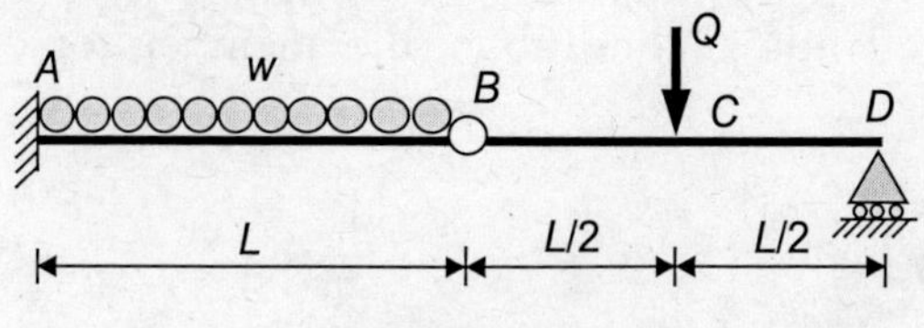

FIGURE E4.22

Solution: For the given loading conditions, the support reactions and the fixed end moment at A can be determined using the equations of equilibrium as follows:

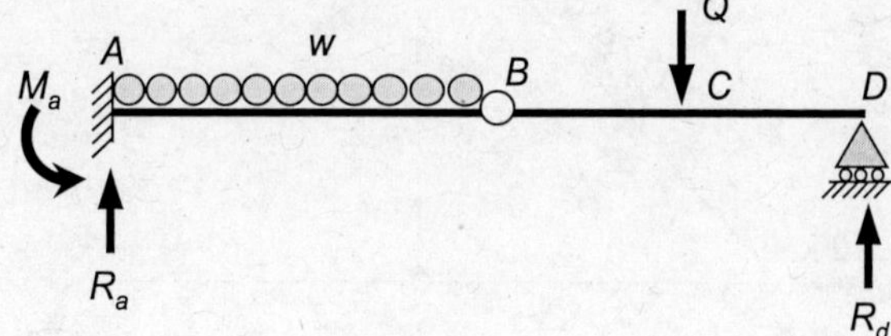

Taking bending moment about the hinge B for the span BD,

$$R_d = \frac{Q}{2}\,(\uparrow)$$

Similarly, for the span AB, $R_a = \dfrac{Q}{2} + wL\,(\uparrow);\ M_a = \dfrac{QL}{2} + \dfrac{wL^2}{2}$

$$M_x = \begin{cases} \dfrac{Q}{2}x & (0 \le x \le L/2)\ \text{for spans DC \& BC} \\[3mm] -\dfrac{Q}{2}x - \dfrac{wx^2}{2} & (0 \le x \le L)\ \ \text{for span BA} \end{cases}$$

To determine the vertical deflection at point B, let's apply a fictitious load P at the same point. The reactions and fixed end moment computed using equilibrium equations are shown in the figure.

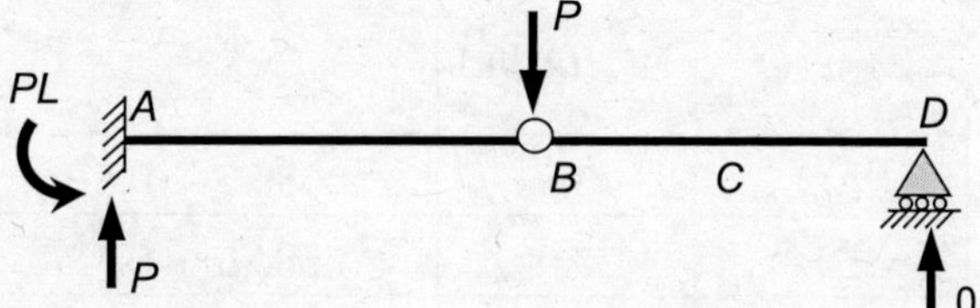

$$M'_x = \begin{cases} 0 & \text{(Span BD)} \\ -Px & \text{(Span BA)} \end{cases}$$

$$\frac{\partial M'_x}{\partial P} = \begin{cases} 0 & \text{(Span BD)} \\ -x & \text{(Span BA)} \end{cases}$$

Deflection at joint B is given by $\Delta_b = \displaystyle\int M\left(\frac{\partial M'}{\partial P}\right)\frac{dx}{EI}$

$$\Delta_b = \int\limits_0^L \frac{Qx}{2}(0)\cdot\frac{dx}{EI} - \int\limits_0^L \left(\frac{Qx}{2} + \frac{wx^2}{2}\right)(-x)\cdot\frac{dx}{EI}$$

$$\Delta_b = \frac{QL^3}{6EI} + \frac{wL^4}{8EI}$$

To determine the slope at point B, let's apply a fictitious bending moment at this hinge joint. The reactions and fixed end moment computed using equilibrium equations are shown in the figure.

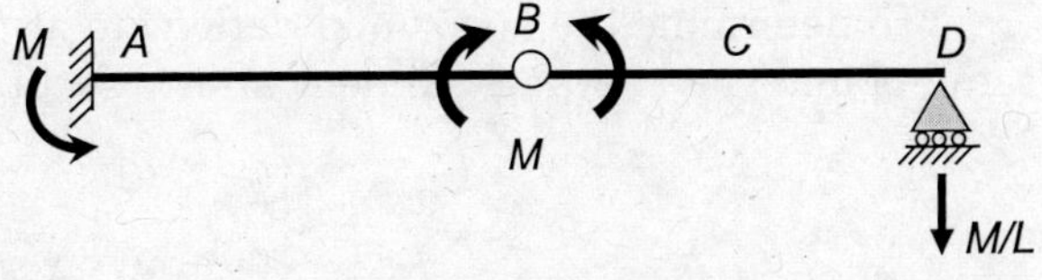

$$M_x'' = \begin{cases} -\dfrac{M}{L}x & \text{(Span DB)} \\[2mm] -M & \text{(Span BA)} \end{cases}$$

$$\frac{\partial M_x''}{\partial M} = \begin{cases} -\dfrac{x}{L} & \text{(Span DB)} \\[2mm] -1 & \text{(Span BA)} \end{cases}$$

Rotation at joint B is given by $\theta_b = \displaystyle\int M\left(\frac{\partial M''}{\partial M}\right)\frac{dx}{EI}$

$$\theta_b = \int_0^L \frac{Qx}{2}\left(-\frac{x}{L}\right)\cdot\frac{dx}{EI} - \int_0^L \left(\frac{Qx}{2}+\frac{wx^2}{2}\right)(-1)\cdot\frac{dx}{EI}$$

$$\therefore \qquad \theta_b = \frac{QL^2}{12EI} + \frac{wL^3}{6EI}$$

EXAMPLE 4.23 Determine the horizontal deflection at joint C of the portal frame shown in Figure E4.23 using *Castigliano's Theorem*. All members of have the same flexural rigidity of EI. Modulus of elasticity (E) of the material is 200 GPa. Moment of inertia (I) of members is 25000 cm^4.

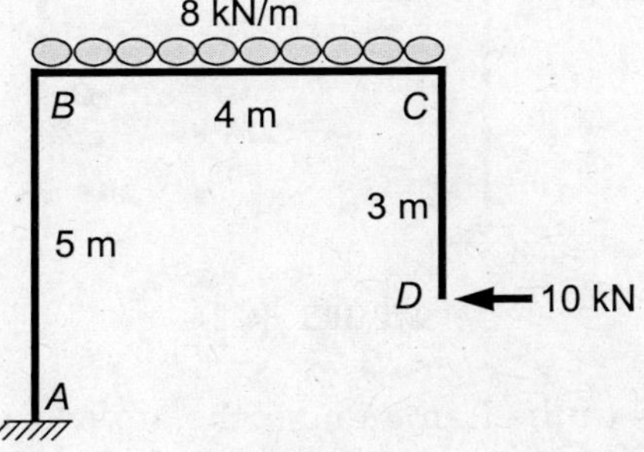

FIGURE E4.23

Solution: Bending moment at any section in different members of the frame due to the applied loading can be written as follows:

$$M_x = \begin{cases} -10x & (0 \le x \le 3) \text{ for span DC} \\[2mm] -30-4x^2 & (0 \le x \le 4) \text{ for span CB} \\[2mm] 10x-94 & (0 \le x \le 5) \text{ for span BA} \end{cases}$$

To determine the horizontal deflection at point B, let's apply a fictitious load P at the same point. The bending moment at any section of three members are as follows:

$$M_x = \begin{cases} 0 & (0 \le x \le 3) \text{ for span DC} \\ 0 & (0 \le x \le 4) \text{ for span CB} \\ -Px & (0 \le x \le 5) \text{ for span BA} \end{cases}$$

$$\frac{\partial M_x}{\partial P} = -x \quad \text{(Span BA)}$$

Deflection at joint B is given by $\Delta_b = \int M\left(\frac{\partial M}{\partial P}\right)\frac{dx}{EI}$

$$\Delta_b = \int_0^5 (10x - 94)(-x) \cdot \frac{dx}{EI} = \frac{758.33}{EI}$$

$$\therefore \quad \Delta_b = \frac{758.33 \times 10^3}{(200 \times 10^6)\,(25000 \times 10^{-8})} \text{ mm} = 15.17 \text{ mm}$$

EXAMPLE 4.24 Determine the horizontal deflection at the hinge joint B of the portal frame shown in Figure E4.24 using *Castigliano's Theorem*. All members of have the same flexural rigidity of *EI*. Modulus of elasticity (E) of the material is 200 GPa. Moment of inertia (I) of members is 25000 cm^4.

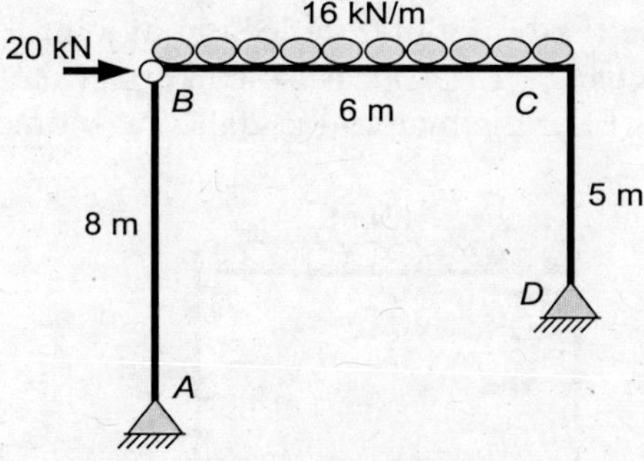

FIGURE E4.24

Solution: Support reactions of the frame under the applied loading can be determined as shown in the figure.

Bending moment at any section of a member can be obtained as follows:

$$M_x = \begin{cases} 0 & (0 \le x \le 8) \text{ for span AB} \\ 31.33x - 8x^2 & (0 \le x \le 6) \text{ for span BC} \\ -20x & (0 \le x \le 5) \text{ for span DC} \end{cases}$$

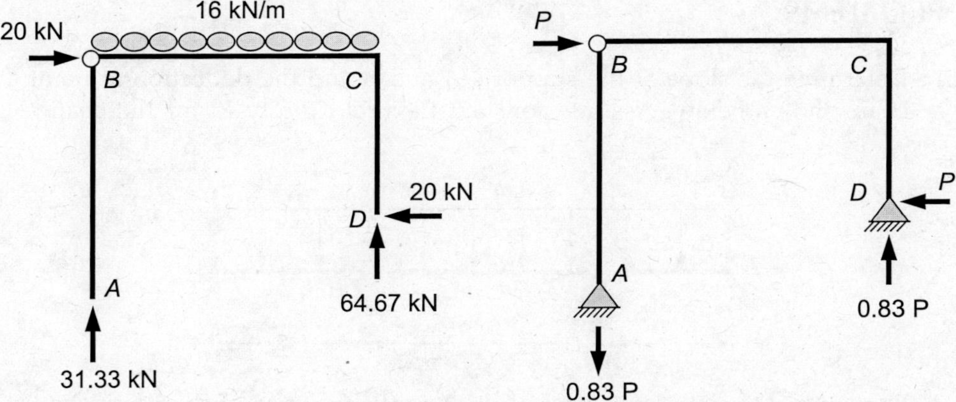

To determine the horizontal deflection at point B, let's apply a fictitious load P at the same point. The bending moment at any section of three members are as follows:

$$M_x = \begin{cases} 0 & (0 \leq x \leq 8) \text{ for span AB} \\ -0.83Px & (0 \leq x \leq 6) \text{ for span BC} \\ -Px & (0 \leq x \leq 5) \text{ for span DC} \end{cases}$$

$$\frac{\partial M_x}{\partial P} = \begin{cases} 0 & (0 \leq x \leq 8) \text{ for span AB} \\ -0.83x & (0 \leq x \leq 6) \text{ for span BC} \\ -x & (0 \leq x \leq 5) \text{ for span DC} \end{cases}$$

Deflection at joint B is given by $\Delta_b = \int M \left(\dfrac{\partial M}{\partial P} \right) \dfrac{dx}{EI}$

$$\Delta_b = \int_0^6 (31.33x - 8x^2)(-0.83x) \cdot \frac{dx}{EI} + \int_0^5 (-20x)(-x) \cdot \frac{dx}{EI}$$

$$\Delta_b = \frac{1112.41 \times 10^3}{(200 \times 10^6)(25000 \times 10^{-8})} \text{ mm} = 22.25 \text{ mm}$$

4.4 SUMMARY

This chapter presents a detailed discussion on various methods of computing deflection and slopes of determinate structures. The same problem can be solved using more than one method. However, the selection of a particular method depends on the type of structure, the complexities involved, and their applicability. For example, the conjugate beam method and double integration method can't be used to solve the displacements of the truss systems. These methods will be extremely useful in solving the indeterminate structures as discussed in the subsequent chapters.

4.5 PROBLEMS

4.1 Determine the slope at the supports *A* and *B* and the deflection at point *C* of the beam shown below. Assume constant flexural rigidity *EI* for the beam.

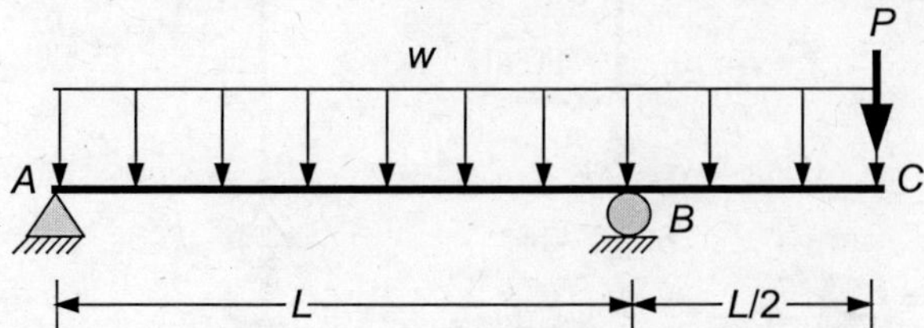

4.2 Determine the slope and the deflection at the free end of the beam shown below. Assume constant flexural rigidity *EI* for the beam.

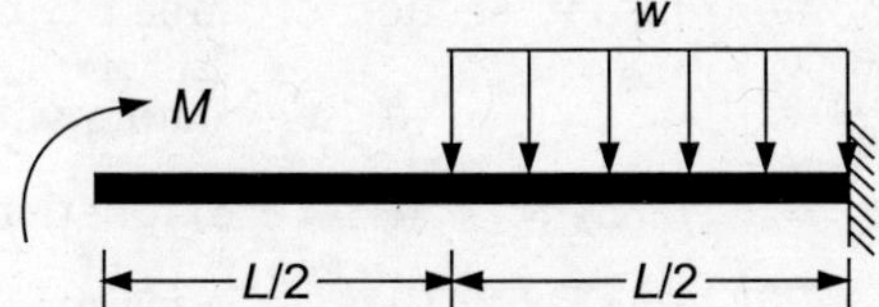

4.3 Determine the slope and the deflection at point A and the deflection at the mid-span of the beam shown below. Assume constant flexural rigidity *EI* for the beam.

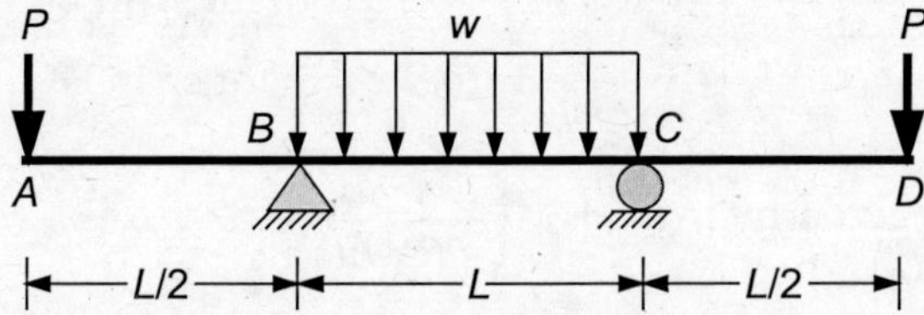

4.4 Determine the deflection at the mid-span of the beam shown below. Assume constant flexural rigidity *EI* for the beam.

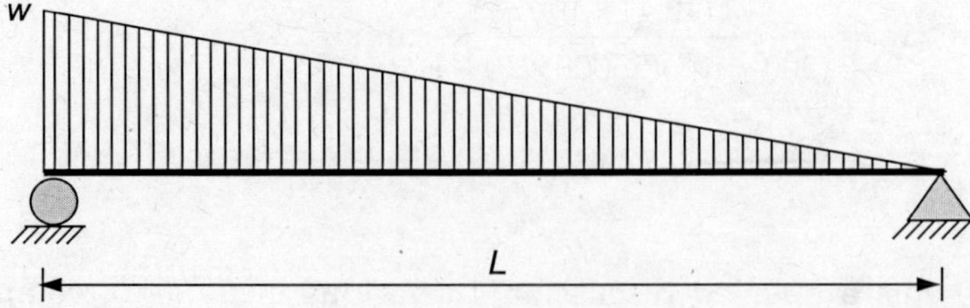

4.5 Determine the vertical deflection at point A of the truss shown below. Assume the value of $AE = 80000$ kN for all members.

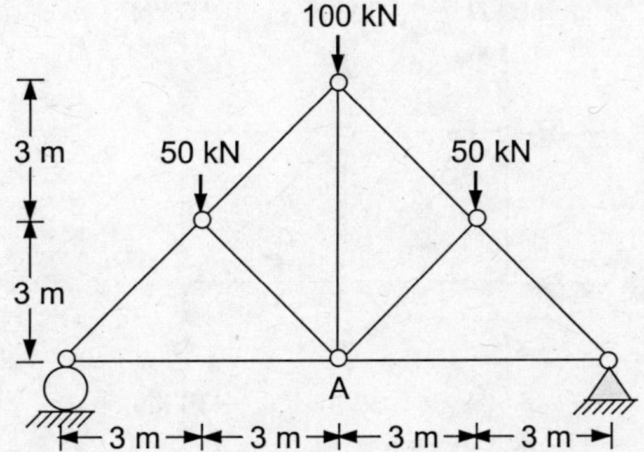

4.6 Determine the horizontal deflection at the support B of the truss shown below. Assume the value of AE = 60000 kN for all members.

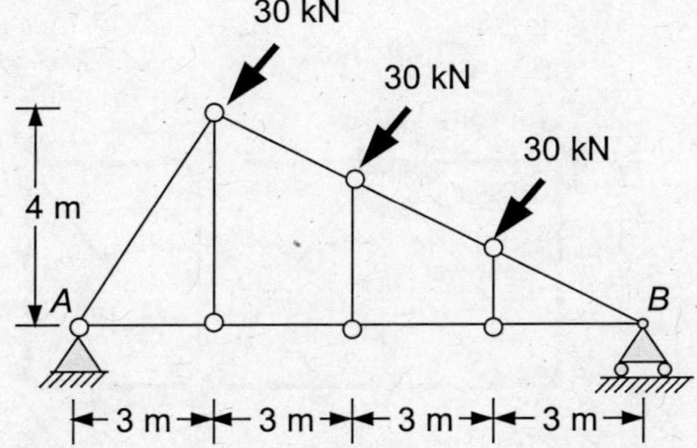

4.7 Determine the vertical deflection and slope at points A and B of the beam shown below. Assume the value of EI = 6000 kNm^2.

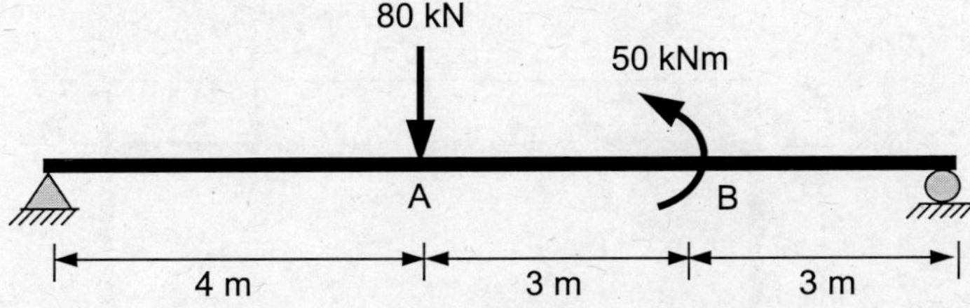

4.8 Determine the horizontal and vertical deflection at point A of the truss shown below. Assume the value of AE = 60000 kN for all members.

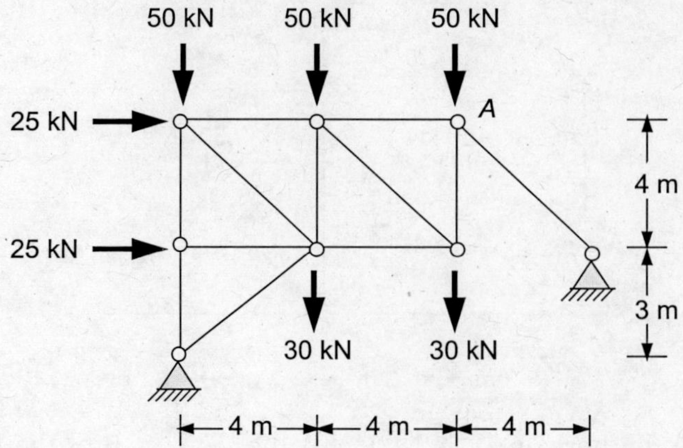

4.9 Determine the horizontal and vertical deflection at point E of the truss shown below. Assume the value of $AE = 60000$ kN for all members.

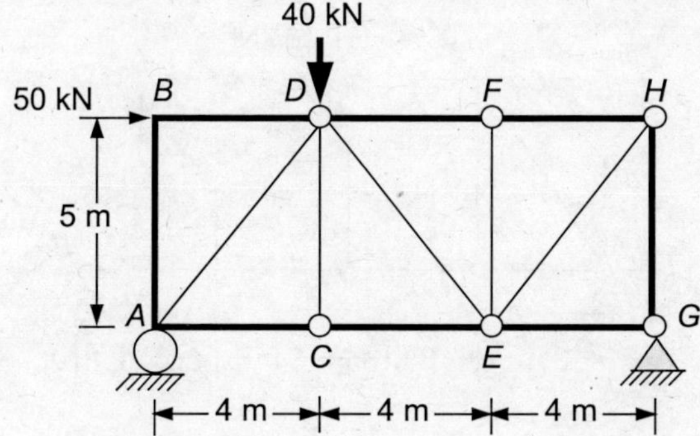

4.10 Determine the horizontal deflection at point C of the frame shown below. Assume the value of $EI = 6000$ kNm^2.

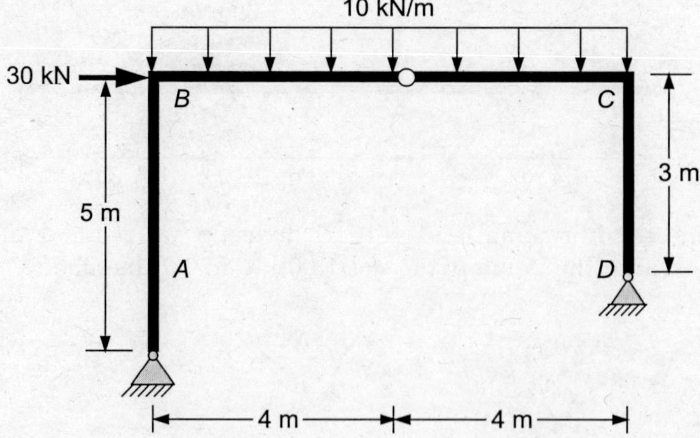

4.11   Determine the horizontal and vertical deflections at point A of the frame shown below. Assume the value of $EI = 10000\ \text{kNm}^2$.

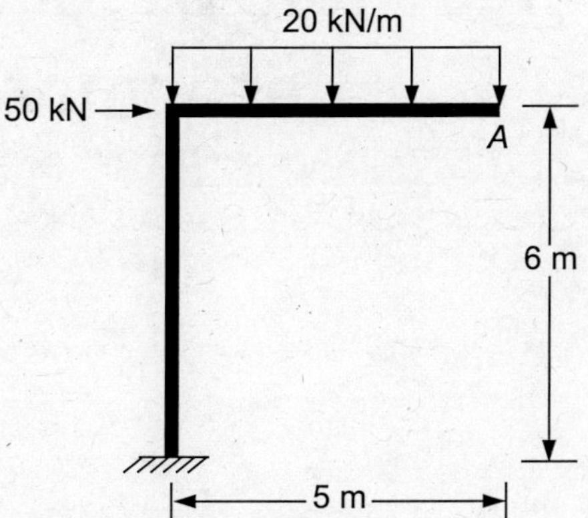

4.12   Determine the horizontal and vertical deflections at point A of the frame shown below. Assume the value of $EI = 10000\ \text{kNm}^2$.

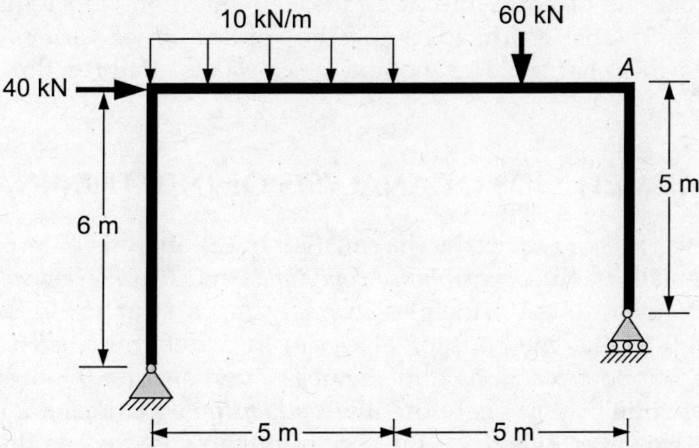

Method of Consistent Deformation

5.1 GENERAL

This chapter presents an introduction to the methods of analysis of indeterminate structures. Since the number of unknowns in an indeterminate structure is higher than the number of available equations of equilibrium, additional conditions are required to analyze this type of structure. The basic principles and the application of various methods of analysis are discussed in this chapter. The method of consistent deformation has been presented in detail.

5.2 GENERAL METHODS OF ANALYSIS OF INDETERMINATE STRUCTURES

Three basic principles are required to be satisfied in the analysis of any structure. These are: (i) *Equilibrium equations*, (ii) *Compatibility conditions*, and (iii) *Force-displacement relationships*. The sequence of using these principles in analyzing a structure depends on the type of structure as well as the methods of analysis. In a determinate structure, all the force unknowns (i.e., support reactions and member internal forces) can be determined using the equations of equilibrium. Therefore, the *equilibrium* equations are used as the first step of the analysis. However, when the number of unknowns exceeds the available equations of equilibrium, the other two principles must be used to determine the force as well as deformation unknowns. The *degree of static indeterminacy* represents the number of *force unknowns* in a structure. Similarly, the *degree of kinematic indeterminacy* (or *degrees of freedom*) depicts the number of *unknown displacements*. The force-displacement equations relate the applied forces to the displacements of the structure as well as the member forces to the member deformations.

The methods of analysis of any indeterminate structure are broadly classified into two categories. These are called as *force (or flexibility) method* and *displacement (or stiffness) method*. The basic concepts of these methods are discussed in the following sections.

Force method

The *force method* was the first method originally developed by James Clerk Maxwell to analyze the indeterminate structures. As the name suggests, the basis of this method is to establish the number of *force unknowns* and to determine these unknowns using *compatibility conditions* and *force-displacement relationships*. Therefore, it is required to compute the *degree of static indeterminacy* of the given structure. The compatibility equations are used to solve these force unknowns. By compatibility equations, it is meant that the displacements (or rotations) at any point of the structure must be consistent with the given support conditions and the member continuity. Once the number of unknowns is determined, the equilibrium equations are utilized to compute all required forces and reactions. Since this method is heavily dependent on the compatibility equations, the force method is also referred to as the *compatibility method* or the *method of consistent deformations*. This chapter deals with this method of analysis.

Displacement method

As the name suggests, the primary unknowns solved in this method are *displacements*, rather than forces. Thus, it is required to compute the degree of kinematic indeterminacy (DKI) to determine the number of unknowns. *The force-displacement relationship* is the first principle to be used in this method. The *compatibility conditions* are utilized in the force-displacement equations. Equilibrium equations are later applied to determine the member forces and support reactions.

In some structures, the number of force unknowns (i.e., DSI) may be higher than the displacement unknowns (i.e., DKI). In such cases, displacement methods are preferred to the force methods to minimize the computational time and efforts, and vice-versa. Hence, it is required to understand both the methods of analysis as presented in this book.

5.3 METHOD OF CONSISTENT DEFORMATION

The primary goal of the *force method* of analysis is to determine the force unknowns in a structure. Therefore, it is required to determine the degree of static indeterminacy of the structure. Depending on the number of force unknowns, the given structure can be divided into a series of stable and determinate structures. This is obtained by removing the extra (*redundant*) reactions and/or internal member forces from the given structure. The determinate structures so obtained by removing these redundant forces but with all imposed external loads are referred to as the *primary structures*. Next, all external loads are removed from these determinate structures and the redundant forces (or moments) are applied as the external load. The deformations corresponding to the redundant forces are computed in both cases. Using the compatibility conditions, the unknown forces can be determined. Once the unknown forces are determined, all support reactions and member forces are then determined. A general procedure of this method has been illustrated in the following section.

5.3.1 General Procedure

Though the method of consistent deformation can be applied to any type of structure, i.e., truss, beam, and frame, a propped cantilever with the uniformly distributed load as shown in Figure 5.1 has been considered for the illustration. The general procedure of this method has been presented as follows:

1. *Determine the DSI of the structure*:   For the propped cantilever shown in the figure, there are three reaction forces and one bending moment. Thus, the value of DSI is one (external). Since no external forces are acting in the horizontal direction, the horizontal reaction at the fixed support is zero.

2. *Selection of redundant forces*:   As the structure is indeterminate to first degree, it is required to identify one redundant force so that one can get a determinate structure by removing the redundant forces. For the propped cantilever, one can choose either the reaction force (R_b) at the support B or the bending moment (M_a) at support A as the redundant.

3. *Primary structure*:   After the redundant force is identified, the primary structure is developed by removing the redundant forces. For the given beam, if the fixed end moment at A is chosen as the redundant, then the primary structure would be a simply supported beam. On the other hand, if the reaction R_b is considered as the redundant, the primary structure would be a cantilever beam as shown in the figure. All external loads are considered in the primary structure. In addition to the primary structures, an additional structure is required to be developed in which the external loads are replaced by the redundant (unknown) forces as shown in the figure. The superimposition of the primary structure and the structure with the redundant forces would thus result in the given structure.

4. *Compatibility conditions*:   The deformations are computed for all primary or released structures in the direction of redundant forces at the points where they act. Compatibility conditions related to the deformations are written corresponding to the redundant forces. These deformations are calculated using any applicable methods of computing deflection. For the propped cantilever beam considered here, the redundant force is R_b and hence, one compatibility condition related to the deflection at B should be written. Let's assume that the deflection at B of the primary structure is Δ_b and that of with the redundant force is Δ'_b. Since the support B is a roller, the deflection at this support must be zero. Accordingly, the compatibility condition is $\Delta_b = \Delta'_b$.

5. *Determination of redundant*:   The unknown redundant force can be determined from the compatibility condition. For the given beam with the flexural rigidity of EI,

$$\Delta_b = \frac{wL^4}{8EI} \quad \text{and} \quad \Delta'_b = \frac{R_b L^3}{3EI} \tag{5.1}$$

$$\because \quad \Delta_b = \Delta'_b \quad \Rightarrow R_b = \frac{3wL}{8} \tag{5.2}$$

Once the reaction force (R_b) at the support B is determined, all support reactions (forces and moments) can now be determined using the equations of equilibrium. Accordingly, the reaction and fixed end moment at support are as follows:

$$R_a + R_b = wL \quad \therefore \ R_a = \frac{5wL}{8} \tag{5.3}$$

$$M_a = \frac{wL^2}{2} - \left(\frac{3wL}{8}\right)L \quad \therefore \ M_a = \frac{wL^2}{8} \tag{5.4}$$

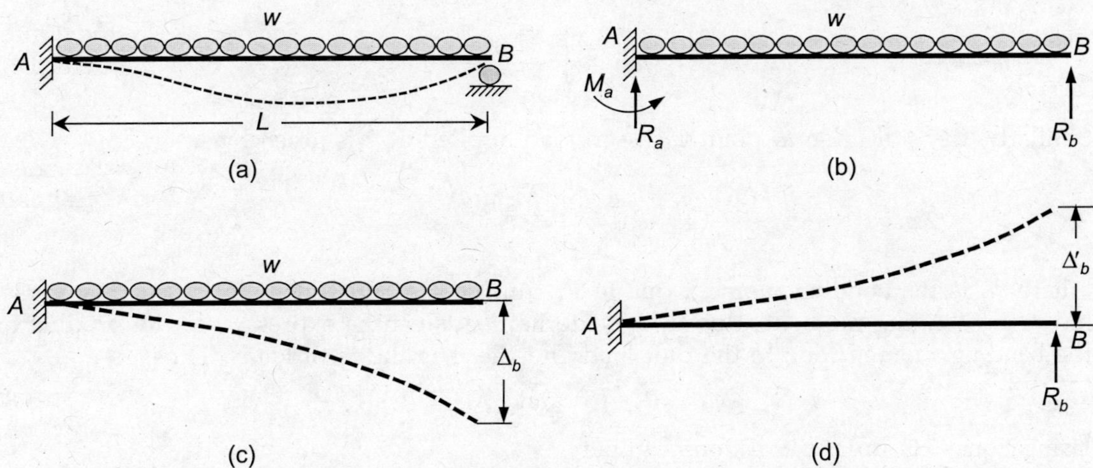

FIGURE 5.1 Analysis of propped cantilever: (a) Beam with given loading, (b) Support reactions, (c) Primary structure, (d) Beam with redundant force only.

5.3.2 Maxwell-Betti's Theorem of Reciprocal Deflections

James Clerk Maxwell (1864) presented a reciprocal theorem that is quite useful in the analysis of statically indeterminate structures. This theorem has been illustrated with the help of the following example.

Consider a simply supported beam, as shown in Figure 5.2(a), subjected to a concentrated load P at point 1 that resulted in the vertical displacement of Δ_{11} at this point. Let's consider any arbitrary point 2 on the beam. The vertical displacement at point 2 due to the load P applied at point 1 is Δ_{21}.

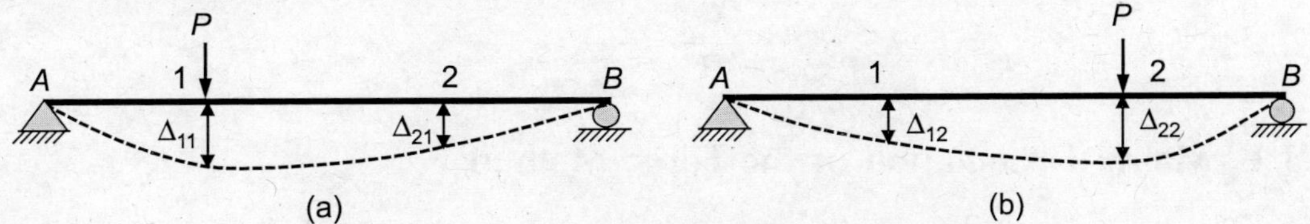

FIGURE 5.2 Maxwell-Betti's reciprocal theorem.

In the second case, the concentrated load P is removed from point 1 and the same load is applied at point 2 as shown in Figure 5.2(b). The vertical displacements at points 1 and 2 due to load P acting at point 2 are Δ_{12} and Δ_{22}, respectively.

Maxwell's theorem of reciprocal deflections states that *the deflection at point 1 due to a load applied at point 2 of a structure is equal to the deflection at point 2 if the same load is applied at point 1.*

i.e.,

$$\Delta_{12} = \Delta_{21} \tag{5.5}$$

To derive this theorem, let's use the method of virtual work to compute these displacements. If the length of the beam having the flexural rigidity of EI is L, the deflection at point 2 due to load applied at 1 is given by

$$\Delta_{21} = \int_0^L \frac{M_1 m_2 \, dx}{EI} \tag{5.6}$$

Similarly, the deflection at point 1 due to load applied at 2 is given by

$$\Delta_{12} = \int_0^L \frac{M_2 m_1 dx}{EI} \tag{5.7}$$

where m_1 is the bending moment due to the unit load applied at point 1 and vice-versa. Now, the bending moments due to the external loads can be expressed as the product of the bending moment due to the unit load and the magnitude of loads. Thus,

$$M_1 = P \cdot m_1 \quad \text{and} \quad M_2 = P \cdot m_2 \tag{5.8}$$

Using Eqn. (5.6) in Eqn. (5.4), one can get

$$\Delta_{21} = \int_0^L \frac{(Pm_1)m_2 dx}{EI} = \int_0^L \frac{m_1(Pm_2)dx}{EI} = \int_0^L \frac{m_1 M_2 dx}{EI} = \Delta_{12} \tag{5.9}$$

As a special case, putting the value of $P = 1$, the above equation can be written as follows:

$$\delta_{12} = \delta_{21} \tag{5.10}$$

where, δ_{21} = deflection at point 2 due to unit load applied at point 1, and δ_{21} = deflection at point 1 due to unit load applied at point 2.

It can be shown that this theorem is valid for rotations as well as displacements of beams and frame structures. This is also valid for the truss structures. The only condition for the validity of this theorem is that the material must be elastic and follow Hooke's law. Betti in 1872 extended this theorem which is stated as follows:

$$\frac{\Delta_{12}}{P_1} = \frac{\Delta_{21}}{P_2} \tag{5.11}$$

5.3.3   Matrix Formulation of the Force Method

The indeterminate structures with more than one redundancy can be solved using the method of consistent deformations using matrix formulations. In order to derive the generalized formulation of this method, let's consider a continuous beam as shown in the following figure. It is a statically indeterminate beam to the second degree. This means that the structure would be statically determinate if two redundant are removed.

Let's consider the vertical reactions at the supports B and C as the two redundant. These two reactions are R_1 and R_2. The primary structure of the given beam is a cantilever beam with all imposed loads and by removing the redundant forces. The resulting structure can be regarded as the superposition of the primary structure with the imposed loads as shown in Figure 5.3(b), and the determinate structure with the redundant forces applied separately as shown in Figures 5.3(c) and (d). Consequently, the deformation of the structure at any point can be determined as the algebraic sum of the deformations at the same of these three structures.

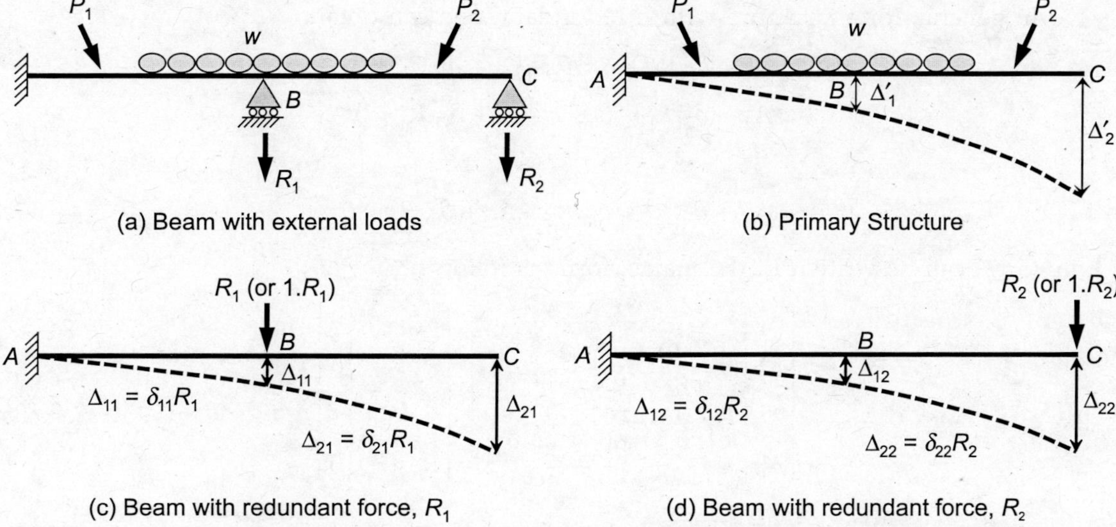

(a) Beam with external loads

(b) Primary Structure

(c) Beam with redundant force, R_1

(d) Beam with redundant force, R_2

FIGURE 5.3　Generalized matrix formulation of force method.

For unyielding supports B and C, the vertical displacements at these supports are zero. Thus, compatibility conditions are as follows:

$$\Delta_1 = 0 \quad \text{and} \quad \Delta_2 = 0 \tag{5.12}$$

where, Δ_1 = deflection at support B in the direction of redundant R_1 and Δ_2 = deflection at support C in the direction of redundant R_2. By the principle of superposition, Eqn. (5.12) can be written as follows:

$$\Delta_1' + \Delta_{11} + \Delta_{12} = 0 \tag{5.13}$$

$$\Delta_2' + \Delta_{21} + \Delta_{22} = 0 \tag{5.14}$$

where, Δ_1' = deflection at support B of the primary structure with all imposed load (See Figure 5.3(b)), Δ_{11} = deflection at point B of the structure with the redundant R_1 applied at the same point (Ref. Figure 5.3(c)), Δ_{12} = deflection at point B of the structure with the redundant R_2 applied at the point C (Ref. Figure 5.3(d)). Other notations can be expressed in a similar manner.

The deflections due to redundant forces, i.e., Δ_{11}, Δ_{12}, ... can be expressed in terms of the deflections due to unit forces, i.e., δ_{11}, δ_{12}, ... times the magnitude of the redundant forces, R_1 and R_2. The deflections, δ_{11}, δ_{12}, ... are also referred to as the *flexibility coefficients*. A flexibility coefficient δ_{ij} is defined as the displacement at the point, i due to a unit action at the point, j assuming that no other actions (forces) are acting at any point. Thus, Eqns. (5.13) and (5.14) can be now written as follows:

$$\Delta_1' + \delta_{11}R_1 + \delta_{12}R_2 = 0 \tag{5.15}$$

$$\Delta_2' + \delta_{21}R_1 + \delta_{22}R_2 = 0 \tag{5.16}$$

where, δ_{11} = deflection at point 1 (support B) due to a unit force acting at point 1, δ_{21} = deflection at point 2 (support C) due to a unit force acting at point 1, and so on.

In general, for a structure with n redundant, one can obtain

$$\Delta_1' + \delta_{11}R_1 + \delta_{12}R_2 + \ldots + \delta_{1n}R_n = 0$$
$$\Delta_2' + \delta_{21}R_1 + \delta_{22}R_2 + \ldots + \delta_{2n}R_n = 0$$
$$\vdots \qquad \vdots \qquad \qquad \vdots \qquad \vdots$$
$$\Delta_n' + \delta_{n1}R_1 + \delta_{n2}R_2 + \ldots + \delta_{nn}R_n = 0$$

(5.17)

Eqn. (5.17) can be written in the matrix form as follows:

$$\begin{Bmatrix} \Delta_1' \\ \Delta_2' \\ \vdots \\ \Delta_n' \end{Bmatrix} + \begin{bmatrix} \delta_{11} & \delta_{12} & \ldots & \delta_{1n} \\ \delta_{21} & \delta_{22} & \ldots & \delta_{2n} \\ \ldots & \ldots & \ldots & \ldots \\ \delta_{n1} & \delta_{n2} & \ldots & \delta_{nn} \end{bmatrix} \begin{Bmatrix} R_1 \\ R_2 \\ \vdots \\ R_n \end{Bmatrix} = \begin{Bmatrix} 0 \\ 0 \\ \vdots \\ 0 \end{Bmatrix}$$

(5.18)

Simply, $$\{\Delta'\} + [F]\{R\} = 0$$ (5.19)

If the points at which the redundant forces acting on the original structure are allowed to undergo some finite deformations, Eqn. (5.19) can be modified as follows:

$$\{\Delta'\} + [F]\{R\} = \{\Delta\}$$

(5.20)

where, $\{\Delta'\}$ is the column matrix representing the displacements at the redundant points of the primary (or released) structure due to the original loads, $[F]$ is the square matrix containing the flexibility coefficients in which each column contains the displacements at the redundant points due to unit forces acting at some definite points, $\{\Delta\}$ is the column matrix represents the actual displacements at the redundant points of the original structure and $\{R\}$ is the column matrix containing the redundant forces (or moments).

5.4 ANALYSIS OF INDETERMINATE BEAMS

The analysis of indeterminate beams under various loading conditions using force method (or method of consistent deformations) has been illustrated by means of the following examples.

EXAMPLE 5.1 Determine the support reactions and the bending moment at point B of the continuous beam shown in Figure E5.1 using the method of consistent deformation. Assume $EI = 200$ MNm^2.

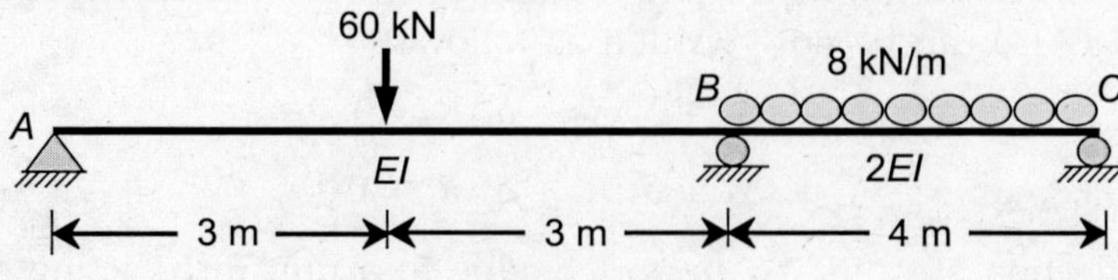

FIGURE E5.1

Solution: The degree of static indeterminacy of the continuous beam is one. This means that removing one reaction force from the beam would make it a determinate structure. Note that the horizontal reaction at the pinned support A is zero as no external forces are acting in the horizontal direction.

Consider the reaction, R_b at the support B is redundant. The primary structure is obtained by removing the support at B as shown in the figure. The compatibility condition for the beam can be written as the net vertical displacement at the support B is zero.

Mathematically, $$\Delta' + \delta R_b = 0$$

where, Δ' = the vertical deflection at point B of the primary structure due to external loads, and δ = the vertical deflection at point B due to unit load applied at the same point.

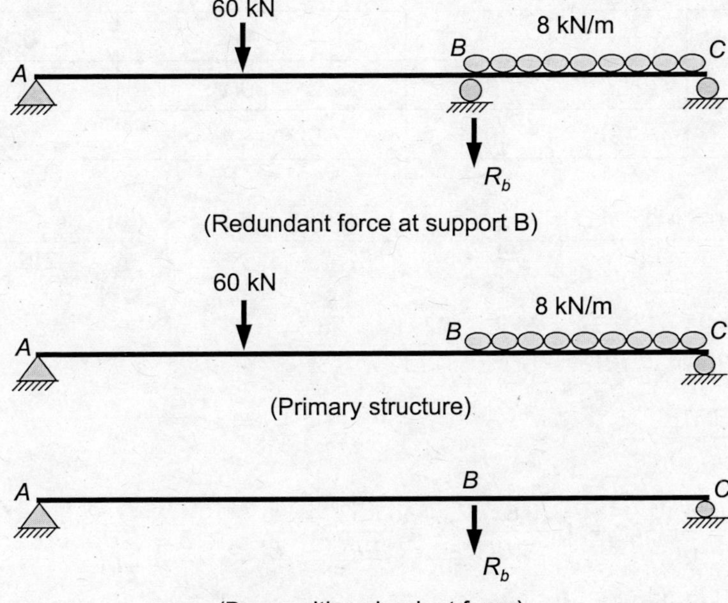

Using the virtual work method, one can write

$$\Delta' = \int \frac{Mm}{EI}\,dx; \quad \delta = \int \frac{m^2}{EI}\,dx$$

In the above expression, M = Bending moment due to the external load, m = Bending moment due to the unit load applied at B in the direction of R_b. The support reactions of beams with external loads and unit load at point B can be determined from the equilibrium equations as shown below. The expressions for M and m in different spans of the beam are summarized in the table.

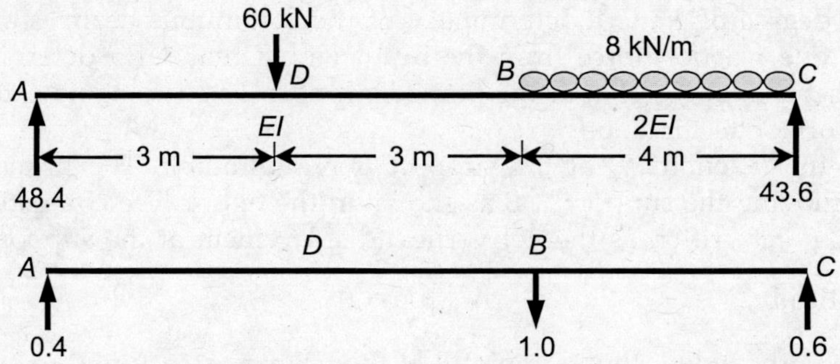

Span	EI	Limit	M	m
AD	EI	0–3	$48.4x$	$0.4x$
DB	EI	3–6	$48.4x - 60(x - 3)$	$0.4x$
BC	2EI	0–4	$43.6x - 4x^2$	$0.6x$

$$\Delta' = \int_0^3 \frac{(48.4x)(0.4x)}{EI}dx + \int_3^6 \frac{(-11.6x + 180)(0.4x)}{EI}dx + \int_0^4 \frac{(43.6x - 4x^2)(0.6x)}{2EI}dx$$

$$= \frac{174.24}{EI} - \frac{292.32}{EI} + \frac{972}{EI} + \frac{279.04}{EI} - \frac{76.8}{EI}$$

$$\therefore \quad \Delta' = \frac{1056.16}{EI}$$

$$\delta = \int_0^6 \frac{(0.4x)^2}{EI}dx + \int_0^4 \frac{(0.6x)^2}{2EI}dx = \frac{11.52}{EI} + \frac{3.84}{EI} \quad \therefore \quad \delta = \frac{15.36}{EI}$$

Using the compatibility condition,

$$\frac{1056.16}{EI} + \frac{15.36}{EI}R_b = 0 \quad \therefore R_b = -68.76 \text{ kN} (\uparrow)$$

The support reactions at supports A and C of the beam can now be determined using the Equilibrium equations as follows:

$$R_a = 20.90 \text{ kN} (\uparrow) \quad \text{and} \quad R_c = 2.34 \text{ kN} (\uparrow)$$

Bending moment at point B, $M_b = (43.6 \times 4 - 8 \times 4 \times 2) \text{ kNm} = 110.4 \text{ kNm}$.

EXAMPLE 5.2   Determine the end moments of the fixed-end beam of flexural rigidity EI subjected to a uniformly distributed load of w. Use the *Method of Consistent Deformations*.

Solution: In general, the degree of static indeterminacy of the beam is two. Because of symmetry in the external loads and support conditions, both supports have the same fixed end moments. Thus, the beam is statically indeterminate to the first degree. Let the fixed end moment at the supports be M.

Using the method of conjugate beam as shown in the figure,

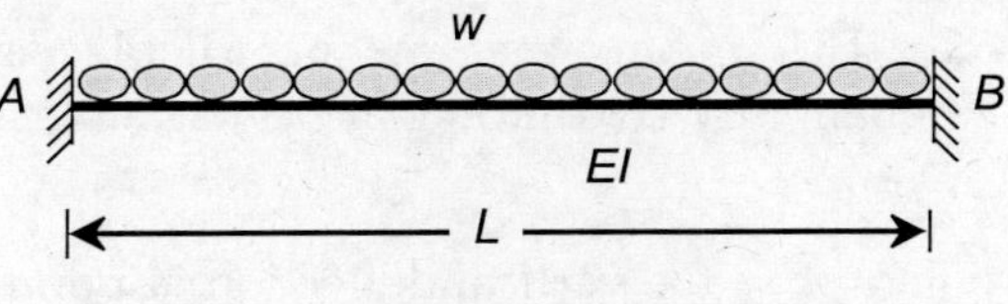

FIGURE E5.2

$$\sum V = 0$$

$$\Rightarrow \frac{2}{3}\left(\frac{wL^2}{8EI}\right)L - \frac{ML}{EI} = 0 \qquad \therefore M = \frac{wL^2}{12}$$

EXAMPLE 5.3 A beam of length L and flexural rigidity EI is fixed at one end and is supported on a spring at the other end. The beam is carrying a concentrated load of P at its mid-span. If the stiffness of the spring is K, determine the spring force using the *method of consistent deformations*.

Solution: The degree of static indeterminacy of the beam is one. Let's assume the force in the spring be R. The same spring force provides the reaction to the beam at C in the opposite direction.

Consider the support reaction (or spring force), R as the redundant. Axial shortening of the spring due to R is

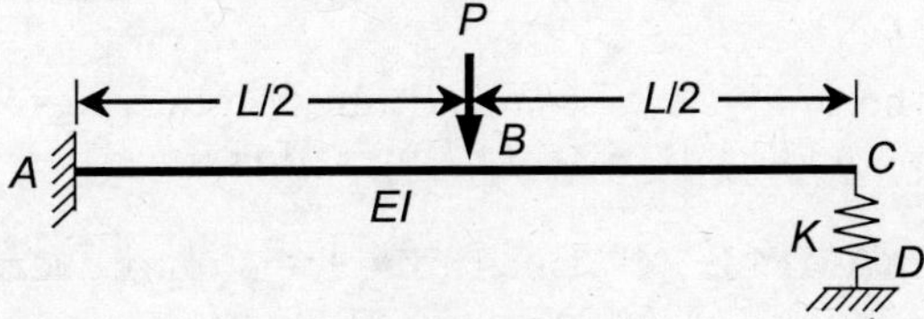

FIGURE E5.3

$$\Delta = \frac{R}{K}$$

This means that the net displacement at the support C should be Δ. Thus, the compatibility condition can be written as follows:

$$\Delta' - \delta R = \Delta$$

where, Δ' = the vertical deflection at point C of the primary structure due to external loads, and δ = the vertical deflection at point C due to unit load applied at the same point.

Using the virtual work method, one can write

$$\Delta' = \int_{L/2}^{L} \frac{(Px)x}{EI} dx = \frac{7\,PL^3}{24\,EI}$$

$$\delta = \int_{0}^{L} \frac{x^2}{EI} dx = \frac{L^3}{3EI}$$

Putting these expressions in the compatibility equation,

$$\frac{7PL^3}{24EI} - \frac{L^3}{3EI} R = \frac{R}{K}$$

$$\therefore \qquad R = \frac{\left(\dfrac{7PL^3}{24EI}\right)}{\left(\dfrac{1}{K} + \dfrac{L^3}{3EI}\right)}$$

5.5   ANALYSIS OF PORTAL FRAMES

The analysis of plane frames under various loading conditions using force method (or method of consistent deformations) has been illustrated by means of the following examples.

EXAMPLE 5.4   Determine the support reactions of the portal frame shown in Figure E5.4 using the *method of consistent deformations*. All members of the frame are of the same flexural rigidity (*EI*). Draw the bending moment diagram (BMD).

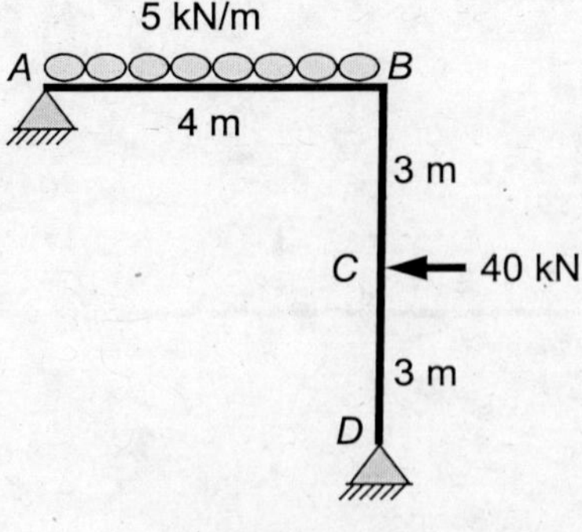

FIGURE E5.4

Solution: Degree of static indeterminacy of the frame = 1. It is required to identify only one redundant force to analyze the structure.

Let's consider the horizontal reaction at the support D is redundant. The primary structure with the external loads and the redundant force, H_d is shown in the figure.

Compatibility condition: $\Delta' - \delta H_d = 0$

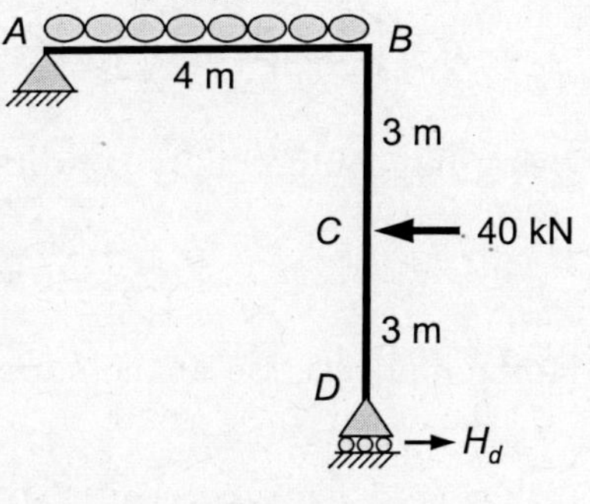

where, Δ' = the horizontal deflection at point D of the primary structure due to external loads, and δ = the horizontal deflection at point D due to unit load applied at the same point.

Let's use the method of virtual work to determine the deflections. The support reactions are determined for the primary structure using the equilibrium conditions.

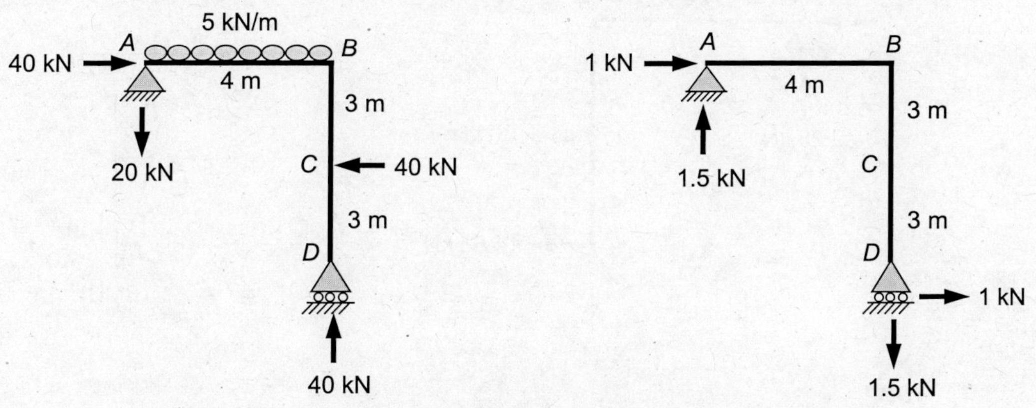

Span	EI	Limit	M	m
AB	EI	0–4	$-(20x + 2.5x^2)$	$1.5x$
BC	EI	3–6	$-40(x-3)$	x
CD	EI	0–3	0	x

$$\Delta' = \int \frac{Mm}{EI}\,dx \quad \Delta' = -\int_0^4 \frac{(20x+2.5x^2)(1.5x)}{EI}\,dx - \int_3^6 \frac{40(x-3)(x)}{EI}\,dx + \int_0^3 \frac{0.(x)}{EI}\,dx$$

$$= -\frac{880}{EI} - \frac{900}{EI} = -\frac{1780}{EI} \quad \therefore \Delta' = -\frac{1780}{EI}$$

$$\delta = \int \frac{m^2}{EI}\,dx$$

$$\delta = \int_0^4 \frac{(1.5x)^2}{EI}\,dx + \int_0^6 \frac{(x)^2}{EI}\,dx = \frac{48}{EI} + \frac{72}{EI} = \frac{120}{EI} \qquad \therefore\ \delta = \frac{120}{EI}$$

Using the compatibility condition,

$$-\frac{1780}{EI} + \frac{120}{EI}H_b = 0 \quad \therefore\ H_b = 14.83 \text{ kN } (\rightarrow)$$

Horizontal reaction at the support A,

$$H_a = 40 - 14.83 \text{ kN} = 25.17 \text{ kN } (\rightarrow)$$

Support reactions can be determined by using equilibrium equations. The frame with support reactions and bending moment diagrams are shown in the following figures.

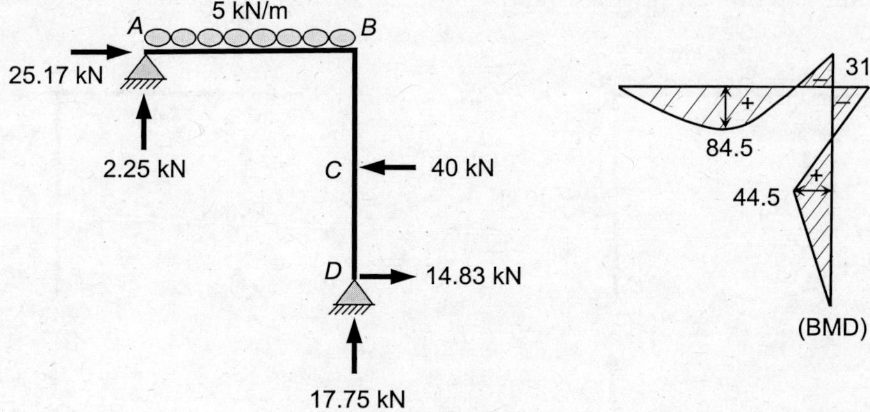

EXAMPLE 5.5   For the plane frame shown in Figure E5.5, determine the reaction forces and fixed end moments at the supports. Assume all the members have the same flexural rigidity.

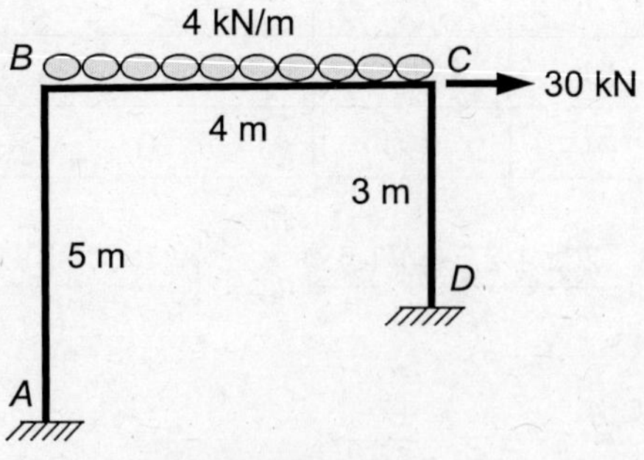

FIGURE E5.5

Solution: Degree of static indeterminacy = 3.

Let the reaction components at the support D, i.e., X_1, X_2, and X_3 are the redundant forces.

Compatibility equation:

$$\{\Delta'\} + [F]\{R\} = \{0\}$$

$$\begin{Bmatrix} \Delta_1' \\ \Delta_2' \\ \Delta_3' \end{Bmatrix} + \begin{bmatrix} \delta_{11} & \delta_{12} & \delta_{13} \\ \delta_{21} & \delta_{22} & \delta_{23} \\ \delta_{31} & \delta_{32} & \delta_{33} \end{bmatrix} \begin{Bmatrix} X_1 \\ X_2 \\ X_3 \end{Bmatrix} = \begin{Bmatrix} 0 \\ 0 \\ 0 \end{Bmatrix}$$

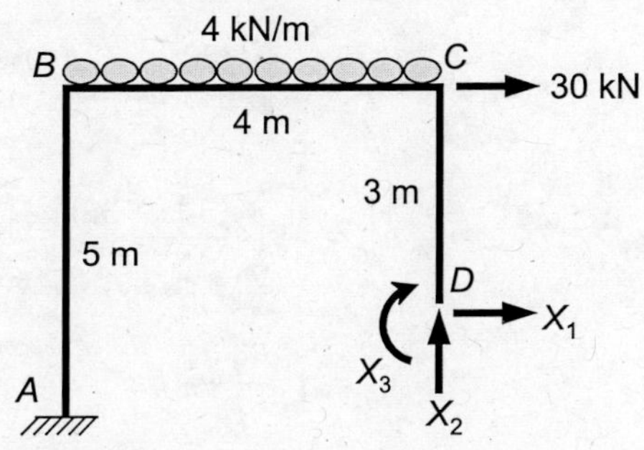

where, Δ_i' = the deflection at point i of the primary structure due to external loads, and δ_{ij} = the deflection at point i due to unit load applied at point j.

The primary structure and the frame with the redundant forces acting only are shown in the following figure. Let's use the method of virtual work to compute the displacements.

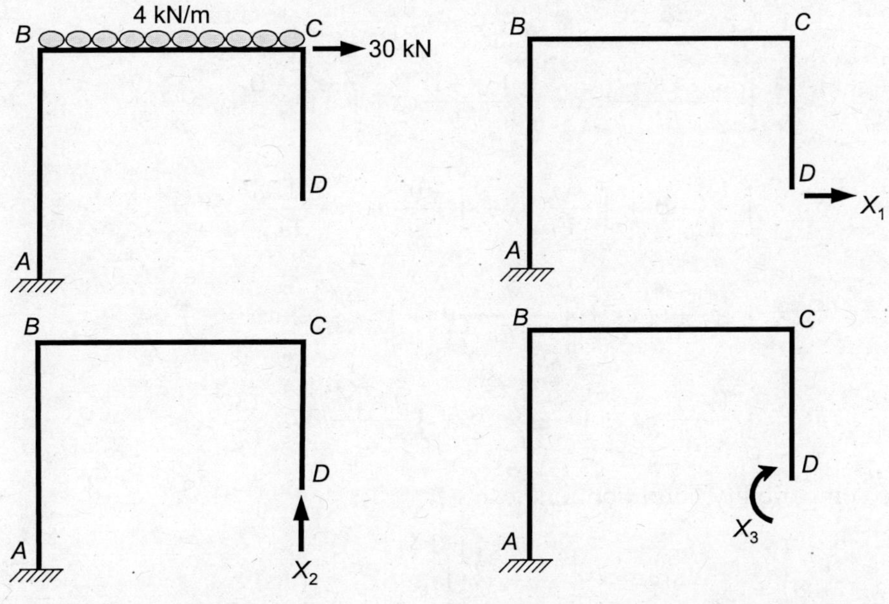

Span	Limit	M	m_1	m_2	m_3
BA	0–5	$-(30x + 32)$	$3 - x$	4	-1
CB	0–4	$-2x^2$	3	x	-1
DC	0–3	0	x	0	-1

$$\Delta_1' = \int \frac{Mm_1}{EI}\,dx; \quad \Delta_2' = \int \frac{Mm_2}{EI}\,dx; \quad \Delta_3' = \int \frac{Mm_3}{EI}\,dx$$

$$\Delta_1' = -\int_0^5 \frac{(30x+32)(3-x)}{EI}\,dx - \int_0^4 \frac{(2x^2)(3)}{EI}\,dx + \int_0^3 \frac{0.(x)}{EI}\,dx = \frac{45}{EI} - \frac{128}{EI} = -\frac{83}{EI}$$

$$\Delta'_2 = -\int_0^5 \frac{(30x+32)(4)}{EI}dx - \int_0^4 \frac{(2x^2)(x)}{EI}dx = -\frac{1980}{EI} - \frac{128}{EI} = -\frac{2108}{EI}$$

$$\Delta'_3 = -\int_0^5 \frac{(30x+32)(-1)}{EI}dx - \int_0^4 \frac{(2x^2)(-1)}{EI}dx = \frac{535}{EI} + \frac{42.67}{EI} = \frac{577.67}{EI}$$

$$\delta_{ij} = \int \frac{m_i m_j}{EI}dx$$

$$\delta_{11} = \int_0^5 \frac{(3-x)^2}{EI}dx + \int_0^4 \frac{(3)^2}{EI}dx + \int_0^3 \frac{(x)^2}{EI}dx = \frac{56.67}{EI}$$

$$\delta_{12} = \int_0^5 \frac{(3-x)(4)}{EI}dx + \int_0^4 \frac{(3)(x)}{EI}dx + \int_0^3 \frac{(x)(0)}{EI}dx = \frac{34}{EI} = \delta_{21}$$

$$\delta_{13} = \int_0^5 \frac{(3-x)(-1)}{EI}dx + \int_0^4 \frac{(3)(-1)}{EI}dx + \int_0^3 \frac{(x)(-1)}{EI}dx = -\frac{19}{EI} = \delta_{31}$$

$$\delta_{22} = \int_0^5 \frac{(4)^2}{EI}dx + \int_0^4 \frac{(x)^2}{EI}dx + \int_0^3 \frac{(0)^2}{EI}dx = \frac{101.33}{EI}$$

$$\delta_{23} = \int_0^5 \frac{(4)(-1)}{EI}dx + \int_0^4 \frac{(x)(-1)}{EI}dx + \int_0^3 \frac{(0)(-1)}{EI}dx = -\frac{28}{EI} = \delta_{32}$$

$$\delta_{33} = \int_0^5 \frac{(-1)^2}{EI}dx + \int_0^4 \frac{(-1)^2}{EI}dx + \int_0^3 \frac{(-1)^2}{EI}dx = \frac{13}{EI}$$

From the compatibility condition, one can get

$$\frac{1}{EI}\begin{Bmatrix} -83 \\ -2108 \\ 577.67 \end{Bmatrix} + \frac{1}{EI}\begin{bmatrix} 56.67 & 34 & -19 \\ 34 & 101.33 & -28 \\ -19 & -28 & 13 \end{bmatrix}\begin{Bmatrix} X_1 \\ X_2 \\ X_3 \end{Bmatrix} = \begin{Bmatrix} 0 \\ 0 \\ 0 \end{Bmatrix}$$

$$\therefore \quad \begin{Bmatrix} X_1 \\ X_2 \\ X_3 \end{Bmatrix} = \begin{Bmatrix} 22.2 \text{ kN} \\ -17.3 \text{ kN} \\ 39.6 \text{ kNm} \end{Bmatrix}$$

The support reactions and fixed end moments can be computed using equations of equilibrium. The final values are shown in the following figure.

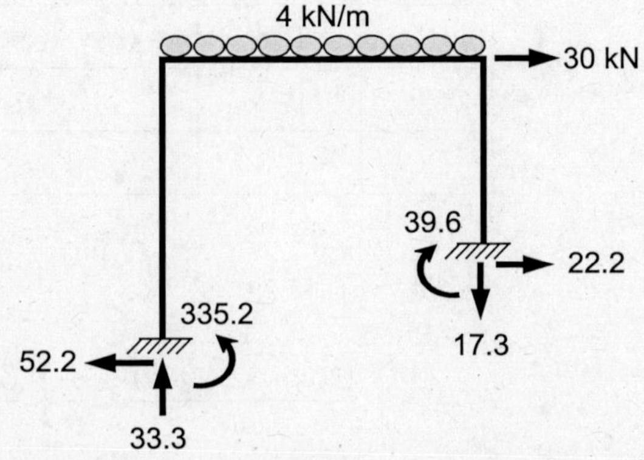

5.6 ANALYSIS OF INDETERMINATE TRUSSES

The analysis of indeterminate truss for different loading conditions using force method (or method of consistent deformations) has been illustrated by means of following examples.

EXAMPLE 5.6 Determine the support reactions and member forces of the truss shown in Figure E5.6. The area (A) of the cross-section of each member is 10 cm^2. The modulus of elasticity (E) is 200 GPa.

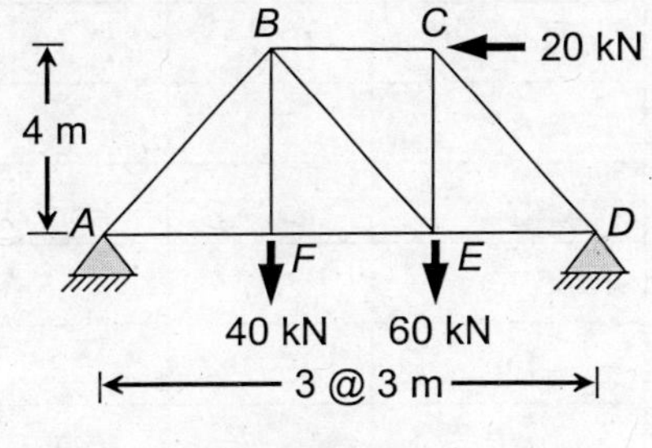

FIGURE E5.6

Solution: Degree of static indeterminacy of the frame = 1 (external indeterminacy)

This indicates that the given truss can be determinate if one component of the support reactions can be removed. Let's assume the horizontal reaction (H_d) at support D be the redundant force. The primary structure with all imposed external loads is shown in the figure.

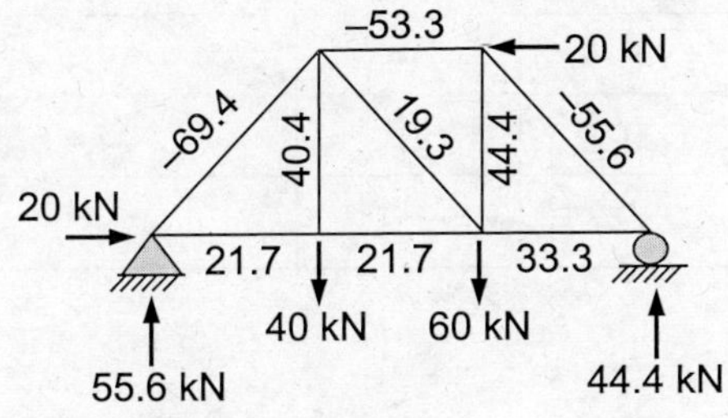

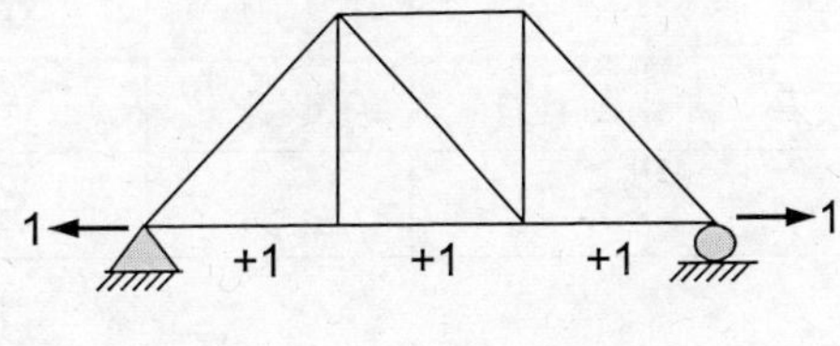

Compatibility equation:

$$\Delta'_d - \delta H_d = 0$$

where, Δ'_d = the horizontal deflection at point D of the primary structure due to the external loads, and δ = the horizontal deflection at point D due to unit load applied at the same point.

Using the method of virtual work gives

$$\sum \frac{SsL}{AE} + H_d \sum \frac{s^2 L}{AE} = 0$$

$$H_d = -\frac{\sum \dfrac{SsL}{AE}}{\sum \dfrac{s^2 L}{AE}}$$

where, S = the internal forces in members of the primary structure due to the external loads, and s = the internal forces in members due to unit load applied at the point D.

Since the value of AE is the same for all members, the above equation can be written as follows:

$$H_d = -\frac{\sum SsL}{\sum s^2L}$$

The calculation of deflections is summarized in the following table.

Member	L (m)	S (kN)	s (kN)	SsL	s^2L	$S + sH_d$ (kN)
AB	5.0	−69.4	0	0	0	−69.4
BC	3.0	−53.3	0	0	0	−53.3
CD	5.0	−55.6	0	0	0	−55.6
DE	3.0	33.3	1	100.0	3	7.7
EF	3.0	21.7	1	65.1	3	−3.9
FA	3.0	21.7	1	65.1	3	−3.9
BF	4.0	40.0	0	0	0	40.0
BE	5.0	19.4	0	0	0	19.4
CE	4.0	44.4	0	0	0	44.4
			Σ	230.2	9.0	

$$H_d = -\frac{230.2}{9.0} = -25.6 \text{ kN}$$

Final member forces are given by $(S + sH_d)$ as computed above and the support reactions of the truss are shown in the figure.

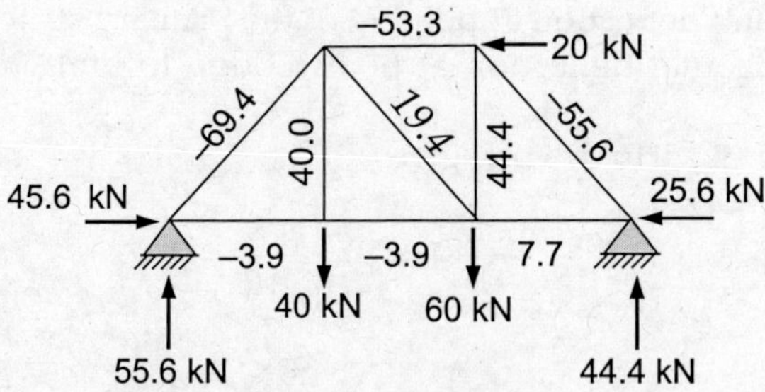

5.7 CASTIGLIANO'S COMPATIBILITY EQUATION

Castigliano's first theorem states that the deflection at any point of a statically determinate structure should be equal to the partial derivative of work done due to the external load by applying a force P at the same point and in the direction in which the deflection is computed. i.e., $\Delta = \dfrac{\partial W_e}{\partial P}$.

Using the compatibility condition for the redundant member, the partial derivative of the total work done with respect to the redundant must be zero. This is referred to as *Castigliano's second theorem* which is mostly applicable to the analysis of the indeterminate structure.

Mathematically,
$$\frac{\partial W}{\partial P} = 0 \tag{5.21}$$

The analysis of statically determinate frames is similar to that of the beams. The internal forces in the frame members consist of axial force, shear force, and bending moment. These forces are determined using the free-body diagrams at any section of the frame. Accordingly, the axial force, the shear force, and the bending moment diagrams are drawn from the frame. While selecting the sign convention for a frame member, it is assumed that the point of observation is within the frame, not outside the frame. The following examples illustrate the procedure to analyze the determinate rigid frames.

EXAMPLE 5.7   Determine the forces in all members of the truss subjected to a horizontal force of 10 kN shown in the figure using the Method of Consistent Deformations.

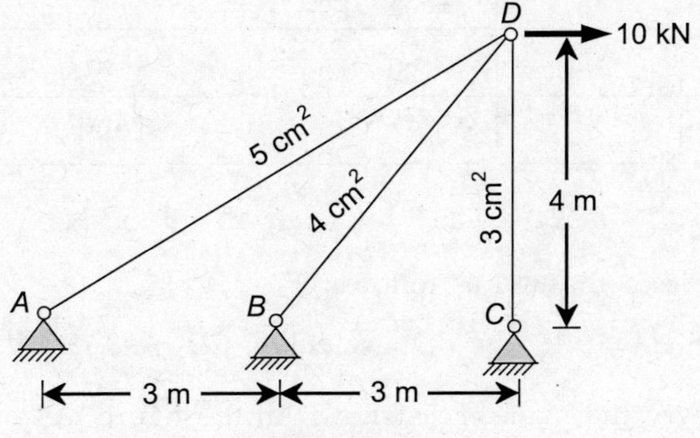

FIGURE E5.7

Solution: Degree of static indeterminacy of the truss = 1.

Let's assume the force in the member *AD* is redundant. Assume the force in this member is *P (tension)*.

Using Castigliano's second theorem, $\dfrac{\partial W}{\partial P} = 0$.

For a plane truss, the above equation can be written as follows: $\displaystyle \sum S\left(\frac{\partial S}{\partial P}\right)\frac{L}{AE} = 0$

The member force S represents the sum of forces in the member of the primary structure (i.e., by removing the redundant member) and that of due to the redundant force P only. It is assumed that all members have same modulus of elasticity E.

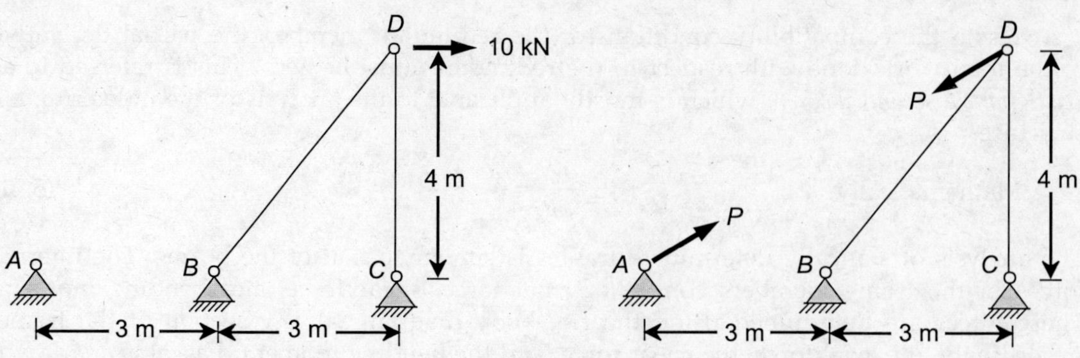

Member	L (m)	A (cm^2)	S	$\partial S/\partial P$	$S\left(\dfrac{\partial S}{\partial P}\right)\dfrac{L}{A}$
AD	7.21	5.0	P	1	$1.44P$
BD	5.0	4.0	$16.7-1.39P$	-1.39	$-29.01+2.41P$
CD	4.0	3.0	$-13.3+0.56P$	0.56	$-9.93+0.42P$
				Total	$-38.94+4.26P$

$$-38.94 + 4.26P = 0 \qquad \therefore\ P = 9.15 \text{ kN}$$

Final member forces are computed as follows:

$$AD = 9.15 \text{ kN};\ BD = 3.98 \text{ kN};\ CD = -8.18 \text{ kN}$$

EXAMPLE 5.8   Analyze the plane truss shown in the figure. Assume all members have the same area (A) of cross-section and modulus of elasticity (E).

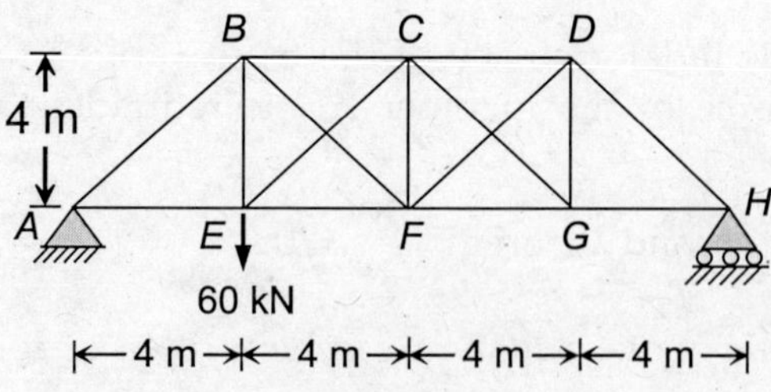

FIGURE E5.8

Solution: Degree of static indeterminacy of the truss is two. Since the truss is internally indeterminate, two members should be considered as redundant. Consider the members CE and CG are redundant and the respective member forces are assumed as X_1 and X_2 as shown in the figure.

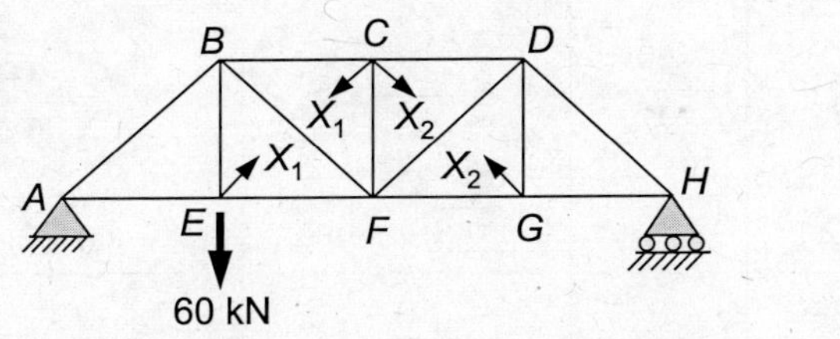 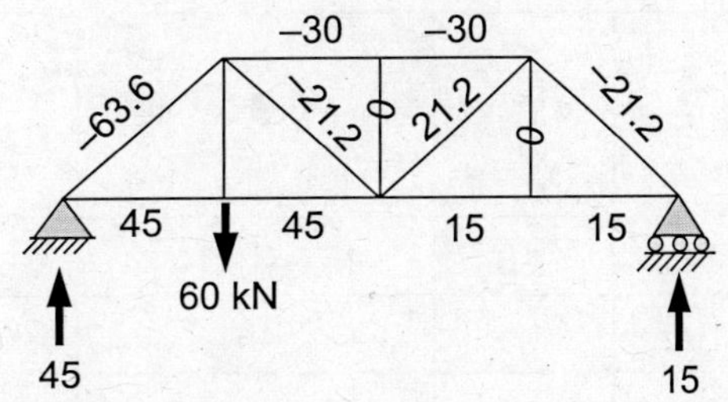

Using the equilibrium equations, one can determine the member forces in the primary structure due to the external loads as computed above.

Using the compatibility condition for the redundant members, one can get two equations, i.e.,

$$\frac{\partial W}{\partial X_1} = 0 \quad \text{and} \quad \frac{\partial W}{\partial X_2} = 0$$

or,

$$\sum S\left(\frac{\partial S}{\partial X_1}\right)\frac{L}{AE} = 0 \quad \text{and} \quad \sum S\left(\frac{\partial S}{\partial X_2}\right)\frac{L}{AE} = 0$$

Since the value of AE is constant for all the members of the truss,

$$\sum S\left(\frac{\partial S}{\partial X_1}\right)L = 0 \quad \text{and} \quad \sum S\left(\frac{\partial S}{\partial X_2}\right)L = 0$$

The member force S represents the sum of forces in the member of the primary structure (i.e., by removing the redundant member) and that of due to the redundant forces (X_1 and X_2) acting at the respective joints independently. These member forces are summarized in the following table.

$$\sum S\left(\frac{\partial S}{\partial X_1}\right)L = -543 + 43.13X_1 + 7.95X_2$$

$$\sum S\left(\frac{\partial S}{\partial X_2}\right)L = 204.6 + 7.95X_1 + 43.13X_2$$

Member	$L(m)$	$S(kN)$	$\dfrac{\partial S}{\partial X_1}$	$\dfrac{\partial S}{\partial X_2}$	$S\left(\dfrac{\partial S}{\partial X_1}\right)L$	$S\left(\dfrac{\partial S}{\partial X_2}\right)L$
AB	5.66	−63.6	0	0	0	0
BC	4.0	$-30 - 1.41X_1$	−1.41	0	$169.2 + 7.95X_1$	0
CD	4.0	$-30 - 1.41X_2$	0	−1.41	0	$169.2 + 7.95X_1$
DH	5.66	−21.2	0	0	0	0
HG	4.0	15	0	0	0	0
GF	4.0	$15 - 1.41X_2$	0	−1.41	0	$-84.6 + 7.95X_1$

Member	$L(m)$	S(kN)	$\dfrac{\partial S}{\partial X_1}$	$\dfrac{\partial S}{\partial X_2}$	$S\left(\dfrac{\partial S}{\partial X_1}\right)L$	$S\left(\dfrac{\partial S}{\partial X_2}\right)L$
FE	4.0	$45 - 1.41X_1$	-1.41	0	$-256.8 + 7.95X_1$	0
EA	4.0	45	0	0	0	0
BE	4.0	$60 - 1.41X_1$	-1.41	0	$-338.4 + 7.95X_1$	0
BF	5.66	$-21.2 + X_1$	1.0	0	$-120 + 5.66X_1$	0
CE	5.66	X_1	1.0	0	$5.66X_1$	0
CF	4.0	$-1.41X_1 - 1.41X_2$	-1.41	-1.41	$7.95X_1 + 7.95X_2$	$7.95X_1 + 7.95X_2$
CG	5.66	X_2	0	1.0	0	$5.66X_2$
FD	5.66	$21.2 + X_2$	0	1.0	0	$5.66X_2$
DG	4.0	$-1.41X_2$	0	-1.41	0	$7.95X_2$

Two simultaneous equations are as follows:

$$-543 + 43.13X_1 + 7.95X_2 = 0$$

$$204.6 + 7.95X_1 + 43.13X_2 = 0$$

On solving, one would get $X_1 = 13.94$ kN   and $X_2 = -7.31$ kN

    Accordingly, the final member forces can be obtained as summarized in the following table.

Member	Final forces (kN)
AB	-63.60
BC	-49.65
CD	-19.69
DH	-21.20
HG	15.00
GF	25.31
FE	25.31
EA	45.00
BE	40.35
BF	-7.26
CE	13.94
CF	-9.34
CG	-7.31
FD	13.89
DG	10.31

EXAMPLE 5.9   A cantilever beam of length 8 m is subjected to a concentrated load of 50 kN at a distance of 4 m from the fixed end and is supported by a bar at the free end. The area of cross-section and moment of inertia of the beam are 80 cm^2 and 6000 cm^4, respectively. The area of the cross-section of the bar is 8 cm^2. Assume that both the beam and the bar have the same modulus of elasticity of 200 GPa. Determine the axial force carried by the bar due to the applied loading.

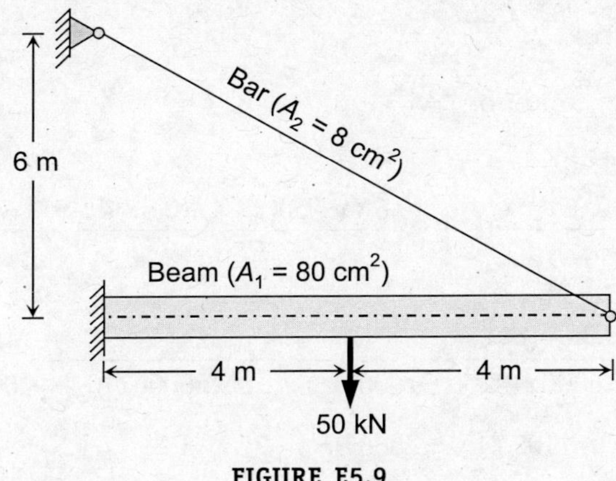

FIGURE E5.9

Solution: The degree of static indeterminacy of the structure is 1. Let's consider the bar force is redundant and assume that the magnitude of this bar force is X.

The value of X can be determined using Castigliano's compatibility conditions, i.e.,

$$\frac{\partial W}{\partial X} = 0$$

where, W is the sum of the work done by the external load and the redundant force. The primary structure with the external loads and the redundant force components are shown in the following figure.

The work done by the bar force, X can be given by

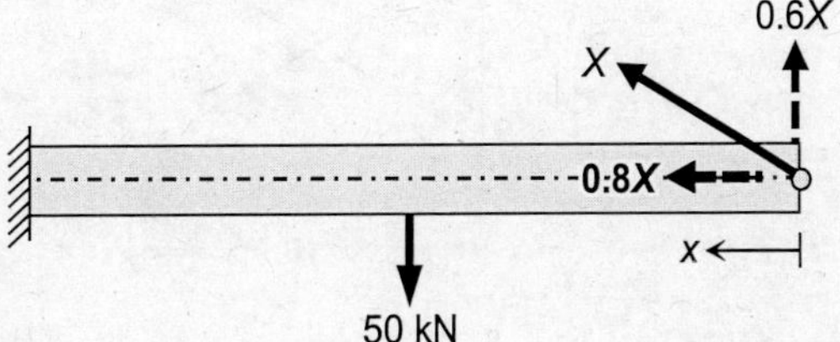

$$W_1 = \frac{X^2 L_2}{2 A_2 E}$$

where, L_2 = Length of bar, A_2 = Area of cross-section of bar.

Work done by the beam due to the external load and the redundant member force,

$$W_2 = \int \frac{M^2 dx}{2EI} + \frac{S^2 L_1}{2 A_1 E}$$

where, L_1 = Length of beam, A_1 = Area of cross-section of beam, and I = Moment of inertia of beam cross-section.

$$W_2 = \int_0^4 \frac{(0.6Xx)^2 dx}{2EI} + \int_4^8 \frac{[0.6Xx - 50(x-4)]^2 dx}{2EI} + \frac{(-0.8X)^2 L_1}{2A_1 E}$$

Total work done by the structure, $W = W_1 + W_2$

$$W = \frac{X^2 L_2}{2A_2 E} + \int_0^4 \frac{(0.6Xx)^2 dx}{2EI} + \int_4^8 \frac{[0.6Xx - 50(x-4)]^2 dx}{2EI} + \frac{(-0.8X)^2 L_1}{2A_1 E}$$

Using the compatibility condition, $\dfrac{\partial W}{\partial X} = 0$

$$\frac{XL_2}{A_2 E} + \int_0^4 \frac{(0.6Xx)(0.6x)\,dx}{EI} + \int_4^8 \frac{[0.6Xx - 50(x-4)](0.6x)dx}{EI} + \frac{(-0.8X)(-0.8)L_1}{A_1 E} = 0$$

$$\frac{X(10)}{(8x10^{-4})} + \frac{0.36X(8)^3}{3x6000x10^{-8}} - \frac{30X(8^3 - 4^3)}{3x6000x10^{-8}} + \frac{120X(8^2 - 4^2)}{2x6000x10^{-8}} + \frac{(0.64X)(8)}{(80x10^{-4})} = 0$$

$$103.71X = 2666.67 \qquad \therefore\ X = 25.17 \text{ kN}$$

EXAMPLE 5.10 A plane truss shown in Figure E5.10 is subjected to a horizontal load of 20 kN at the joint C. During the fabrication, the member BC was short by 10 mm and was forced in the position. If the area of cross-section each member is 100 mm², determine the forces in all the members of the truss. Assume, $E = 200$ GPa.

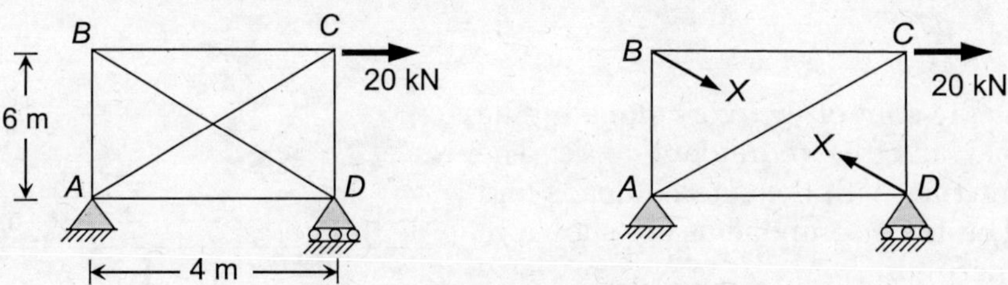

FIGURE E5.10

Solution: The degree of static indeterminacy of the structure is one. Since the truss is internally indeterminate, we have to select the internal member force as the redundant. Consider the force in the member BD is redundant and assume that the magnitude of this force is X.

Since the truss is subjected to forces due to the external loads and the lack of fit, the total forces can be computed by superimposing these effects considered independently.

Compatibility equation: $\Delta_1 + \Delta_2 + \delta X = 0$

where, Δ_1 = displacement between the points B and D due to the external loads, Δ_2 = displacement between the points B and D due to the lack of fit effects, δ = displacement between the points B and D due to unit loads applied these two points.

$$\Delta_1 = \sum \frac{SsL}{AE}; \quad \Delta_2 = \sum s\Delta'; \quad \delta = \sum \frac{s^2 L}{AE}$$

where, S = member forces due to external loads, s = member forces due to unit loads applied at joints B and D, and Δ' = Displacement due to lack of fit. Putting these expressions in the compatibility equation, the redundant force can be determined as follows:

$$X = -\frac{\sum \dfrac{SsL}{AE} + \sum s\Delta'}{\sum \dfrac{s^2 L}{AE}}$$

The calculation of redundant force and final member forces are summarized in the following table.

Member	L (m)	S(kN)	s(kN)	Δ′ (m)	SsL	s²L	sΔ′	Final force (S+sX)
AB	3.0	0.0	−0.6	0	0	1.08	0	5.28
BC	4.0	0.0	−0.8	−0.01	0	2.56	0.008	7.04
CD	3.0	−15.0	−0.6	0	27	1.08	0	−9.72
DA	4.0	0.0	−0.8	0	0	2.56	0	7.04
AC	5.0	25.0	1.0	0	125	5.0	0	16.02
BD	5.0	0.0	1.0	0	0	5.0	0	−8.80
				Total	152	17.28	0.008	

$$X = -8.8 \text{ kN}$$

5.8 PROBLEMS

5.1 Determine the support reactions of the beam shown below subjected to a triangular loading using the Method of Consistent Deformations. Assume constant flexural rigidity *EI*.

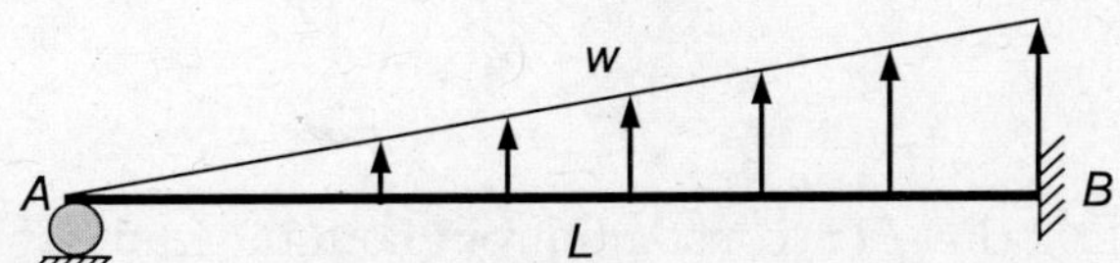

5.2 Determine the support reactions of the beam shown below subjected to a triangular loading using the Method of Consistent Deformations. Assume constant flexural rigidity *EI*.

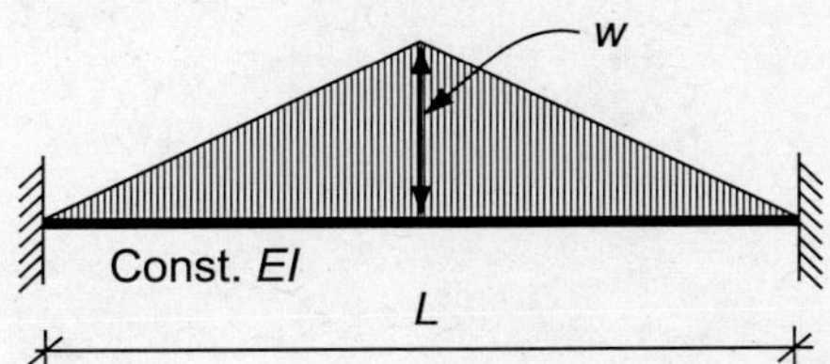

5.3 Determine the support reactions of the beam shown below subjected to a uniformly distributed loading using the Method of Consistent Deformations. Draw the bending moment and shear force diagrams. Assume constant flexural rigidity EI.

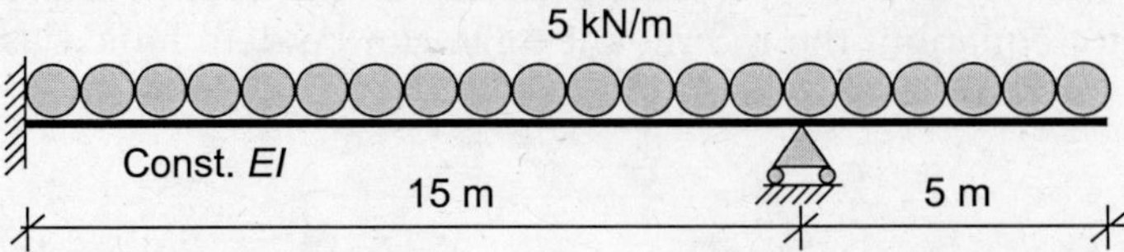

5.4 Determine the support reactions and bending moments at the supports of the beam shown below if support A moves upward by 30 mm. Use the method of consistent deformations. Assume constant flexural rigidity of the beam.

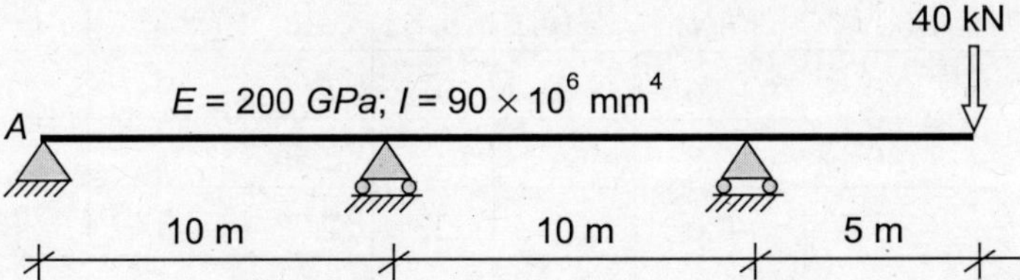

5.5 Determine the support reactions and bending moments at the supports of the frame shown below using the method of consistent deformations.

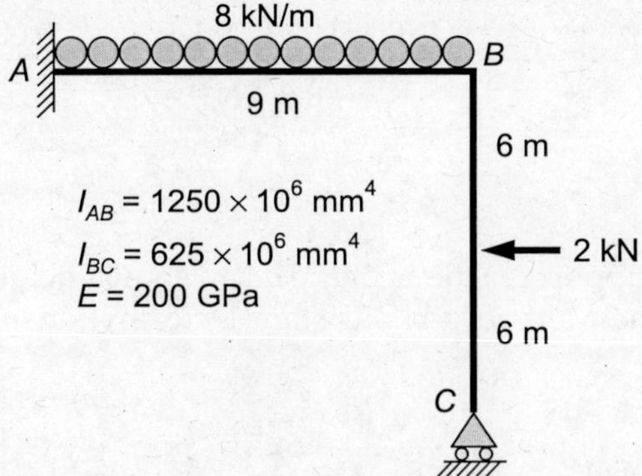

5.6 If the flexural rigidity EI is constant throughout the frame shown below, determine the support reactions using the method of consistent deformations. Draw the bending moment diagram for the frames.

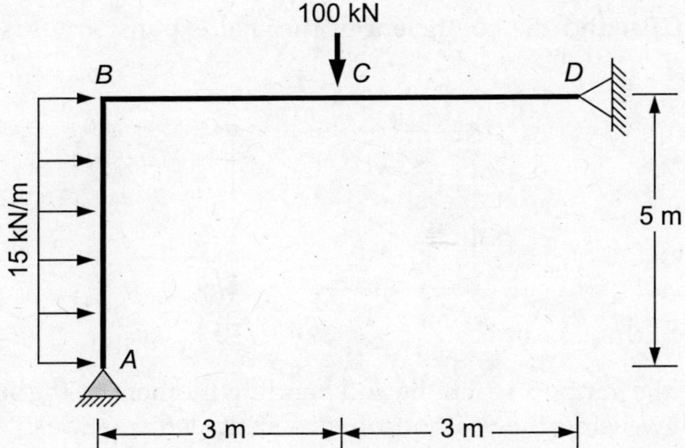

5.7 Determine the support reactions at A and D using the method of consistent deformations. Draw the bending moment diagram for the frames.

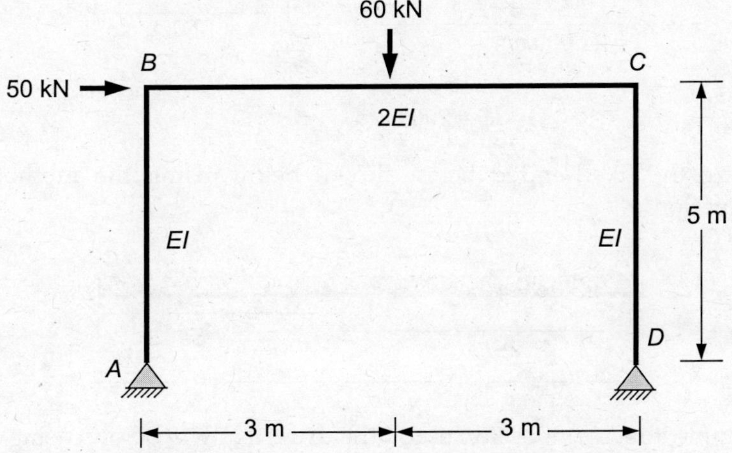

5.8 Determine the member forces of the truss shown below using the method of consistent deformations if the member BC is short in length by 10 mm and forced into position. Assume the value of AE = 8000 kN for all members.

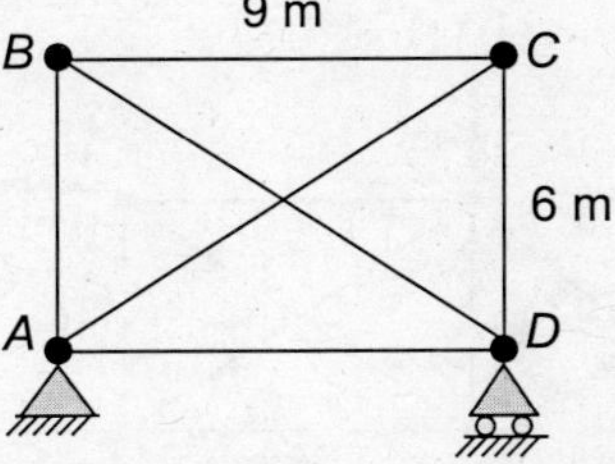

5.9 Determine the member forces of the truss shown below using the method of consistent deformations if the member BC is subjected to a temperature change of 60°C. Assume the area of cross-section (A) is 100 mm², the modulus of elasticity

(E) is 200 GPa, and the coefficient of thermal expansion (a) is $6 \times 10^{-6}/°C$ for all members.

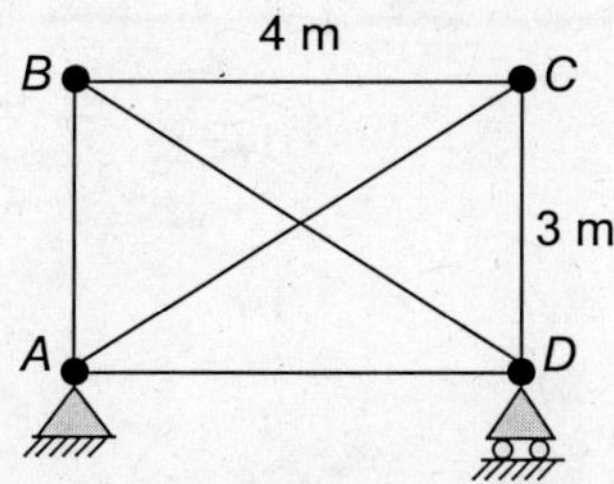

5.10   Determine the support reactions and bending moment at the supports of the beam shown below using the method of consistent deformations.

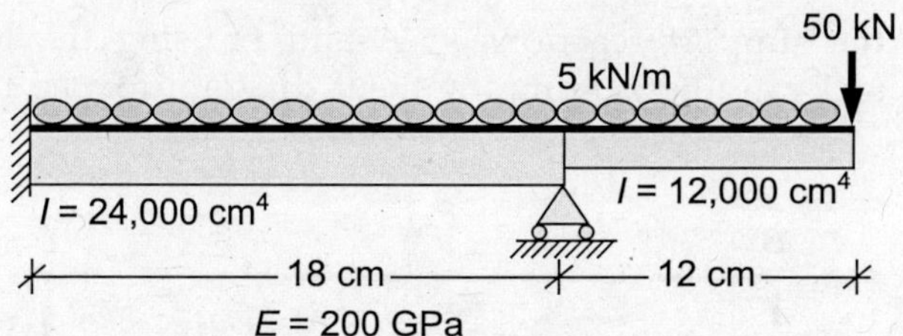

5.11   Analyze the fixed-ended beam shown below using the method of consistent deformations.

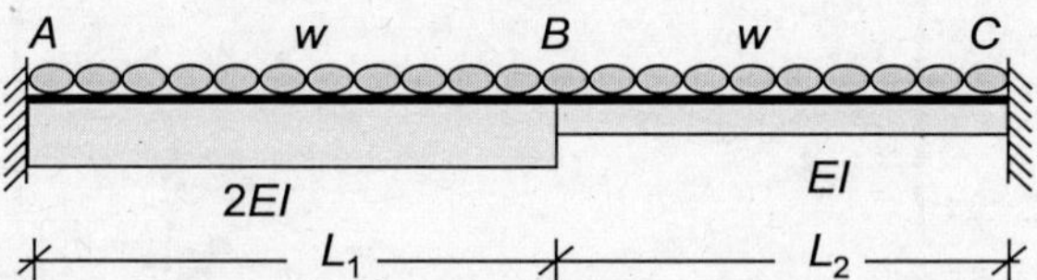

5.12   Determine the member forces of the truss shown below using the method of consistent deformations if the member BF is subjected to a temperature change of 50°C. Assume the area of cross-section (A) is 200 mm^2, the modulus of elasticity (E) is 200 GPa, and the coefficient of thermal expansion (a) is $6 \times 10^{-6}/°C$ for all members.

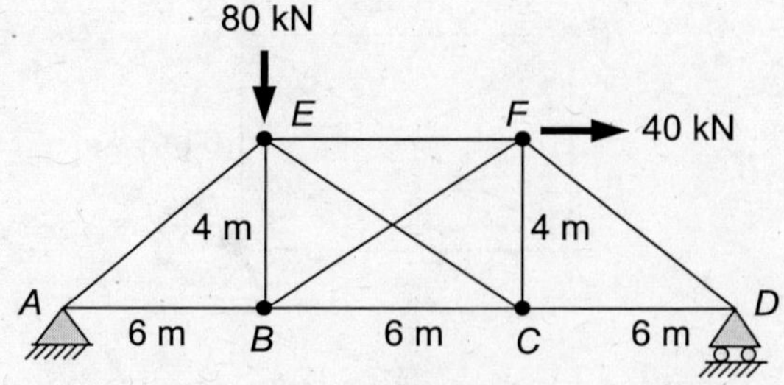

5.13   Analyze the beam using the method of consistent deformations. Draw the bending moment and shear force diagrams.

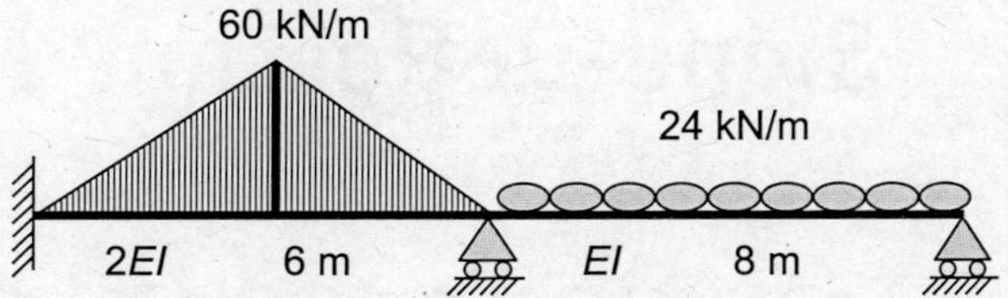

5.14   Analyze the structure shown below using the method of consistent deformations.

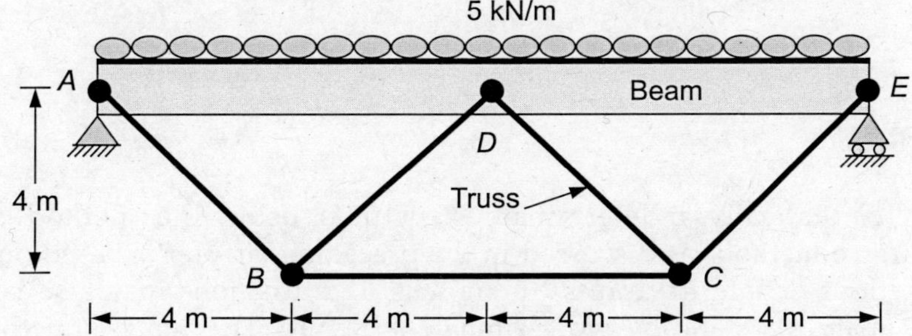

Beam: $E = 200$ GPa; $I = 30,000$ cm^4
Truss: $E = 200$ GPa; $A = 800$ cm^2

5.15   Analyze the structure shown below using the method of consistent deformations.

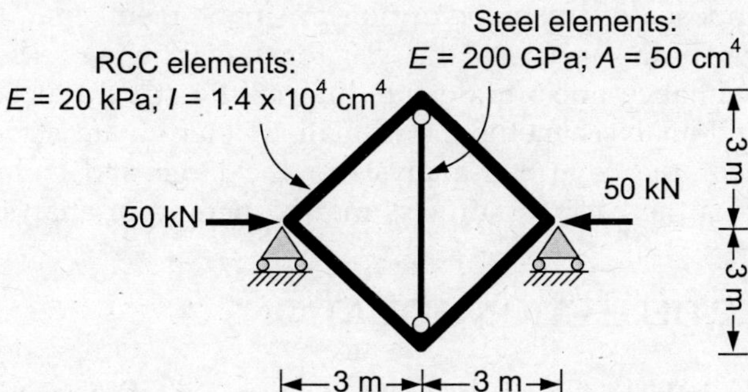

Slope-Deflection Method

6.1 GENERAL

The analysis of statically indeterminate structures using the method of consistent deformation (force method) as discussed in the preceding chapter is based on determining the force (reactions or internal forces) unknowns. Deformations may also be used as the unknowns to analyze a statically indeterminate structure. The analysis methods in which the displacements (or rotations) are considered as the basic unknowns are referred to as *displacement methods*. The *Slope-deflection method* is one of the displacement methods, which is based on determining the deflections and rotations at various joints of a structure. The end moments are computed for each member using the force-displacement relationships.

George Maney (1915) developed the *slope-deflection method* for the analysis of beams and rigid frames composed of prismatic or non-prismatic members. This method is one of the important concepts in the structural analysis for the following reasons: (i) A knowledge of this method can enhance one's understanding of the structural behaviour, (ii) It is a simple and easy hand-analysis method for small indeterminate structure, (iii) It can be easily programmed for the computer-analysis of the large and complex structures, and (iv) This method forms the basis of stiffness matrix method of analysis of structures.

6.2 BASIC SLOPE-DEFLECTION EQUATIONS

The basis of the slope-deflection method lies in the *slope-deflection equations*, which relate the end moments of each member in terms of end rotations and deflections of that member. In nutshell, these equations represent the *force-displacement relationship* which is one of three principles of structural analysis. The compatibility conditions are then utilized in these equations followed by the use of moment equilibrium at the joints to finally determine the member forces and the support reactions.

In order to derive the basic slope-deflection equations, let's consider a continuous beam (or part of the rigid frame) subjected to arbitrary loadings and the support settlement as shown in Figure 6.1. The beam is assumed to be prismatic having the constant flexural rigidity of *EI*. However, this method can be applied to the non-prismatic members as well. Consider the interior span *ab*, which under the given loading condition would experience the end moments in addition to the end rotations and deflections. Figure 6.2 shows the deformed shape of the member *ab* along with the end rotations, end moments, and the relative displacement.

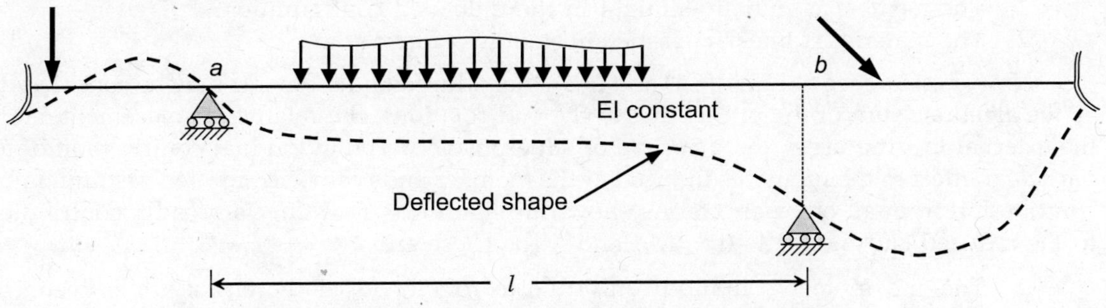

FIGURE 6.1   Continuous beam with arbitrary loading.

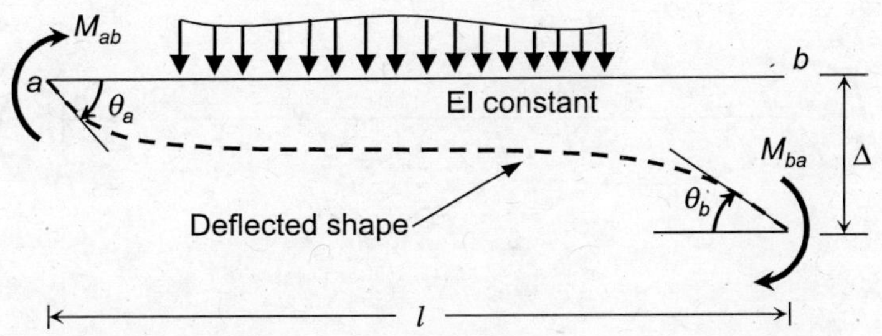

FIGURE 6.2   Deformed shape of beam with end moments.

Let's assume that the induced end moments of the member *ab* are M_{ab} and M_{ba}. Similarly, the end rotations are assumed as θ_a and θ_b, and the relative vertical displacement between the endpoints *a* and *b* is Δ. The induced end moments are a function of the elastic deformations/distortions at both ends as well as the applied external loads. Thus,

$$M_{ab} = f(\theta_a, \theta_b, \Delta, load) \tag{6.1}$$

$$M_{ba} = g(\theta_a, \theta_b, \Delta, load) \tag{6.2}$$

To derive the above expressions, it is necessary to consider the following sign conventions for the slope and deflections:

1. The *moments* at member ends are considered *positive* when acting in the *clockwise* direction and vice-versa.
2. The *rotation* at the ends of members is considered *positive* when the tangent to the elastic deformed force at the ends rotates *clockwise* from its original (undeformed) position.
3. The *relative deflection* between the ends of a member is considered *positive* when it corresponds to the *clockwise* rotation of the member (i.e., the straight-line joining ends of the elastic line).

Accordingly, all end moments, end rotations, and relative deflection as shown in Figure 6.2 are considered positive. In addition to the above sign convention, the following assumptions are necessary to derive the slope-deflection equations.

1. The member is initially straight in the unloaded configuration.
2. The material is linearly elastic and follows Hooke's law.

Now, in reference to Eqns. (6.1) and (6.2), the induced end moments may be considered as the algebraic sum of the effects due to the end rotations, the relative displacement, and the external loading using the principle of superposition. The individual contribution of a particular effect to the moments induced at the member ends can be computed assuming no contribution from all other effects. As shown in Figure 6.3, four effects broadly contribute to the induced end moments (i.e., M_{ab} and M_{ba}). These are:

1. The end moments induced only due to rotation θ_a at the end a while keeping the other end b fixed.
2. The end moments induced only due to rotation θ_b at the end b while keeping the other end a fixed.

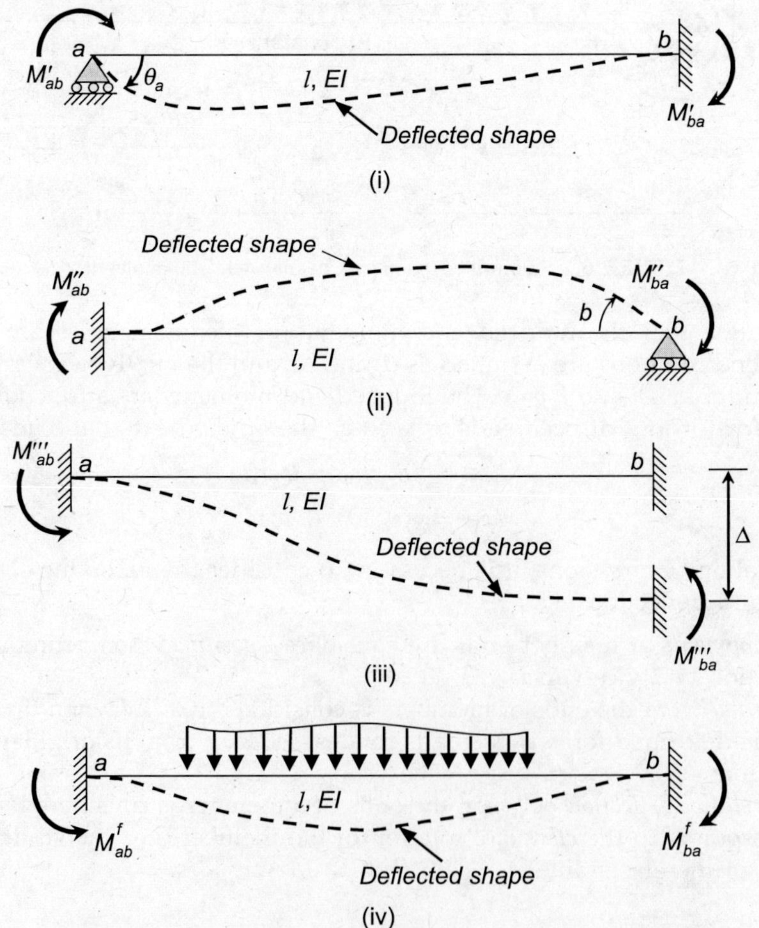

FIGURE 6.3   End moments of beam due to all possible effects.

3. The end moments induced only due to the relative displacement Δ between the ends of the member assuming no end rotations.

4. The end moments induced only due to the external loads assuming no end rotations and zero relative displacement.

$$M_{ab} = M'_{ab} + M''_{ab} + M'''_{ab} + M^{f}_{ab} \tag{6.3}$$

$$M_{ba} = M'_{ba} + M''_{ba} + M'''_{ba} + M^{f}_{ba} \tag{6.4}$$

Let's derive the end moments induced due to these four actions acting independently.

1. Figure 6.4 shows the beam with applicable boundary conditions when it is subjected to end rotation, θ_a only. The moment applied at the end a to induce a rotation θ_a at the same end keeping the other end fixed is M'_{ab}. As a result, the fixed end induced at end b is M'_{ba}. To determine these moments, let's use the conjugate beam method. Figure 6.4 also shows the conjugate beam with M/EI as the loading. The support reaction at the end a of the conjugate beam is equal to the end rotation θ_a of the real beam.

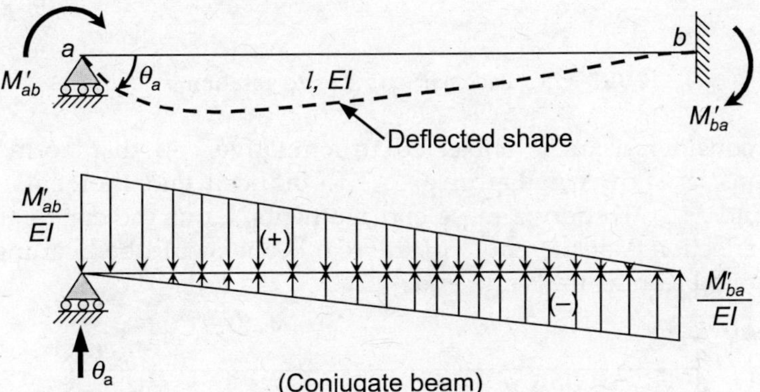

FIGURE 6.4   End moments due to rotation at end A.

Using the equilibrium conditions for the conjugate beam,

$$\sum M_a = 0 \left(\frac{M'_{ab}l}{2EI} \right)\left(\frac{l}{3} \right) - \left(\frac{M'_{ba}l}{2EI} \right)\left(\frac{2l}{3} \right) = 0 \tag{6.5}$$

$$\sum M_b = 0 \ \theta_a l - \left(\frac{M'_{ab}l}{2EI} \right)\left(\frac{2l}{3} \right) + \left(\frac{M'_{ba}l}{2EI} \right)\left(\frac{l}{3} \right) = 0 \tag{6.6}$$

From Eq. (6.5), $$M'_{ba} = \frac{M'_{ab}}{2} \tag{6.7}$$

Substituting Eq. (6.7) in Eq. (6.6),

$$M'_{ab} = \frac{4EI\theta_a}{l} \tag{6.8}$$

$$M'_{ba} = \frac{2EI\theta_a}{l} \tag{6.9}$$

2. Consider the beam ab subjected to a rotation θ_b at end b keeping the end b fixed as shown in Figure 6.5. The corresponding applied moment at the end b is M''_{ba} and the induced moment at the end a is M''_{ab}. Following the same procedure as discussed above,

$$M''_{ab} = \frac{M''_{ba}}{2} \tag{6.10}$$

$$M''_{ab} = \frac{2EI\theta_b}{l} \tag{6.11}$$

$$M''_{ba} = \frac{4EI\theta_b}{l} \tag{6.12}$$

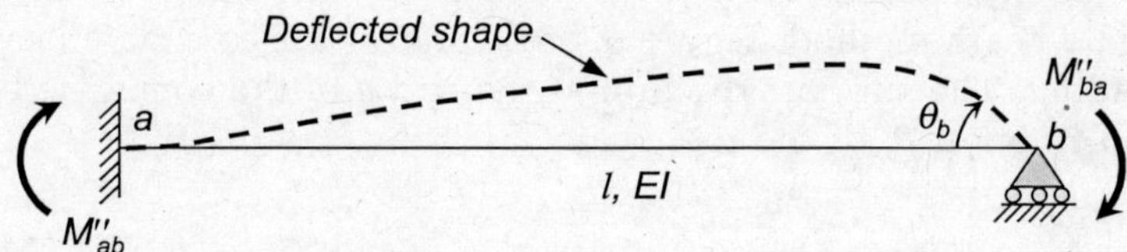

FIGURE 6.5 End moments due to rotation at end B.

3. Now consider beam is subjected to a relative end displacement without end rotations as shown in Figure 6.6. The induced end moments are assumed as M'''_{ab} and M'''_{ba}. To derive these end moments, let us the conjugate beam method. The deflection Δ at the end b of the real beam would be bending moment in the conjugate beam at the same end.

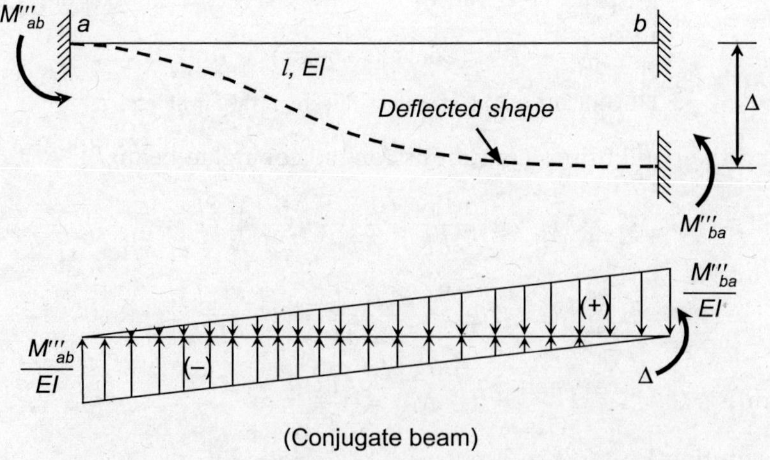

FIGURE 6.6 End moments due to relative end displacement.

Considering the force equilibrium of the conjugate beam, $\dfrac{M'''_{ab}l}{2EI} = \dfrac{M'''_{ba}l}{2EI}$

$$\therefore \qquad M'''_{ab} = M'''_{ba} = M \qquad (6.13)$$

Moment equilibrium of the conjugate beam would give

$$\Delta + \left(\frac{Ml}{2EI}\right)\left(\frac{l}{3}\right) - \left(\frac{Ml}{2EI}\right)\left(\frac{2l}{3}\right) = 0$$

$$\therefore \qquad M = \frac{6EI\Delta}{l^2} \qquad (6.14)$$

It is worth noting that the directions of both M'''_{ab} and M'''_{ba} are opposite to those of M'_{ab}, M'_{ba}, M''_{ab} and M''_{ba}. Thus,

$$\therefore \qquad M'''_{ab} = M'''_{ab} = -\frac{6EI\Delta}{l^2} \qquad (6.15)$$

4. Consider the beam with external loads with both ends are fixed (i.e., no end rotations and relative displacements). The induced fixed end moments depend on the magnitude and type of external loads. The fixed end moments are M^f_{ab} and M^f_{ba}. The directions of these fixed moments are often opposite to each other.

   The final induced end moments can be obtained by substituting the expressions of four different components in Eqs. (6.3) and (6.4).

$$M_{ab} = \pm M^f_{ab} + \frac{4EI\theta_a}{l} + \frac{2EI\theta_b}{l} - \frac{6EI\Delta}{l^2}$$

$$M_{ba} = \pm M^f_{ba} + \frac{2EI\theta_a}{l} + \frac{4EI\theta_b}{l} - \frac{6EI\Delta}{l^2}$$

Rearranging the terms, one can get

$$M_{ab} = \pm M^f_{ab} + \frac{2EI}{l}\left(2\theta_a + \theta_b - \frac{3\Delta}{l}\right) \qquad (6.16)$$

$$M_{ba} = \pm M^f_{ba} + \frac{2EI}{l}\left(\theta_a + 2\theta_b - \frac{3\Delta}{l}\right) \qquad (6.17)$$

Eqs. (6.16) and (6.17) are the basic slope-deflection equations for a general deformed member of uniform cross-section. These equations provide the end moments in terms of both end rotations (i.e., θ_a and θ_b), the relative displacement (Δ) between two ends, and the externally applied loads on the member.

If we consider $K = \dfrac{I}{l}$ and $R = \dfrac{\Delta}{l}$, the slope-deflection equations can be written as follows:

$$M_{ab} = \pm M^f_{ab} + 2EK(2\theta_a + \theta_b - 3R) \qquad (6.18)$$

$$M_{ba} = \pm M^f_{ba} + 2EK(\theta_a + 2\theta_b - 3R) \qquad (6.19)$$

where, K = *Stiffness factor* of the member and R = *Rigid-body rotation* of the member.

The fixed end moments (i.e., M^f_{ab} and M^f_{ba}) for the members with various loading conditions are summarized in Table 6.1.

TABLE 6.1   Fixed end moments of fixed-ended beams

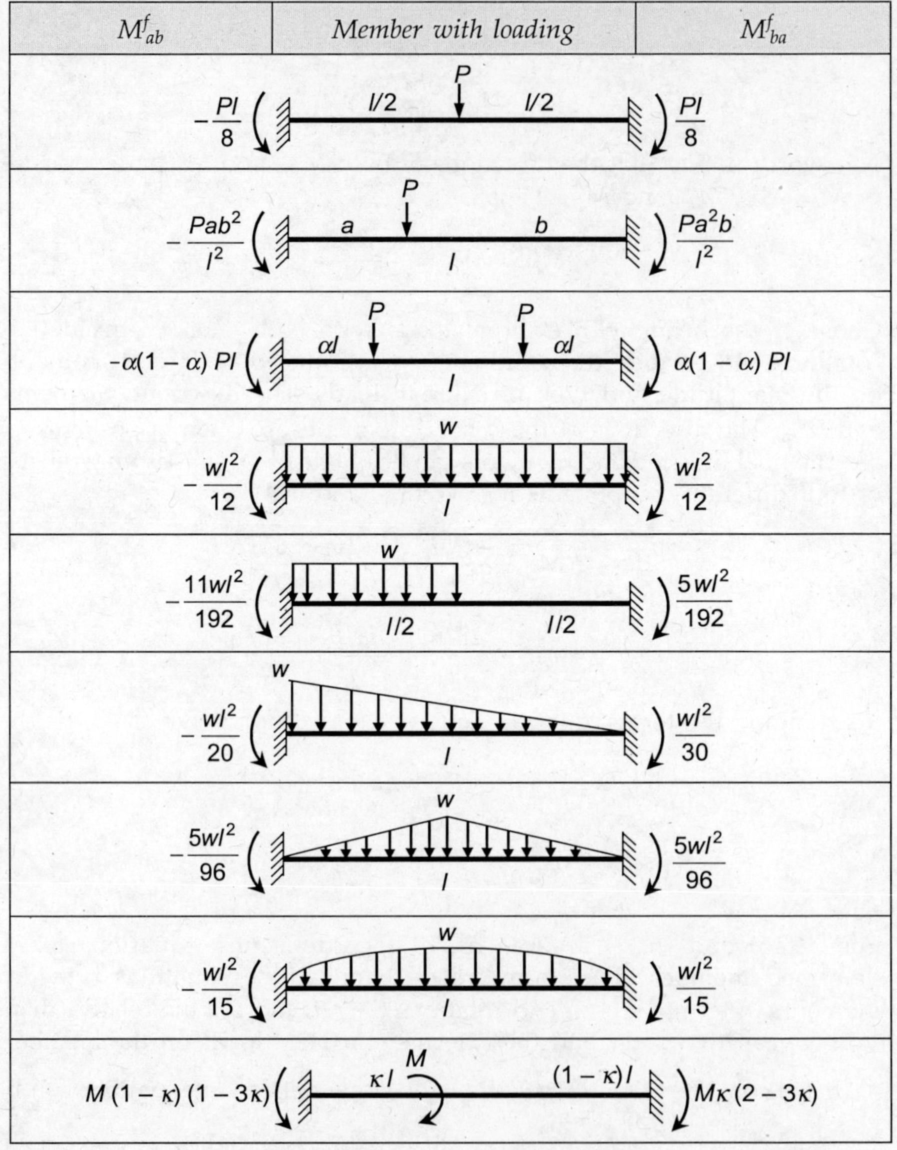

M^f_{ab}	Member with loading	M^f_{ba}
$-\dfrac{Pl}{8}$	P at $l/2$, $l/2$	$\dfrac{Pl}{8}$
$-\dfrac{Pab^2}{l^2}$	P at a, b, span l	$\dfrac{Pa^2b}{l^2}$
$-\alpha(1-\alpha)\,Pl$	P at αl, P at αl, span l	$\alpha(1-\alpha)\,Pl$
$-\dfrac{wl^2}{12}$	w over span l	$\dfrac{wl^2}{12}$
$-\dfrac{11wl^2}{192}$	w over $l/2$, $l/2$	$\dfrac{5wl^2}{192}$
$-\dfrac{wl^2}{20}$	w (triangular) over span l	$\dfrac{wl^2}{30}$
$-\dfrac{5wl^2}{96}$	w (triangular peak) over span l	$\dfrac{5wl^2}{96}$
$-\dfrac{wl^2}{15}$	w (parabolic) over span l	$\dfrac{wl^2}{15}$
$M(1-\kappa)(1-3\kappa)$	M at κl, $(1-\kappa)l$	$M\kappa(2-3\kappa)$

6.3   MODIFIED SLOPE-DEFLECTION EQUATIONS

The basic slope-deflection equations can be modified for the members with one simple support (i.e., hinge or roller). Instead of assuming the simple support as the fixed support and later setting the final bending moment at the simple support to zero, the modified slope-deflection equation can be adopted to reduce the computational cost and time. The modified slope-deflection equation for a member with simple support has been derived in the following section.

Consider a member AB with hinge support at end B subjected to any external loads as shown in Figure 6.7. Using the basic slope-deflection equations, the end moments can be written as follows:

$$M_{ab} = \pm M_{ab}^{f} + \frac{2EI}{l}\left(2\theta_a + \theta_b - \frac{3\Delta}{l}\right) \tag{6.20}$$

$$M_{ba} = \pm M_{ba}^{f} + \frac{2EI}{l}\left(\theta_a + 2\theta_b - \frac{3\Delta}{l}\right) \tag{6.21}$$

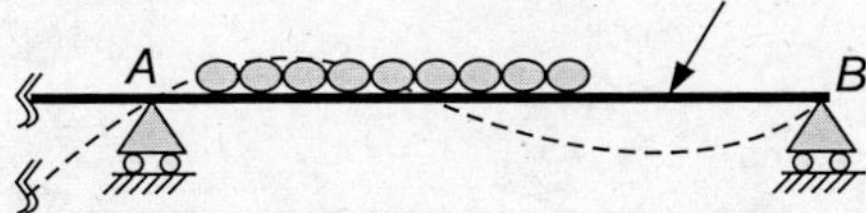

FIGURE 6.7 Beam with one end hinged.

Since the end B is supported on simple support, $M_{ba} = 0$. Further, if the end B is not assumed to be fixed but considered as simple support as it is, the value of the fixed end moment must be zero, i.e., $M_{ba}^{f} = 0$. Substituting these values in the Eq. (6.21), one would get

$$0 = 0 + \frac{2EI}{l}\left(\theta_a + 2\theta_b - \frac{3\Delta}{l}\right)$$

$$\theta_b = -\frac{1}{2}\left(\theta_a - \frac{3\Delta}{l}\right) \tag{6.22}$$

Substituting Eq. (6.22) in Eq. (6.20), the end moment at A can be obtained as follows:

$$M_{ab} = \pm M_{ab}^{f'} + \frac{3EI}{l}\left(\theta_a - \frac{\Delta}{l}\right) \tag{6.23}$$

where, $M_{ab}^{f'}$ is the fixed end moment at end A of the beam having the end B supported on a hinge/roller. Therefore, the fixed end moment as listed in Table 6.1 should not be used in Eqn. (6.23). However, one can get the corresponding fixed end moments by simply using the analogy between the applied moment and the induced fixed end moment as derived in Eqn. (6.10). This states that if a moment M is applied at one end, the moment to be induced at the other fixed end would be half of the applied moment, i.e., $M/2$. Accordingly, one can easily derive the value of the fixed end moment of a beam with a simply-supported end from a fixed-ended beam with the same external loading by adding an equal and opposite bending moment at the hinged end and adding half of this value to the induced fixed end moment at the other end. This is illustrated by means of an example given below.

Consider a beam having one end fixed and the other end being pinned subjected to a concentrated load at its mid-span as shown in Figure 6.8. This beam can be assumed a combination of a fixed-ended beam with the given load and a beam with an end moment applied at one end. The end moments of the fixed-ended beam are already highlighted in Table 6.1. The fixed end moment at B of the fixed-ended beam is $Pl/8$. Since the end B is a pined end, the net bending moment must be zero. Hence, an equal and opposite bending moment of $-Pl/8$ is applied at the end B in the second case. The moment induced at end A due to this applied bending moment would be $-Pl/16$. Thus, the final bending moment at end A, $M_{ab}^{f'}$ would be $-3Pl/16$.

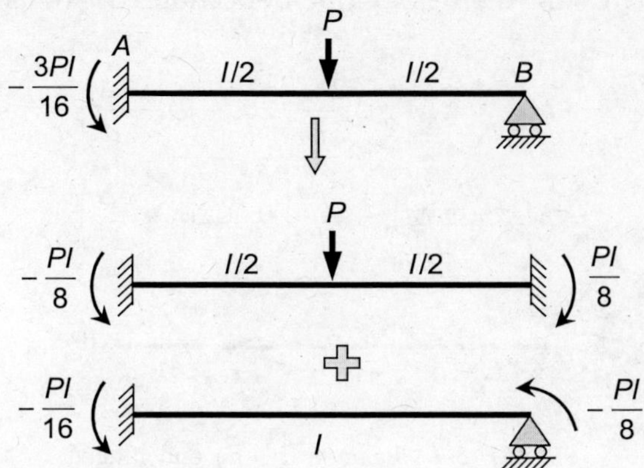

FIGURE 6.8 Computation of fixed end moment of beam with hinged end.

Table 6.2 shows the bending moment at the fixed end of the beam having one end pinned and subjected to various loading conditions.

TABLE 6.2 Fixed end moments of beams with one simple support

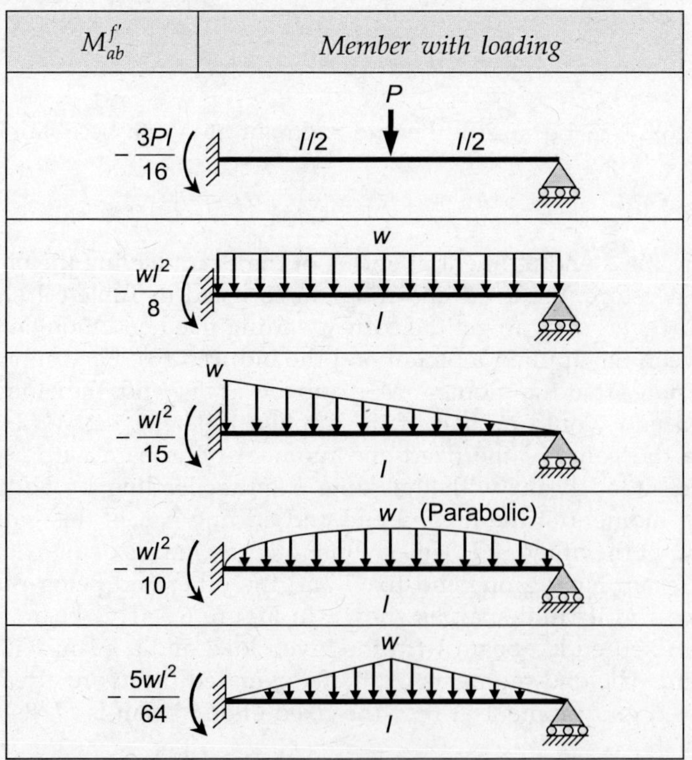

6.4   PROCEDURE OF ANALYSIS BY USING SLOPE-DEFLECTION METHOD

The general procedure of the conventional slope-deflection method has been presented as follows:

- Split the entire structure into a series of members between the joints and/or supports.
- Assume both ends of all members so formed as fixed. Determine the fixed end moments at both ends of all members for the given loading.
- Write the slope-deflection equations for each member. If there are n number of members, the total number of slope-deflection equations would be $2n$.
- Use the displacement computability conditions in the slope deflection equations depending on the boundary conditions.
- Using equilibrium equations at joints, the unknown rotations (i.e., slopes) and relative displacements can be determined.
- Once the values of all slopes and deflections are computed, the end moments can be computed by substituting these values in the slope-deflection equations.

6.5   ANALYSIS OF INDETERMINATE BEAMS

The analysis of indeterminate beams under various loading conditions using the slope-deflection method has been illustrated in the following examples.

EXAMPLE 6.1   Determine the support reactions and the end moments of the continuous beam shown in Figure E6.1 using the Slope-deflection method. Assume $EI = 200$ MNm^2.

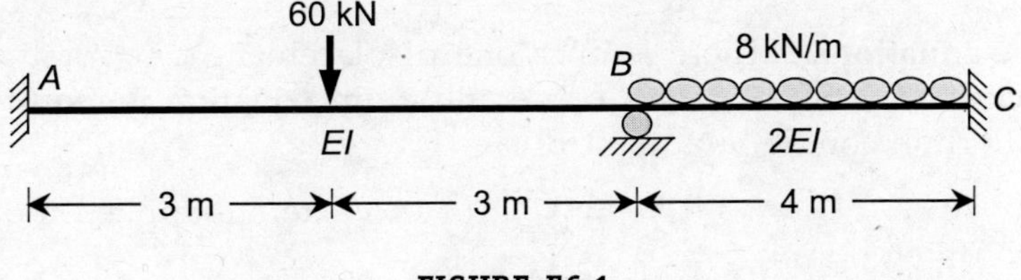

FIGURE E6.1

Solution: **(a) Fixed end moments:**

Span AB:

$$M_{ab}^{f} = -\frac{Pl}{8} = -\frac{60 \times 6}{8}\ \text{kNm} = -45\ \text{kNm}$$

$$M_{ba}^{f} = \frac{Pl}{8} = \frac{60 \times 6}{8}\ \text{kNm} = 45\ \text{kNm}$$

Span BC:

$$M^f_{bc} = -\frac{wl^2}{12} = -\frac{8 \times 4^2}{12} \text{ kNm} = -10.67 \text{ kNm}$$

$$M^f_{cb} = \frac{wl^2}{12} = \frac{8 \times 4^2}{12} \text{ kNm} = 10.67 \text{ kNm}$$

(b) Compatibility Conditions:

$$\theta_a = 0; \quad \theta_c = 0 \qquad \text{(Because of fixed supports)}$$

$$\Delta_{ab} = 0; \quad \Delta_{bc} = 0 \qquad \text{(Because of no support settlements)}$$

(c) Slope-deflection Equations:
Span AB:

$$M_{ab} = -45 + \frac{2EI}{6}\left(2\theta_a + \theta_b - \frac{3\Delta_{ab}}{6}\right) = -45 + \frac{EI\theta_b}{3} \tag{1}$$

$$M_{ba} = 45 + \frac{2EI}{6}\left(\theta_a + 2\theta_b - \frac{3\Delta_{ab}}{6}\right) = 45 + \frac{2EI\theta_b}{3} \tag{2}$$

Span BC:

$$M_{bc} = -10.67 + \frac{2(2EI)}{4}\left(2\theta_b + \theta_c - \frac{3\Delta_{bc}}{4}\right) = -10.67 + 2EI\theta_b \tag{3}$$

$$M_{cb} = 10.67 + \frac{2(2EI)}{4}\left(\theta_b + 2\theta_c - \frac{3\Delta_{bc}}{4}\right) = 10.67 + EI\theta_b \tag{4}$$

(d) Equilibrium equations: There is only one displacement (i.e., rotation) unknown in the slope deflection equations. Hence, one equilibrium equation is required to solve this unknown. Equilibrium condition at joint B is

$$M_b = 0 \qquad (\because \text{No external applied moment at joint B})$$

$$M_{ba} + M_{bc} = 0 \tag{5}$$

Substituting Eqs. (2) and (3) in Eq. (5),

$$34.33 + \frac{8EI\theta_b}{3} = 0 \qquad \therefore EI\theta_b = -12.88 \tag{6}$$

Putting the value of $EI\theta_b$ in Eqs. (1)–(4), the final end moments can be determined as follows:

$$M_{ab} = -49.29 \text{ kNm}; \quad M_{ba} = -M_{bc} = 36.41 \text{ kNm}; \quad M_{cb} = -2.21 \text{ kNm}$$

(e) Support reactions: To compute the support reactions, consider the free body diagram of each span with external loads and end moments. As shown in the following figure, the positive end moments are shown clockwise, and the negative end moments are shown anti-clockwise as per the considered sign conventions of this method. Let the reactions at supports A, B, and C are R_a, R_b and R_c, respectively.

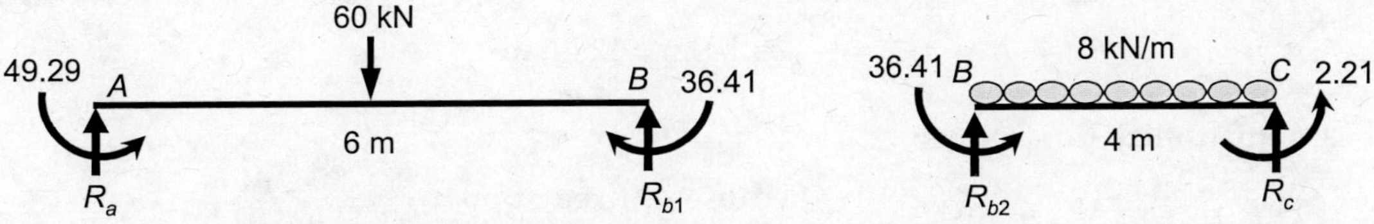

Span *AB*: Taking moment about *A*,

$$6R_{b1} + 49.29 - 36.41 - 60 \times 3 = 0 \quad \therefore R_{b1} = 27.85 \text{ kN}$$

Force equilibrium, $R_{b1} + R_a = 60 \quad \therefore R_a = 32.15$ kN

Span *BC*: Taking moment about *B*,

$$4R_c + 2.21 + 36.41 - 8 \times 4 \times 2 = 0 \quad \therefore R_c = 6.35 \text{ kN}$$

Force equilibrium, $R_{b2} + R_c = 8 \times 4 = 32 \; kN \quad \therefore R_{b2} = 25.66$ kN

Thus, support reaction at *B*, $R_b = R_{b1} + R_{b2} = 53.51$ kN.

EXAMPLE 6.2 Analyze the two-span beam shown in Figure E6.2 using the Slope-deflection method. Draw the bending moment diagram and shear force diagram. Assume the constant flexural rigidity of *EI* for both spans.

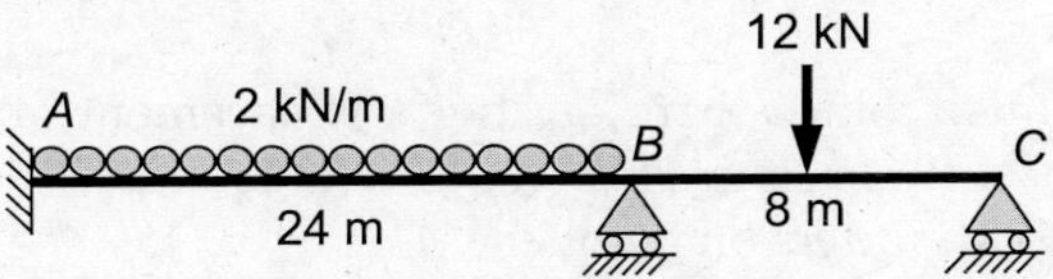

FIGURE E6.2

Solution: This problem can be solved by using basic slope-deflection equations as well as the modified slope-deflection equations. Let's use the basic slope-deflection equations.

(a) Fixed end moments:

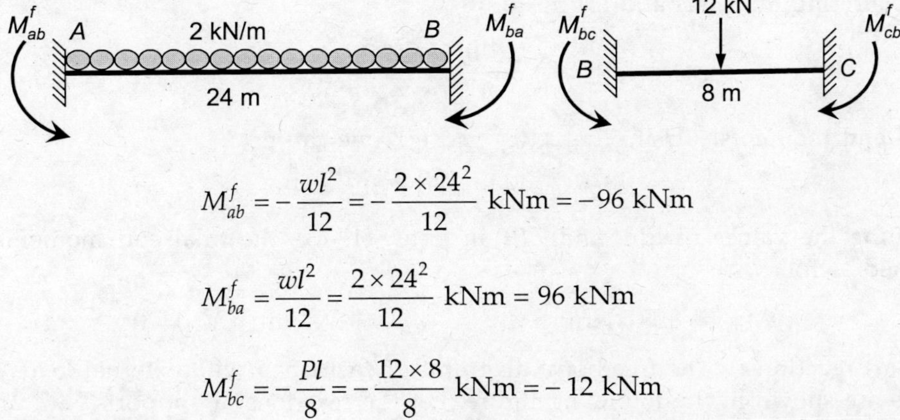

$$M_{ab}^f = -\frac{wl^2}{12} = -\frac{2 \times 24^2}{12} \text{ kNm} = -96 \text{ kNm}$$

$$M_{ba}^f = \frac{wl^2}{12} = \frac{2 \times 24^2}{12} \text{ kNm} = 96 \text{ kNm}$$

$$M_{bc}^f = -\frac{Pl}{8} = -\frac{12 \times 8}{8} \text{ kNm} = -12 \text{ kNm}$$

$$M_{cb}^f = \frac{Pl}{8} = \frac{12 \times 8}{8} \text{ kNm} = 12 \text{ kNm}$$

(b) Compatibility Conditions:

$$\theta_a = 0 \quad \text{(Because of fixed support)}$$

$$\Delta_{ab} = 0; \quad \Delta_{bc} = 0 \quad \text{(Because of no support settlements)}$$

(c) Slope-deflection Equations:
Span AB:

$$M_{ab} = -96 + \frac{2EI}{24}\left(2\theta_a + \theta_b - \frac{3\Delta_{ab}}{24}\right) = -96 + \frac{EI\theta_b}{12} \tag{1}$$

$$M_{ba} = 96 + \frac{2EI}{24}\left(\theta_a + 2\theta_b - \frac{3\Delta_{ab}}{24}\right) = 96 + \frac{EI\theta_b}{6} \tag{2}$$

Span BC:

$$M_{bc} = -12 + \frac{2EI}{8}\left(2\theta_b + \theta_c - \frac{3\Delta_{bc}}{8}\right) = -12 + \frac{EI}{4}(2\theta_b + \theta_c) \tag{3}$$

$$M_{cb} = 12 + \frac{2EI}{8}\left(\theta_b + 2\theta_c - \frac{3\Delta_{bc}}{8}\right) = 12 + \frac{EI}{4}(\theta_b + 2\theta_c) \tag{4}$$

(d) Equilibrium equations: Since there are two displacement/rotation unknowns in the slope-deflection equations, it is required to write two equations of equilibrium.
Since the end C is supported on a roller, $M_{cb} = 0$.
Using Eqn. (4), $(\theta_b + 2\theta_c)$

$$\therefore \qquad \theta_b + 2\theta_c = -\frac{48}{EI} \tag{5}$$

Equilibrium condition at joint B, $M_b = 0$

$$M_{ba} + M_{bc} = 0 \tag{6}$$

Substituting Eqns. (2) and (3) in Eq. (6),

$$84 + \frac{2EI\theta_b}{3} + \frac{EI\theta_c}{4} = 0 \tag{7}$$

(e) Final end moments: Using Eqns. (5) and (7), one can get

$$EI\theta_b = -144 \quad \text{and} \quad EI\theta_c = 48$$

Putting the values of $EI\theta_b$ and $EI\theta_c$ in Eqns. (1)–(3), the final end moments can be determined as follows:

$$M_{ab} = -108 \text{ kNm}; \quad M_{ba} = -M_{bc} = 72 \text{ kNm}; \quad M_{cb} = 0$$

(f) Support reactions: The free-body diagrams of each span with external loads and end moments are shown in the following figure. Let the reactions at supports A, B and C are R_a, R_b and R_c, respectively.

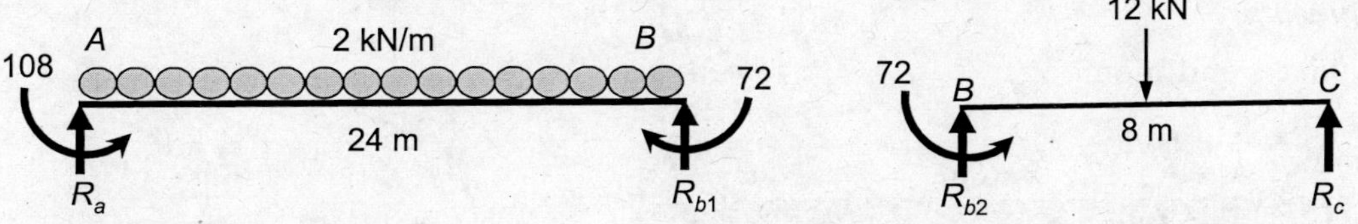

Span AB:

Taking moment about A,

$$24R_{b1} + 108 - 72 - 2 \times 24 \times 12 = 0 \quad \therefore R_{b1} = 22.5 \text{ kN}$$

Force equilibrium,

$$R_{b1} + R_a = (24 \times 2)\text{kN} = 48 \text{ kN}; \quad \therefore R_a = 25.5 \text{ kN}$$

Span BC:

Taking moment about B,

$$8R_c + 72 - 12 \times 4 = 0 \quad \therefore R_c = -3 \text{ kN}$$

Force equilibrium, $R_{b2} + R_c = 12 \text{ kN} \quad \therefore R_{b2} = 15 \text{ kN}$

Support reaction at B, $R_b = R_{b1} + R_{b2} = 37.5 \text{ kN}$

Shear force diagram (SFD) and bending moment diagram (BMD) of the beam are shown in the following figure.

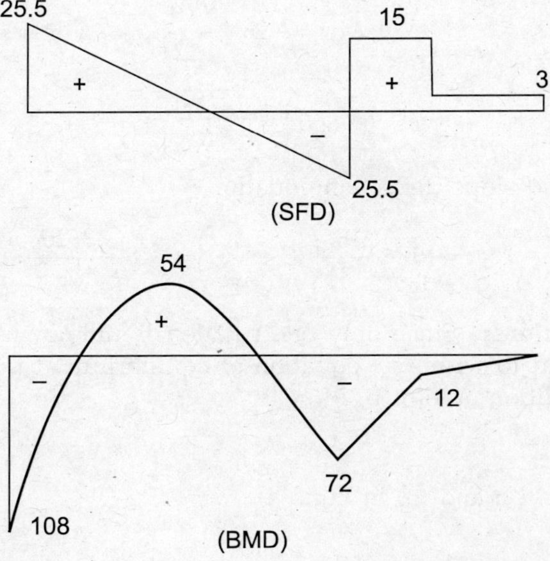

EXAMPLE 6.3 Analyze the two-span beam shown in Figure E6.2 using the modified slope-deflection equations.

Solution:

(a) Fixed end moments:

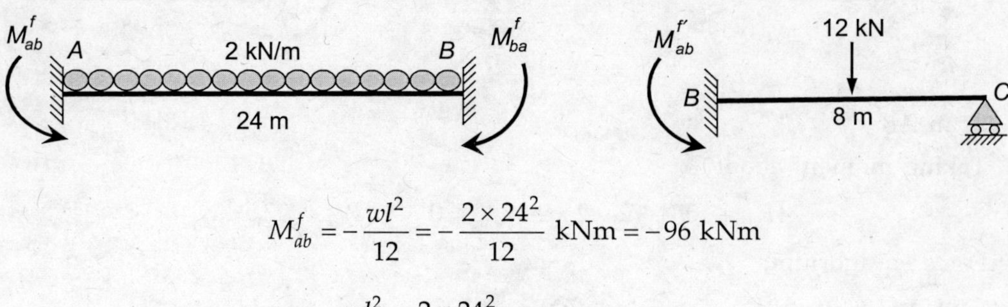

$$M_{ab}^{f} = -\frac{wl^2}{12} = -\frac{2 \times 24^2}{12} \ \text{kNm} = -96 \ \text{kNm}$$

$$M_{ba}^{f} = \frac{wl^2}{12} = \frac{2 \times 24^2}{12} \ \text{kNm} = 96 \ \text{kNm}$$

$$M_{bc}^{f} = -\frac{3Pl}{16} = -\frac{3 \times 12 \times 8}{16} \ \text{kNm} = -18 \ \text{kNm}$$

(b) Compatibility Conditions:

$$\theta_a = 0 \quad \text{(Because of fixed support)}$$

$$\Delta_{ab} = 0; \quad \Delta_{bc} = 0 \quad \text{(Because of no support settlements)}$$

(c) Slope-deflection Equations:
Span *AB*:

$$M_{ab} = -96 + \frac{EI\theta_b}{12} \tag{1}$$

$$M_{ba} = 96 + \frac{EI\theta_b}{6} \tag{2}$$

Span *BC* (modified slope deflection equation):

$$M_{bc} = -18 + \frac{3EI}{8}\left(\theta_b - \frac{\Delta_{bc}}{8}\right) = -18 + \frac{3EI\theta_b}{8} \tag{3}$$

(d) Equilibrium equations: Since only one rotation is unknown in the slope-deflection equations, it is required to write one equation of equilibrium at the joint.
Equilibrium condition at joint B, $M_b = 0$

$$M_{ba} + M_{bc} = 0 \tag{4}$$

Substituting Eqns. (2) and (3) in Eqn. (4),

$$78 + \frac{13EI\theta_b}{24} = 0 \tag{5}$$

$$\therefore \qquad EI\theta_b = -144$$

(e) Final end moments: Putting the values of $EI\theta_b$ in Eqns. (1)-(3), the final end moments can be determined as follows:

$$M_{ab} = -\,108 \text{ kNm}; \quad M_{ba} = -M_{bc} = 72 \text{ kNm}; \quad M_{cb} = 0$$

The same values of end moments have also been obtained earlier.

EXAMPLE 6.4   Analyze the continuous beam shown in Figure E6.4 using the slope-deflection equations. Draw the bending moment and shear force diagrams.

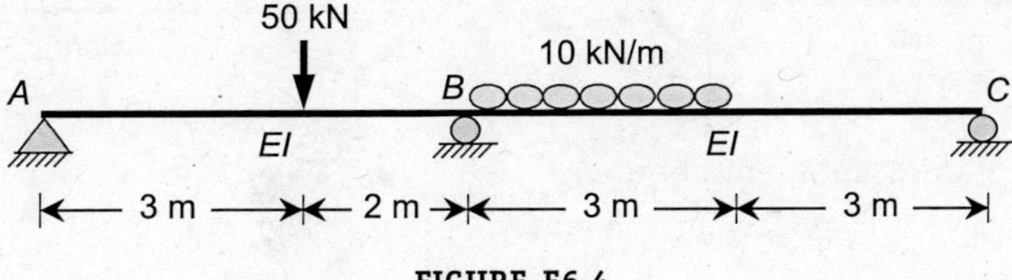

FIGURE E6.4

Solution: Since the ends A and C are simply supported, let's use the modified slope-deflection equations to analyze the beam.

(a) Fixed end moments:

$$M_{ba}^{f'} = \frac{Pa^2b}{l^2} + \frac{1}{2}\left(\frac{Pab^2}{l^2}\right) = \left(\frac{50 \times 3^2 \times 2}{5^2}\right) + \frac{1}{2}\left(\frac{50 \times 3 \times 2^2}{5^2}\right) = 48 \text{ kNm}$$

$$M_{bc}^{f'} = -\frac{11wl^2}{192} - \frac{1}{2}\left(\frac{5wl^2}{192}\right) = -\frac{27wl^2}{384} = -\frac{27 \times 10 \times 6^2}{384} \text{ kNm} = -25.31 \text{ kNm}$$

(b) Compatibility Conditions:

$$\Delta_{ab} = 0; \quad \Delta_{bc} = 0 \quad \text{(Because of no support settlements)}$$

(c) Modified Slope-deflection Equations:

$$M_{ba} = 48 + \frac{3EI\theta_b}{5} \tag{1}$$

$$M_{bc} = -\,25.31 + \frac{3EI\theta_b}{6} \tag{2}$$

(d) Equilibrium equations: Equilibrium condition at joint B, $M_b = 0$

$$M_{ba} + M_{bc} = 0 \tag{3}$$

Substituting Eqs. (1) and (2) in Eq. (3),

$$22.69 + 1.1EI\theta_b = 0$$

$$\therefore \qquad EI\theta_b = -20.63 \tag{4}$$

(e) Final end moments: Putting the values of $EI\theta_b$ in Eqs. (1)-(2), the final end moments can be determined as follows:

$$M_{ab} = 0; \quad M_{ba} = -M_{bc} = 35.62 \text{ kNm}; \quad M_{cb} = 0$$

(f) Support reactions: Free body diagrams of both spans with applicable end moments re shown in the following figures.

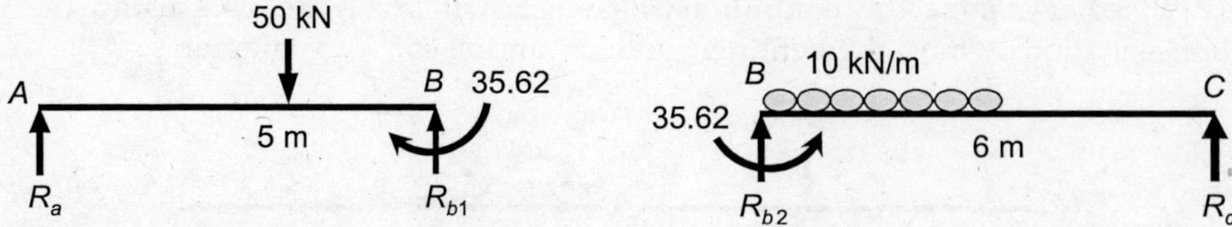

Span AB: Taking moment about A,

$$5R_{b1} - 35.62 - 50 \times 3 = 0 \quad \therefore R_{b1} = 37.12 \text{ kN}$$

Force equilibrium,

$$R_{b1} + R_a = 50 \text{ kN} \quad \therefore R_a = 12.88 \text{ kN}$$

Span BC: Taking moment about B,

$$6R_c + 35.62 - 10 \times 3 \times 1.5 = 0 \quad \therefore R_c = 1.56 \text{ kN}$$

Force equilibrium, $R_{b2} + R_c = 30 \text{ kN} \quad \therefore R_{b2} = 28.44 \text{ kN}$

Support reaction at B, $R_b = R_{b1} + R_{b2} = 65.56 \text{ kN}$

Shear force diagram (SFD) and bending moment diagram (BMD) of the beam are shown in the following figure.

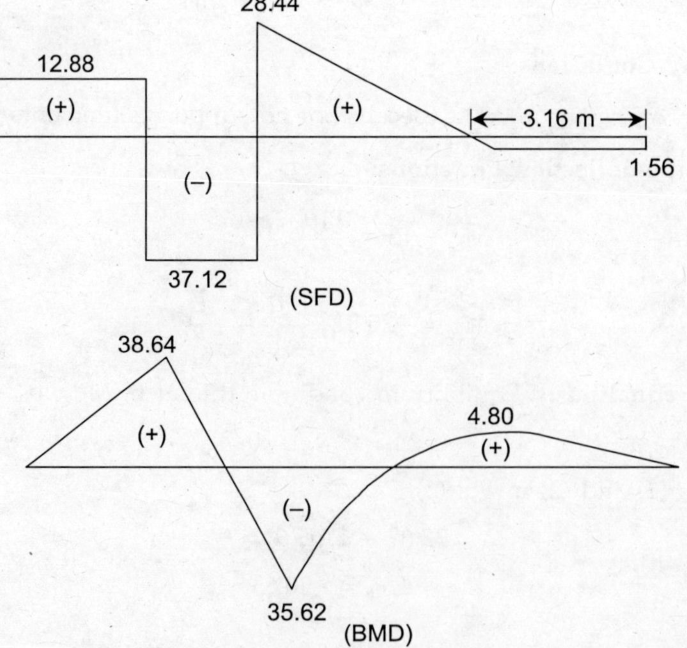

EXAMPLE 6.5 Re-analyze the continuous beam shown in Figure E6.4 using the slope-deflection equations if the support B settles by 25 mm. Assume $E = 200$ GPa and $I = 40,000$ cm^4.

Solution: As computed earlier, the fixed end moments are

$$M_{ba}^{f'} = 48 \text{ kNm}; \quad M_{bc}^{f'} = -25.31 \text{ kNm}$$

(a) Modified Slope-deflection Equations:

$$M_{ba} = 48 + \frac{3EI}{5}\left(\theta_b - \frac{0.025}{5}\right) \tag{1}$$

$$M_{bc} = -25.31 + \frac{3EI}{6}\left(\theta_b + \frac{0.025}{6}\right) \tag{2}$$

(b) Equilibrium equations: Equilibrium condition at joint B, $M_b = 0$

$$M_{ba} + M_{bc} = 0 \tag{3}$$

Substituting Eqs. (1) and (2) in Eq. (3),

$$22.69 - 73.33 + 1.1EI\theta_b = 0$$

$$\therefore \qquad EI\theta_b = 46.04 \tag{4}$$

(c) Final end moments: Putting the values of $EI\theta_b$ in Eqs. (1)–(2), the final end moments can be determined as follows:

$$M_{ab} = 0; \quad M_{ba} = -M_{bc} = -164.38 \text{ kNm}; \quad M_{cb} = 0$$

(d) Support reactions: Free body diagrams of both spans with applicable end moments are shown in the following figures.

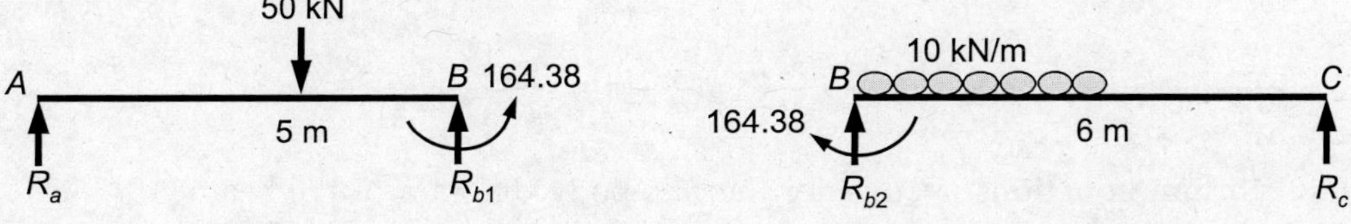

Span AB: Taking moment about A,

$$5R_{b1} + 164.38 - 50 \times 3 = 0 \quad \therefore R_{b1} = -2.88 \text{ kN}$$

Force equilibrium,

$$R_{b1} + R_a = 50 \text{ kN} \quad \therefore R_a = 52.88 \text{ kN}$$

Span BC: Taking moment about B,

$$6R_c - 164.38 - 10 \times 3 \times 1.5 = 0 \quad \therefore R_c = 34.90 \text{ kN}$$

Force equilibrium, $R_{b2} + R_c = 30$ kN $\quad \therefore R_{b2} = -4.90$ kN

Support reaction at B, $R_b = R_{b1} + R_{b2} = -7.78$ kN

EXAMPLE 6.6 Determine the spring force and fixed end moment of a beam AB supported on a spring and subjected to a uniformly distributed load of w per unit length. The spring constant is k and the flexural rigidity of the beam is EI. Use the *Slope-deflection method*.

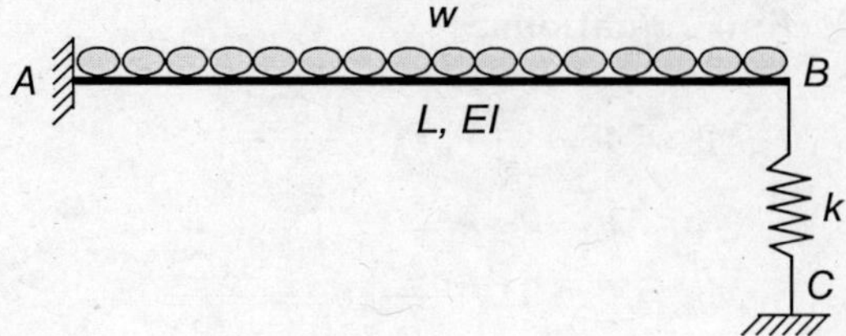

FIGURE E6.6

Since the bending moment at B is zero, let's use the modified slope-deflection equation to analyze the beam.

(a) Fixed end moments:

$$M_{ab}^{f'} = -\left(\frac{wL^2}{12}\right)1.5 = -\frac{wL^2}{8}$$

(b) Compatibility Conditions:

$$\theta_a = 0 \quad \text{(Because of the fixed end)}$$

$$\Delta_{ab} = \frac{R}{k}, \text{ where } R \text{ is the force in the spring.}$$

Assume the R is compressive such that the displacement at B, Δ_{ab} is downward.

(c) Modified Slope-deflection Equations:

$$M_{ab} = M_{ab}^{f'} + \frac{3EI}{L}\left(\theta_a - \frac{\Delta}{L}\right) = -\frac{wL^2}{8} - \frac{3EIR}{kL^2} \tag{1}$$

(d) Equilibrium equations: Consider the free-body diagram of the beam AB.

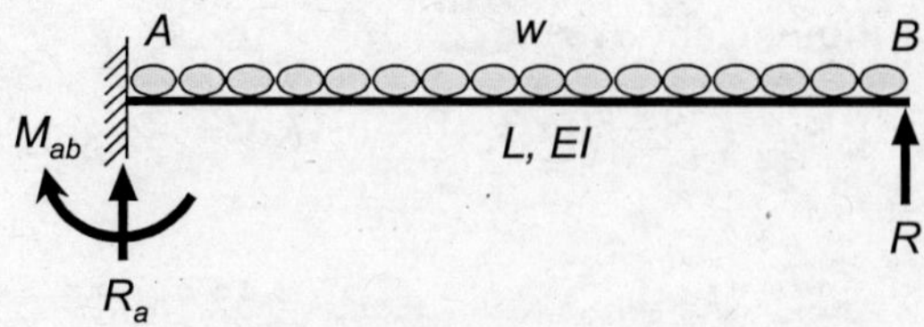

Taking the moment equilibrium about end A,

$$R = \frac{wL}{2} + \frac{M_{ab}}{L} \tag{2}$$

Substituting Eq. (1) in Eq. (2) and rearranging the terms,

$$R = \frac{wL}{2} - \frac{wL}{8} - \frac{3EIR}{kL^3}$$

$$\therefore \qquad R = \frac{3wL}{8} \left(\frac{1}{1 + \dfrac{3EI}{kL^3}} \right) \qquad\qquad (3)$$

(e) Special cases

If spring constant, $k = 0$ (i.e., *Cantilever beam*), $R = 0$; $M_{ab} = -\dfrac{wL^2}{2}$

If spring constant, $k = \infty$ (i.e., *Propped-cantilever beam*)

$$R = \frac{3wL}{8}; \quad M_{ab} = -\frac{wL^2}{8}$$

EXAMPLE 6.7 Analyze the continuous beam shown in Figure 6.7 using the *Slope-deflection method*. Draw the bending moment and shear force diagrams.

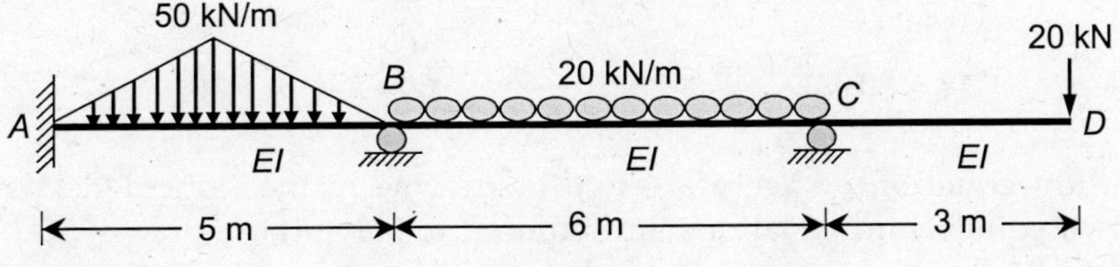

FIGURE E6.7

Solution: Though there are three spans in the beam, one could easily get the variation in the bending moment and shear force in the overhanging span *CD*. Thus, the given beam can be idealized as a two-span member shown below for computing the end moment and reactions at the supports.

Bending moment and shear force at *C* of the member *CD* due to a point load of 20 kN are 60 kNm and 20 kN. These forces are applied at the end C of the beam as shown in the following figure. It is worth mentioning that such an idealization has been carried out only for the purpose of solving the beam using the Slope-deflection method. However, the bending moment and shear force diagrams must be drawn for the entire beam consisting of the three spans.

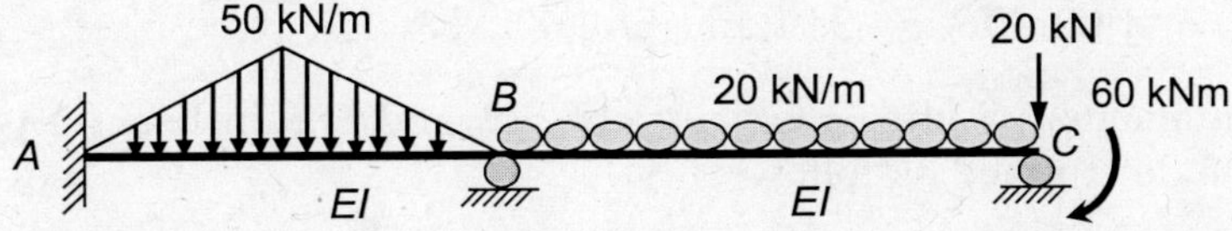

(a) Fixed end moments:

$$M_{ab}^{f} = -M_{ba}^{f} = -\frac{5wL^2}{96} = -\frac{5\times 50 \times 5^2}{96}\,\text{kNm} = -65.1\,\text{kNm}$$

$$M_{bc}^{f} = -M_{cb}^{f} = -\frac{wL^2}{12} = -\frac{20\times 6^2}{12}\,\text{kNm} = -60.0\,\text{kNm}$$

(b) Compatibility Conditions:

$$\theta_a = 0 \text{ (Because of the fixed end)}$$

$$\Delta_{ab} = \Delta_{bc} = 0 \text{ (Because of unyielding supports)}$$

(c) Slope-deflection Equations:

$$M_{ab} = -65.1 + \frac{2EI}{5}\left(2\theta_a + \theta_b - \frac{3\Delta_{ab}}{5}\right) = -65.1 + \frac{2EI\theta_b}{5} \tag{1}$$

$$M_{ba} = 65.1 + \frac{2EI}{5}\left(\theta_a + 2\theta_b - \frac{3\Delta_{ab}}{5}\right) = 65.1 + \frac{4EI\theta_b}{5} \tag{2}$$

$$M_{bc} = -60 + \frac{2EI}{6}\left(2\theta_b + \theta_c - \frac{3\Delta_{bc}}{6}\right) = -60 + \frac{EI}{3}(2\theta_b + \theta_c) \tag{3}$$

$$M_{cb} = 60 + \frac{2EI}{6}\left(\theta_b + 2\theta_c - \frac{3\Delta_{bc}}{6}\right) = 60 + \frac{EI}{3}(\theta_b + 2\theta_c) \tag{4}$$

(d) Equilibrium equations: There are two unknowns in the slope-deflection equations. Therefore, two equilibrium equations are required to solve these unknowns.

Equilibrium condition at joint B, $M_b = 0$

$$M_{ba} + M_{bc} = 0 \tag{5}$$

Substituting Eqs. (2) and (3) in Eq. (5),

$$5.1 + 1.47EI\theta_b + 0.33EI\theta_c = 0 \tag{6}$$

Now, consider the free-body diagram at joint C as shown in the figure.

Moment equilibrium at the end C, $M_{cb} = 60$

Using Eq. (4),

$$60 + \frac{EI}{3}(\theta_b + 2\theta_c) = 60$$

$$0.33EI\theta_b + 0.67EI\theta_c = 0 \tag{7}$$

Solving Eqs. (6) and (7), $EI\theta_b = -3.91$ and $EI\theta_c = 1.96$.

(e) Final end moments: Putting the values of $EI\theta_b$ and $EI\theta_b$ in Eqs. (1)–(4), the final end moments can be determined as follows:

$$M_{ab} = -66.67\,\text{kNm}; \quad M_{ba} = -M_{bc} = 61.97\,\text{kNm}; \quad M_{cb} = 60\,\text{kNm}$$

(f) Support reactions: Free body diagrams of all spans with applicable end moments are shown in the following figures.

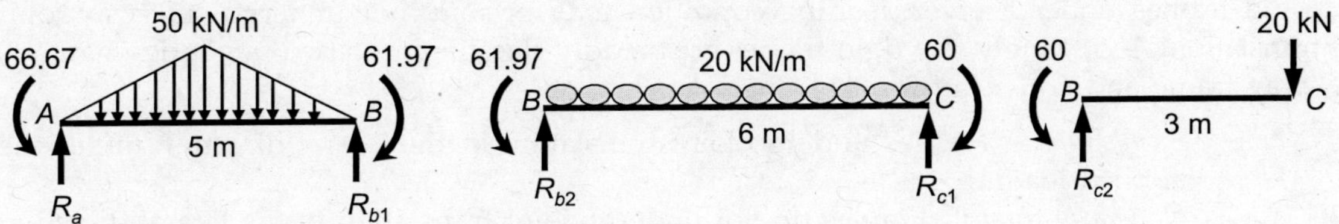

Span *AB*: Taking moment about *A*,

$$5R_{b1} + 66.67 - 61.97 - \frac{50 \times 5^2}{2 \times 2} = 0 \quad \therefore R_{b1} = 61.56 \text{ kN}$$

Force equilibrium,

$$R_{b1} + R_a = \frac{50 \times 5}{2} \text{ kN} = 125 \text{ kN} \quad \therefore R_a = 63.44 \text{ kN}$$

Span *BC*: Taking moment about B,

$$6R_{c1} - 60 + 61.97 - 20 \times 6 \times 3 = 0 \quad \therefore R_{c1} = 59.67 \text{ kN}$$

Force equilibrium,

$$R_{b2} + R_{c1} = 20 \times 6 \text{ kN} = 120 \text{ kN} \quad \therefore R_{b2} = 60.33 \text{ kN}$$

Span *CD*: Force equilibrium, $R_{c2} = 20$ kN
Final support reactions:

$$R_a = 63.44 \text{ kN}$$

$$R_b = R_{b1} + R_{b2} = 121.89 \text{ kN}$$

$$R_c = R_{c1} + R_{c2} = 79.67 \text{ kN}$$

Shear force diagram (SFD) and bending moment diagram (BMD) of the beam are shown in the following figure.

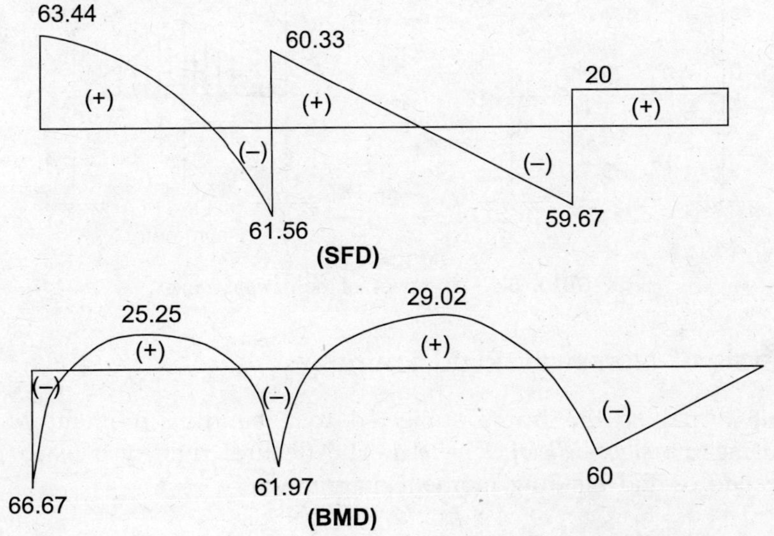

6.6 ANALYSIS OF RIGID FRAMES

Rigid frames under a given loading condition may or may not undergo side-way joint translation. Accordingly, the rigid frames are broadly classified into two categories, namely, sway frame and non-sway frame.

1. *Sway frame*: Frames undergo joint translation in the lateral direction under the applied loading
2. *Non-sway frame*: Frames do not undergo joint translation in the lateral direction under the applied loading

Figure 6.9 shows the examples of the non-sway frames. The lateral displacement (swat) of joints may be prevented by means of the supports. Sometimes, if the loading and boundary conditions are symmetric, the frames may not undergo the lateral joint displacements as shown in the figure.

Unless stated otherwise, all members of the rigid frames are considered as axially inextensible. Depending on the sway or non-sway frame, the step-by-step procedure of the slope-deflection method may vary in the analysis. The analysis of rigid frames has been illustrated by means of the following examples.

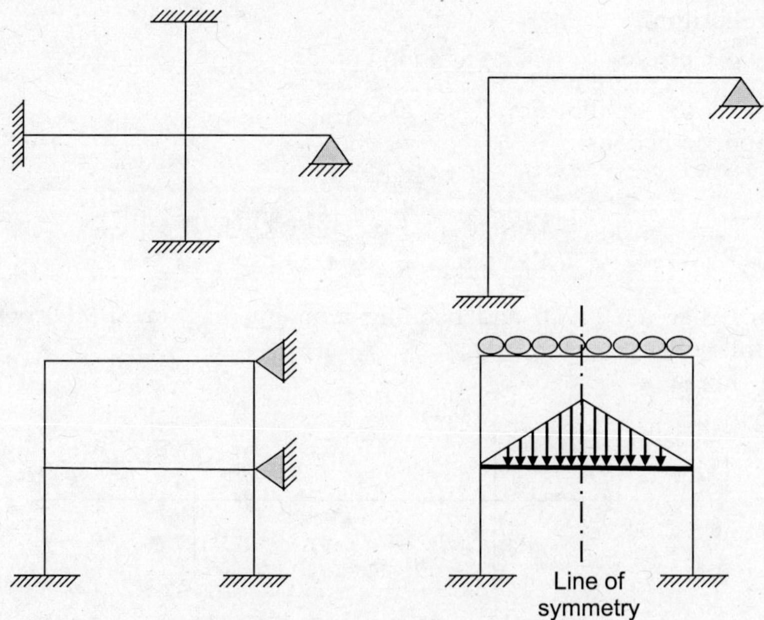

FIGURE 6.9 Examples of non-sway frames.

6.6.1 Analysis of Non-sway Rigid Frames

EXAMPLE 6.8 Analyze the frame subjected to a bending moment M as shown in Figure E6.8 using the *slope-deflection method*. The flexural rigidity of each member is EI. Draw the shear force and bending moment diagrams.

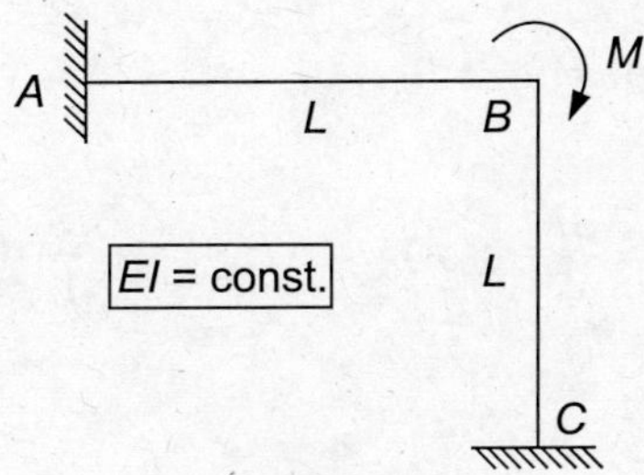

FIGURE E6.8

Solution: Since the frame is a non-sway frame, the joint translations in the lateral directions of both members are zero.

Both members are not subjected to any external loading. Thus, all fixed end moments are zero, it is worth mentioning that the external moment M is applied at the joint and hence, it should not be considered in the computation of the fixed end moments.

Fixed end moments:

$$M_{ab}^{f} = M_{ba}^{f} = M_{bc}^{f} = M_{cb}^{f} = 0$$

Compatibility conditions:

$$\theta_a = \theta_c = 0 \ \text{(Because of fixed supports)}$$

$$\Delta_{ab} = \Delta_{bc} = 0 \ \text{(Because of no joint translations)}$$

Slope-deflection Equations:

$$M_{ab} = \frac{2EI\theta_b}{L} ; \ M_{ba} = \frac{4EI\theta_b}{L} ; \ M_{bc} = \frac{4EI\theta_b}{L} ; \ M_{cb} = \frac{2EI\theta_b}{L}$$

Equilibrium Equations: Joint equilibrium at B, $M_{ba} + M_{bc} = M$
Using slope-deflection equations,

$$\frac{4EI\theta_b}{L} + \frac{4EI\theta_b}{L} = M \quad \therefore EI\theta_b = \frac{ML}{8}$$

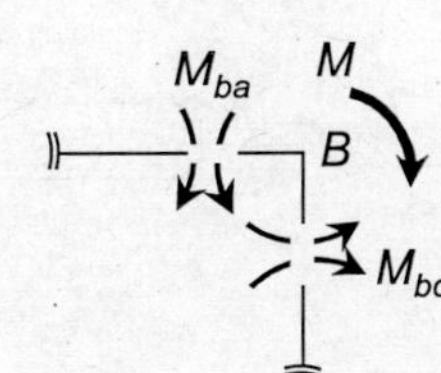

Final end moments:
Putting the value of $EI\theta_b$ in the slope-deflection equations,

$$M_{ab} = M_{cb} = \frac{M}{4} ; \quad M_{ba} = M_{bc} = \frac{M}{2}$$

Support reactions: The free-body diagrams of all members are shown in the following figures.

Span AB:

Taking moment about B, $V_a L + \dfrac{M}{4} + \dfrac{M}{2} = 0$

$$\therefore \quad V_a = -\frac{3M}{4L} \ (\downarrow) \quad \therefore V_c = \frac{3M}{4L} \ (\uparrow)$$

Span *BC*:   Taking moment about *B*, $H_c L - \dfrac{M}{4} - \dfrac{M}{2} = 0$

$$\therefore \qquad H_c = \frac{3M}{4L}\,(\rightarrow) \qquad \therefore\ H_a = -\frac{3M}{4L}\,(\leftarrow)$$

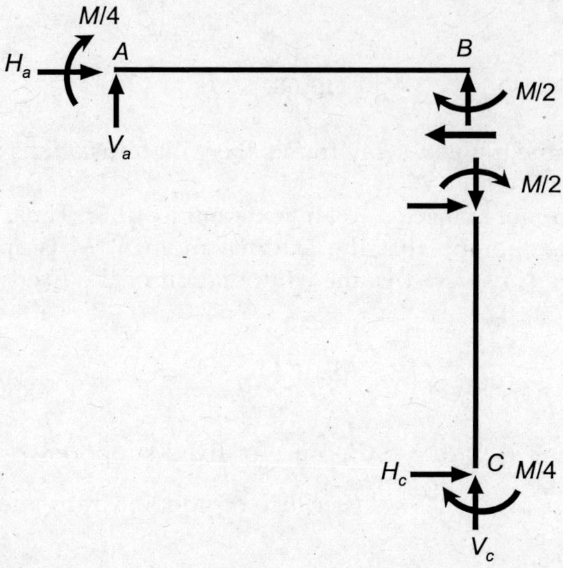

Shear force and bending moment diagrams of the frame are shown below:

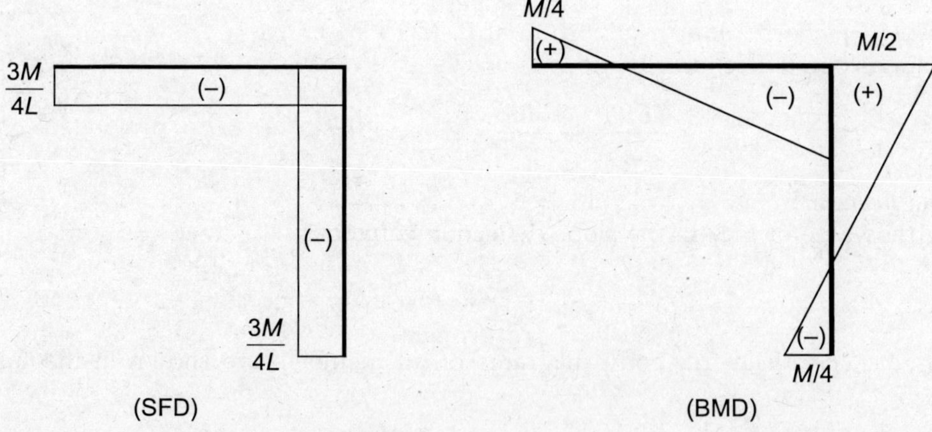

EXAMPLE 6.9   Analyze the frame shown in Figure E6.9 using the *slope-deflection method*. The flexural rigidity of each member is *EI*. Draw the shear force and bending moment diagrams.

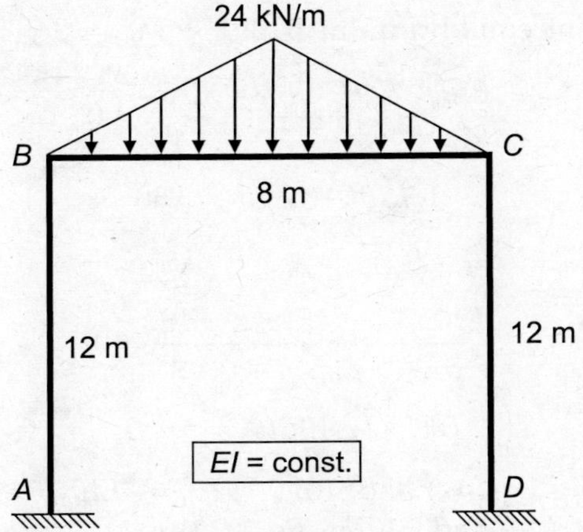

FIGURE E6.9

Solution: Since the loading and boundary conditions are symmetric about the vertical axis, the given frame is a non-sway frame, i.e., the joint translation in the lateral direction is zero.

Fixed end moments:

$$M_{ab}^f = M_{ba}^f = M_{cd}^f = M_{dc}^f = 0$$

$$M_{bc}^f = -M_{cb}^f = -\frac{5 \times 24 \times 8^2}{96}\,\text{kNm} = -80\ \text{kNm}$$

Compatibility conditions:

$$\theta_a = \theta_d = 0 \ \text{(Because of fixed supports)}$$

$$\Delta_{ab} = \Delta_{bc} = \Delta_{cd} = 0 \ \text{(Because of no joint translations)}$$

Slope-deflection Equations:

$$M_{ab} = \frac{2EI}{12}(\theta_b) = \frac{EI\theta_b}{6} \tag{1}$$

$$M_{ba} = \frac{2EI}{12}(2\theta_b) = \frac{EI\theta_b}{3} \tag{2}$$

$$M_{bc} = -80 + \frac{2EI}{8}(2\theta_b + \theta_c) = -80 + \frac{EI\theta_b}{2} + \frac{EI\theta_c}{4} \tag{3}$$

$$M_{cb} = 80 + \frac{2EI}{8}(\theta_b + 2\theta_c) = 80 + \frac{EI\theta_b}{4} + \frac{EI\theta_c}{2} \tag{4}$$

$$M_{cd} = \frac{2EI}{12}(2\theta_c) = \frac{EI\theta_c}{3} \tag{5}$$

$$M_{dc} = \frac{2EI}{12}(\theta_c) = \frac{EI\theta_c}{6} \tag{6}$$

Equilibrium Equations: Joint equilibrium at B, $M_{ba} + M_{bc} = 0$

$$\frac{EI\theta_b}{3} - 80 + \frac{EI\theta_b}{2} + \frac{EI\theta_c}{4} = 0$$

$$10EI\theta_b + 3EI\theta_c = 960 \tag{7}$$

Joint equilibrium at C, $M_{cb} + M_{cd} = 0$

$$\frac{EI\theta_c}{3} + 80 + \frac{EI\theta_b}{4} + \frac{EI\theta_c}{2} = 0$$

$$3EI\theta_b + 10EI\theta_c = -960 \tag{8}$$

Final end moments: Solving Eqs. (7) and (8), $\therefore$ $EI\theta_b = -EI\theta_c = -137.14$
Putting the values of $EI\theta_b$ and $EI\theta_c$ in Eqs. (1)–(6),

$$M_{ab} = -22.86 \text{ kNm}; \ M_{ba} = -M_{bc} = 45.71 \text{ kNm}$$

$$M_{dc} = 22.86 \text{ kNm}; \ M_{cd} = -M_{cb} = -45.71 \text{ kNm}$$

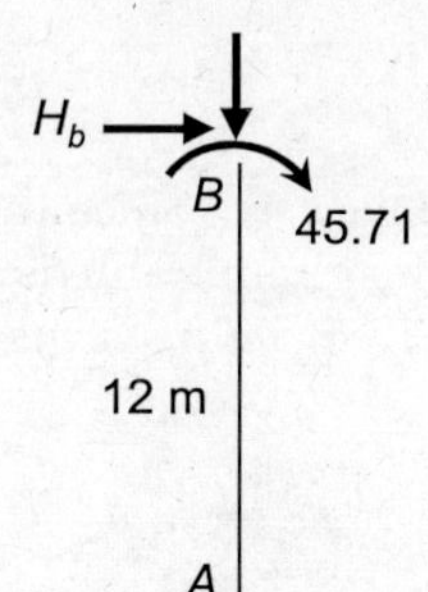

Support reactions: The free-body diagrams of all members are shown
in the following figures.
  Span *AB*:   Taking moment about B,

$$12 \times H_a + 22.86 - 45.71 = 0 \quad \therefore \quad H_a = 1.90 \text{ kN}(\rightarrow)$$

Force equilibrium, $\qquad\qquad$ $H_a + H_b = 0$

$\therefore$ $\qquad\qquad\qquad$ $H_b = -1.90 \text{ kN} (\leftarrow)$

Span *BC*:   Taking moment about C,

$$12 \times H_d - 22.86 + 45.71 = 0 \quad \therefore \quad H_d = -1.90 \text{ kN}(\leftarrow)$$

Force equilibrium,

$$H_c + H_d = 0 \quad \therefore \quad H_c = 1.90 \text{ kN}(\rightarrow)$$

Span *CD*:   Taking moment about C,

$$8 \times V_c - 45.71 + 45.71 - \frac{1}{2}(24)(8)(4) = 0$$

$\therefore$ $\qquad\qquad\qquad$ $V_c = 48 \text{ kN}(\uparrow)$

Force equilibrium,

$$V_b + V_c = \frac{1}{2}(24)(8) \text{ kN} = 96 \text{ kN}$$

$\therefore$ $\quad$ $V_b = 48 \text{ kN}(\uparrow)$

Shear force and bending moment diagrams
of the frame are shown below:

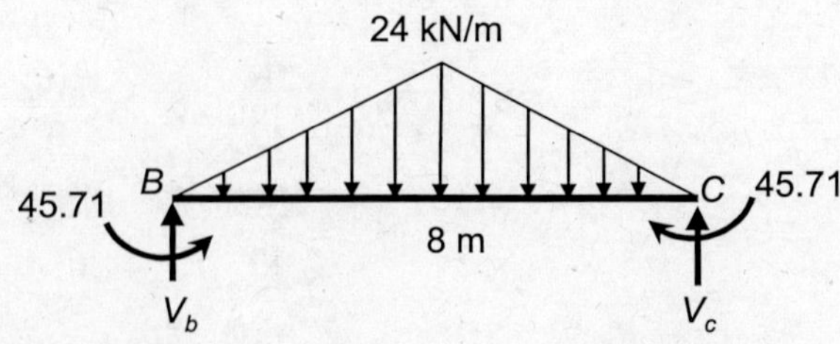

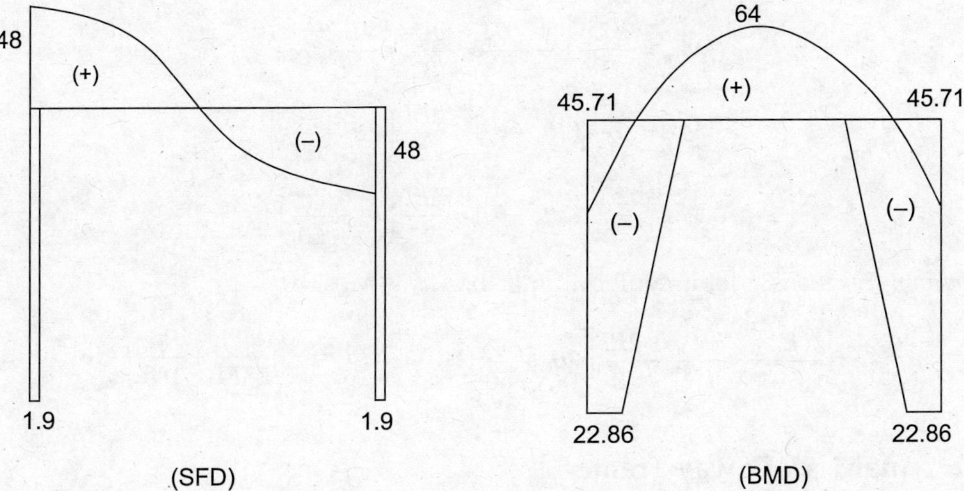

EXAMPLE 6.10   Determine the rotation of joint B of the frame shown in Figure E6.10 using the *slope-deflection method,* if the support C rotates by $\pi/60$ radian (clockwise). The flexural rigidity of each member is *EI*. Draw the bending moment diagram.

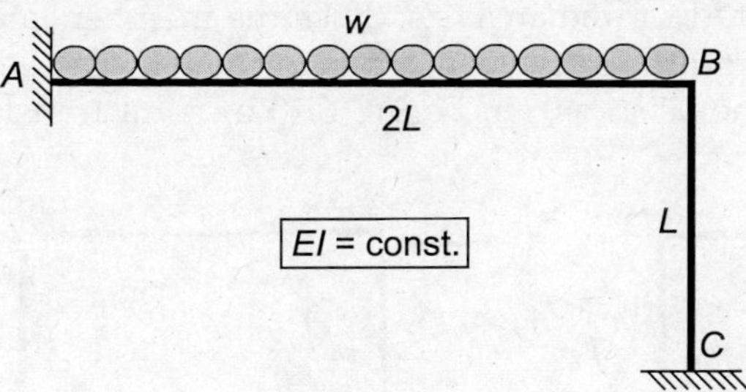

FIGURE E6.10

Solution: *Fixed end moments:*

$$M_{ab}^f = -\,M_{ba}^f = -\frac{w(2L)^2}{12} = -\frac{wL^2}{3}$$

$$M_{bc}^f = M_{cb}^f = 0$$

Compatibility conditions:

$$\theta_a = 0; \quad \theta_c = \frac{\pi}{60}$$

$$\Delta_{ab} = \Delta_{bc} = 0 \ \text{(Because of no joint translations)}$$

Slope-deflection Equations:

$$M_{ab} = -\frac{wL^2}{3} + \frac{2EI\theta_b}{2L} = -\frac{wL^2}{3} + \frac{EI\theta_b}{L} \tag{1}$$

$$M_{ba} = \frac{wL^2}{3} + \frac{4EI\theta_b}{2L} = \frac{wL^2}{3} + \frac{2EI\theta_b}{L} \qquad (2)$$

$$M_{bc} = \frac{2EI}{L}(2\theta_b + \theta_c) = \frac{2EI}{L}\left(2\theta_b + \frac{\pi}{60}\right) \qquad (3)$$

$$M_{cb} = \frac{2EI}{L}(2\theta_b + \theta_c) = \frac{2EI}{L}\left(\theta_b + \frac{2\pi}{60}\right) \qquad (4)$$

Equilibrium Equations: Joint equilibrium at B, $M_{ba} + M_{bc} = 0$

$$\frac{wL^2}{3} + \frac{2EI\theta_b}{L} + \frac{2EI}{L}\left(2\theta_b + \frac{\pi}{60}\right) = 0 \quad \therefore \; \theta_b = -\left(\frac{wL^3}{18EI} + \frac{\pi}{180}\right)$$

6.6.2   Analysis of Sway Frames

Sway frames are those in which the joint translations in the lateral direction are not prevented under the applied loading. For instance, Figure 6.10 shows some examples of the sway frames. Under the application of a load, the joint a is translated to a point a' such that the joint displacement aa' is Δ. The joint displacement normal to the longitudinal axes of columns is only considered for the analysis. Since the members are axially inextensible, the member ab would undergo the same translation at the joint b, i.e., $aa' = bb' = \Delta$. In the case of a two-bay portal frame as shown in Figure 6.10(b), joint translations $aa' = bb' = cc' = \Delta$.

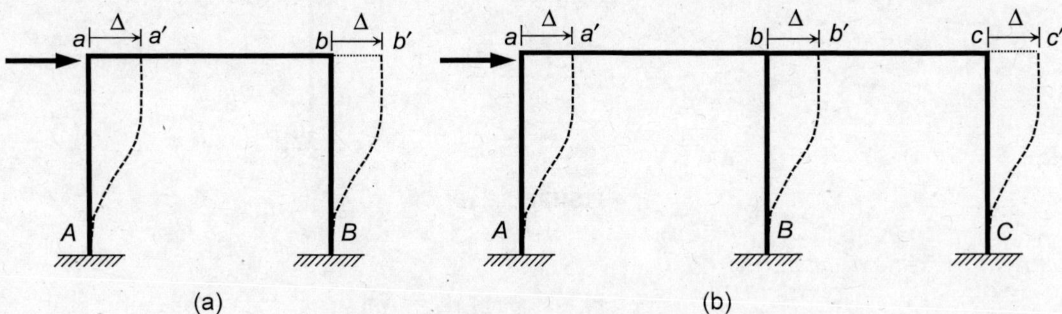

(a)   (b)

FIGURE 6.10   Examples of sway frames.

Now consider a portal frame shown in Figure 6.11 in which one member is inclined at an angle θ to the horizontal. The qualitative deflected shape of the frame is shown by the dashed lines. Under the applied load, the joints a and b would undergo translation as shown in the figure. The displacement of joint a relative to the end A is assumed as $aa' = \Delta_1$ which would be perpendicular to the member Aa. Similarly, the joint b would translate to a point b' in a direction perpendicular to the member Bb such that the relative displacement $bb' = \Delta_3$. The relative perpendicular displacement between ends of a' and b' of the member ab is $b'b'' = \Delta_2$. All these joint displacements are interrelated and can be computed using the sine law as follows:

$$\frac{\Delta_1}{\sin\theta} = \frac{\Delta_2}{\sin(90^\circ - \theta)} = \frac{\Delta_3}{\sin 90^\circ} \qquad (6.24)$$

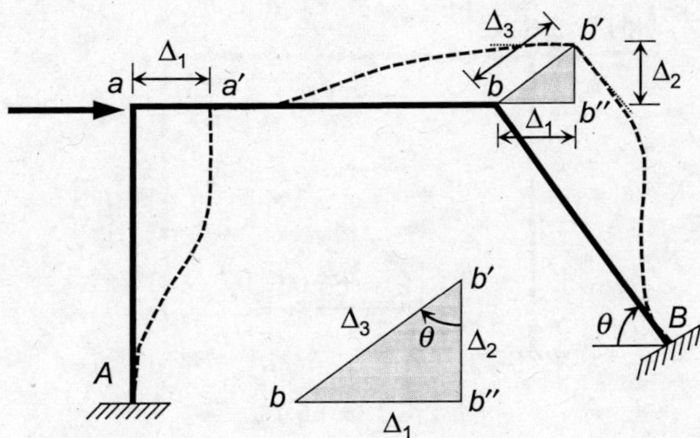

FIGURE 6.11 Gable frame subjected to lateral load.

Now consider a single-bay gable frame subjected to a lateral loading as shown in Figure 6.12. The relative joint displacement of member Aa is $aa' = \Delta_1$. This displacement is perpendicular to the member Aa. Similarly, the relative joint displacement of member Bb is $bb' = \Delta_3$. The relative joint displacement, $b'b'' = \Delta_2$ of member ab can be obtained by joining the points b' and b'' such that $bb' = \Delta_1$. All these displacements can be determined as follows:

$$\frac{\Delta_1}{\sin\theta_2} = \frac{\Delta_2}{\sin(\theta_1 + \theta_2)} = \frac{\Delta_3}{\sin\theta_1} \tag{6.25}$$

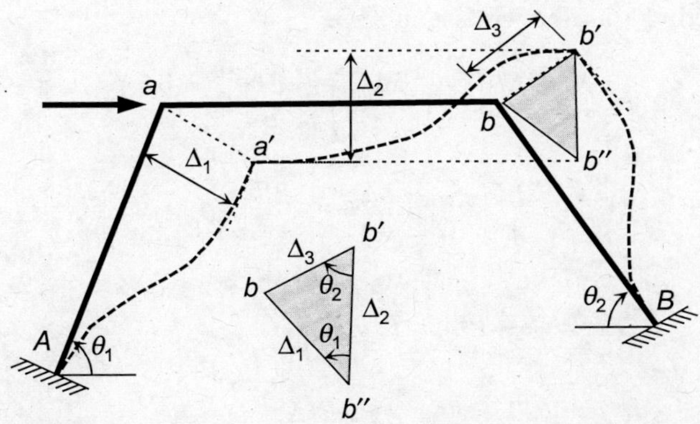

FIGURE 6.12 Deflected shape of gable frame under lateral loading.

This procedure has been illustrated by means of the following examples.

EXAMPLE 6.11 Analyze the frame shown in Figure E6.11 using the *slope-deflection method*. The flexural rigidity of each member is *EI*. Draw the bending moment diagram.

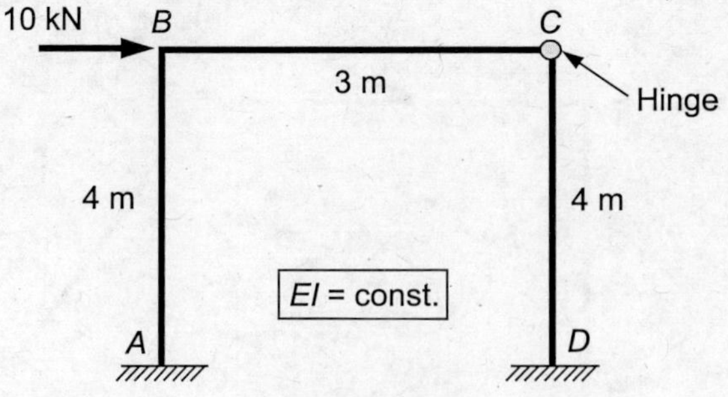

FIGURE E6.11

Solution: The qualitative deflected shape of the frame is shown below. The joint displacements at *B* and *C* would be the same, i.e., $Bb = Cc = D$.

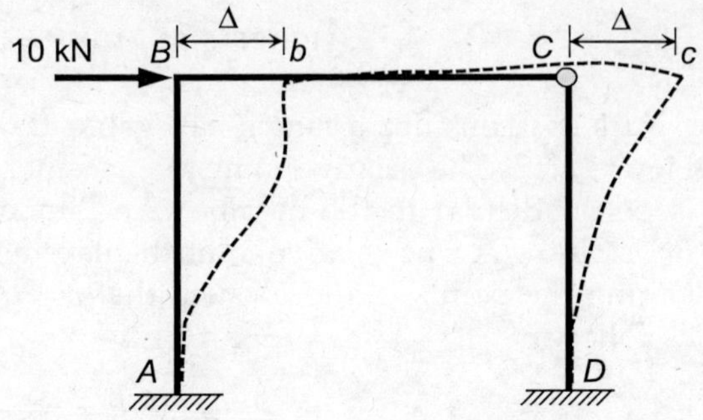

Fixed end moments:

$$M^f_{ab} = M^f_{ba} = M^f_{bc} = M^f_{cb} = M^f_{cd} = M^f_{dc} = 0$$

Compatibility conditions:

$$\theta_a = \theta_d = 0$$

Slope-deflection Equations: Since joint C is a hinge, the modified slope-deflections can be used for the connecting members BC and CD.

$$M_{ab} = \frac{2EI}{4}\left(\theta_b - \frac{3\Delta_{ab}}{4}\right) = \frac{EI}{2}\left(\theta_b - \frac{3\Delta}{4}\right) \tag{1}$$

$$M_{ba} = \frac{2EI}{4}\left(2\theta_b - \frac{3\Delta_{ab}}{4}\right) = \frac{EI}{2}\left(2\theta_b - \frac{3\Delta}{4}\right) \tag{2}$$

$$M_{bc} = \frac{3EI}{3}\left(\theta_b\right) = EI\theta_b \tag{3}$$

$$M_{dc} = \frac{3EI}{4}\left(-\frac{\Delta}{4}\right) = -\frac{3EI\Delta}{16} \tag{4}$$

Equilibrium Equations: Joint equilibrium at B, $M_{ba} + M_{bc} = 0$

$$\frac{EI}{2}\left(2\theta_b - \frac{3\Delta}{4}\right) + EI\theta_b = 0 \qquad \therefore \ \theta_b = \frac{3\Delta}{16} \tag{5}$$

Another equilibrium can be obtained by considering the force equilibrium of the frame. Let's consider the free-body diagram of all members as below:

Taking moment about B,

$$4 \times H_a - M_{ab} - M_{ba} = 0 \quad \therefore H_a = \frac{M_{ab} + M_{ba}}{4}$$

Taking moment about C,

$$4 \times H_d - M_{dc} = 0$$

$$\therefore \qquad H_d = \frac{M_{dc}}{4}$$

Shear equilibrium of the frame,

$$H_a + H_d + 10 = 0$$

$$\frac{M_{ab} + M_{ba}}{4} + \frac{M_{dc}}{4} + 10 = 0$$

$$\frac{EI}{2}\left(\theta_b - \frac{3\Delta}{4}\right) + \frac{EI}{2}\left(2\theta_b - \frac{3\Delta}{4}\right) - \frac{3EI\Delta}{16} + 40 = 0$$

$$24EI\theta_b - 15EI\Delta + 640 = 0 \qquad\qquad (6)$$

Solving Eq. (5) and (6), $EI\theta_b = 11.43$; $EI\Delta = 60.95$

Final bending moments:

$$M_{ab} = -17.14 \text{ kNm}$$

$$M_{ba} = -M_{bc} = -11.43 \text{ kNm}$$

$$M_{dc} = -11.43 \text{ kNm}$$

$$M_{cd} = M_{cb} = 0$$

The bending moment diagram of the frame is shown below:

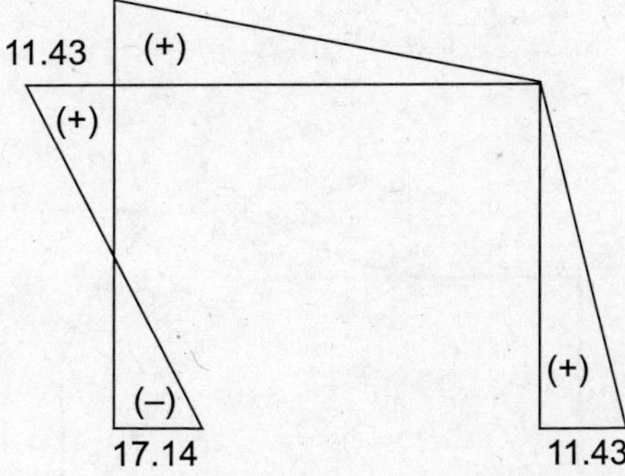

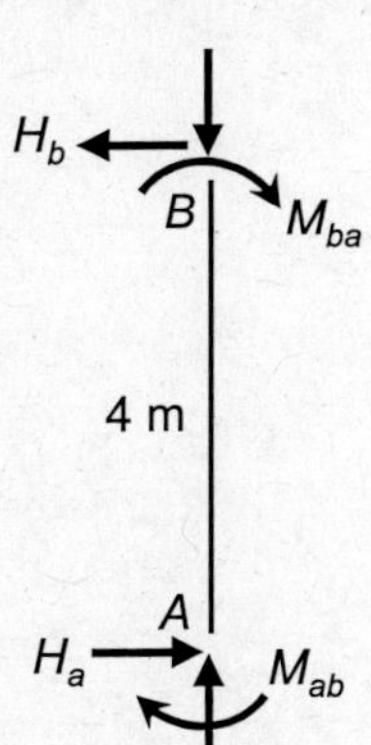

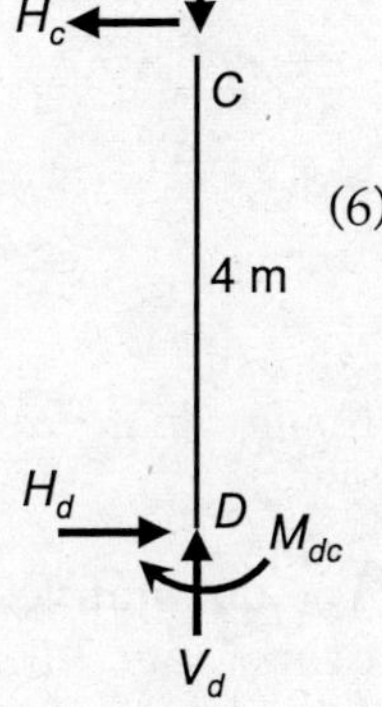

EXAMPLE 6.12   Analyze the frame shown in Figure E6.12 using the *slope-deflection method*. Draw the bending moment diagram.

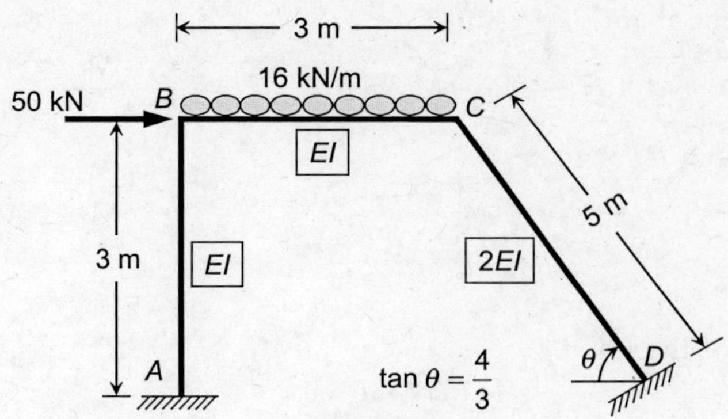

FIGURE E6.12

Solution: *Fixed end moments:*

$$M_{ab}^{f} = M_{ba}^{f} = M_{cd}^{f} = M_{dc}^{f} = 0$$

$$M_{bc}^{f} = - M_{cb}^{f} = - \frac{16 \times 3^2}{12}\,\text{kNm} = -12\ \text{kNm}$$

Compatibility conditions:

$$\theta_a = \theta_d = 0$$

The qualitative deflected shape of the frame is shown below. Let the relative displacement of member *AB*, *BC*, and *CD* be Δ_1, Δ_2 and Δ_3, respectively. The relationship between these displacements can be obtained using the joint-displacement triangle.

$$\frac{\Delta_1}{\sin\theta} = \frac{\Delta_2}{\sin(90-\theta)} = \frac{\Delta_3}{\sin 90}$$

Let $\Delta_1 = \Delta$; then, $\Delta_2 = \Delta \cot\theta = \dfrac{3}{4}\Delta;\ \ \Delta_3 = \dfrac{\Delta}{\sin\theta} = \dfrac{5}{4}\Delta$

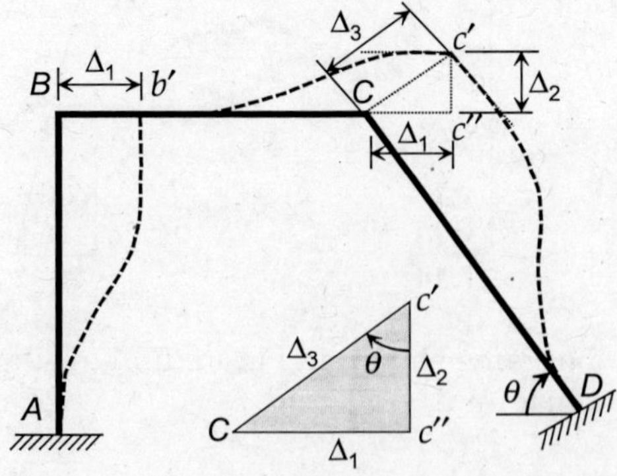

Slope-deflection Equations:

$$M_{ab} = \frac{2EI}{3}\left(\theta_b - \frac{3\Delta_1}{3}\right) = \frac{2EI}{3}(\theta_b - \Delta) \tag{1}$$

$$M_{ba} = \frac{2EI}{3}\left(2\theta_b - \frac{3\Delta_1}{3}\right) = \frac{2EI}{3}(2\theta_b - \Delta) \tag{2}$$

$$M_{bc} = -12 + \frac{2EI}{3}\left(2\theta_b + \theta_c - \frac{3(-\Delta_2)}{3}\right) = -12 + \frac{2EI}{3}\left(2\theta_b + \theta_c + \frac{3\Delta}{4}\right) \tag{3}$$

$$M_{cb} = 12 + \frac{2EI}{3}\left(\theta_b + 2\theta_c - \frac{3(-\Delta_2)}{3}\right) = 12 + \frac{2EI}{3}\left(\theta_b + 2\theta_c + \frac{3\Delta}{4}\right) \tag{4}$$

$$M_{cd} = \frac{2(2EI)}{5}\left(2\theta_c - \frac{3\Delta_3}{5}\right) = \frac{4EI}{5}\left(2\theta_c - \frac{3\Delta}{4}\right) \tag{5}$$

$$M_{dc} = \frac{2(2EI)}{5}\left(\theta_c - \frac{3\Delta_3}{5}\right) = \frac{4EI}{5}\left(\theta_c - \frac{3\Delta}{4}\right) \tag{6}$$

Equilibrium Equations:

Joint equilibrium at B, $M_{ba} + M_{bc} = 0$

$$\frac{2EI}{3}(2\theta_b - \Delta) - 12 + \frac{2EI}{3}\left(2\theta_b + \theta_c + \frac{3\Delta}{4}\right) = 0$$

$\therefore \qquad\qquad 16\,EI\theta_b + 4EI\theta_c - EI\Delta = 72 \tag{7}$

Joint equilibrium at C, $M_{cb} + M_{cd} = 0$

$$12 + \frac{2EI}{3}\left(\theta_b + 2\theta_c + \frac{3\Delta}{4}\right) + \frac{4EI}{5}\left(2\theta_c - \frac{3\Delta}{4}\right) = 0$$

$\therefore \qquad\qquad 20EI\theta_b + 88EI\theta_c - 3EI\Delta + 360 = 0 \tag{8}$

Since there are three unknowns to be solved, the third equation can be obtained by taking the overall equilibrium of the frame.

Taking moment about O,

$$4 \times 50 + (7)H_a + (10)H_d - M_{ab} - M_{dc} - \frac{16 \times 3^2}{2} = 0$$

$$\left(\frac{M_{ab} + M_{ba}}{3}\right) + (10)\left(\frac{M_{dc} + M_{cd}}{5}\right) - M_{ab} - M_{dc} - 128 = 0$$

Using Eqs. (1), (2), (5) and (6),

$$EI\theta_b + EI\theta_c - 1.06EI\Delta + 45.5 = 0 \tag{9}$$

Solving Eqs. (7), (8) and (9),

$$EI\theta_b = 8.52; \quad EI\theta_c = -4.37; \quad EI\Delta = 46.84$$

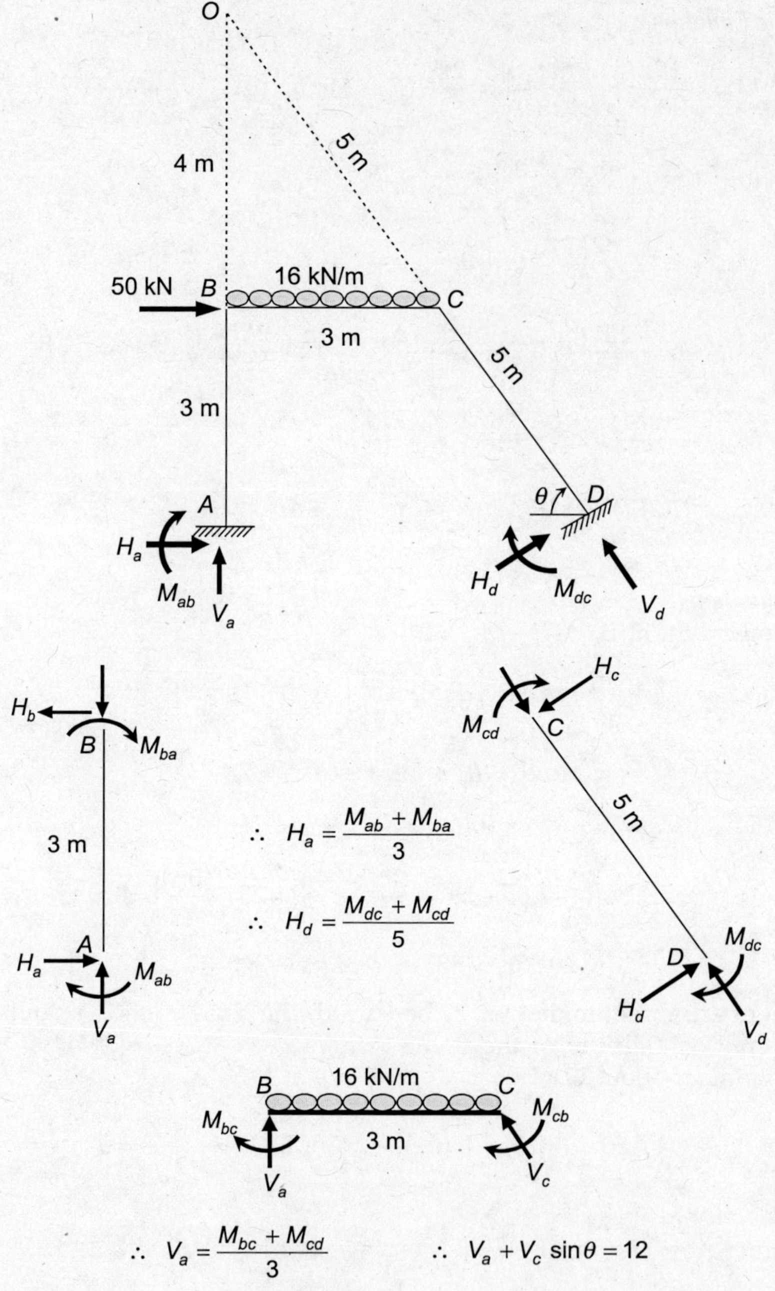

$$\therefore \ H_a = \frac{M_{ab} + M_{ba}}{3}$$

$$\therefore \ H_d = \frac{M_{dc} + M_{cd}}{5}$$

$$\therefore \ V_a = \frac{M_{bc} + M_{cd}}{3} \qquad \therefore \ V_a + V_c \sin\theta = 12$$

Final bending moments:

$$M_{ab} = -25.55 \ \text{kNm}$$

$$M_{ba} = -M_{bc} = -19.87 \ \text{kNm}$$

$$M_{cd} = -M_{cb} = -35.27 \ \text{kNm}$$

$$M_{dc} = -31.6 \ \text{kNm}$$

The bending moment diagram of the frame is shown in the figure.

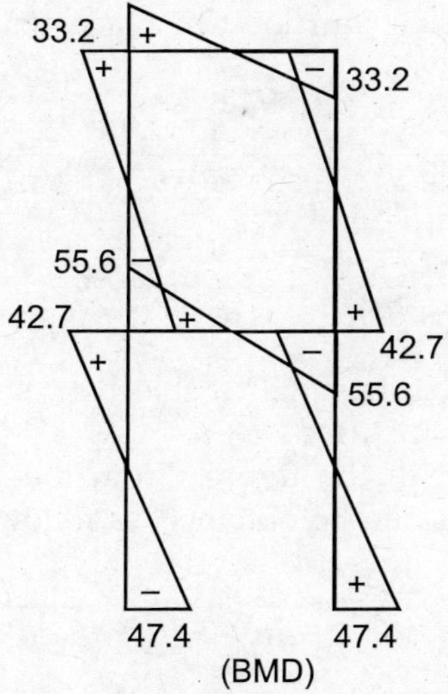

6.7   ANALYSIS OF RIGID FRAMES WITH TWO JOINT TRANSLATIONS

The analysis of indeterminate rigid frames with more than one unknown joint translation can be solved using the same procedure as discussed above. The joint-displacement triangles are required to be developed to develop a relationship between the joint displacements. Some examples of rigid frames with two degrees of freedom of joint translation are discussed in the following sections.

Figure 6.13 shows a single-bay gable bent frame subjected to a vertical load. The deflected shape of the frame is shown in the figure. Under the load, the joint B is translated to b' by a distance of Δ_1. Similarly, the relative displacement of member DE is Δ_2. These displacements are drawn at the joint C by means of the parallel lines BC and CD of the

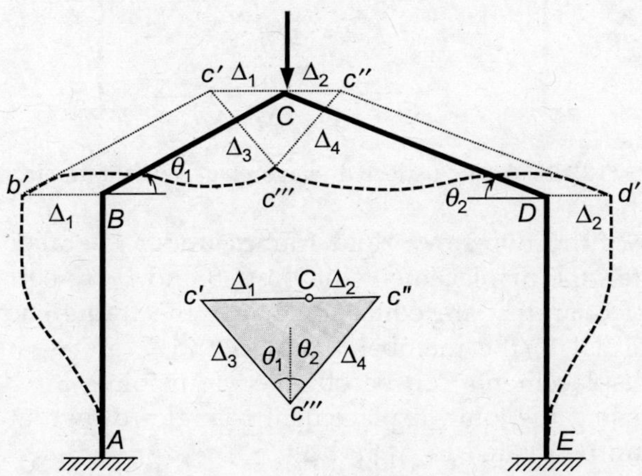

FIGURE 6.13 Deflected shape of gable bent frame subjected to a vertical load.

respective lengths. The relative displacement of joints B and C is Δ_3, and that of C and D is Δ_4. The joint-displacement triangle is drawn at C in which $Cc' = \Delta_1$, $Cc'' = \Delta_2$, $c'c''' = \Delta_3$ and $c''c''' = \Delta_4$. Using sine law, all four joint displacements can be expressed in terms of any two displacements as follows:

$$\frac{(\Delta_1 + \Delta_2)}{\sin(\theta_1 + \theta_2)} = \frac{\Delta_3}{\sin(90 - \theta_2)} = \frac{\Delta_4}{\sin(90 - \theta_1)} \tag{6.26}$$

or
$$\frac{(\Delta_1 + \Delta_2)}{\sin(\theta_1 + \theta_2)} = \frac{\Delta_3}{\cos\theta_2} = \frac{\Delta_4}{\cos\theta_1} \tag{6.27}$$

Figure 6.14 shows a rigid frame with gable bent subjected to a lateral load. From the deflected shape, $Bb' = Cc' = \Delta_1$, $Dd' = Cc'' = \Delta_2$, $c'c''' = \Delta_3$ and $c''c''' = \Delta_4$. The joint displacement triangle is shown in the figure. From the triangle $c'c''c'''$, the relationship between the joint displacements can be obtained as follows:

$$\frac{(\Delta_1 - \Delta_2)}{\sin(\theta_1 + \theta_2)} = \frac{\Delta_3}{\sin(90 - \theta_2)} = \frac{\Delta_4}{\sin(90 - \theta_1)} \tag{6.28}$$

or,
$$\frac{(\Delta_1 + \Delta_2)}{\sin(\theta_1 + \theta_2)} = \frac{\Delta_3}{\cos\theta_2} = \frac{\Delta_4}{\cos\theta_1} \tag{6.29}$$

Thus, four joint displacements can be expressed in terms of any two joint displacements.

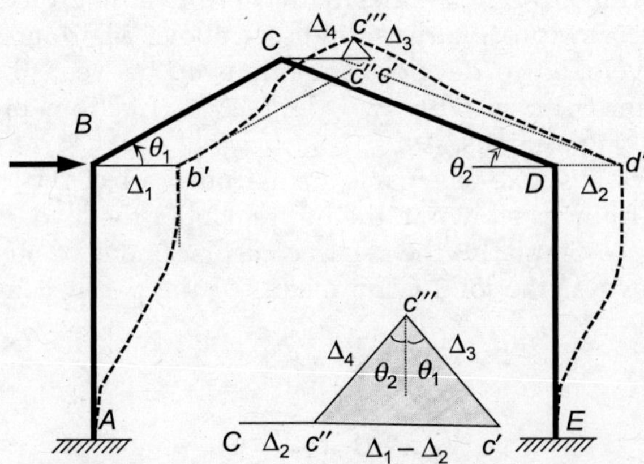

FIGURE 6.14   Deflected shape of gable bent frame.

Figure 6.15 shows the two-story rigid frames under lateral loading. For the frame shown in Figure 6.15(a), the displacements of joints B and E are equal to Δ_1. Similarly, the joint displacements of C and D are equal to Δ_2. For the frame shown in Figure 6.15(b), the displacements normal to the member AC at B and C are Δ_1 and Δ_2, respectively. A total of six relative displacements can be obtained four joints B, C, D, and E for the six members. However, using the joint-displacement triangles drawn at joints D and E, these joint displacements can be related to Δ_1 and Δ_2.

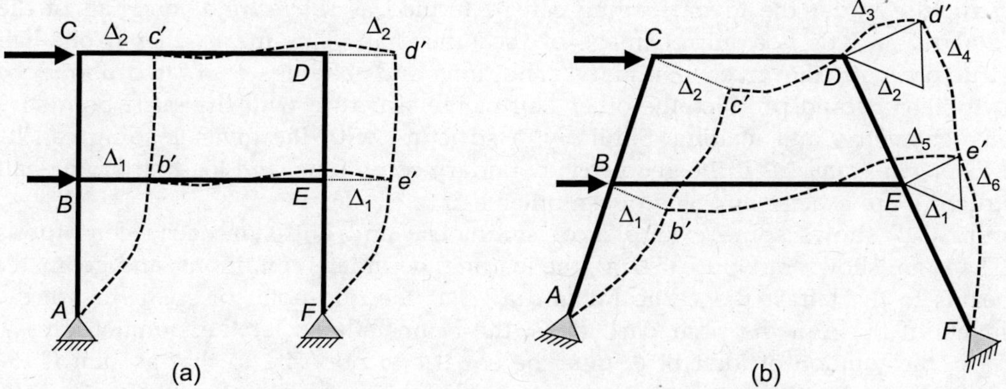

FIGURE 6.15　Two-story rigid frames under lateral loading.

The procedure can now be extended to a two-bay frame shown in Figure 6.16. The joint-displacement diagram shows that all the displacements can be expressed using the sine law for two triangles having one common base. This procedure can be extended to any number of bays. These joint displacements should be used in the slope-deflections equations for the respective members.

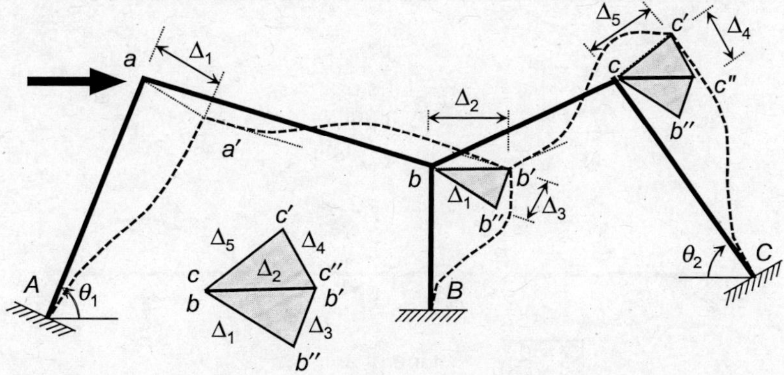

FIGURE 6.16　Deflectedshapeof two-bay rigid frame under lateral load.

6.8　ANALYSIS OF RIGID FRAMES WITH SYMMETRY AND ANTI-SYMMETRY

The analysis of a beam or rigid frame by using the slope-deflection method is simple and can be carried out by the hand calculation when there are few numbers of unknown deformations (i.e., degree of kinematic indeterminacy). As the number of unknown deformations is increased, the number of simultaneous equations to be solved is also increasing. Sometimes, the geometry, boundary conditions, and loading of structures are such that their behaviour may be classified into symmetric or anti-symmetric cases. In such cases, the number of simultaneous equations is reduced to a great extent since some of the deformations at the joints are known beforehand. Therefore, it is necessary to understand the behaviour of symmetric and anti-symmetric structures.

A structure under the given loading can be termed as *symmetric* if one-half of the structure about an axis is a mirror-image of the other half. This means that if one-half of the structure with the given boundary conditions and loadings is rotated about the line of symmetry would produce the other half of the structure with the same boundary conditions, geometry, and loading. Similarly, a structure with the given loading can be referred to as *anti-symmetric* if the geometry, boundary conditions, and loading of one-half of the structure are exactly opposite of the other half.

Figure 6.17 shows some examples of symmetric and anti-symmetric continuous beams. The beam shown in Figure 6.17(a), the loading, boundary conditions, and geometry of the beams to the left of the vertical axis drawn at the mid-point of span BC are the mirror-image of those in the right part. Thus, the beam falls under the symmetric case. In this case, the rotation of joint B, θ_b must be equal and opposite to that at joint C, θ_c, i.e., $\theta_b = -\theta_c$. Similarly, the rotations at joints A and D are equal and opposite, i.e., $\theta_a = -\theta_d$.

The beam shown in Figure 6.17(b) is an example of an anti-symmetric case since the boundary conditions and geometry of both halves of the beam are exactly with equal and opposite loading acting on these spans. In such case, the rotations of joint B and C would be equal and same directions. Thus, $\theta_b = \theta_c$ and $\theta_a = \theta_d$.

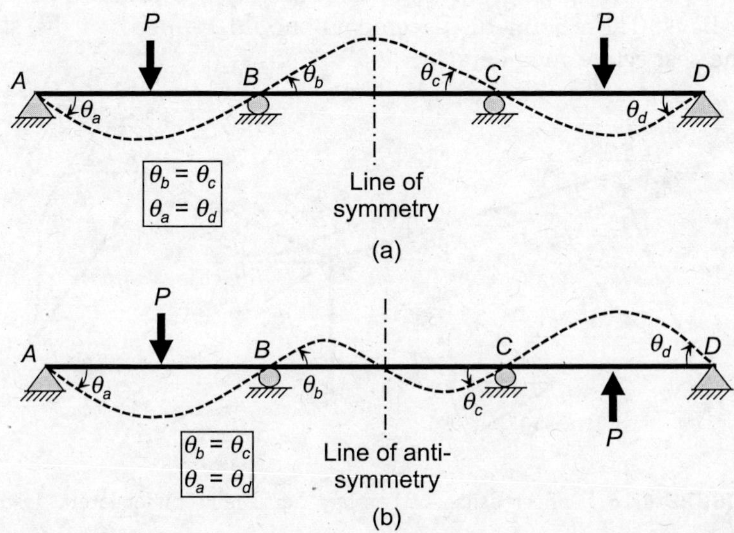

FIGURE 6.17   Deflected shape of beam with (a) symmetric and (b) anti-symmetric loadings.

Figure 6.18 shows the examples of the symmetric and anti-symmetric frames in which lateral sway is prevented. For the symmetric case, $\theta_b = -\theta_c$. and for the antisymmetric case, $\theta_b = \theta_c$. The same symmetric and anti-symmetric concepts can be applied to the rigid frames with lateral sway as well depending on the geometry, boundary, and loading conditions.

Since the relationships between the deformations at the joints are known beforehand, the same should be used in the respective slope-deflection equations. For example, the generalized slope-deflection equations for member BC can be written as follows:

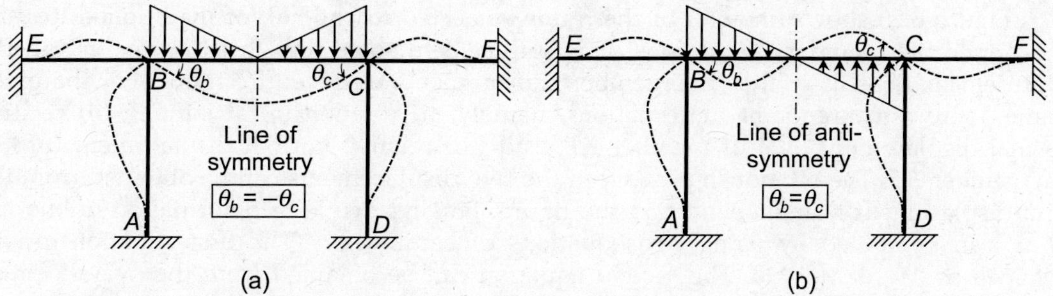

FIGURE 6.18 Deflected shape of frame with (a) symmetric and (b) anti-symmetric loadings.

For symmetric loading,

$$M_{bc} = \pm M_{bc}^{f} + \frac{2EI}{l}\left(2\theta_b + (-\theta_b) - \frac{3\Delta}{l}\right) \tag{6.30}$$

or

$$M_{bc} = \pm M_{bc}^{f} + \frac{2EI}{l}\left(\theta_b - \frac{3\Delta}{l}\right) \tag{6.31}$$

For anti-symmetric loading,

$$M_{bc} = \pm M_{bc}^{f} + \frac{2EI}{l}\left(2\theta_b + (+\theta_b) - \frac{3\Delta}{l}\right) \tag{6.32}$$

or

$$M_{bc} = \pm M_{bc}^{f} + \frac{6EI}{l}\left(\theta_b - \frac{\Delta}{l}\right) \tag{6.33}$$

For beams or rigid frames with symmetric or anti-symmetric loadings, only one-half of the structure can be solved using slope-deflection equations. Figure 6.19 shows a rigid frame with symmetrical loading and boundary conditions and the deflected shape of the frame. The joint displacement at B is equal to that of joint D and assumed to be Δ. Accordingly, $Cc' = c'c'' = \Delta$. The relative joint displacements of members BC and CD can be obtained by drawing the perpendicular lines at point c' and c''. This results in the vertical displacement of joint C without rotation. This, the support at C can be considered as a guided roller.

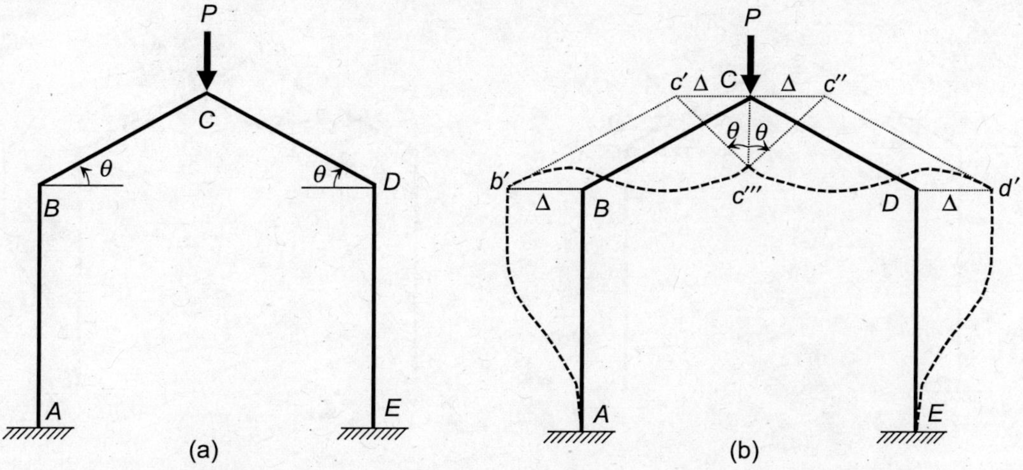

FIGURE 6.19 Symmetric frame along with deflected shape.

Figure 6.20 shows one-half of the frame subjected to one-half of the applied loading and applicable boundary conditions. It is sufficient to solve this structure to find all the joint deformations as well as the member end moments of the entire frame. For the given frame, there are three joint deformations, namely, (i) rotation (q) at joint B, (ii) relative normal displacement (Δ_{ab}) of member AB, and (iii) relative normal displacement (Δ_{bc}) of the member BC. The relationship between the two displacements can be obtained from the joint-displacement triangle shown in the figure. Finally, two joint deformation unknowns (θ, Δ) can be solved by using two equations of equilibrium. The first equation can be obtained as $M_{ba} + M_{bc} = 0$. The second equation can be obtained from the overall force/moment equilibrium of the frame.

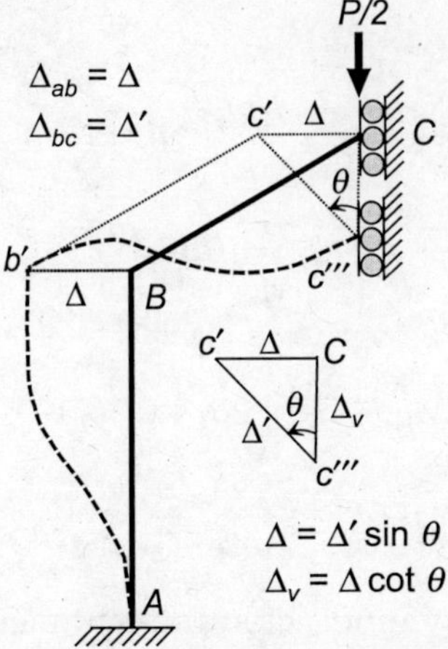

FIGURE 6.20   One-half of the symmetric frame.

Figure 6.21 shows an example of a rigid frame with anti-symmetric loading. As shown in the deflected shape, the vertical displacement at joint C would be zero. However, the joint C would translate in the horizontal direction along with rotation. Thus, the joint C can be considered as roller support. This frame can be solved by considering only half of the frame.

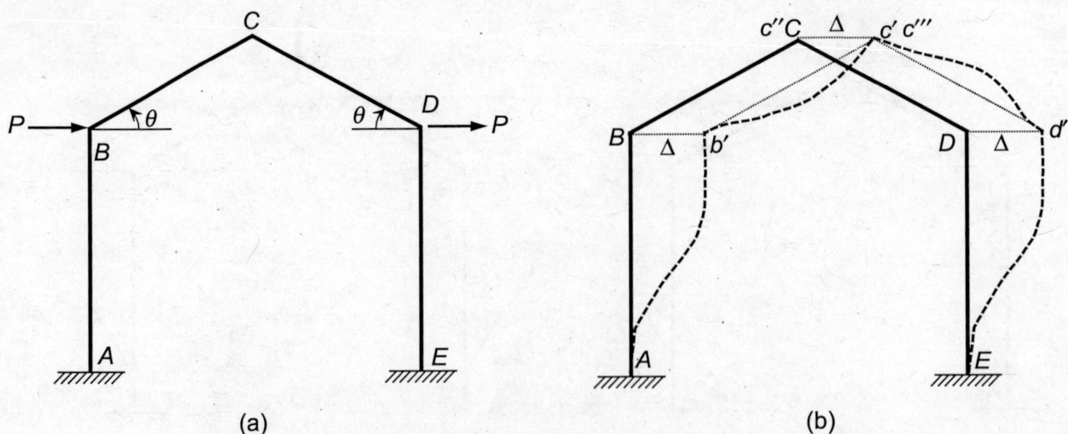

FIGURE 6.21   Anti-symmetric frame along with deflected shape.

Figure 6.22 shows the half of the frame with applicable support and loading conditions. The relationship between the joint displacements of members can be obtained using the joint-displacement triangle. Accordingly, there will be three unknown joint deformations, namely, (i) the rotation (θ_b) of joint B, (ii) the rotation (θ_c) of joint C, and (iii) the joint displacement (Δ). Using the modified slope-deflection equations, three joint deformations can be reduced to two unknowns only, which can be obtained using two equations of equilibrium. One equation can be written in terms of the moment equilibrium at joint B. The other equation can be obtained by considering the overall force/moment equilibrium of the frame.

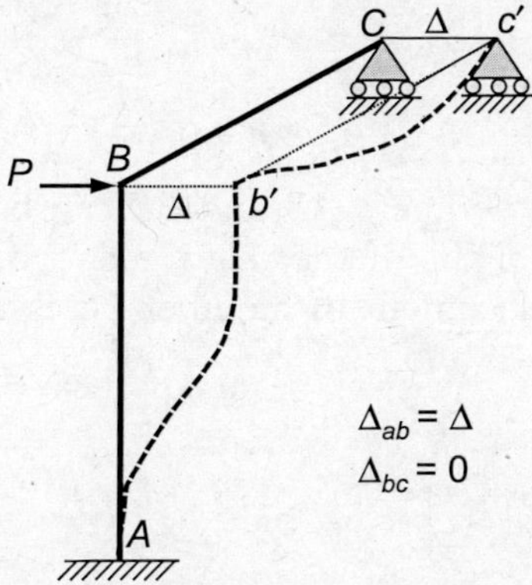

FIGURE 6.22 One-half of the anti-symmetric frame.

The following examples illustrate the application of these concepts in rigid frames.

EXAMPLE 6.13 Analyze the frame shown in Figure E6.13 using the *slope-deflection method*. Draw the bending moment diagram.

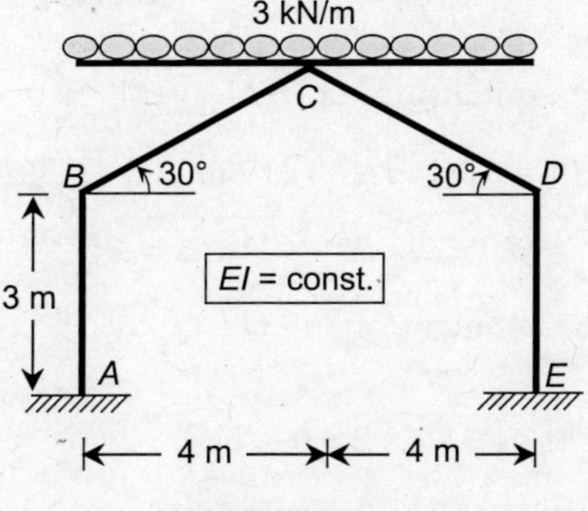

FIGURE E6.13

Solution: The degree of kinematic indeterminacy of the frame is five. This means that five unknown joint deformations are required to solve if the conventional slope-deflection method is used. However, the frame is symmetric for the given member configuration, supports, and loading conditions. Thus, all end moments of the frame can be obtained by solving only one-half of the frame.

The part of frame ABC with the applied loading and support conditions is shown in the following figure. The deflected shape of the frame is also shown in the figure.

Fixed end moments:

$$M_{ab}^f = M_{ba}^f = 0$$

$$M_{bc}^f = -M_{cb}^f = -\frac{3 \times 4^2}{12}\ \text{kNm} = -4\ \text{kNm}$$

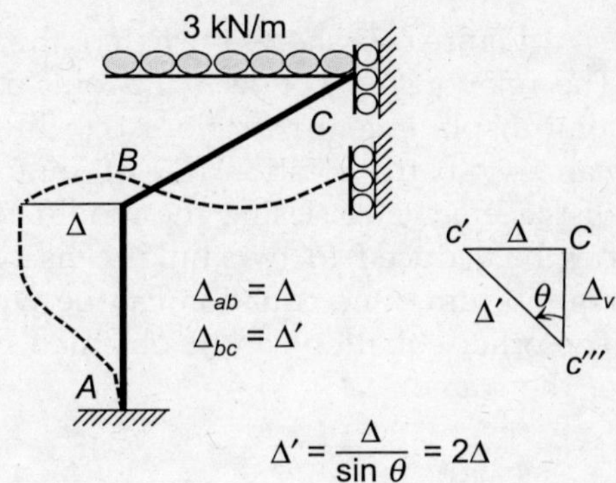

Compatibility conditions:

$$\theta_a = \theta_c = 0$$

The joint displacements of the members AB and BC can be related to each other through the joint-displacement triangle. Thus, if the relative end displacement of the member AB is Δ, then the relative end displacement of member BC is $\Delta' = 2\Delta$.

Slope-deflection Equations:

$$M_{ab} = \frac{2EI}{3}\left(\theta_b + \frac{3\Delta}{3}\right) = 0.667(EI\theta_b + EI\Delta) \tag{1}$$

$$M_{ba} = \frac{2EI}{3}\left(2\theta_b + \frac{3\Delta}{3}\right) = 1.333EI\theta_b + 0.667EI\Delta \tag{2}$$

$$M_{bc} = -4 + \frac{2EI}{4.62}\left(2\theta_b - \frac{3\Delta'}{4.62}\right) = -4 + 0.867EI\theta_b - 0.562EI\Delta \tag{3}$$

$$M_{cb} = -4 + \frac{2EI}{4.62}\left(\theta_b - \frac{3\Delta'}{4.62}\right) = -4 + 0.433EI\theta_b - 0.562EI\Delta \tag{4}$$

Equilibrium Equations: Joint equilibrium at B, $M_{ba} + M_{bc} = 0$

$$1.333EI\theta_b + 0.667EI\Delta - 4 + 0.867EI\theta_b - 0.562EI\Delta = 0$$

$\therefore$
$$2.2\,EI\theta_b + 0.11EI\Delta = 4 \tag{5}$$

For overall equilibrium, the moment about O must be zero.

$$H_a(5.31) - M_{ab} - M_{cd} - 3 \times 4 \times 2 = 0$$

$$\frac{(M_{ab} + M_{ba})}{3}(5.31) - M_{ab} - M_{cd} - 24 = 0$$

$\therefore$
$$2.43\,EI\theta_b + 2.26\,EI\Delta = 28 \tag{6}$$

Solving Eqs. (5), and (6), $EI\theta_b = 1.267$; $EI\Delta = 11.02$

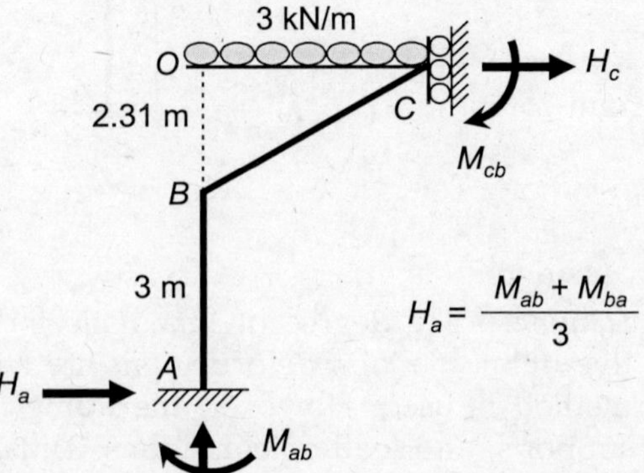

Final bending moments:

$$M_{ab} = 8.19 \text{ kNm}; \ M_{ba} = -M_{bc} = 9.05 \text{ kNm}; \ M_{cb} = -1.64 \text{ kNm}$$

The end bending moments of members in the other half of the frame are equal and opposite to the corresponding values as computed below:

$$M_{ed} = -8.19 \text{ kNm}; \ M_{de} = -M_{dc} = -9.05 \text{ kNm}; \ M_{cd} = 1.64 \text{ kNm}$$

The bending moment diagram of the frame is shown below:

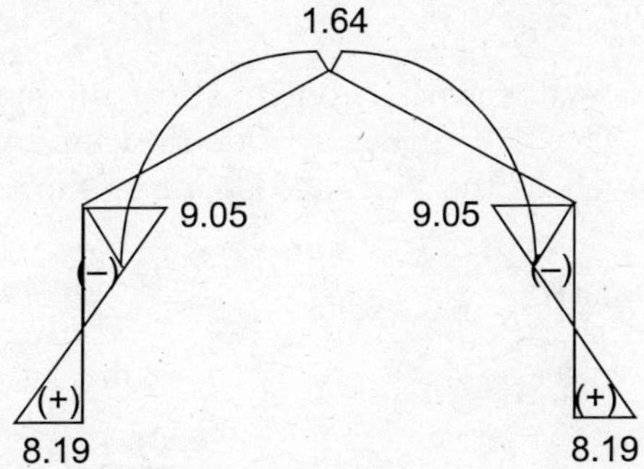

EXAMPLE 6.14 Analyze the frame shown in Figure E6.14 using the *slope-deflection method*. Draw the bending moment diagram.

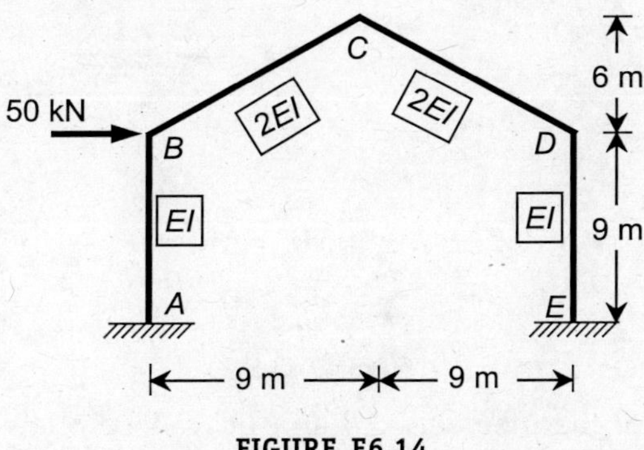

FIGURE E6.14

Solution: The degree of kinematic indeterminacy of the frame is five as there are three joint rotations and two independent joint (sway) displacements. However, this frame can be solved by using the principle of superposition considering the symmetric and anti-symmetric loading cases. The given loading can be split into two parts, namely, *symmetric* and *anti-symmetric* as shown below. The final bending moments of the frame can be considered as the sum of the bending moments in the frames with symmetric and anti-symmetric loading cases.

Let's determine the member end moments for the frames with symmetric and anti-symmetric loading cases.

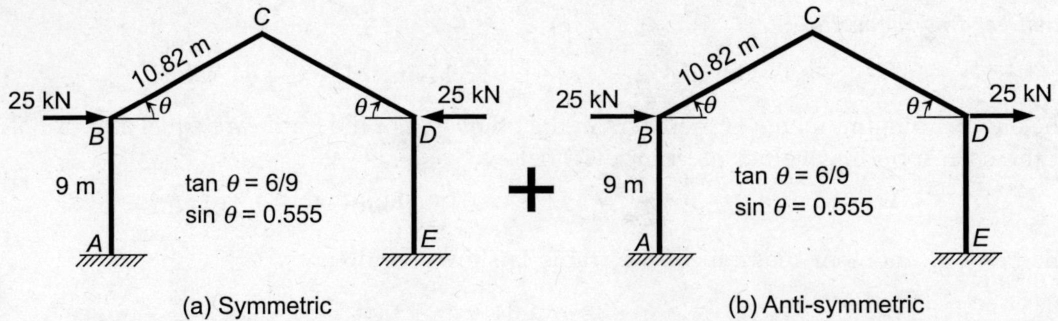

(a) Symmetric   (b) Anti-symmetric

(a) Symmetric loading: As stated earlier, only half of the symmetric loading frame can be analyzed for simplicity. The half of the symmetric frame with the applicable loading and support conditions as well as the deflected shape is shown below:

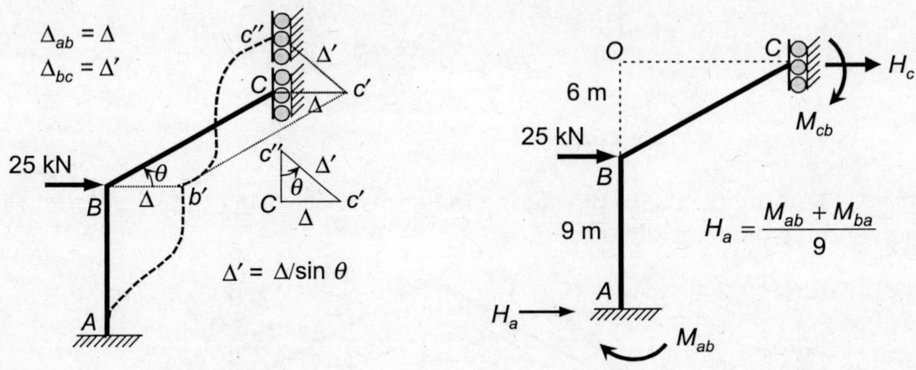

Fixed end moments:

$$M_{ab}^{f} = M_{ba}^{f} = M_{bc}^{f} = M_{cb}^{f} = 0$$

Compatibility conditions:

$$\theta_a = \theta_c = 0$$

The joint displacements of the members AB and BC can be related to each other through the joint-displacement triangle. Thus, if the relative end displacement of the member AB is Δ, then the relative end displacement of member BC is $\Delta' = 1.81\Delta$.

Slope-deflection Equations:

$$M_{ab} = \frac{2EI}{9}\left(\theta_b - \frac{3\Delta}{9}\right) = 0.222EI\theta_b - 0.074EI\Delta \tag{1}$$

$$M_{ba} = \frac{2EI}{9}\left(2\theta_b - \frac{3\Delta}{9}\right) = 0.444EI\theta_b - 0.074EI\Delta \tag{2}$$

$$M_{bc} = \frac{2(2EI)}{10.82}\left(2\theta_b + \frac{3\Delta'}{10.82}\right) = 0.739EI\theta_b + 0.186EI\Delta \tag{3}$$

$$M_{cb} = \frac{2(2EI)}{10.82}\left(\theta_b + \frac{3\Delta'}{10.82}\right) = 0.37EI\theta_b + 0.186EI\Delta \tag{4}$$

Equilibrium Equations: Joint equilibrium at B, $M_{ba} + M_{bc} = 0$

$$0.444EI\theta_b - 0.074EI\Delta + 0.739EI\theta_b + 0.186EI\Delta = 0$$

$$\therefore \qquad 10.56EI\theta_b + EI\Delta = 0 \tag{5}$$

For overall equilibrium, the moment about O must be zero.

$$H_a(15) - M_{ab} - M_{cd} + (25)(6) = 0$$

$$\frac{(M_{ab} + M_{ba})}{9}(15) - M_{ab} - M_{cd} - 150 = 0$$

$$\therefore \qquad 0.52EI\theta_b - 0.31EI\Delta + 150 = 0 \tag{6}$$

Solving Eqs. (5), and (6), $EI\theta_b = -39.54$; $EI\Delta = 417.54$

Final bending moments:

$$M_{ab} = -39.7 \text{ kNm}; \ M_{ba} = -M_{bc} = -48.4 \text{ kNm}; \ M_{cb} = 63.1 \text{ kNm}$$

The end bending moments of members in the other half of the frame are equal and opposite to the corresponding values as computed below:

$$M_{ed} = 39.7 \text{ kNm}; \ M_{de} = -M_{dc} = 48.4 \text{ kNm}; \ M_{cd} = -63.1 \text{ kNm}$$

(b) Anti-symmetric loading: The half of the anti-symmetric frame with the applicable loading and support conditions as well as the deflected shape is shown below:

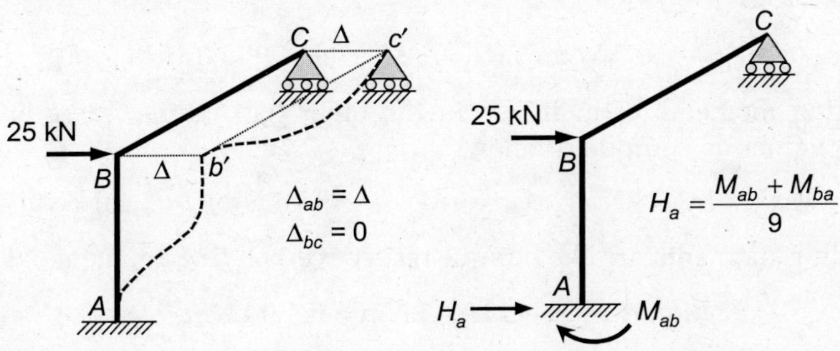

Fixed end moments:

$$M_{ab}^f = M_{ba}^f = M_{bc}^f = M_{cb}^f = 0$$

Compatibility condition: $\theta_a = 0$.

The joint displacements of the members AB and BC can be related to each other through the joint-displacement triangle. Thus, if the relative end displacement of the member AB is Δ, then the relative end displacement of member BC is $\Delta' = 0$.

Modified Slope-deflection Equations:

$$M_{ab} = \frac{2EI}{9}\left(\theta_b - \frac{3\Delta}{9}\right) = 0.222EI\theta_b - 0.074EI\Delta \tag{7}$$

$$M_{ba} = \frac{2EI}{9}\left(2\theta_b - \frac{3\Delta}{9}\right) = 0.444EI\theta_b - 0.074EI\Delta \tag{8}$$

$$M_{bc} = \frac{3(2EI)}{10.82}(\theta_b) = 0.555EI\theta_b \tag{9}$$

$$M_{cb} = 0 \tag{10}$$

Equilibrium Equations: Joint equilibrium at B, $M_{ba} + M_{bc} = 0$

$$0.444EI\theta_b - 0.074EI\Delta + 0.555EI\theta_b = 0$$

$$\therefore \qquad EI\theta_b - 0.074EI\Delta = 0 \tag{11}$$

For overall horizontal shear equilibrium,

$$H_a + 25 = 0$$

$$\frac{(M_{ab} + M_{ba})}{9} + 25 = 0$$

$$\therefore \qquad 0.667EI\theta_b - 0.148EI\Delta + 225 = 0 \tag{12}$$

Solving Eqs. (11), and (12), $EI\theta_b = 168.67$; $EI\Delta = 2279.26$

Final bending moments:

$$M_{ab} = -131.3 \text{ kNm}; \ M_{ba} = -M_{bc} = -96.9 \text{ kNm}; \ M_{cb} = 0$$

The end bending moments of members in the other half of the frame are equal to the corresponding values as computed below:

$$M_{ed} = -131.3 \text{ kNm}; \ M_{de} = -M_{dc} = -93.9 \text{ kNm}; \ M_{cd} = 0 \text{ kNm}$$

The final bending moments are the sum of respective bending moments at the joints.

$$M_{ab} = (-39.7 - 131.3) \text{ kNm} = 171.0 \text{ kNm}$$

$$M_{ba} = -M_{bc} = (-48.4 - 96.9) \text{ kNm} = -142.3 \text{ kNm}$$

$$M_{cb} = -M_{cd} = (63.1 + 0) \text{ kNm} = 63.1 \text{ kNm}$$

$$M_{de} = -M_{dc} = (48.4 - 93.9) \text{ kNm} = -45.5 \text{ kNm}$$

$$M_{ed} = -91.6 \text{ kNm}$$

The bending moment diagram of the frame is shown below:

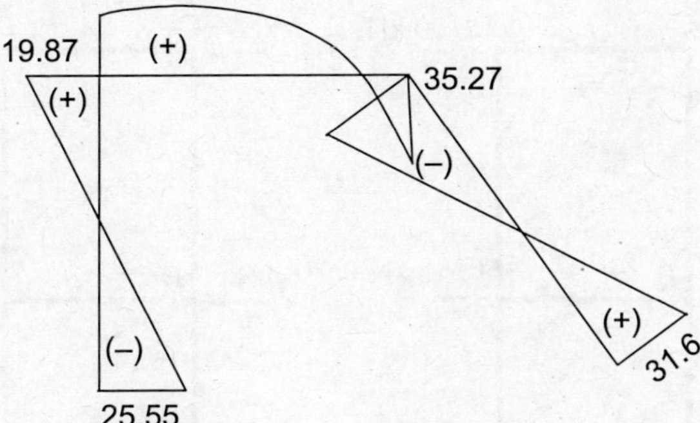

EXAMPLE 6.15 Analyze the two-story frame shown in Figure E6.15 using the *slope-deflection method*. Draw the bending moment diagram.

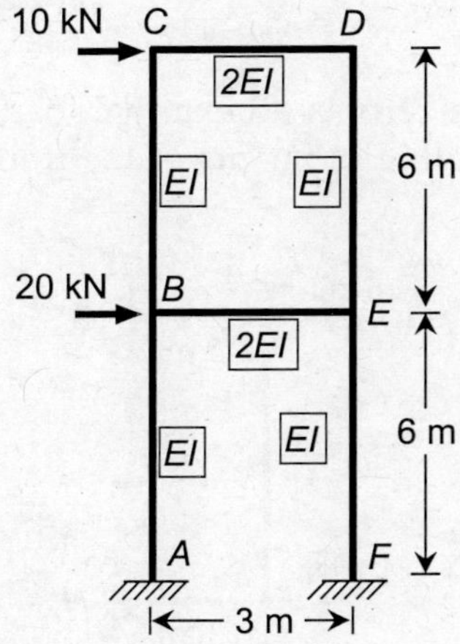

FIGURE E6.15

Solution: The degree of kinematic indeterminacy of the frame is six as there are four joint rotations and two independent joint (sway) displacements. However, this frame can be solved by using the principle of superposition considering the *symmetric* and *anti-symmetric* loading cases.

The given frame can be assumed to be a combination of one symmetric case and one antisymmetric case as shown below. The final bending moments of the frame can be considered as the sum of the bending moments in the frames with symmetric and anti-symmetric loading cases.

Let's determine the member end moments for the frames with symmetric and anti-symmetric loading cases.

(a) Symmetric loading: As stated earlier, only half of the symmetric loading frame can be analyzed for simplicity. Half of the symmetric frame with the applicable loading and support conditions is shown below. It can be seen that the bending moments at the ends

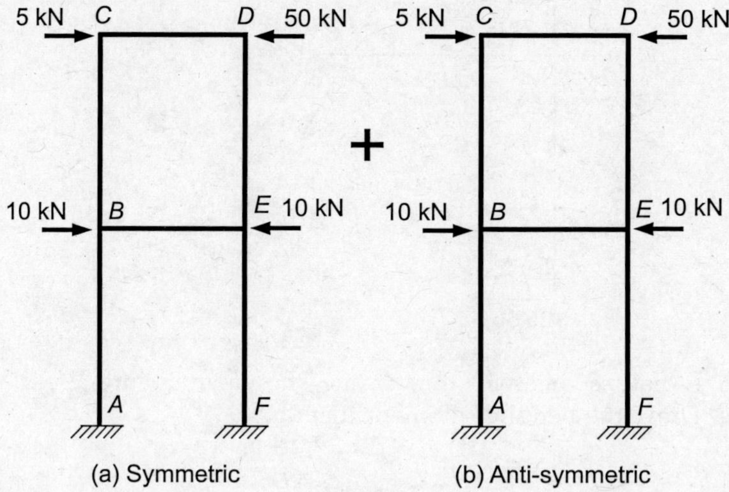

(a) Symmetric   (b) Anti-symmetric

of all members of this frame are zero as all external loads are applied at the joints along the supports. Thus, the final bending moments of the frame are equal to those of the anti-symmetric loading cases.

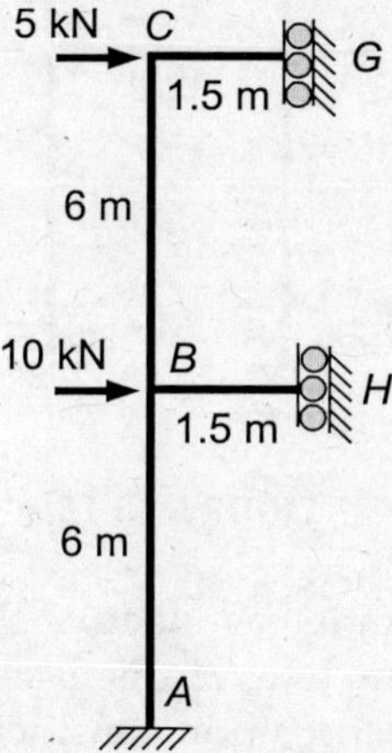

(b) Anti-symmetric loading: The half of the anti-symmetric frame with the applicable loading and support conditions as well as the deflected shape is shown below:

Fixed end moments:

$$M_{ab}^{f} = M_{ba}^{f} = M_{bh}^{f} = M_{bc}^{f} = M_{cb}^{f} = M_{cg}^{f} = 0$$

Compatibility condition: $\theta_a = 0.$

The joint displacements of the members AB and BC are assumed as Δ_1 and Δ_2, respectively.

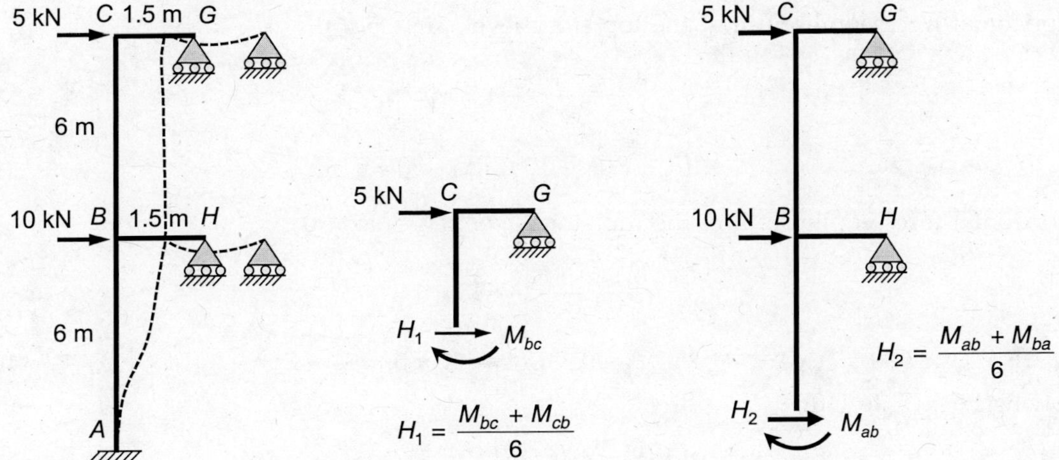

Modified Slope-deflection Equations:

$$M_{ab} = \frac{2EI}{6}\left(\theta_b - \frac{3\Delta_1}{6}\right) = 0.33EI\theta_b - 0.17EI\Delta_1 \tag{1}$$

$$M_{ba} = \frac{2EI}{6}\left(2\theta_b - \frac{3\Delta_1}{6}\right) = 0.67EI\theta_b - 0.17EI\Delta_1 \tag{2}$$

$$M_{bh} = \frac{3(2EI)}{1.5}(\theta_b) = 4EI\theta_b \tag{3}$$

$$M_{bc} = \frac{2EI}{6}\left(2\theta_b + \theta_c - \frac{3\Delta_2}{6}\right) = 0.67EI\theta_b + 0.33EI\theta_c - 0.17EI\Delta_2 \tag{4}$$

$$M_{cb} = \frac{2EI}{6}\left(\theta_b + 2\theta_c - \frac{3\Delta_2}{6}\right) = 0.33EI\theta_b + 0.67EI\theta_c - 0.17EI\Delta_2 \tag{5}$$

$$M_{cg} = \frac{3(2EI)}{1.5}(\theta_c) = 4EI\theta_c \tag{6}$$

Equilibrium Equations: There are four unknowns and therefore, four equations of equilibrium are required to solve these simultaneous equations.

Joint equilibrium at B, $M_{ba} + M_{bc} + M_{bh} = 0$

$$0.67EI\theta_b - 0.17EI\Delta_1 + 0.67EI\theta_b + 0.33EI\theta_c - 0.17EI\Delta_2 + 4EI\theta_b = 0$$

$$\therefore \quad 5.33EI\theta_b + 0.33EI\theta_b - 0.17EI\Delta_1 - 0.17EI\Delta_2 = 0 \tag{7}$$

Joint equilibrium at C, $M_{cb} + M_{cg} = 0$

$$0.33EI\theta_b + 0.67EI\theta_c - 0.17EI\Delta_2 + 4EI\theta_c = 0$$

$$\therefore \quad 0.33EI\theta_b + 4.67EI\theta_c - 0.17EI\Delta_2 = 0 \tag{8}$$

Horizontal force equilibrium at the top story level, $H_1 + 5 = 0$

$$\frac{(M_{bc} + M_{cb})}{6} + 5 = 0$$

$$EI\theta_b + EI\theta_c - 0.34 EI\Delta_2 + 30 = 0 \qquad (9)$$

Horizontal force equilibrium at the top story level, $H_2 + 15 = 0$

$$\frac{(M_{ab} + M_{ba})}{6} + 15 = 0$$

$$\therefore \qquad EI\theta_b - 0.33 EI\Delta_1 + 90 = 0 \qquad (10)$$

Solving Eqs. (7) to (10),

$$EI\theta_b = 13.9 \,;\; EI\theta_c = 4.16 \,;\; EI\Delta_1 = 312 \,;\; EI\Delta_2 = 144$$

Final bending moments: Putting the above values in the slope-deflection equations,

$$M_{ab} = -47.4 \text{ kNm}; \; M_{ba} = -42.7 \text{ kNm}; \; M_{bh} = 55.6 \text{ kNm}$$

$$M_{bc} = -13.4 \text{ kNm}; \; M_{cb} = -16.6 \text{ kNm}; \; M_{cg} = 16.6 \text{ kNm}$$

The bending moments at the ends of members in the other half of the frame are equal to the corresponding values as computed below:

$$M_{fe} = -47.4 \text{ kNm}; \; M_{ef} = -42.7 \text{ kNm}; \; M_{eh} = 55.6 \text{ kNm}$$

$$M_{ed} = -13.4 \text{ kNm}; \; M_{de} = -16.6 \text{ kNm}; \; M_{dg} = 16.6 \text{ kNm}$$

The bending moment diagram of the frame is shown below:

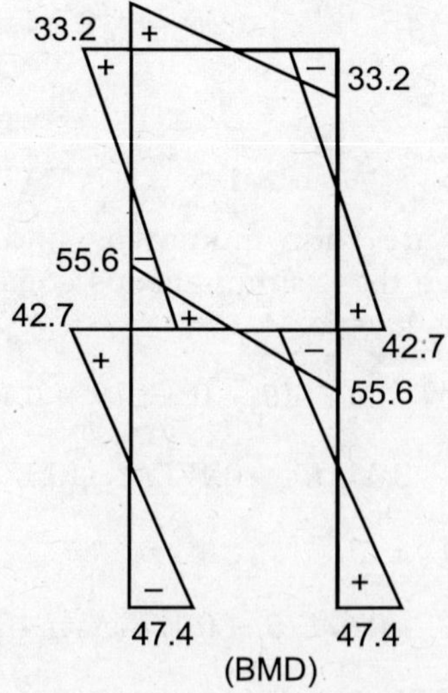

6.9 PROBLEMS

6.1 Determine the reactions and bending moments at the supports of the beam shown below using the Slope-Deflection method. Draw the shear force and bending moment diagrams.

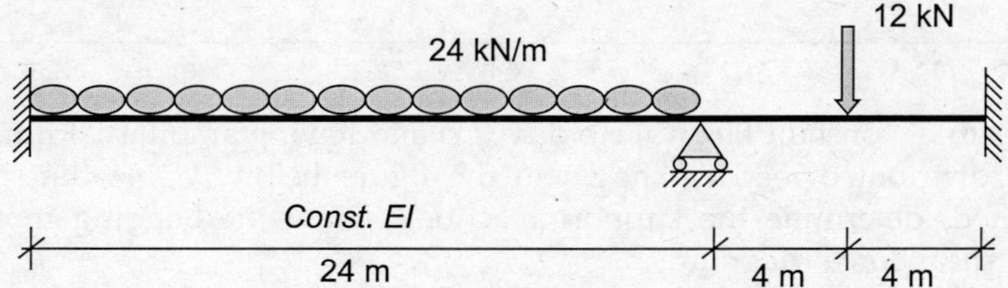

6.2 Determine the reactions and bending moments at the supports of the beam shown below using the Slope-Deflection method. Draw the shear force and bending moment diagrams.

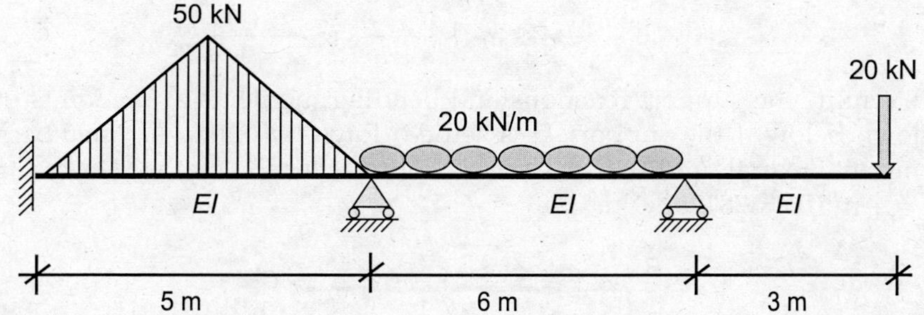

6.3 Determine the reactions and the bending moments at the supports of the beam supported on a spring at one end as shown below using the Slope-deflection method. The value of spring constant, $k = 10$ kN/mm. Check your solutions if (i) the value of k is zero, and (ii) the value of k is infinity.

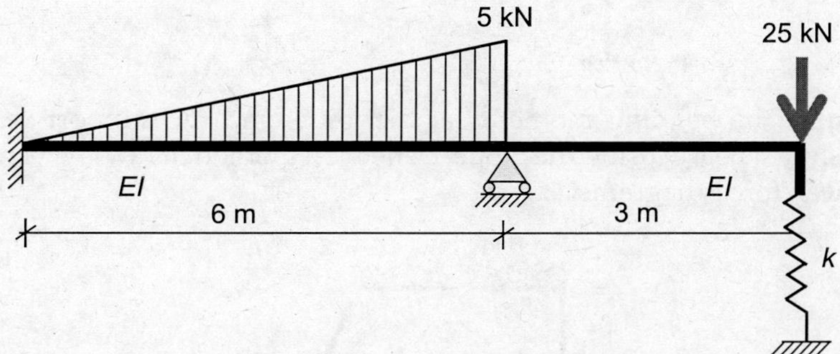

6.4 Determine the reactions and the bending moments at the supports of the beam as shown below using the Slope-deflection method. The support B is settled down by 50 mm and the support C is rotated clockwise by 0.02 radian. Assume the value of $EI =1000$ kNm^2.

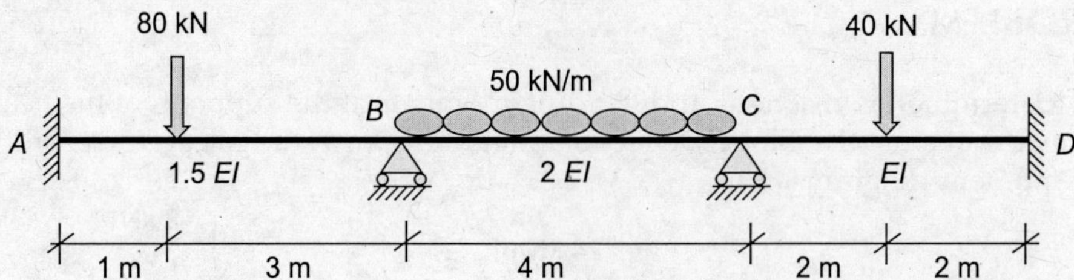

6.5 A beam of constant flexural rigidity (EI) and having an internal hinge is subjected to a distributed load as shown in the figure below. Using the slope-deflection method, determine the support reactions. Draw the bending moment diagram and shear force diagram.

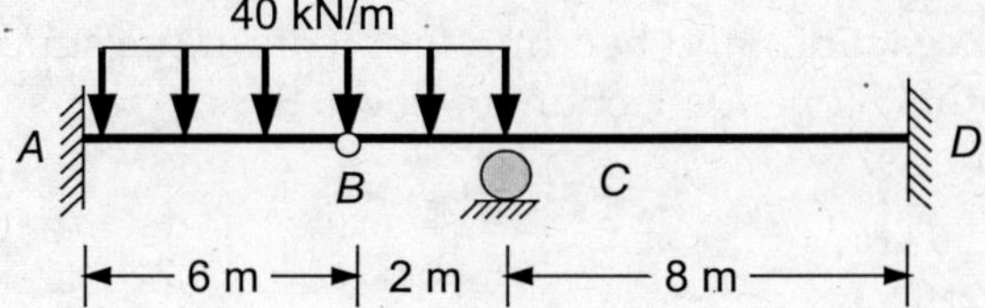

6.6 Determine the support reactions and bending moments at the joints of the frame shown below if the support D is settled down by $0.01L$. All members are of the constant flexural rigidity (EI). Use the Slope-Deflection method. Take $w = 10$ kN/m; $L = 4$ m; $EI = 78125$ kNm^2.

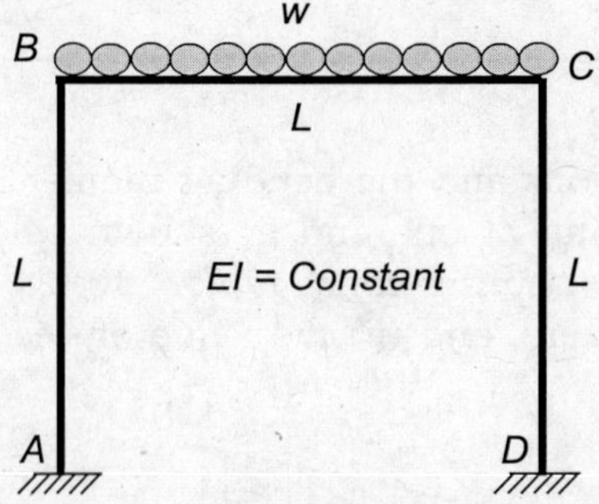

6.7 Determine the bending moments at the joints and the support reactions of the frame shown below using the Slope-Deflection method. Draw the bending moment and shear force diagrams.

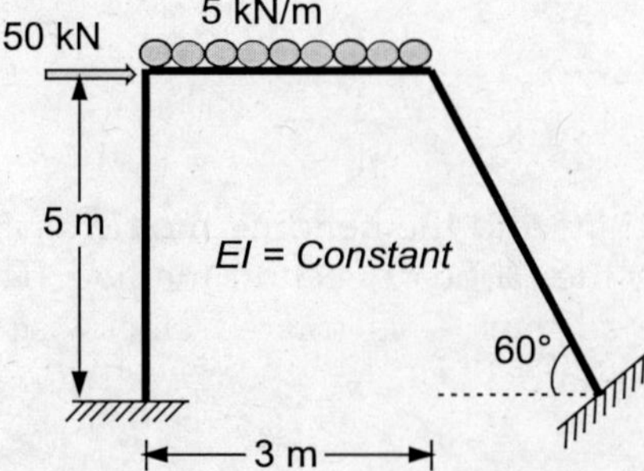

6.8 Determine the bending moments at the joints and the support reactions of the frame shown below using the Slope-Deflection method. Draw the bending moment and shear force diagrams.

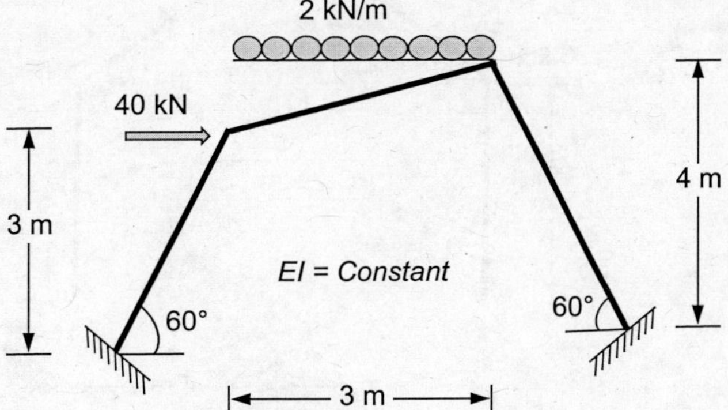

6.9 Determine the bending moments at the joints and the support reactions of the beam shown below using the Slope-Deflection method. Determine the vertical displacement at the hinge if the value of $EI=6336$ kNm^2. Draw the bending moment and shear force diagrams.

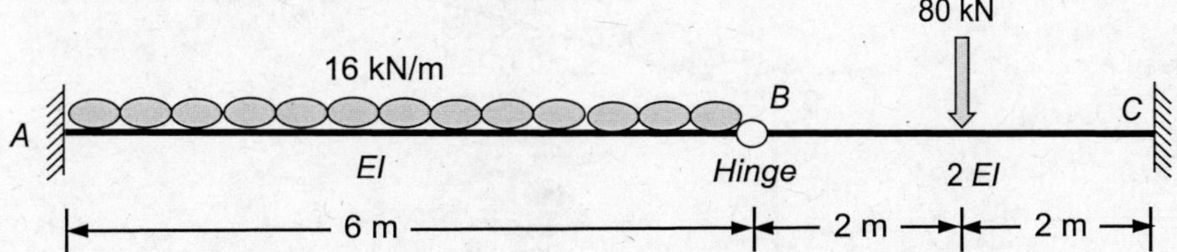

6.10 Determine the bending moments at the joints and the support reactions of the frame shown below using the Slope-Deflection method. Draw the bending moment and shear force diagrams.

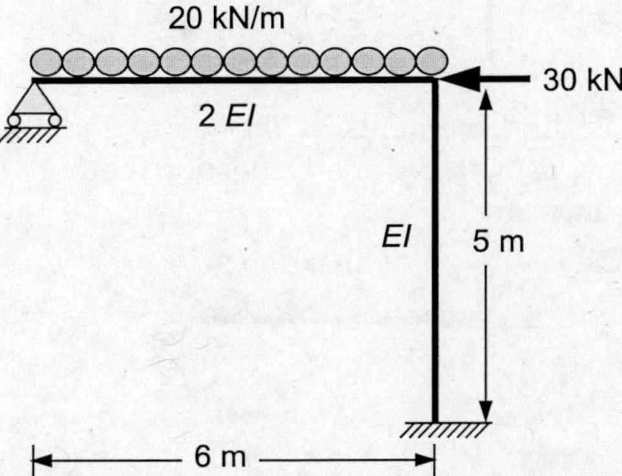

6.11 Determine the bending moments at the joints and the support reactions of the frame shown below using the Slope-Deflection method. Draw the bending moment and shear force diagrams.

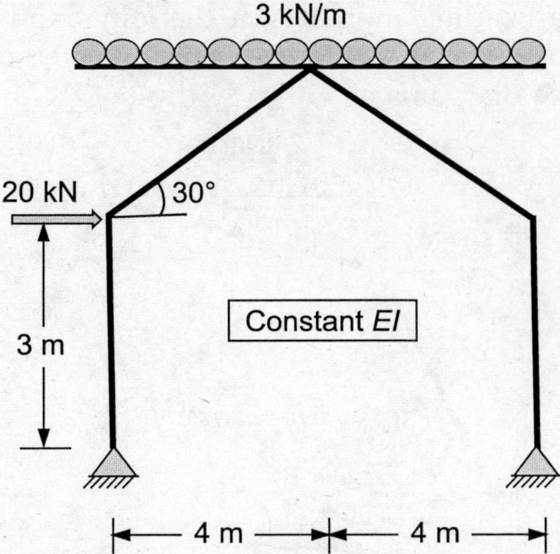

6.12 Determine the bending moments at the joints and the support reactions of the frame shown below using the Slope-Deflection method. Draw the bending moment and shear force diagrams.

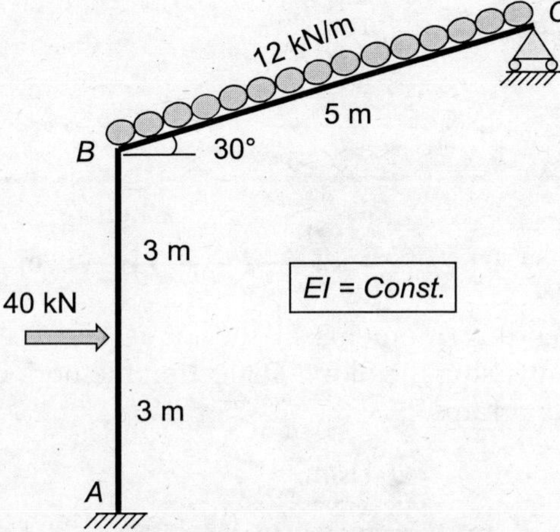

6.13 Determine the bending moments at the joints and the support reactions of the frame shown below using the Slope-Deflection method. Draw the bending moment and shear force diagrams.

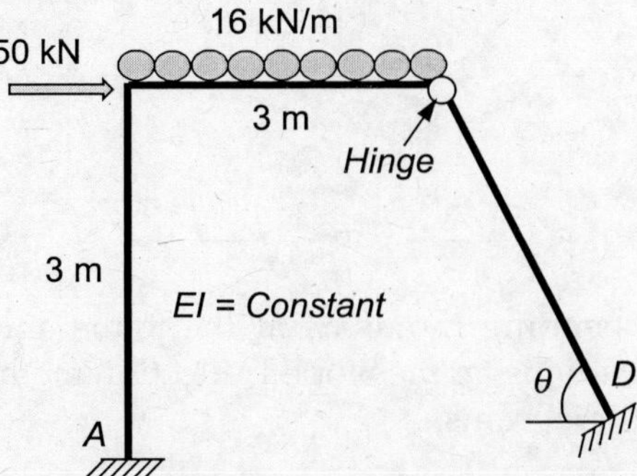

6.14 Determine the bending moments at the joints and the support reactions of the frame shown below using the Slope-Deflection method. Draw the bending moment and shear force diagrams.

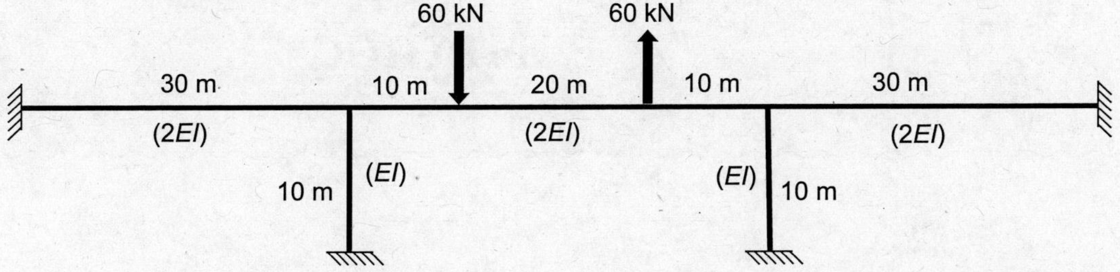

6.15 Determine the bending moments at the joints and the support reactions of the frame shown below using the Slope-Deflection method. Draw the bending moment and shear force diagrams.

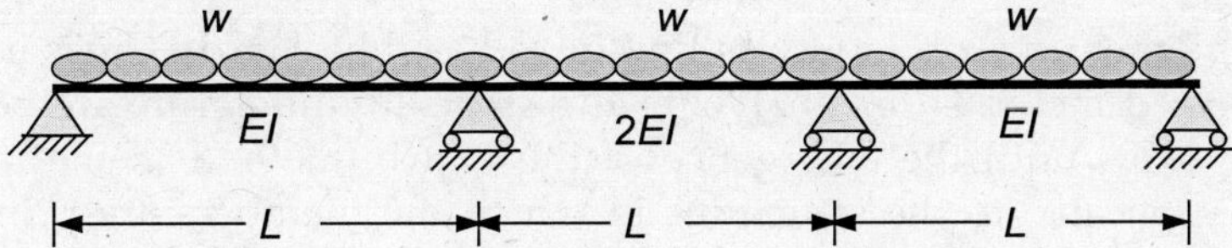

Moment-Distribution Method

7.1 GENERAL

The method of moment-distribution was first introduced by Hardy Cross in 1924. This was one of the preferred methods for analyzing the statically indeterminate beams and plane frames in early 1930s until the emergence of the matrix method in the late 1950s. Thus, this method is applicable to the structures in which the members are primarily subjected to bending moments. The process of moment distribution is initiated from the basic slope-deflection equation. Instead of solving the simultaneous equations as followed in the slope-deflection method, the final end moments are obtained by a quick and simple iteration procedure. Thus, the moment-distribution method is also one of the displacement methods.

The moments acting at the ends of members are the algebraic sum of four actions (or effects) as follows:

1. The fixed end moments generated due to the applied loads of members with no end rotations and relative displacements (i.e., the fixed-ended members),
2. The moments due to the rotation at near end of the members without applied loads keeping the far end being fixed,
3. The moments due to the rotation at the far end of the members without applied loads keeping the near end fixed, and
4. The moments due to the relative displacement between the ends of members without applied loads keeping both ends being fixed.

The moment distribution method is a successive correction procedure in which the errors in the end moments are minimized in the subsequent steps. The method begins with the assumption that all joints are fixed supports, which permits to compute the fixed-end moments due to applied loads on the members. This process in which the joint rotations and the relative joint translation of the members considered to be zero is called as *locking* the joints. Since all joints are not truly fixed and may be free to rotate, the restraint is released at a joint that is free to rotate. This process of called as *unlocking* the joint. This would induce the additional end moments at the released end and may get carried over to the other end. Then, the rotated joint is reclamped and the restraint at the other joint is released to consider the effect of the member deformations on the fixed end moments.

This rotated joint is then reclamped. Joints may be successively released or reclamped, one by one, as many times, as required. As the process of locking and unlocking of joints is repeated, the corrections to the end moments become smaller and smaller. The process is continued until the errors in the moment equilibrium at the joints are sufficiently small to be neglected. This process is essentially one of the successive approximations which can be carried out to any degree of accuracy desired. The convergence to an exact solution can be achieved by moment distribution if enough iterative cycles are performed. However, only a few iterations cycles have shown to produce a solution with an acceptable accuracy.

At present, the moment distribution method continues to be used for preliminary analysis and design. This method also serves to check the accuracy of results obtained from computer analysis. In fact, the moment distribution method of analysis is considered as efficient and, in some cases, more economical than a computer analysis for small-scale structures. Nevertheless, this method is equally applicable in the analysis of continuous beams and multi-story rigid frames.

7.2 FUNDAMENTAL CONCEPTS

The basic assumptions used in the moment distribution method are as follows:

- All members of the structure behave elastically, i.e., Hooke's law is applicable.
- All members are primarily subjected to bending actions.
- Axial deformations of the members are negligible.
- Rigid body rotations occur at the joints of the members.
- All members are prismatic.

The sign conventions used in the slope-deflection methods are also valid in the moment distribution method. All clockwise internal end moments and rotations are considered as positive and vice-versa.

The main parameters used in the moment-distribution method are presented as follows:

7.2.1 Fixed End Moments

The application of the method of moment-distribution requires knowledge of the moments developed at the ends of the fixed-ended beams subjected to the applied loads. These moments are called as the fixed end moments (FEMs), often denoted by M^f. The fixed end moments of a member subjected to different types of loadings have already been discussed in previous chapter.

7.2.2 Stiffness and Distribution Factor

The *stiffness* (or more specifically, the *rotational stiffness*) of a member of uniform section (constant *EI*) is defined as the moment required to produce a unit rotation at one end of the member keeping the other end as fixed.

Consider a beam of length l and flexural rigidity *EI*. The end b of the beam is fixed, and the end a is allowed to rotate. There is no external load acting on the beam.

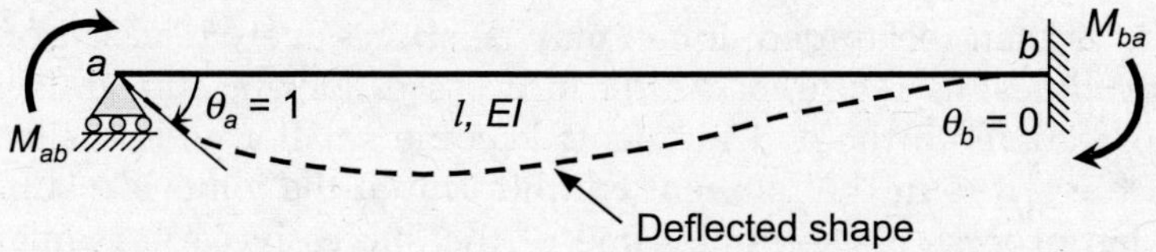

FIGURE 7.1   Beam with end rotation at hinge support.

The end moment required at end *a* to cause rotation θ_a at the same end keeping $\theta_b = 0$ can be determined by the slope-deflection equation as follows:

$$M_{ab} = M_{ab}^f + \frac{2EI}{l}\left(2\theta_a + \theta_b - \frac{3\Delta}{l}\right) = \frac{2EI}{l}(2\theta_a) = \frac{4EI\theta_a}{l} \tag{7.1}$$

Stiffness (or *rotational stiffness*) of the member is the moment required to cause a unit rotation at end *a* (i.e., $\theta_a = 1$) and can be denoted as S.

Thus,
$$S = \frac{M_{ab}}{\theta_a} = \frac{4EI}{l} = 4EK \tag{7.2}$$

Stiffness factor,
$$K = \frac{I}{l} \tag{7.3}$$

Now, let us consider a rigid frame composed of four members as shown in Figure 7.2(a). One end of each member is fixed, whereas the other end is rigidly connected at joint *o* whose translation is prevented. If an external clockwise moment, *M* is applied at the joint *o*, it would cause the joint to rotate clockwise by an angular deformation, θ as shown in Figure 7.2(b). Since the joint *o* is rigid, the tangent drawn to the elastic curves of each member at the connected end rotates by the same angle, θ.

The applied moment *M* is resisted by the four members meeting at the joint. As shown in Figure 7.2(c), the resisting moments M_{oa}, M_{ob}, M_{oc}, and M_{od} induced at the ends of four members must balance the effect of the external moment, *M*.

Moment equilibrium at the joint *o*,

$$M = M_{oa} + M_{ob} + M_{oc} + M_{od} \tag{7.4}$$

Using the slope-deflection equations,

$$M_{oa} = \frac{4EI_{oa}\theta}{l_{oa}} = 4EK_{oa}\theta; \quad M_{ob} = \frac{4EI_{ob}\theta}{l_{ob}} = 4EK_{ob}\theta$$

$$M_{oc} = \frac{4EI_{oc}\theta}{l_{oc}} = 4EK_{oc}\theta; \quad M_{od} = \frac{4EI_{od}\theta}{l_{od}} = 4EK_{od}\theta$$

$$M = 4EK_{oa}\theta + 4EK_{ob}\theta + 4EK_{oc}\theta + 4EK_{od}\theta$$

or
$$M = 4E\theta(K_{oa} + K_{ob} + K_{oc} + K_{od}) = 4E\theta(\Sigma K)$$

$\therefore$
$$\theta = \frac{M}{4E(\Sigma K)} \tag{7.5}$$

where, $\Sigma K = K_{oa} + K_{ob} + K_{oc} + K_{od}$

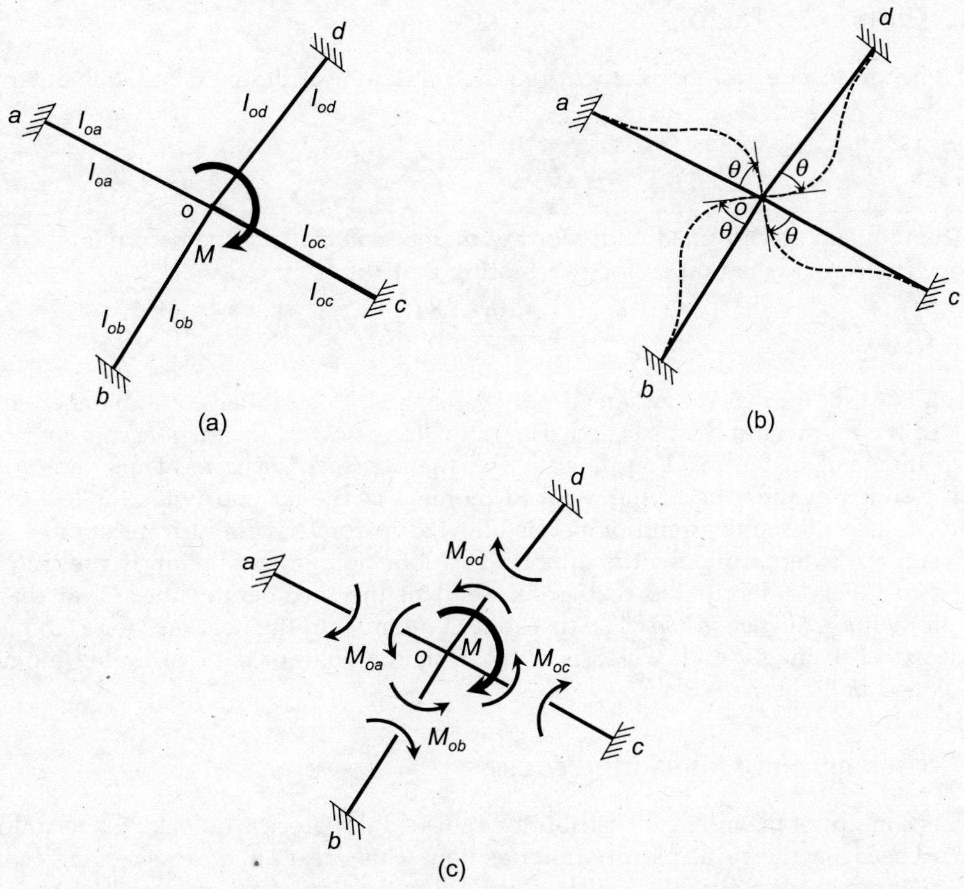

FIGURE 7.2 Frame composed of four members subjected to a joint moment.

Accordingly,

$$M_{oa} = \frac{K_{oa}}{\sum K} M = D_{oa} M; \quad M_{ob} = \frac{K_{ob}}{\sum K} M = D_{ob} M$$

$$M_{oc} = \frac{K_{oc}}{\sum K} M = D_{oc} M; \quad M_{od} = \frac{K_{od}}{\sum K} M = D_{od} M$$

where, $D_{ij} = \dfrac{K_{ij}}{\sum K}$ = *Distribution factor.*

Thus, the *distribution factor* (D_{ij}) for a member at joint is defined as the ratio of the stiffness factor of that member to the sum of the stiffness factors of all members meeting at the same joint. This shows that a moment applied at a joint is resisted by the members meeting at the same joint in proportion to their *distribution factors*. It is worth mentioning that the distribution factors are determined based on the relative stiffness, rather than absolute stiffness, of the members.

7.2.3   Carry-over Factor

In reference to Figure 7.1, the moment required at end *a* of the member *ab* is determined as follows:

$$M_{ab} = \frac{4EI\theta_a}{l}$$

The moment at end fixed end, *b* of the member *ab* can also be determined by using the slope-deflection equation as follows (noting that $\theta_b = 0$):

$$M_{ba} = \frac{2EI\theta_a}{l} = \left(\frac{1}{2}\right)M_{ab} \tag{7.6}$$

Equation (7.6) shows that the moment induced at the far end (fixed) of a member equal to *one-half* of the moment at the near end. The ratio (1/2) is called the *carry-over factor*, and the induced moments at the fixed end is called as the *carry-over moments*. Thus, the *carry-over factor* is defined as the ratio of the induced moment at the far end (which is fixed) to the applied moment at the near end, which is allowed to rotate without translation.

When the external moment is applied to a joint whose translation is prevented, the resisting member is distributed to the near end of the members connected at that joint based on their *distribution factors*. The distributed moment to the near end for each member is carried over to the far end, which is equal to the one-half of the distributed moment at the near end and is same sign.

7.2.4   Locking and Unlocking Process

The basic concept of the moment distribution relies on the process of locking and unlocking of joints based on the principle of superposition. This means that any bending moment added to any joint of a structure due to locking process is later eliminated by unlocking. To explain this behaviour, consider the rigid frame shown in the Figure 7.3. The member *ao* is subjected to a concentrated load of 40 kN at its mid-span.

1. At the first instance, let's assume that the end *o* of the loaded member *ao* is locked (or fixed) as shown in Figure 7.3(b). The fixed end moment at both fixed ends of the member *ao* can be determined as follows:

    $$M_{ao}^f = -M_{bo}^f = -\frac{PL}{8} = -\frac{(40)(3)}{8} \text{ kNm} = -15 \text{ kNm}$$

    At this stage, all other members are not subjected to any end moments. Considering the equilibrium condition at the joint *o*, $\Sigma M_0 = 0$. This requires that the locking moment at the joint *o*, $M = 15$ kNm, which is clockwise as shown in Figure 7.3(b).
2. Now, release the joint *o* from the locking position. For equilibrium, an equal and opposite moment should be applied at the joint *o*. As shown in Figure 7.3(c), the unlocking moment, $M = -15$ kNm. This unlocking moment now must be distributed among all members connected at this joint.

The same process of locking and unlocking of joints must be applied if other members are subjected to the external loadings.

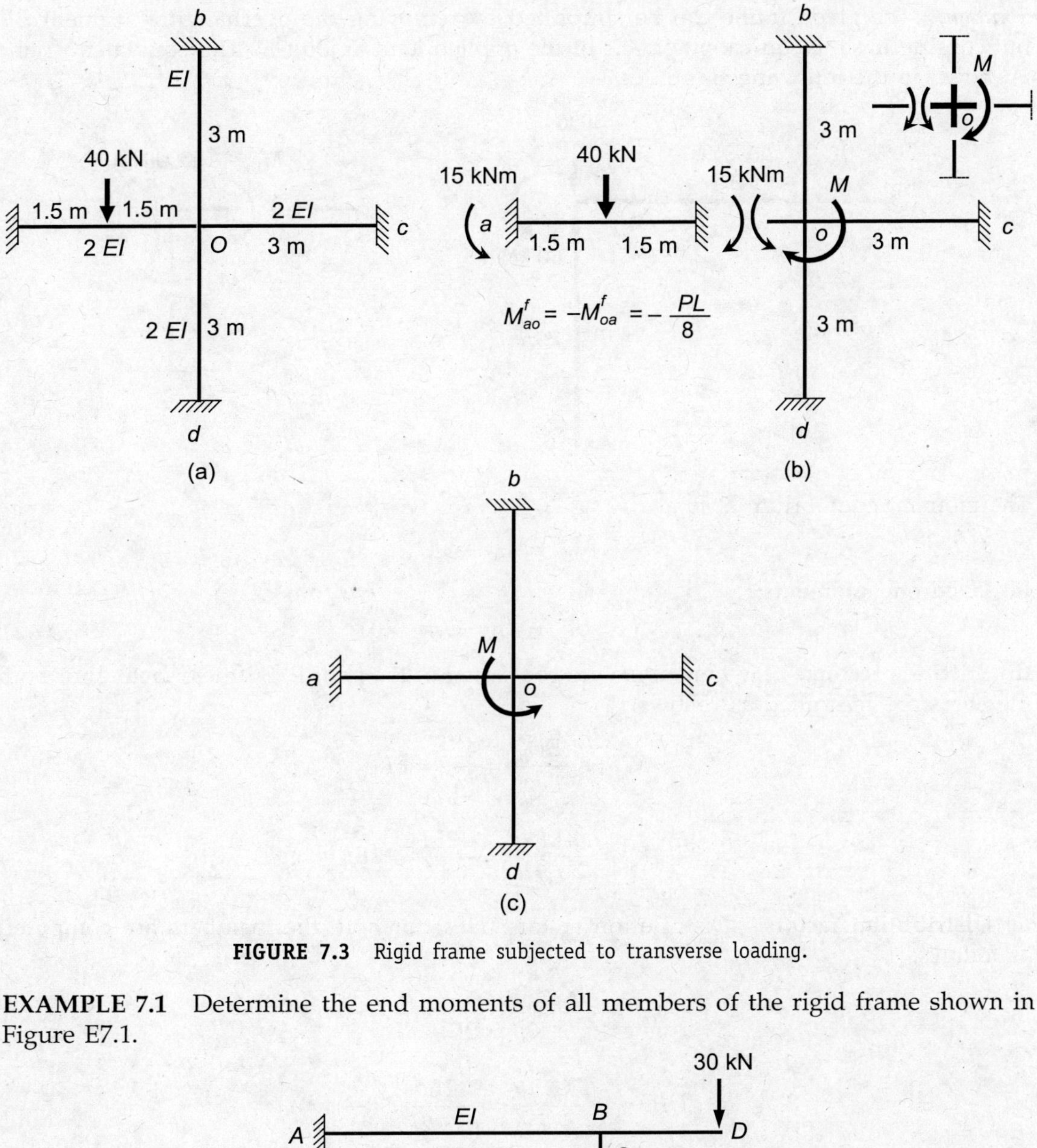

$$M^f_{ao} = -M^f_{oa} = -\frac{PL}{8}$$

FIGURE 7.3 Rigid frame subjected to transverse loading.

EXAMPLE 7.1 Determine the end moments of all members of the rigid frame shown in Figure E7.1.

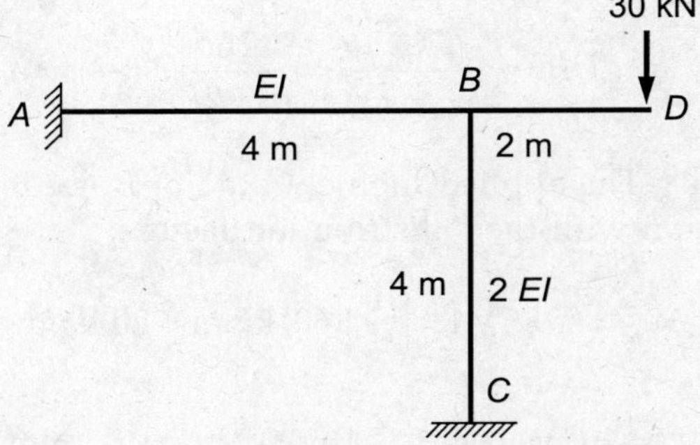

FIGURE E7.1

Solution: The given frame can be simplified by removing the overhanging segment *BD* but considering the equivalent effects of the applied load at joint *D*. The equivalent frame is shown in the following figure.

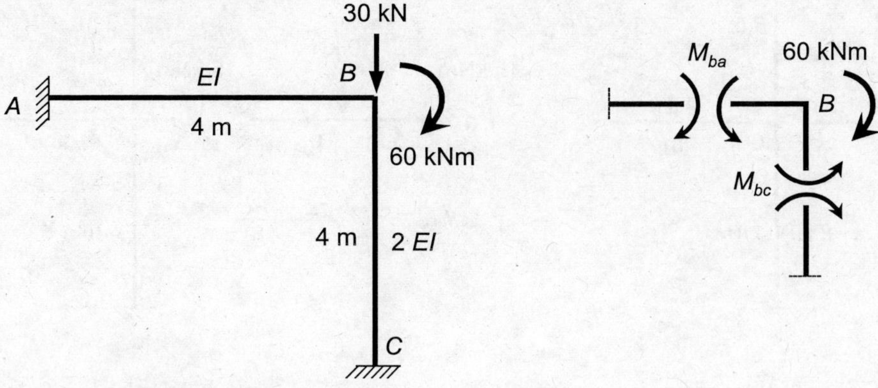

The moment equilibrium at joint *B*,

$$M_{ba} + M_{bc} = 60$$

(a) Fixed end moments:

$$M_{ab}^{f} = M_{ba}^{f} = M_{bc}^{f} = M_{cb}^{f} = 0$$

(b) Stiffness factors: Two members are connected at the joint B. Stiffness factors for both members are computed as follows:

$$K_{ba} = \frac{4(EI)_{ba}}{L_{ba}} = \frac{4EI}{4} = EI$$

$$K_{bc} = \frac{4(EI)_{bc}}{L_{bc}} = \frac{4(2EI)}{4} = 2EI$$

(c) Distribution factors: Distribution factors (*DFs*) for both the members are computed as follows:

$$DF_{ba} = \frac{K_{ba}}{K_{ba} + K_{bc}} = \frac{EI}{EI + 2EI} = \frac{1}{3}$$

$$DF_{bc} = \frac{K_{bc}}{K_{ba} + K_{bc}} = \frac{2EI}{EI + 2EI} = \frac{2}{3}$$

(d) Moment distribution: The applied moment at joint *B* needs to be distributed among two members in proportion with their distribution factors.

$$M_{ba} = (DF_{ba})M = \left(\frac{1}{3}\right)(60)\,\text{kNm} = 20\ \text{kNm}$$

$$M_{bc} = (DF_{bc})M = \left(\frac{2}{3}\right)(60)\,\text{kNm} = 40\ \text{kNm}$$

(e) Carry-over moment:

$$M_{ab} = \left(\frac{1}{2}\right)M_{ba} = \left(\frac{1}{2}\right)(20)\ \text{kNm} = 10\ \text{kNm}$$

$$M_{cb} = \left(\frac{1}{2}\right)M_{bc} = \left(\frac{1}{2}\right)(40)\ \text{kNm} = 20\ \text{kNm}$$

It should be noted that $M_{bd} = -(2)(30)\ \text{kNm} = -60\ \text{kNm}$

The end moments of all members are shown in the following figures.

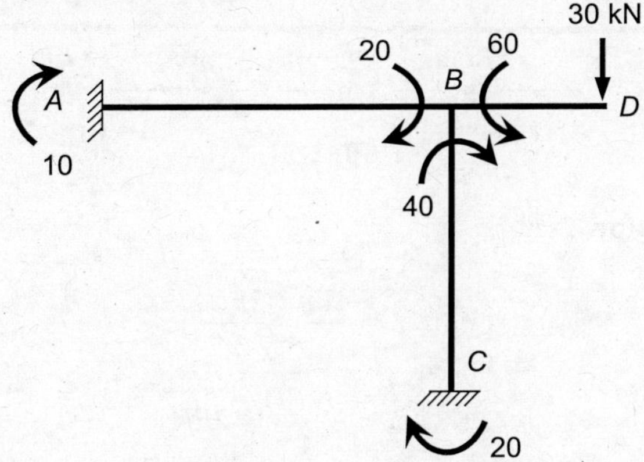

7.3 PROCEDURE OF MOMENT-DISTRIBUTION METHOD

The general procedure of moment-distribution method can be summarized as follows:

1. Determine the distribution factors for the members at all joints where more than one member is connected.
2. Lock all joints: Compute fixed end moments (FEMs) for each member due to applied loadings.
3. Unlock (release) one joint which is free to rotate.
4. Determine the unbalanced moment at the released joint.
5. Compute the balanced moment for each member at the released joint using the distribution factors.
6. Carry over one-half of the balanced moment to the joint supports at the opposite ends of members.
7. Lock the released joint in its new position and repeat steps (3)–(4) at another supported joint that is free to rotate.
8. Since the moment carried over to a joint induces additional unbalanced moment, repeat steps (5) and (6) to get the balanced moment and the carry over moments.
9. Continue the process of balancing and carry-over at all the joints until the value of unbalanced moments becomes very small, which can be considered to be negligible.
10. Compute the final end moments of members by adding all the balanced moments at the joints.

7.4   BEAMS WITHOUT SUPPORT SETTLEMENT

The procedure summarized in the preceding section can be directly adopted to analyze the beams without support settlement as illustrated in the following examples.

EXAMPLE 7.2   Determine the end moments of the continuous beam shown in the Figure E7.2 using Moment-distribution method.

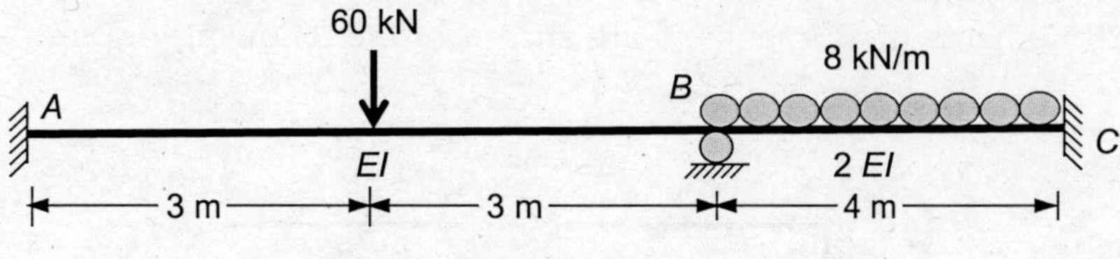

FIGURE E7.2

Solution: **(a) Distribution factors:**
Stiffness factors:

$$K_{ba} = \frac{4EI}{6} = \frac{2EI}{3}$$

$$K_{bc} = \frac{4(2EI)}{4} = 2EI$$

Distribution factors:

$$DF_{ba} = \frac{K_{ba}}{K_{ba} + K_{bc}} = 0.25$$

$$DF_{bc} = \frac{K_{bc}}{K_{ba} + K_{bc}} = 0.75$$

(b) Fixed end moments:

$$M_{ab}^{f} = -\frac{Pl}{8} = -\frac{60 \times 6}{8} \text{ kNm} = -45 \text{ kNm}$$

$$M_{ba}^{f} = \frac{Pl}{8} = \frac{60 \times 6}{8} \text{ kNm} = 45 \text{ kNm}$$

$$M_{bc}^{f} = -\frac{wl^2}{12} = -\frac{8 \times 4^2}{12} \text{ kNm} = -10.67 \text{ kNm}$$

$$M_{cb}^{f} = \frac{wl^2}{12} = \frac{8 \times 4^2}{12} \text{ kNm} = 10.67 \text{ kNm}$$

(c) Moment-distribution: The moment-distribution process is usually carried out in a tabular format as shown below:

Joint	A	B		C	Remarks
Member	AB	BA	BC	CB	
DF		0.25	0.75		Distribution factor
FEM	−45.0	45.0	−10.67	10.67	Fixed end moment
Distribution moment		−8.58	−25.75		Balance moment = −34.33
Carry-over moment	−4.43			−12.88	Carry-over factor = 0.5
Final moment	−49.43	36.42	−36.42	−2.21	Sum of all moments

Thus, final end moments are:

$$M_{ab} = -49.43 \text{ kNm}; \ M_{ba} = -M_{bc} = 36.42 \text{ kNm}; \ M_{cb} = -2.21 \text{ kNm}$$

Notes:
- The same end bending moments for the beam have been obtained in Example E6.1 using the Slope-deflection method.
- Moment-distribution process is simple and less computational expensive.

EXAMPLE 7.3 Analyze the two-span beam shown in Figure E7.3 using the Moment-distribution method. Draw the bending moment diagram and shear force diagram. Assume the constant flexural rigidity of *EI* for both the spans.

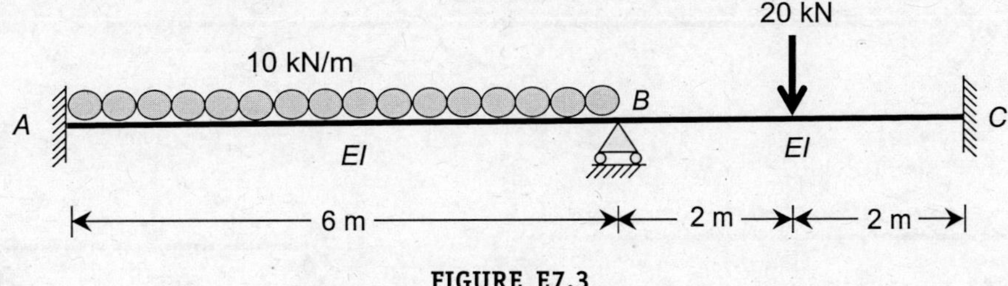

FIGURE E7.3

Solution: **(a) Distribution factors:**
Stiffness factors:

$$K_{ba} = \frac{4EI}{6} = \frac{2EI}{3}; \ K_{bc} = \frac{4EI}{4} = EI$$

Distribution factors:

$$DF_{ba} = \frac{K_{ba}}{K_{ba} + K_{bc}} = 0.4; \ DF_{bc} = \frac{K_{bc}}{K_{ba} + K_{bc}} = 0.6$$

(b) Fixed end moments:

$$M_{ab}^{f} = -M_{ba}^{f} = -\frac{wl^2}{12} = -\frac{10 \times 6^2}{12} \text{ kNm} = -30 \text{ kNm}$$

$$M_{bc}^{f} = -M_{cb}^{f} = -\frac{Pl}{8} = -\frac{20 \times 4}{8} \text{ kNm} = -10 \text{ kNm}$$

(c) Moment-distribution: The moment-distribution process is usually carried out in a tabular format as shown below:

Joint	A	B		C	Remarks
Member	AB	BA	BC	CB	
DF		0.4	0.6		Distribution factor
FEM	−30.0	30.0	−10.0	10.0	Fixed end moment
Distribution moment		−8.0	−12.0		Balance moment = −20
Carry-over moment	−4.0			−6.0	Carry-over factor = 0.5
Final moment	−34.0	−22.0	−22.0	−4.0	Sum of all moments

Thus, the final end moments are

$$M_{ab} = -34.0 \text{ kNm}; \ M_{ba} = -M_{bc} = 22.0 \text{ kNm}; \ M_{cb} = -4.0 \text{ kNm}$$

The beam with the final end moments are shown in the following figure. The support reactions can be determined using the free-body diagram of each span.

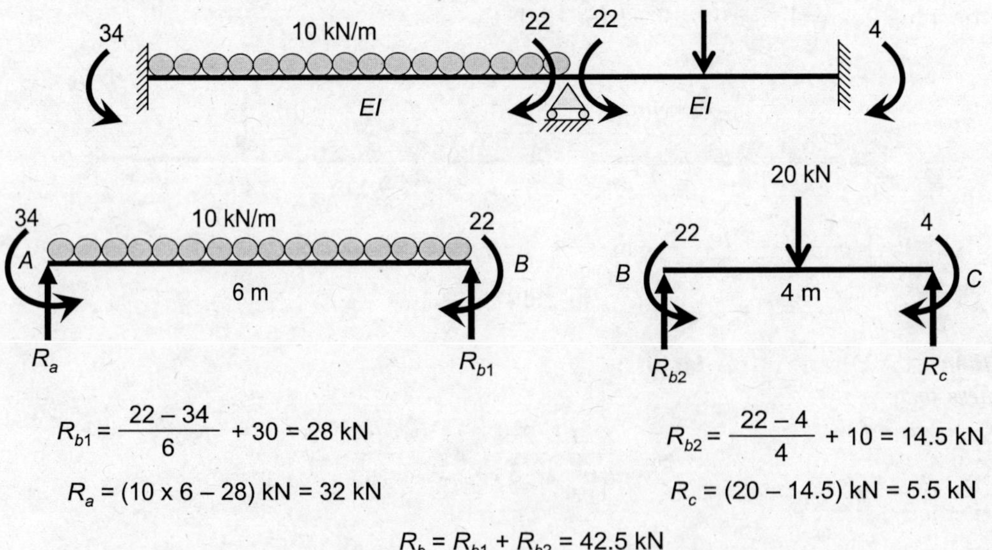

$$R_{b1} = \frac{22 - 34}{6} + 30 = 28 \text{ kN}$$

$$R_a = (10 \times 6 - 28) \text{ kN} = 32 \text{ kN}$$

$$R_{b2} = \frac{22 - 4}{4} + 10 = 14.5 \text{ kN}$$

$$R_c = (20 - 14.5) \text{ kN} = 5.5 \text{ kN}$$

$$R_b = R_{b1} + R_{b2} = 42.5 \text{ kN}$$

The bending moment and shear force diagrams are shown below:

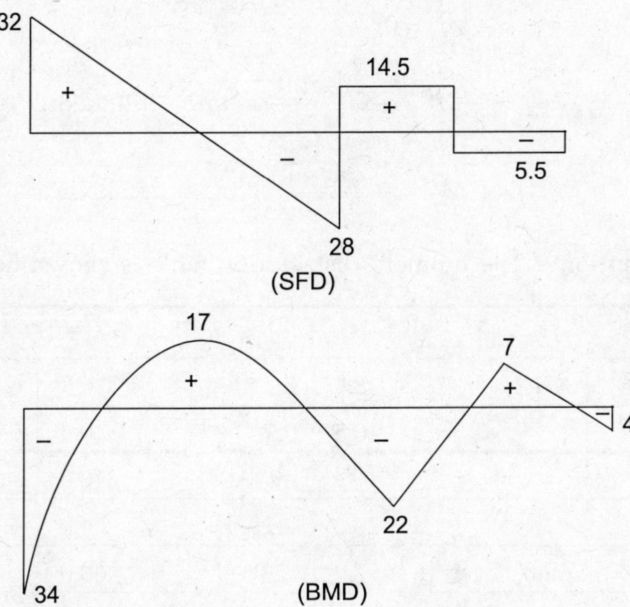

EXAMPLE 7.4   Analyze the continuous beam shown below using the moment-distribution method.

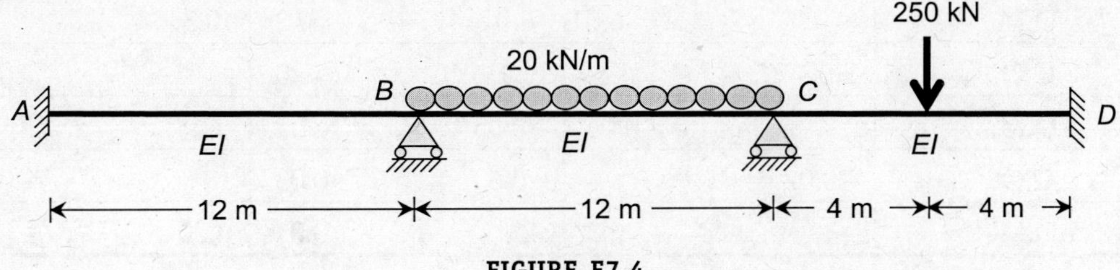

FIGURE E7.4

Solution: **(a) Distribution factors:**

Stiffness factors:

$$K_{ba} = \frac{4EI}{12} = \frac{EI}{3}; \quad K_{bc} = \frac{4EI}{12} = \frac{EI}{3}$$

$$K_{cb} = \frac{4EI}{12} = \frac{EI}{3}; \quad K_{cd} = \frac{4EI}{8} = \frac{EI}{2}$$

Distribution factors:

At joint *B*,
$$DF_{ba} = \frac{K_{ba}}{K_{ba} + K_{bc}} = 0.5; \quad DF_{bc} = \frac{K_{bc}}{K_{ba} + K_{bc}} = 0.5$$

At joint *C*,
$$DF_{cb} = \frac{K_{cb}}{K_{cb} + K_{cd}} = 0.4; \quad DF_{cd} = \frac{K_{cb}}{K_{cb} + K_{cd}} = 0.6$$

(b) Fixed end moments:

$$M_{ab}^{f} = -M_{ba}^{f} = 0$$

$$M_{bc}^{f} = -M_{bc}^{f} = -\frac{20 \times 12^2}{12} \text{ kNm} = -240 \text{ kNm}$$

$$M_{cd}^{f} = -M_{dc}^{f} = -\frac{250 \times 8}{8} \text{ kNm} = -250 \text{ kNm}$$

(c) Moment-distribution: The moment-distribution table is shown below:

Joint	A		B		C	D
Member	AB	BA	BC	CB	CD	DC
DF		0.5	0.5	0.4	0.6	
FEM			−240.0	240.0	−250.0	250.0
D.M.		120.0	120.0	4.0	6.0	
C.O.	60.0		2.0	60.0		3.0
D.M.		−1.0	−1.0	−24.0	−36.0	
C.O.	−0.5		−12.0	−0.5		−18.0
D.M.		6.0	6.0	0.2	0.3	
C.O.	3.0		0.1	3.0		0.15
D.M.		−0.05	−0.05	−1.2	−1.8	
C.O.	−0.02		−0.6	−0.03		−0.90
D.M.		0.3	0.3	0.01	0.02	
Final moment	62.5	125.2	−125.2	281.5	−281.5	234.3

The final end moments are:

$$M_{ab} = 62.5 \text{ kNm}; \; M_{ba} = -M_{bc} = 125.2 \text{ kNm}$$

$$M_{cb} = -M_{cd} = 281.5 \text{ kNm}; \; M_{dc} = 234.3 \text{ kNm}$$

Support reactions can be determined by considered the free-body diagram of each span with applicable end moment and external loading.

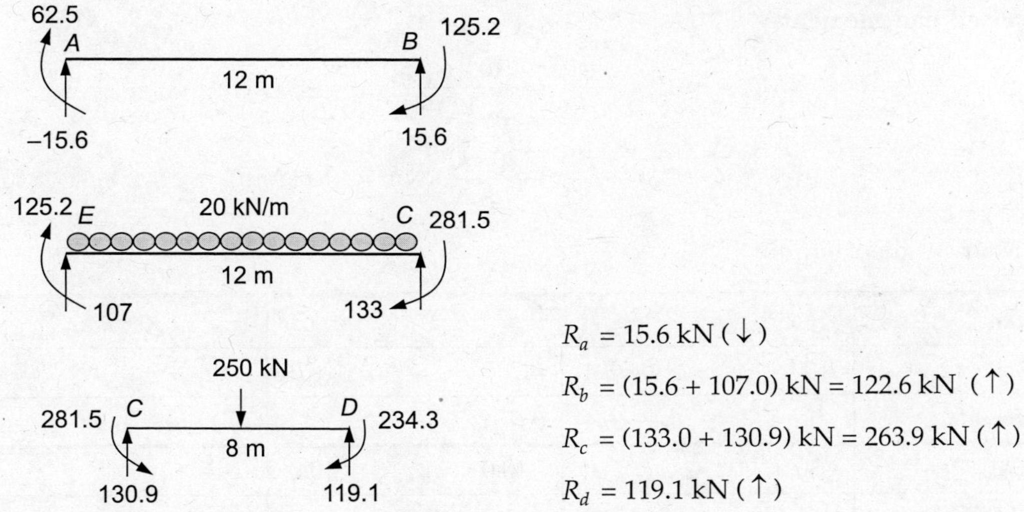

$R_a = 15.6 \text{ kN} (\downarrow)$

$R_b = (15.6 + 107.0) \text{ kN} = 122.6 \text{ kN} (\uparrow)$

$R_c = (133.0 + 130.9) \text{ kN} = 263.9 \text{ kN} (\uparrow)$

$R_d = 119.1 \text{ kN} (\uparrow)$

The shear force and bending moment diagrams are shown below:

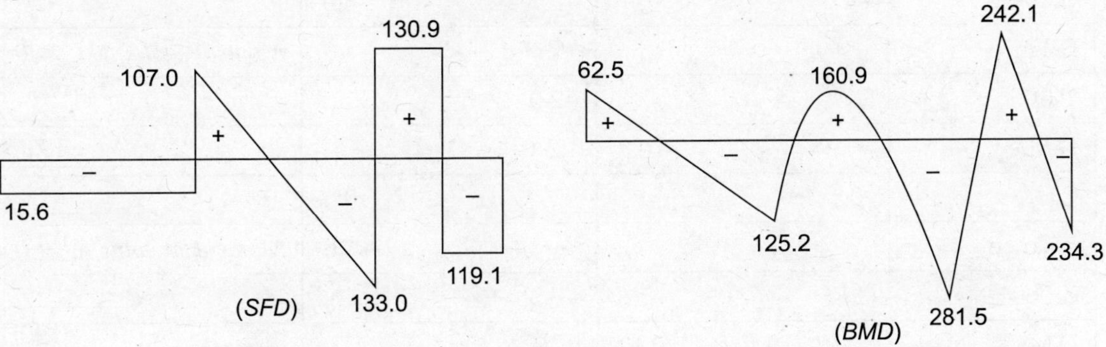

EXAMPLE 7.5   Analyze the continuous beam shown below using the moment-distribution method.

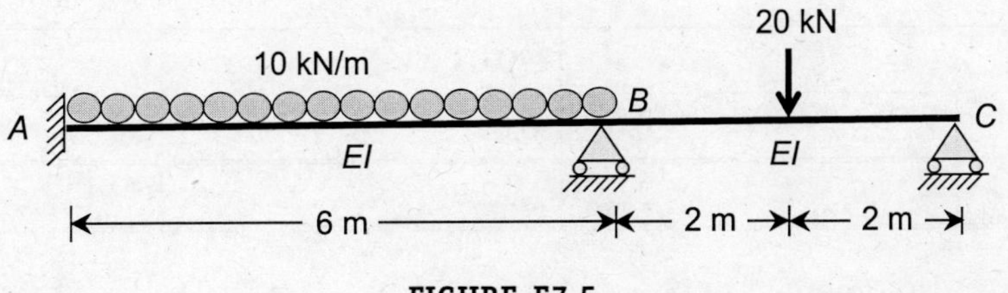

FIGURE E7.5

Solution:
(a) Distribution factors:

$$DF_{ba} = \frac{(4EI/6)}{(4EI/6) + (4EI/4)} = 0.4; \quad DF_{bc} = \frac{(4EI/4)}{(4EI/6) + (4EI/4)} = 0.6$$

(b) Fixed end moments:

$$M_{ab}^{f} = -M_{ba}^{f} = -\frac{wl^2}{12} = -\frac{10 \times 6^2}{12} \text{ kNm} = -30 \text{ kNm}$$

$$M_{bc}^{f} = -M_{cb}^{f} = -\frac{Pl}{8} = -\frac{20 \times 4}{8} \text{ kNm} = -10 \text{ kNm}$$

(c) Moment-distribution:

Joint	A		B	C	Remarks
Member	AB	BA	BC	CB	
DF		0.4	0.6		
FEM	−30.0	30.0	−10.0	10.0	
D.M.		−8.0	−12.0		
C.O.	−4.0			−6.0	
Balance				−4.0	*Moment at joint, C = 0*
C.O.			−2.0		
D.M.		0.8	1.2		
C.O.	0.4			0.6	
Balance				−0.6	*Moment at joint, C = 0*
C.O.			−0.3		
D.M.		0.12	0.18		
C.O.	0.06			0.09	
Balance				−0.09	*Moment at joint, C = 0*
			−0.03		
C.O.		0.01	0.02		
Final moment	−33.5	22.9	−22.9	0	

The final end moments are:

$$M_{ab} = -33.5 \text{ kNm}; \quad M_{ba} = -M_{bc} = 22.9 \text{ kNm}; \quad M_{cb} = 0$$

EXAMPLE 7.6 Analyze the continuous beam shown below using the moment-distribution method.

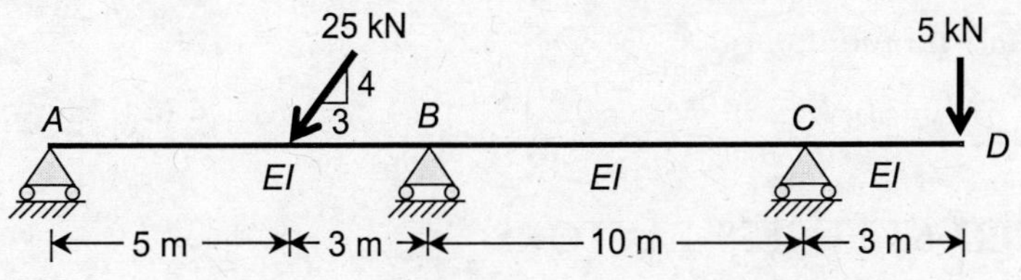

FIGURE E7.6

Solution:

(a) Distribution factors:

$$DF_{ba} = \frac{(4EI/8)}{(4EI/8)+(4EI/10)} = 0.56$$

$$DF_{bc} = \frac{(4EI/10)}{(4EI/8)+(4EI/10)} = 0.44$$

(b) Fixed end moments:

$$M_{ab}^f = -\frac{(25 \times 0.8)(5)(3)^2}{8^2} = -14.1 \text{ kNm}$$

$$M_{ba}^f = \frac{(25 \times 0.8)(5)^2(3)}{8^2} = 23.4 \text{ kNm}$$

$$M_{bc}^f = -M_{cb}^f = 0$$

$$M_{cd}^f = -(5)(3) \text{ kNm} = -15 \text{ kNm}$$

(c) Moment-distribution

Joint	A		B		C	
Member	AB	BA	BC	CB	CD	
DF		0.56	0.44			
FEM	−14.1	23.4			−15.0	
D.M.	14.1	−13.1	−10.3	15.0	Balance at A & C	
C.O.	−6.5	7.1	7.5	−5.2		
D.M.	6.5	−8.2	−6.4	5.2		
C.O.	−4.1	3.2	2.6	−3.2		
D.M.	4.1	−3.2	−2.6	3.2		
C.O.	−1.6	2.1	1.6	−1.3		
Balance	1.6	−2.1	−1.6	1.3		
Final moment	0	9.2	−9.2	15.0	−15.0	

The final end moments are:

$$M_{ab} = 0; \ M_{ba} = -M_{bc} = 9.2 \text{ kNm}; \ M_{cb} = -M_{cd} = 15.0 \text{ kNm}$$

7.5   MODIFIED STIFFNESS FACTORS

In the conventional procedure of moment distribution, all joints of a structure are locked at the beginning and the stiffness factors are computed accordingly. However, in many cases, the moment or deformation conditions at the joints are known a priori. For example, in case of the simple end supports (pins or rollers), the final bending moments must be zero. Similarly, in case of symmetric (or anti-symmetric) cases, the rotations both ends of the members are equal with opposite (or same) directions. In such cases, the modified stiffness factors may be used to reduce the number of iterations in the moment distribution. The modified stiffness factors for these special cases are derived in the following sections.

7.5.1   Members with Simple End Supports

Since the simple end supports are assumed to be fixed in the conventional moment distribution process, the distributed balanced moments are carried over to these joints. Later, the additional moments of equal magnitude but opposite sign are added to order to make the zero final moments at these simple end supports. This procedure has been adopted in previous examples. However, these redundant steps of adding the balanced moment to make the bending moment zero at the simple supports can be eliminated by adopting the modified stiffness factors. The modification permits the simple end supports to remain unlocked after the moments are balanced so that no additional carry over moments are transferred to these joints.

Consider a simply supported beam of length l and flexural rigidity EI as shown in the Figure 7.4. Assume that a bending moment M_{ab} is applied at the joint at which induces a rotation of θ_a at the same joint. The bending moment at joint b, $M_{ba} = 0$. The relation between M_{ab} and θ_a can be obtained using the modified slope-deflection equation as discussed previously.

$$M_{ab} = \frac{3EI}{L}\left(\theta_a - \frac{\Delta}{L}\right) = \frac{3EI}{L}\left(\theta_a - \frac{0}{L}\right) = \frac{3EI}{L}\theta_a \tag{7.7}$$

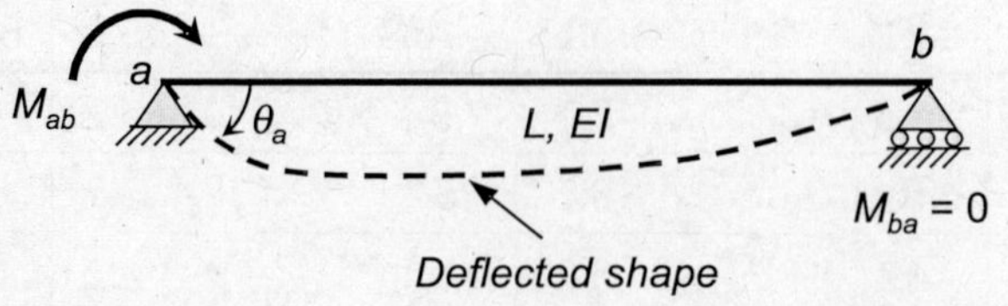

FIGURE 7.4   Simply supported beam with applied end rotation at one support.

The *stiffness* of the beam with a simple support at the far end can be determined as the ratio of the moment required to induce a unit rotation at the near end. Thus,

$$S' = \frac{M_{ab}}{\theta_a} = \frac{3EI}{L} \tag{7.8}$$

Recall that the *stiffness factor* (K) of a beam having the far end as fixed is I/L. Thus, the *modified stiffness factor* (K') for a beam having the far end a simple support is 3/4 times the stiffness factor of the beam with fixed end, i.e.,

$$K' = \frac{3}{4}K \tag{7.9}$$

The *modified stiffness factor* (K') may be used to compute the distribution factors for members connected at a joint. If the modified stiffness factor is used for a beam with simple end support, the modified fixed end moment (M_f') must be used instead of the fixed end moment (M_f).

7.5.2 Members with Symmetric Loading

A beam or frame under symmetric loading has the symmetric joint deformations and bending moments about the axis of symmetry. For the member shown in the Figure 7.6, it is assumed that the loading is symmetric. The end rotations as well as the end moment are equal and opposite in nature.

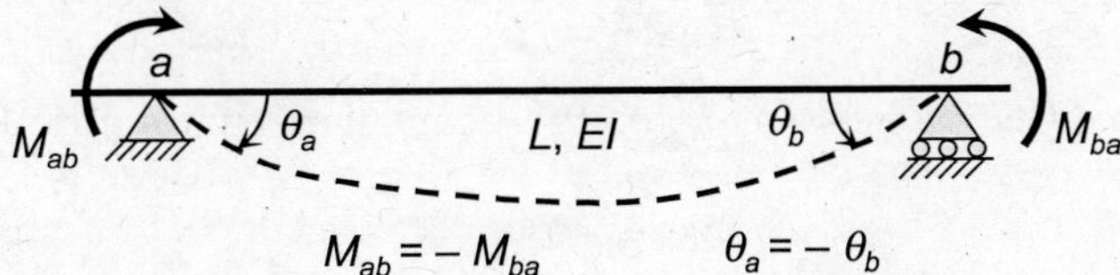

FIGURE 7.5 Beam with symmetric loading.

The moment-rotation relationship can be obtained using the slope deflection equation.

$$M_{ab} = \frac{2EI}{L}(2\theta_a - \theta_a) = \frac{2EI}{L}\theta_a \tag{7.10a}$$

$$M_{ba} = \frac{2EI}{L}(2\theta_b - \theta_b) = \frac{2EI}{L}\theta_b \tag{7.10b}$$

Thus, the *stiffness* for the beam with symmetric loading,

$$K' = \frac{M_{ab}}{\theta_a} = \frac{M_{ba}}{\theta_b} = \frac{2EI}{L} \tag{7.11}$$

Accordingly, the *modified stiffness factor* for the beam with symmetric loading is

$$K' = \frac{2}{4}K = \frac{1}{2}K \tag{7.12}$$

7.5.3 Members with Anti-symmetric Loading

A beam or frame under anti-symmetric loading has the equal joint deformations and bending moments about the axis of symmetry as shown in the Figure 7.7.

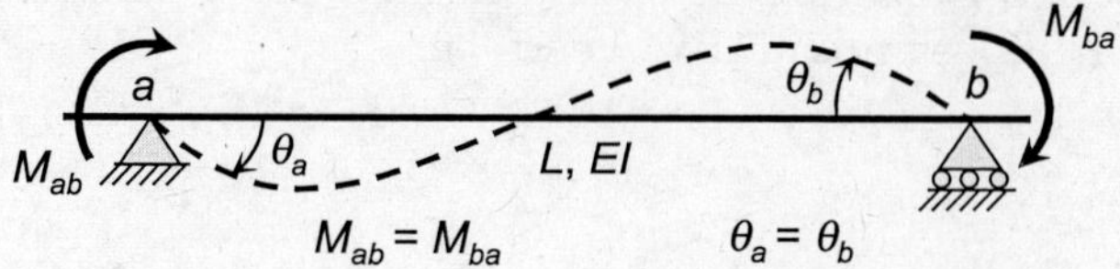

FIGURE 7.7 Beam with anti-symmetric loading.

The moment-rotation relationship can be obtained using the slope deflection equation.

$$M_{ab} = \frac{2EI}{L}(2\theta_a + \theta_a) = \frac{6EI}{L}\theta_a \tag{7.13a}$$

$$M_{ba} = \frac{2EI}{L}(2\theta_b + \theta_b) = \frac{6EI}{L}\theta_b \tag{7.13b}$$

Thus, the *stiffness* for the beam with anti-symmetric loading,

$$K' = \frac{M_{ab}}{\theta_a} = \frac{M_{ba}}{\theta_b} = \frac{6EI}{L} \tag{7.14}$$

Accordingly, the *modified stiffness factor* for the beam with symmetric loading is

$$K' = \frac{6}{4}K = \frac{3}{2}K \tag{7.15}$$

The *stiffness* and *stiffness factors* for members with various end conditions are summarized in Table 7.1.

TABLE 7.1 Stiffness Factors of Members with different End Conditions

End condition	Stiffness	Stiffness factor
Fixed	$\dfrac{4EI}{L}$	K
Pinned	$\dfrac{3EI}{L}$	$\dfrac{3}{4}K$
Symmetric	$\dfrac{2EI}{L}$	$\dfrac{1}{2}K$
Anti-symmetric	$\dfrac{6EI}{L}$	$\dfrac{3}{2}K$

7.6 BEAMS WITH SUPPORT SETTLEMENT

Most civil engineering structures rest on soil or rock foundations which can deform and possibly result in the differential settlement of supports. This relative displacement of supports also develops additional bending moment in the structural members. The procedure to consider this effect in the moment distribution process has been explained in the following section.

Consider a prismatic beam with both ends fixed against rotation. Assume that the support a remains stationary while the support b undergoes a settlement of Δ as shown in Figure 7.8.

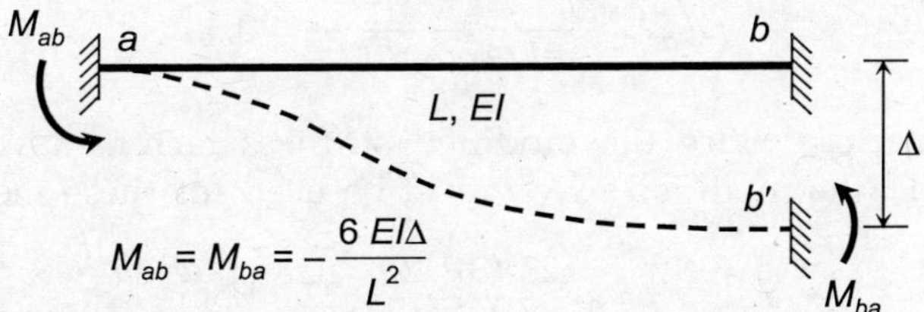

FIGURE 7.8 Beam with relative end displacement.

In a moment-distribution method, the fixed end moments due to the external loads are computed separately assuming no support settlements. The bending moment due to the support settlement is then determined without taking the external loads into account. Both these moments are algebraically added using the principle of superposition. The moment-distribution is carried out for the combined moment at the joints. Similar process is adopted for members with the end rotations as well.

The following examples illustrates the moment distribution analysis of structures with various end conditions.

EXAMPLE 7.7 Analyze the continuous beam shown below using the moment-distribution method. Draw the bending moment diagram.

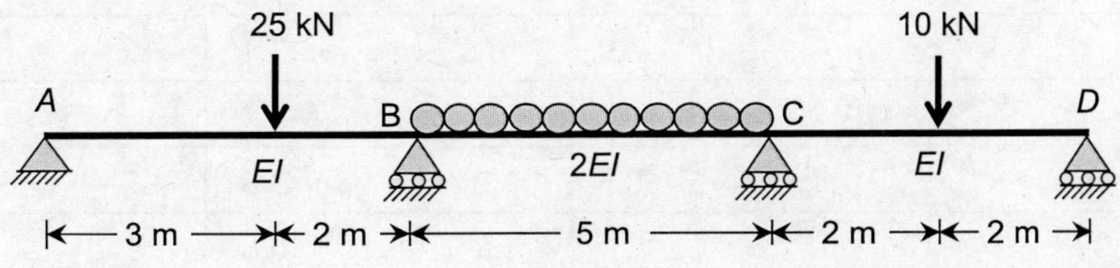

FIGURE E7.7

Solution:

(a) Distribution factors: Since the far ends of members AB and CD are pinned supports, the modified stiffness factors are used to compute the distribution factors for both these members. However, since the bending moments at both ends of the member BC are non-zero, the stiffness factor applicable to the fixed end member should be used.

At joint B,

$$DF_{ba} = \frac{(3EI/5)}{(3EI/5)+(8EI/5)} = 0.27$$

$$DF_{bc} = \frac{(8EI/5)}{(3EI/5)+(3EI/5)} = 0.73$$

At joint C,

$$DF_{cb} = \frac{(8EI/5)}{(8EI/5)+(3EI/4)} = 0.68$$

$$DF_{cd} = \frac{(3EI/4)}{(8EI/5)+(3EI/4)} = 0.32$$

(b) Fixed end moments: Since the modified stiffness factors have been used for the members AB and CD, the modified FEMs must be used for these members.

$$M_{ba}^{f'} = 1.5 \times \frac{(25)(3)^2(2)}{5^2} = 27.0 \text{ kNm}$$

$$M_{bc}^{f} = -M_{cb}^{f} = -\frac{(6)(5)^2}{12} = -12.5 \text{ kNm}$$

$$M_{cd}^{f'} = -1.5 \times \frac{(10)(4)}{8} \text{ kNm} = -7.5 \text{ kNm}$$

(c) Moment-distribution:

Joint	A	B			C	D
Member	AB	BA	BC	CB	CD	DC
DF		0.27	0.73	0.68	0.32	
FEM		27.0	−12.5	12.5	−7.5	
D.M.		−3.9	−10.6	−3.4	−1.6	
C.O.			−1.7	−5.3		
D.M.		0.5	1.2	3.6	1.7	
C.O.			1.8	0.6		
D.M.		−0.5	−1.3	−0.4	−0.2	
C.O.			−0.2	−0.7		
D.M.		0.1	0.1	0.5	0.2	
Final moment	0	23.2	−23.2	7.4	−7.4	0

The final end moments are:

$$M_{ab} = 0; \ M_{ba} = -M_{bc} = 23.2 \text{ kNm}; \ M_{cb} = -M_{cd} = 7.4 \text{ kNm}; \ M_{dc} = 0$$

The bending moment diagram of the beam is shown below:

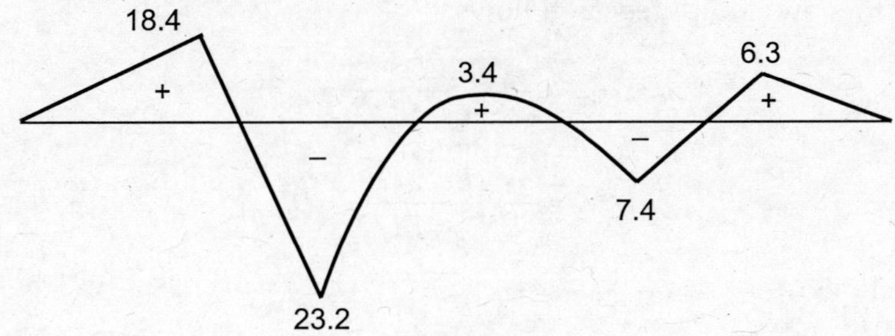

EXAMPLE 7.8 Determine the end moments of the continuous beam shown in Figure E7.8.

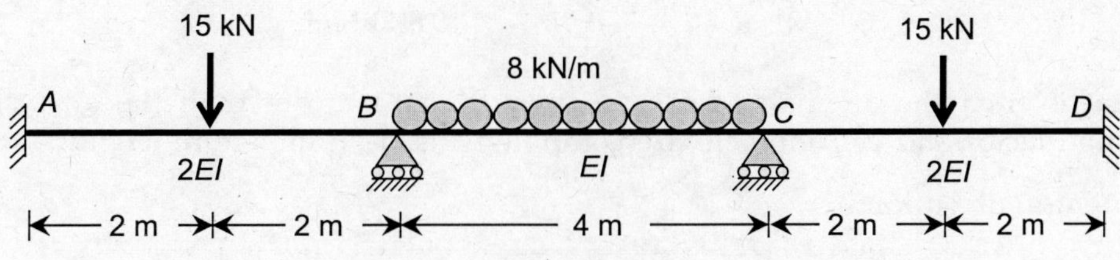

FIGURE E7.8

Solution: For the given support conditions and external loading, it can be seen that the loading on beam is symmetric. The qualitative deformed shape of the beam is shown below. Since the end moments and joint deformations of members on one side of the centre line would be equal and opposite to those of the other side. Thus, it is sufficient to analyze only one-half of the beam considering the symmetric case. In this example, segment *AE* of the beam is considered in the moment-distribution analysis.

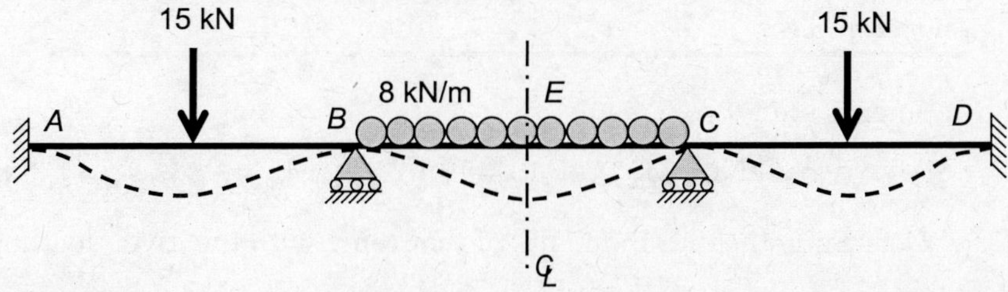

(a) Distribution factors: Since the far end of the member AB is fixed, the stiffness of this member is given by

$$S_{ba} = \frac{4(EI)_{ba}}{L_{ba}} = \frac{4(2EI)}{4} = 2EI$$

For the member BC, the end rotations are symmetric. Thus, the stiffness of member BC is

$$S_{bc} = \frac{4(EI)_{bc}}{L_{bc}} = \frac{2(EI)}{4} = \frac{EI}{2}$$

Distribution factors are computed as follows:

$$DF_{ba} = \frac{(2EI)}{(2EI) + (EI/2)} = 0.8$$

$$DF_{bc} = \frac{(EI/2)}{(2EI) + (EI/2)} = 0.2$$

(b) Fixed end moments:

$$M_{ab}^{f} = -M_{ba}^{f} = -\frac{(15)(4)}{8} = -7.5 \text{ kNm}$$

$$M_{bc}^{f} = -\frac{(8)(4)^2}{12} \text{ kNm} = -10.67 \text{ kNm}$$

Note that even if the one-half of the member BC is considered in the analysis, both distribution factor and end moment are computed based on the entire length.

(c) Moment-distribution:

Joint	A	B		E
Member	AB	BA	BC	EB
DF		0.8	0.2	
FEM	–7.5	7.5	–10.7	
D.M.		2.6	0.6	
C.O.	1.3			
Final moment	–6.2	10.1	–10.1	

The final end moments are:

$$M_{ab} = -M_{dc} = -6.2 \text{ kNm}; \ M_{ba} = -M_{bc} = 10.1 \text{ kNm}; \ M_{cb} = -M_{cd} = 10.1 \text{ kNm}$$

EXAMPLE 7.9 Determine the end moments of the frame with the given loading as shown in Figure E7.9.

Solution: For the given support conditions and external loading, it can be seen that the loading on beam is anti-symmetric. Thus, it is sufficient to analyze only one-half of the beam considering the anti-symmetric case.

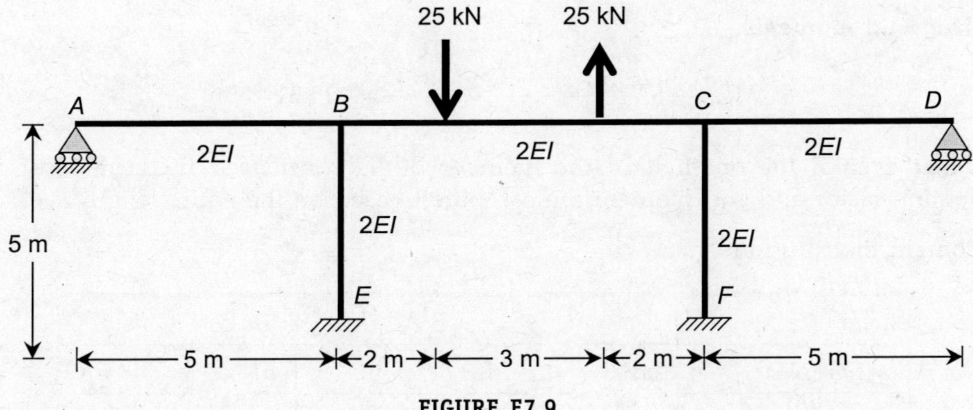

FIGURE E7.9

(a) Distribution factors:

Modified stiffness member *AB* is $S_{ba} = \dfrac{3(EI)_{ba}}{L_{ba}} = \dfrac{3(2EI)}{5} = 1.2EI$

Stiffness of member *BE* is $S_{be} = \dfrac{4(EI)_{be}}{L_{be}} = \dfrac{4(2EI)}{5} = 1.6EI$

Modified stiffness member *BC* is $S_{bc} = \dfrac{6(EI)_{ba}}{L_{ba}} = \dfrac{6(2EI)}{7} = 1.71EI$

Distribution factors are computed as follows:

$$DF_{ba} = \frac{(1.2EI)}{(1.2EI) + (1.6EI) + (1.71EI)} = 0.27$$

$$DF_{be} = \frac{(1.6EI)}{(1.2EI) + (1.6EI) + (1.71EI)} = 0.35$$

$$DF_{bc} = \frac{(1.71EI)}{(1.2EI) + (1.6EI) + (1.71EI)} = 0.38$$

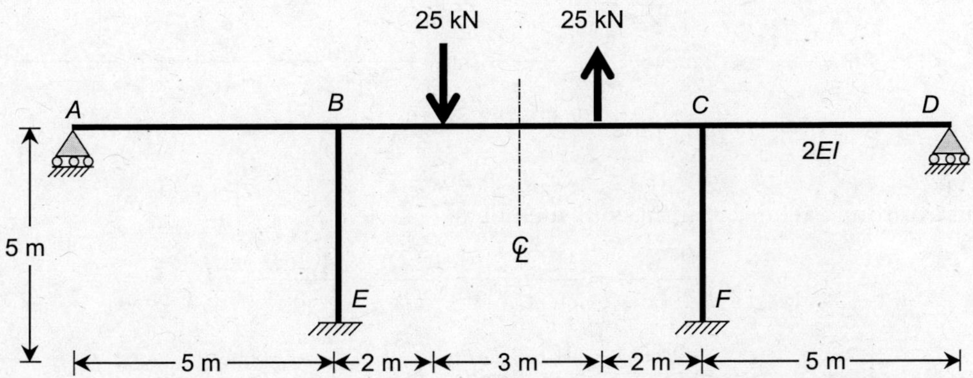

(b) Fixed end moments:

$$M_{bc}^f = -\frac{(25)(2)(5)^2}{7^2} + \frac{(25)(5)(2)^2}{7^2} = -15.31 \text{ kNm}$$

Note that even if the one-half of the member BC is considered in the analysis, both distribution factor and end moment are computed based on the entire length.

(c) Moment-distribution:

Joint	A	B			E
Member	AB	BA	BC	BE	EB
DF		0.27	0.38	0.35	
FEM			−15.31		
D.M.		4.13	5.82	5.36	
C.O.					2.68
Final moment	0	4.13	−9.49	5.36	2.68

The final end moments are:

$$M_{ab} = M_{dc} = 0; \ M_{ba} = M_{cd} = 4.13 \text{ kNm}; \ M_{bc} = M_{cb} = -9.49 \text{ kNm}$$

$$M_{be} = M_{cf} = 5.36 \text{ kNm}; \ M_{eb} = M_{fc} = 2.68 \text{ kNm}$$

EXAMPLE 7.10   Determine the end moments of the continuous beam shown in Figure E7.10 if the supports B and C settle by 30 mm and 50 mm, respectively. $EI = 8000 \text{ kNm}^2$.

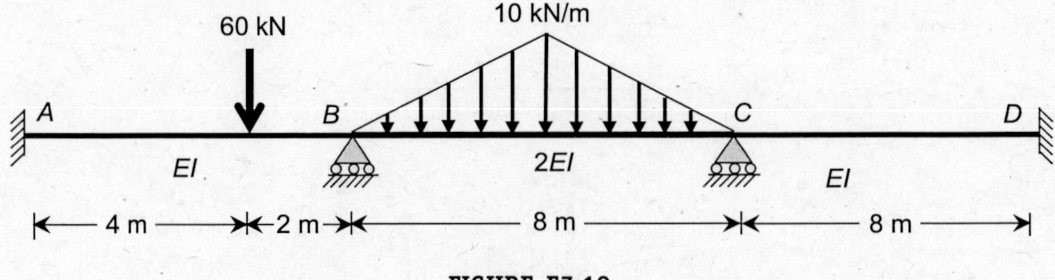

FIGURE E7.10

Solution:

(a) Distribution factors: Stiffness of members

$$S_{ba} = \frac{4(EI)_{ba}}{L_{ba}} = \frac{4EI}{6}; \ S_{bc} = \frac{4(EI)_{bc}}{L_{bc}} = \frac{4(2EI)}{8} = EI$$

$$S_{cd} = \frac{4(EI)_{cd}}{L_{cd}} = \frac{4EI}{8} = 0.5EI$$

Distribution factors for members at joints B and C:

$$DF_{ba} = \frac{(4EI/6)}{(4EI/6)+(EI)} = 0.4 \; ; \; DF_{bc} = \frac{(EI)}{(4EI/6)+(EI)} = 0.6$$

$$DF_{cb} = \frac{(EI)}{(EI)+(0.5EI)} = 0.67 \; ; \; DF_{cd} = \frac{(0.5EI)}{(EI)+(0.5EI)} = 0.33$$

(b) Fixed end moments: Fixed end moments of members are computed by adding the moments due to the applied loads and the support settlements.

$$M_{ab}^{f} = -\frac{(60)(4)(2)^2}{6^2} - \frac{(6)(8000)(0.03)}{8^2} = -66.7 \text{ kNm}$$

$$M_{ba}^{f} = \frac{(60)(4)^2(2)}{6^2} - \frac{(6)(8000)(0.03)}{8^2} = 13.3 \text{ kNm}$$

$$M_{bc}^{f} = -\frac{(5)(40)(8)^2}{96} - \frac{(6)(8000)(0.02)}{8^2} = -148.3 \text{ kNm}$$

$$M_{cb}^{f} = \frac{(5)(40)(8)^2}{96} - \frac{(6)(8000)(0.02)}{8^2} = 118.3 \text{ kNm}$$

$$M_{cd}^{f} = M_{dc}^{f} = \frac{(6)(8000)(0.05)}{8^2} = 37.5 \text{ kNm}$$

(c) Moment-distribution:

Joint	A	B		C		D
Member	AB	BA	BC	CB	CD	DC
DF		0.4	0.6	0.67	0.33	
FEM	−66.7	13.3	−148.3	118.3	37.5	37.5
D.M.		54.0	81.0	−104.4	−51.4	
C.O.	27.0		−52.2	40.5		−25.7
D.M.		20.9	31.3	−27.1	−13.4	
C.O.	10.5		−13.6	15.7		−6.7
D.M.		5.4	8.2	−10.5	−5.2	
C.O.	2.7		−5.3	4.1		−2.6
D.M.		2.1	3.2	−2.7	−1.4	
C.O.	1.1		−1.4	1.6		−0.7
D.M.		0.6	0.8	−1.1	−0.5	
C.O.	0.3		−0.6	0.4		−0.3
D.M.		0.2	0.4	−0.3	−0.1	
C.O.	0.1					
Final moment	−25.0	96.5	−96.5	34.5	−34.5	1.5

The final end moments are:

$$M_{ab} = -25 \text{ kNm}; \quad M_{ba} = -M_{bc} = 96.5 \text{ kNm}$$

$$M_{cb} = -M_{cd} = 34.5 \text{ kNm}; \quad M_{dc} = 1.5 \text{ kNm}$$

7.7 FRAMES WITHOUT SIDESWAY

Frames would not undergo sidesway if they are restrained against sidesway or they have symmetric loading and geometry. For these frames, the moment-distribution analysis procedure explained earlier is also applicable. This has been illustrated in the following examples.

EXAMPLE 7.11 Find the end moments of the frame shown in Figure E7.11 if the support C is rotated by 0.0015 radians clockwise. $EI = 5000 \text{ kNm}^2$.

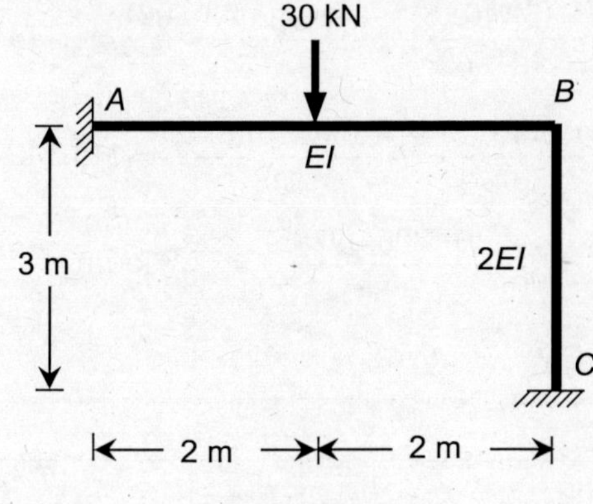

FIGURE E7.11

Solution:
(a) Distribution factors:

$$DF_{ba} = \frac{(4EI/4)}{(4EI/4) + (8EI/3)} = 0.27$$

$$DF_{bc} = \frac{(8EI/3)}{(4EI/4) + (8EI/3)} = 0.73$$

(b) Fixed end moments:

$$M_{ab}^{f} = -M_{ba}^{f} = -\frac{(30)(4)}{8} = -15.0 \text{ kNm}$$

$$M_{bc}^{f} = 0$$

$$M_{cb}^{f} = \frac{4(EI)_{bc}\theta_c}{L_{bc}} = \frac{(4)(2 \times 5000)(0.0015)}{3} \text{ kNm} = 20.0 \text{ kNm}$$

(c) Moment-distribution:

Joint	A	B		C
Member	AB	BA	BC	CB
DF	0.27		0.73	
FEM	−15.0	15.0	20.0	
C.O.			10.0	
D.M.		−6.8	−18.2	
C.O.	−3.4			−9.1
Final moment	−18.4	8.2	−8.2	10.9

The final end moments are:

$$M_{ab} = -18.4 \text{ kNm}; \ M_{ba} = -M_{bc} = 8.2 \text{ kNm}; \ M_{cb} = 10.9 \text{ kNm}$$

EXAMPLE 7.12 Determine the end moments of the rigid frame shown in Figure E7.12 using the moment-distribution method. All members are of constant flexural rigidity (*EI*).

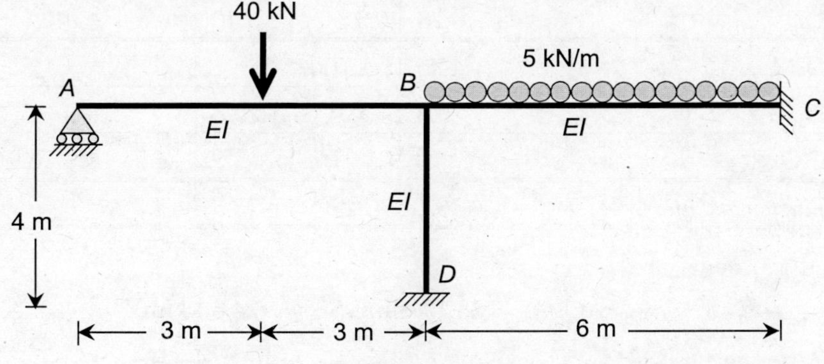

FIGURE E7.12

Solution:

(a) Distribution factors: Stiffness of members

$$S_{ba} = \frac{3(EI)_{ba}}{L_{ba}} = \frac{3EI}{6} = \frac{EI}{2}$$

$$S_{bc} = \frac{4(EI)_{bc}}{L_{bc}} = \frac{4(EI)}{6} = \frac{2EI}{3}$$

$$S_{bd} = \frac{4(EI)_{bd}}{L_{bd}} = \frac{4EI}{4} = EI$$

Distribution factors for members at the joint B

$$DF_{ba} = \frac{(EI/2)}{(EI/2) + (2EI/3) + (EI)} = 0.23$$

$$DF_{bc} = \frac{(2EI/3)}{(EI/2) + (2EI/3) + (EI)} = 0.31$$

$$DF_{bd} = \frac{(EI)}{(EI/2) + (2EI/3) + (EI)} = 0.46$$

(b) Fixed end moments:

$$M^f_{ba} = 1.5 \times \frac{(40)(6)}{8} = 45 \text{ kNm}$$

$$M^f_{bc} = -M^f_{cb} = -\frac{(5)(6)^2}{12} = -15 \text{ kNm}$$

(c) Moment-distribution:

Joint		A	B			C	D
Member	AB	BA	BC		BD	CB	DB
DF		0.23	0.31		0.46		
FEM		45.0	−15.0				
D.M.		−6.9	−9.3		−13.8		
C.O.						−4.7	−6.9
Final moment	0	38.1	−24.3		−13.8	−4.7	−6.9

The final end moments are:

$$M_{ab} = 0; \ M_{ba} = 38.1 \text{ kNm}; \ M_{bc} = -24.3 \text{ kNm}$$

$$M_{bd} = -13.8 \text{ kNm}; \ M_{cb} = -4.7 \text{ kNm}; \ M_{db} = -6.9 \text{ kNm}$$

EXAMPLE 7.13 Determine the end moments of the rigid frame shown in Figure E7.13 using the moment-distribution method. All members are of constant flexural rigidity (EI).

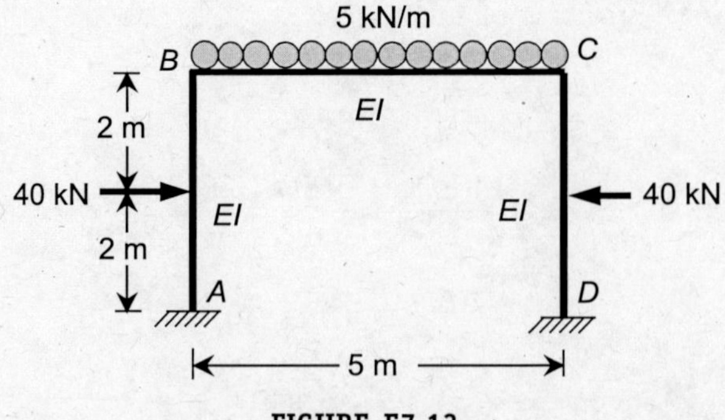

FIGURE E7.13

Solution: The geometry and loading of the frame are symmetric. Hence, the stiffness factors of the member BC can be computed considering symmetric case. Further, it is sufficient to analyze only one-half of the frame.

(a) Distribution factors:

Stiffness of member AB, $S_{ba} = \dfrac{4(EI)_{ba}}{L_{ba}} = \dfrac{4EI}{4} = EI$

Modified stiffness of member BC, $S_{bc} = \dfrac{2(EI)_{bc}}{L_{bc}} = \dfrac{2EI}{5}$

Distribution factors for members at the joint B

$$DF_{ba} = \frac{(EI)}{(EI)+(2EI/5)} = 0.71$$

$$DF_{bc} = \frac{(2EI/5)}{(EI)+(2EI/5)} = 0.29$$

(b) Fixed end moments:

$$M^f_{ab} = -M^f_{ba} = -\frac{(40)(4)}{8}\,\text{kNm} = -20\ \text{kNm}$$

$$M^f_{bc} = -\frac{(5)(5)^2}{12}\,\text{kNm} = -10.4\ \text{kNm}$$

(c) Moment-distribution:

Joint	A	B	
Member	AB	BA	BC
DF		0.71	0.29
FEM	−20.0	20.0	−10.4
D.M.		−6.8	−2.8
C.O.	−3.4		
Final moment	−23.4	13.2	−13.2

The final end moments are:

$$M_{ab} = -23.4\ \text{kNm}; \quad M_{ba} = -M_{bc} = 13.2\ \text{kNm}$$

$$M_{dc} = 23.4\ \text{kNm}; \quad M_{cd} = -M_{cb} = -13.2\ \text{kNm}$$

7.8   FRAMES WITH SIDESWAY

Many frames in practice undergo joint translations (sidesway) under the application of external loadings. However, the magnitudes of these translations are not known a priori.

As a result, the actual fixed end moments of members due to the relative joint translations cannot be computed for moment distribution analysis. Though it is possible to first perform displacement analyses, this process would be tedious and practically not possible in the context of the moment distribution analysis.

Moment-distribution analysis of sway-frames are carried out in broadly in two steps, namely, (i) non-sway analysis, and (ii) sway analysis. Non-sway analysis involves the distribution of fixed end moments for the frame due to applied loading with sway prevented. Sway analysis is carried out for the frames with assume fixed end moments due to joint translations without applied loading. The final end moments are computed by superposing the moments obtained from the non-sway analysis and sway analysis using appropriate corrections. This procedure has been illustrated as follows:

Consider a single-story single-bay rigid frame subjected to vertical and horizontal forces as shown in Figure 7.9. As evident from the deflected shape, the frame is sway frame in which the joint translations can be observed at joints B and C. Considering the members are axially inextensible, both joints would undergo the same translation, Δ.

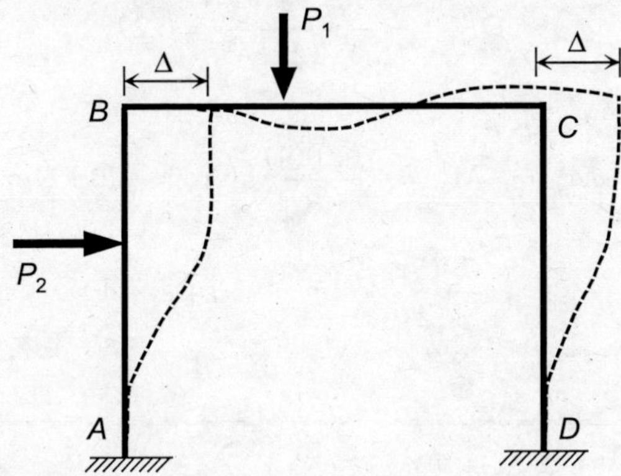

FIGURE 7.9 Portal frame with side-sway displacement

The analysis of this rigid frame should be carried out in two steps in the moment distribution method. Figure 7.10(a) shows the frame with all external loads assuming an artificial roller support at joint C in order to prevent the joint translations. The moment-distribution analysis is first carried out for this non-sway frame to determine the end moments of all members. Using the end moments obtained from the analysis and the applied loads, the horizontal shear forces can be obtained for both vertical members (i.e., *AB* and *CD*). These horizontal forces would not in equilibrium with the applied loads in the horizontal loads as the effect of sway (joint translation) has not been considered in the moment-distribution analysis at this stage. The unbalanced horizontal shear would result in the reaction force, R at the artificial roller support at the joint C.

Next, assume that the frame is subjected to a horizontal force, R at the joint C. Since there is no actual force acting in this direction at the same joint, the applied load should be equal in magnitude but opposite in sign. When the force R is applied to the frame, one would expect the translations of joints *B* and *C*. The moment-distribution analysis can then be conducted following the case of support settlements as illustrated in some examples earlier. Though one could compute the exact magnitude of this joint translation, this process would be very time-consuming and tedious. Instead, the moment-distribution analysis in the second stage (i.e., sway analysis) is carried out assuming any value of joint translation (D_1) as shown in the Figure 7.10(b). At the end of the moment distribution analysis, one would get the value of horizontal force R_1 from the horizontal shear equilibrium. Since for the equilibrium R must be equal to R_1, all end moments obtained in the sway analysis must be corrected in the ratio of R_1/R to get the end moments corresponding to the actual horizontal reaction of R. The final end moments are obtained by summing up the end moments obtained from the non-sway analysis and the sway analysis.

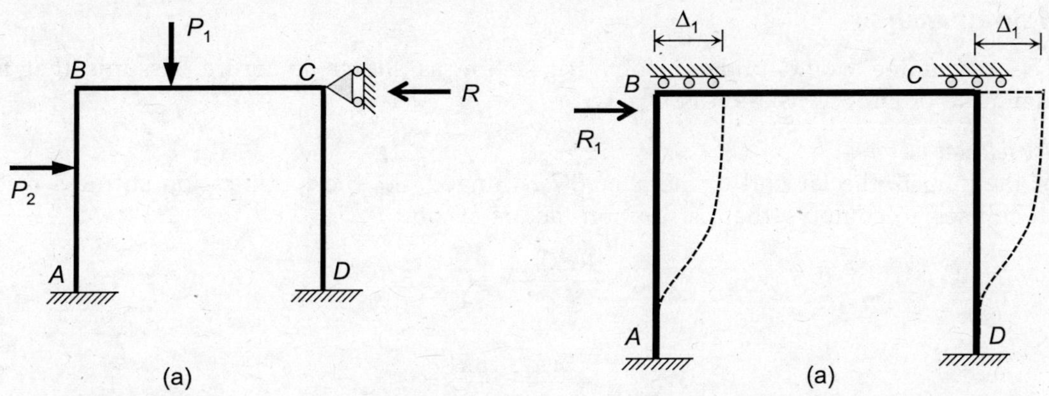

FIGURE 7.10   (a) Assumed frame for non-sway analysis, (b) Frame with side sway.

The moment-distribution analysis of the sway frames including the gable ends has been illustrated in the following examples.

EXAMPLE 7.14   Analyze the frame shown in Figure E7.14 using *moment-distribution method*. The joint C is a hinged. Flexural rigidity of each member is *EI*.

Solution: The frame can undergo side-sway under the applied loading. Thus, the moment-distribution analysis is required to be carried out in two-steps, i.e., non-sway analysis and sway analysis. The frames considered for the non-sway and sway analyses are shown in the following figure.

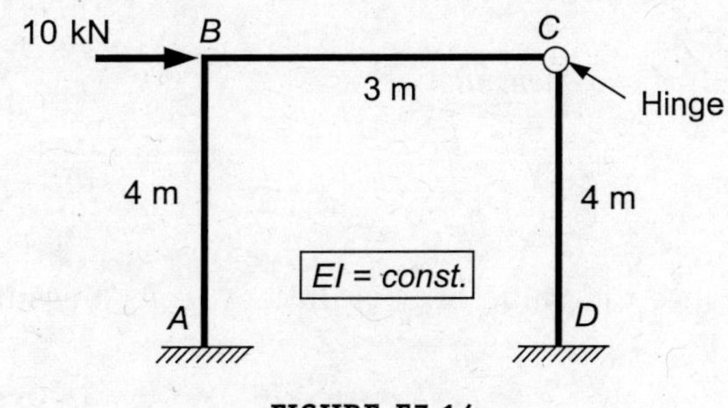

FIGURE E7.14

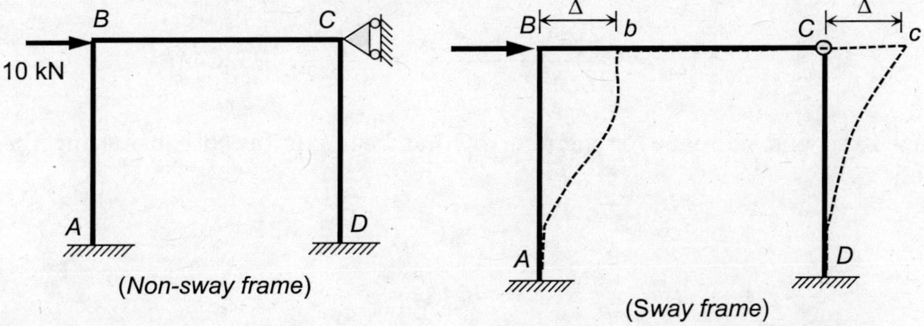

(a) Non-sway analysis

For the non-sway analysis, the joint translation should be prevented using a restraint at joint C in the horizontal direction. The moment-distribution analysis should be carried out for the frame with all applied loads.

One could note that the external load of 10 kN is applied at the joints. Hence, there would be no fixed end moments in any members of the non-sway frame in this case. Thus, end moments of the non-sway frame are zero.

(b) Sway analysis

The sway frame would undergo joint translation as shown in figure. Assume that the magnitude of joint translation is D.

Distribution factors:
For the joint B, the far end of member BC is hinged. So, the modification stiffness factor can be used to compute the distribution factors at joint B.

$$S_{ba} = \frac{4(EI)_{ba}}{L_{ba}} = \frac{4EI}{4} = EI$$

$$S_{bc} = \frac{3(EI)_{bc}}{L_{bc}} = \frac{3EI}{3} = EI$$

Distribution factors at the joint B

$$DF_{ba} = \frac{(EI)}{(EI)+(EI)} = 0.5 = DF_{bc}$$

Fixed end moments:

$$M_{ab}^{f} = M_{ba}^{f} = -\frac{6EI\Delta}{(L_{ab})^2} = -\frac{6EI\Delta}{4^2}$$

Since the value of D is unknown, let's assume that $\dfrac{6EI\Delta}{4^2} = 10 \text{ kNm}$. Thus,

$$M_{ab}^{f} = M_{ba}^{f} = -10 \text{ kNm}$$

$$M_{bc}^{f} = 0$$

$$M_{dc}^{f} = -\frac{6EI\Delta}{(L_{dc})^2} + \frac{1}{2}\frac{6EI\Delta}{(L_{dc})^2} = -\frac{3EI\Delta}{4^2} = -5 \text{ kNm}$$

Note that fixed end moment for member BC has been computed considering the far end is hinged.

Moment-distribution

Joint	A	B		C		D
Member	AB	BA	BC	CB	CD	DC
DF		0.5	0.5			
FEM	−10.0	−10.0				−5.0
D.M.		5.0	5.0			
C.O.	2.5					
Final moment	−7.5	−5.0	5.0	0	0	−5.0

For the end moments computed above, next step is to determine the horizontal shear. For this consider the free-body diagrams of each vertical member separately as shown below.

Equilibrium Equations:
Taking moment about B,

$$4 \times H_a - M_{ab} - M_{ba} = 0 \qquad \therefore H_a = \frac{M_{ab} + M_{ba}}{4} = \frac{-7.5 - 5.0}{4} \text{ kN} = -3.12 \text{ kN} (\leftarrow)$$

Taking moment about C,

$$4 \times H_d - M_{dc} = 0 \qquad \therefore H_d = \frac{M_{dc}}{4} = \frac{-5}{4} \text{ kN} = -1.25 \text{ kN} (\leftarrow)$$

Total horizontal shear, $\therefore H = H_a + H_d = (-3.12 - 1.25) \text{ kN} = -4.37 \text{ kN} (\leftarrow)$

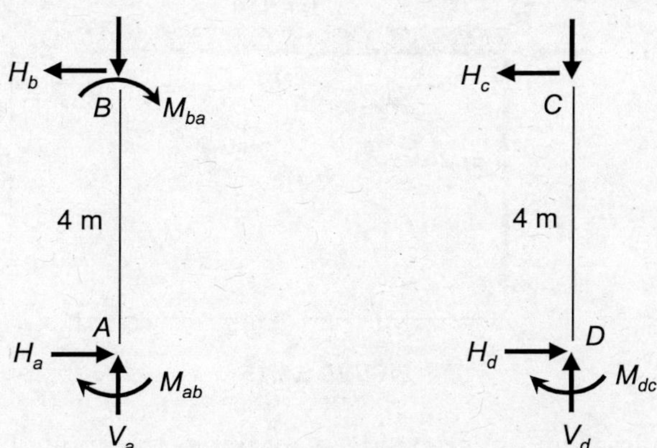

Considering the overall equilibrium of the frame,

$$R' = H = 4.37 \text{ kN}$$

However, the actual value of R' should be 10 kN. Therefore, the computed end moments should be corrected by a factor of (10/4.37).

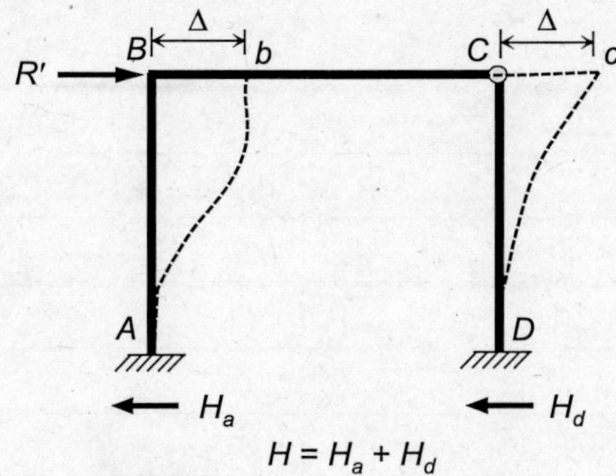

(c) Final bending moments

$$M_{ab} = \left(\frac{10.0}{4.37}\right)(-7.5) \text{ kNm} = -17.17 \text{ kNm}$$

$$M_{ba} = -M_{bc} = \left(\frac{10.0}{4.37}\right)(-5) \text{ kNm} = -11.45 \text{ kNm}$$

$$M_{cb} = M_{cd} = 0$$

$$M_{dc} = \left(\frac{10.0}{4.37}\right)(-5) \text{ kNm} = -11.45 \text{ kNm}$$

EXAMPLE 7.15   Analyze the frame shown in Figure E7.15 using *moment-distribution method*.

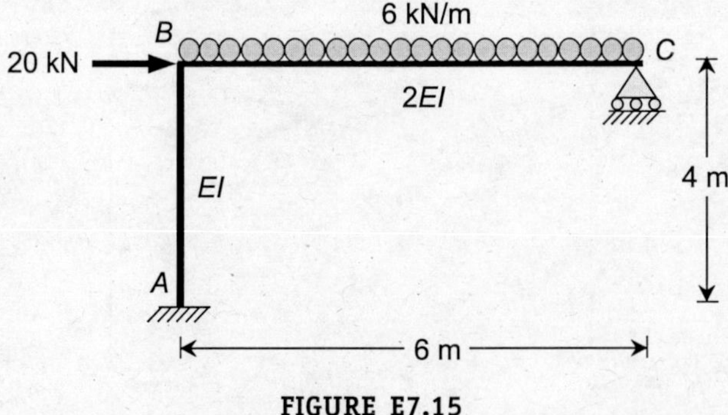

FIGURE E7.15

Solution: Since the frame can undergo side-sway under the applied loading, the moment-distribution analysis has to be carried out in two-steps, i.e., non-sway analysis and sway analysis.

(a) Non-sway analysis

For the non-sway analysis, the joint translation should be prevented using a restraint at joint *C* in the horizontal direction as shown in the figure below. The moment-distribution analysis should be carried out for the frame with all applied loads.

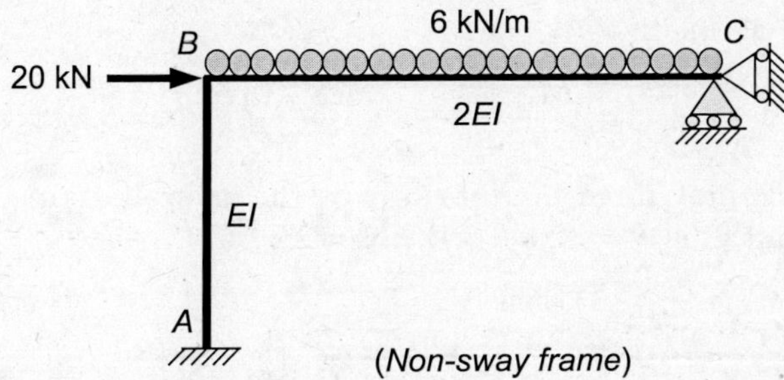

Distribution factors: For the joint B, the far end of member BC is hinged. So, the modification stiffness factor can be used to compute the distribution factors at joint B.

$$S_{ba} = \frac{4(EI)_{ba}}{L_{ba}} = \frac{4EI}{4} = EI$$

$$S_{bc} = \frac{3(EI)_{bc}}{L_{bc}} = \frac{6EI}{6} = EI$$

Distribution factors at the joint B

$$DF_{ba} = \frac{(EI)}{(EI)+(EI)} = 0.5 = DF_{bc}$$

Fixed end moments

$$M^f_{ab} = M^f_{ba} = 0$$

$$M^f_{bc} = -1.5 \times \frac{(6)(6)^2}{12} \text{ kNm} = -27.0 \text{ kNm}$$

Note that the fixed end moment for member BC is computed considering the far end as hinged.

Moment-distribution

Joint	A	B		C
Member	AB	BA	BC	CB
DF		0.5	0.5	
FEM			−27.0	
D.M.		13.5	13.5	
C.O.	6.8			
Final moment	6.8	13.5	−13.5	0

Horizontal shear: Determine the horizontal reaction at the joint A considering the free-body diagrams.

Taking moment about B,

$$4 \times H_a - M_{ab} - M_{ba} = 0 \quad \therefore \ H_a = \frac{M_{ab} + M_{ba}}{4} = \frac{6.8 + 13.5}{4} \text{kN} = 5.1 \text{ kN}(\rightarrow)$$

Considering the horizontal force equilibrium of the overall frame, the reaction at the restraint C is $R = (20 + 5.1)$ kN $= 25.1$ kN.

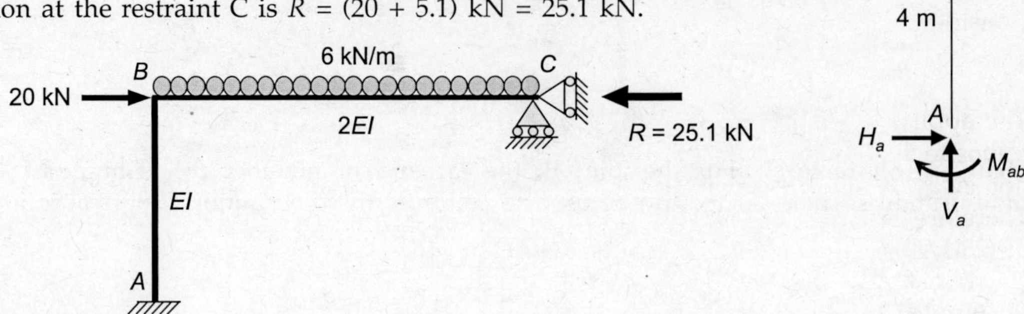

(b) Sway analysis

The sway frame would undergo joint translation as shown in figure. Assume that the magnitude of joint translation is D.

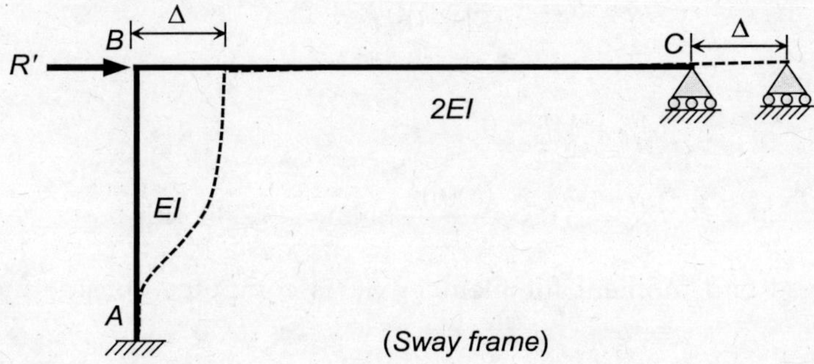

Fixed end moments

$$M_{ab}^f = M_{ba}^f = -\frac{6EI\Delta}{(L_{ab})^2} = -\frac{6EI\Delta}{4^2} = -10 \text{ kNm } (say)$$

$$M_{bc}^f = 0$$

Moment-distribution

Joint	A	B		C
Member	AB	BA	BC	CB
DF		0.5	0.5	
FEM	−10.0	−10.0		
D.M.		5.0	5.0	
C.O.	2.5			
Final moment	−7.5	−5.0	5.0	0

Horizontal shear

Taking moment about B,

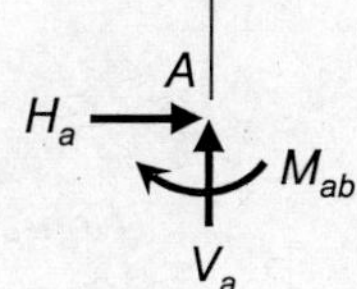

$$4 \times H_a - M_{ab} - M_{ba} = 0 \quad \therefore \quad H_a = \frac{M_{ab} + M_{ba}}{4} = \frac{-7.5 - 5.0}{4} \text{ kN} = -3.12 \text{ kN} \;(\leftarrow)$$

Considering the overall equilibrium of the frame,

$$R' = H_a = 3.12 \text{ kN}$$

The applied force (R') to cause the joint translation should be equal and opposite to the reaction (R) obtained at the assumed restraint in the non-sway frame. However, the actual value of R' should be 25.1 kN. Therefore, the computed end moments should be corrected using a factor of (25.1/3.12).

(c) Final bending moments

The final end moments can be obtained by adding the end moments obtained from the non-sway and sway analyses.

$$M_{ab} = 6.8 + \left(\frac{25.1}{3.12}\right)(-7.5) = -53.5 \text{ kNm}$$

$$M_{ba} = -M_{bc} = 13.5 + \left(\frac{25.1}{3.12}\right)(-5.0) = -26.7 \text{ kNm}$$

$$M_{cb} = 0$$

EXAMPLE 7.16 Analyze the frame shown in Figure E7.16 using *moment-distribution method*. All members are if constant flexural rigidity (EI).

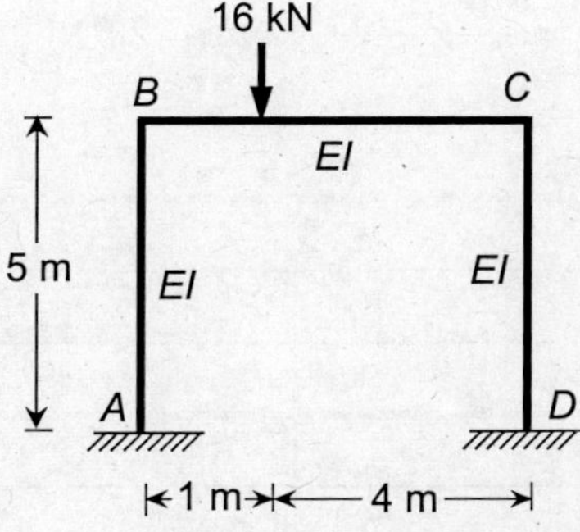

FIGURE E7.16

Solution: The moment-distribution analysis of the frame is carried out in two-steps, i.e., non-sway analysis and sway analysis. The non-sway and sway frames are shown below.

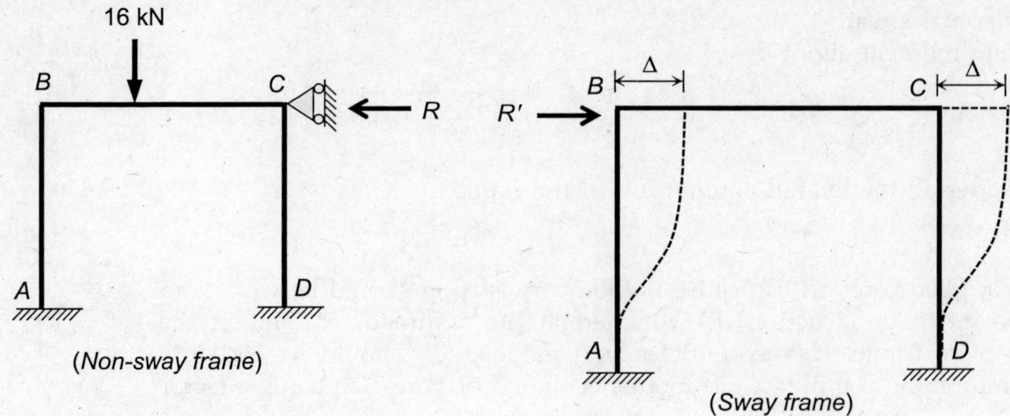

(a) Non-sway analysis

The moment-distribution analysis of the non-sway frame is presented below:

Distribution factors:

$$DF_{ba} = \frac{(4EI/5)}{(4EI/5)+(4EI/5)} = 0.5 = DF_{bc}$$

$$DF_{cb} = \frac{(4EI/5)}{(4EI/5)+(4EI/5)} = 0.5 = DF_{cd}$$

Fixed end moments

$$M^f_{ab} = M^f_{ba} = M^f_{cd} = M^f_{dc} = 0$$

$$M^f_{bc} = -\frac{(16)(1)(4)^2}{5^2} \text{ kNm} = -10.2 \text{ kNm}$$

$$M^f_{cb} = \frac{(16)(1)^2(4)}{5^2} \text{ kNm} = 2.6 \text{ kNm}$$

Moment-distribution

Joint	A	B		C		D
Member	AB	BA	BC	CB	CD	DC
DF		0.5	0.5	0.5	0.5	
FEM			−10.2	2.6		
D.M.		5.1	5.1	−1.3	−1.3	
C.O.	2.6		−0.6	2.6		−0.7
D.M.		0.3	0.3	−1.3	−1.3	
C.O.	0.2		−0.6	0.2		−0.7
D.M.		0.3	0.3	−0.1	−0.1	
Final moment	2.8	5.7	−5.7	2.7	−2.7	−1.4

Horizontal shear: Determine the horizontal reactions at the joints A and D considering the free-body diagrams.

Taking moment about B,

$$5 \times H_a - M_{ab} - M_{ba} = 0 \quad \therefore \; H_a = \frac{M_{ab} + M_{ba}}{5}$$

$$H_a = \frac{2.8 + 5.7}{5} \text{ kN} = 1.7 \text{ kN} (\rightarrow)$$

Taking moment about C,

$$5 \times H_d - M_{cd} - M_{dc} = 0$$

$$\therefore \qquad H_d = \frac{M_{cd} + M_{dc}}{5}$$

$$\therefore \qquad H_d = \frac{-2.7 - 1.4}{5} \text{ kN} = -0.8 \text{ kN} (\leftarrow)$$

Reaction at the support C,

$$R = H_a - H_d = (1.7 - 0.8) \text{ kN} = 0.9 \text{ kN} (\leftarrow)$$

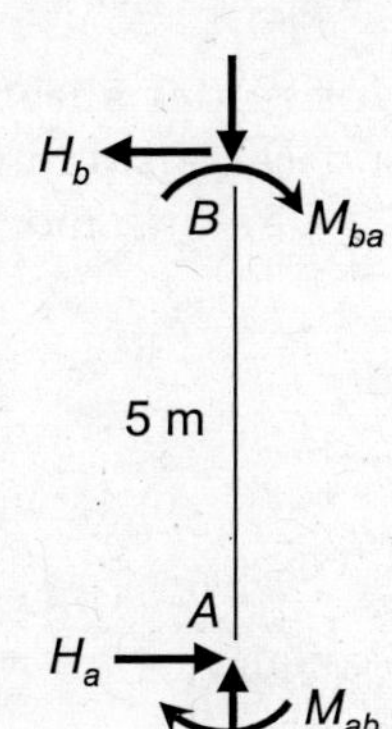

(b) Sway analysis

The moment-distribution analysis of the sway frame is presented below:

Fixed end moments

$$M_{ab}^f = M_{ba}^f = -\frac{6EI\Delta}{5^2} = -10 \text{ kNm } (say)$$

$$M_{bc}^f = M_{cb}^f = 0$$

$$M_{cd}^f = M_{dc}^f = -\frac{6EI\Delta}{5^2} = -10 \text{ kNm}$$

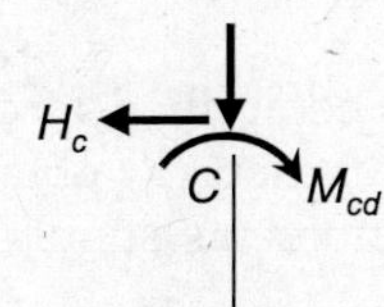

Moment-distribution

Joint	A	B		C		D
Member	AB	BA	BC	CB	CD	DC
DF		0.5	0.5	0.5	0.5	
FEM	−10.0	−10.0			−10.0	−10.0
D.M.		5.0	5.0	5.0	5.0	
C.O.	2.5		2.5	2.5		2.5
D.M.		−1.2	−1.2	−1.2	−1.2	
C.O.	−0.6		−0.6	−0.6		−0.6
D.M.		0.3	0.3	0.3	0.3	
C.O.	0.2		0.2	0.2		0.2
D.M.		−0.1	−0.1	−0.1	−0.1	
Final moment	−7.9	−6.0	6.0	6.0	−6.0	−7.9

Horizontal shear: Determine the horizontal reactions at the joints A and D considering the free-body diagrams.

Taking moment about B,

$$5 \times H_a - M_{ab} - M_{ba} = 0 \quad \therefore \quad H_a = \frac{M_{ab} + M_{ba}}{5}$$

$$H_a = \frac{-7.9 - 6.0}{5}\,\text{kN} = -2.8\ \text{kN}(\leftarrow)$$

Similarly, $H_d = \dfrac{-7.9 - 6.0}{5}\,\text{kN} = -2.8\ \text{kN}(\leftarrow)$

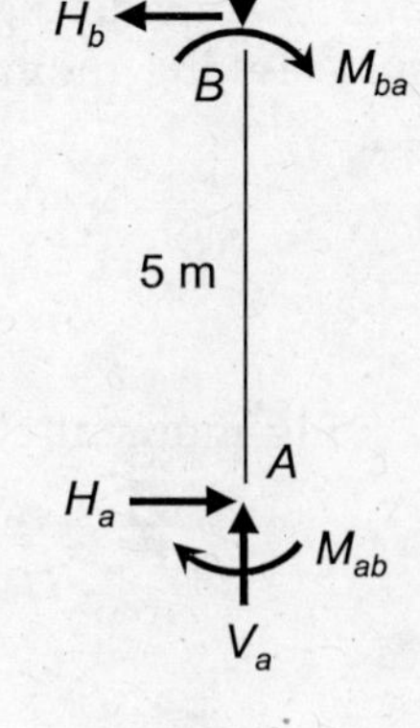

Considering the overall equilibrium of the frame,

$$R' = H_a + H_d = 5.6\ \text{kN}(\rightarrow)$$

The applied force (R') to cause the joint translation should be equal and opposite to the reaction (R) obtained at the assumed restraint in the non-sway frame. The computed end moments should be corrected using a factor of $(0.9/5.6)$.

(c) *Final bending moments*

The final end moments can be obtained by adding the end moments obtained from the non-sway and sway analyses.

$$M_{ab} = 2.8 + \left(\frac{0.9}{5.6}\right)(-7.9) = 1.5\ \text{kNm}$$

$$M_{ba} = -M_{bc} = 5.7 + \left(\frac{0.9}{5.6}\right)(-6) = 4.7\ \text{kNm}$$

$$M_{cb} = -M_{cd} = 2.7 + \left(\frac{0.9}{5.6}\right)(6) = 3.7\ \text{kNm}$$

$$M_{dc} = -1.4 + \left(\frac{0.9}{5.6}\right)(-7.9) = -2.7\ \text{kNm}$$

EXAMPLE 7.17   Analyze the frame shown in Figure E7.17 using *moment-distribution method*.

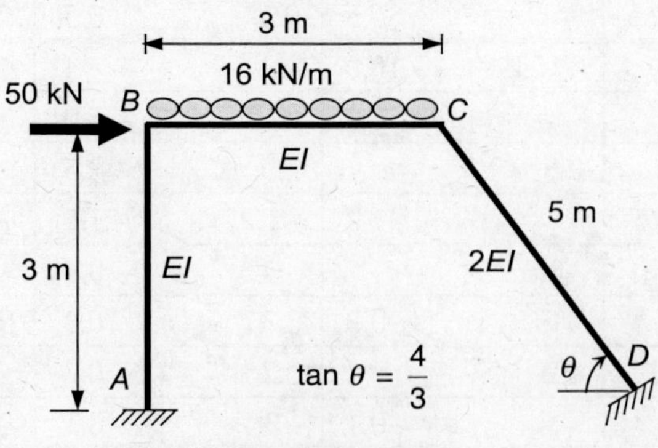

FIGURE E7.17

Solution: The analysis of the frame is carried out in two steps and presented below.

(a) Non-sway analysis

The non-sway frame with joint translation prevented is shown in the figure below.

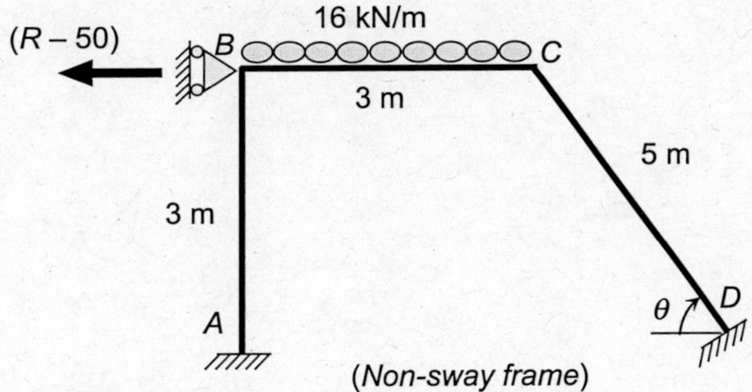

Distribution factors

$$DF_{ba} = \frac{(4EI/3)}{(4EI/3)+(4EI/3)} = 0.5; \quad DF_{bc} = 1 - 0.50 = 0.50$$

$$DF_{cb} = \frac{(4EI/3)}{(4EI/3)+(8EI/5)} = 0.45; \quad DF_{cd} = 1 - 0.45 = 0.55$$

Fixed end moments

$$M^f_{ab} = M^f_{ba} = M^f_{cd} = M^f_{dc} = 0$$

$$M^f_{bc} = -M^f_{cb} = -\frac{16 \times 3^2}{12}\,\text{kNm} = -12\ \text{kNm}$$

Moment distribution

Joint	A	B		C		D
Member	AB	BA	BC	CB	CD	DC
DF		0.5	0.5	0.45	0.55	
FEM			−12.0	12.0		
D.M.		6.0	6.0	−5.4	−6.6	
C.O.	3.0		−2.7	3.0		−3.3
D.M.		1.3	1.3	−1.3	−1.7	
C.O.	0.7		−0.6	0.6		−0.9
D.M.		0.3	0.3	−0.3	−0.3	
C.O.	0.1					−0.1
Final moment	3.8	7.6	−7.6	8.6	−8.6	−4.3

Determine the horizontal reaction at the restraint B:　Consider the free body diagrams of the members AB and CD. The shear forces normal to the longitudinal axes of members are determined as follows:

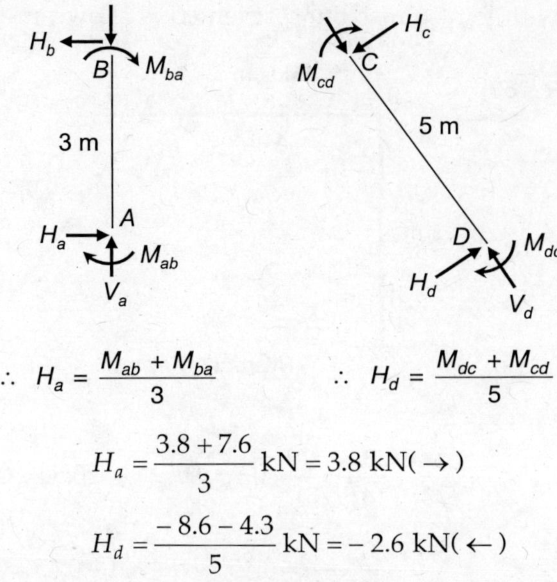

$$\therefore \; H_a = \frac{M_{ab} + M_{ba}}{3} \qquad \therefore \; H_d = \frac{M_{dc} + M_{cd}}{5}$$

$$H_a = \frac{3.8 + 7.6}{3}\, \text{kN} = 3.8\ \text{kN}(\rightarrow)$$

$$H_d = \frac{-8.6 - 4.3}{5}\, \text{kN} = -2.6\ \text{kN}(\leftarrow)$$

The horizontal reaction, R at the support B can be determined from the overall equilibrium of the frame. To simplify the calculation, this reaction can be found by taking the bending moment about point O obtained by extrapolating the members AB and CD. Accordingly,

$$(R-50)(4) - H_a(7) - H_d(10) + M_{ab} + M_{dc} + (16)(3)(1.5) = 0$$

$$(R-50)(4) - (3.8)(7) + (2.6)(10) + 3.8 - 4.3 + 72 = 0$$

$$\therefore \qquad R = 32.3\ \text{kN}\ (\leftarrow)$$

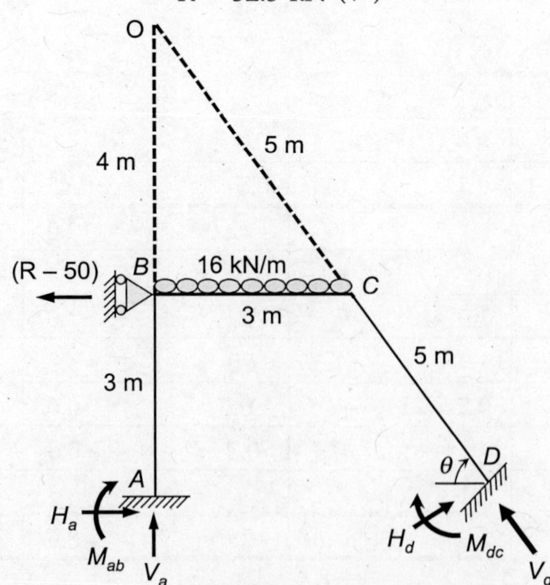

(b) Sway analysis

The sway frame with the assumed deflected shape is shown in the following figure. The relative deflections of the members can be related to each other through joint-displacement triangle as discussed in the previous chapter.

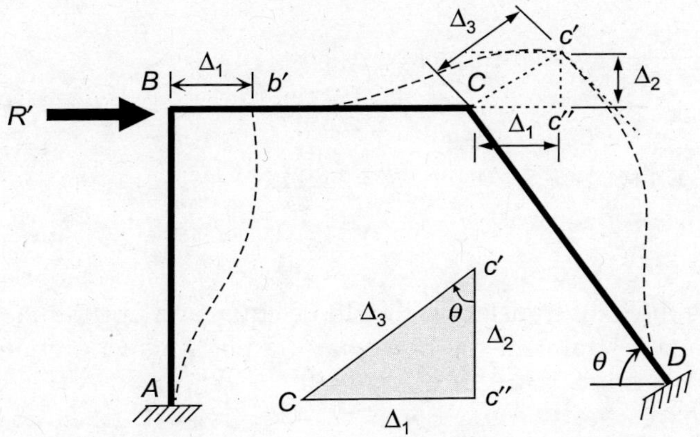

Fixed end moments

$$M_{ab}^{f} = M_{ba}^{f} = -\frac{6EI\Delta_1}{3^2} = -\frac{6EI\Delta}{3^2} = -10 \text{ kNm (say)}$$

$$M_{bc}^{f} = M_{cb}^{f} = \frac{6EI\Delta_2}{3^2} = \frac{6EI\Delta}{3^2}\left(\frac{3}{4}\right) = 7.5 \text{ kNm}$$

$$M_{cd}^{f} = M_{dc}^{f} = -\frac{6(2EI)\Delta_3}{5^2} = -\frac{6EI\Delta}{3^2}\left(\frac{2\times 5}{4}\right)\left(\frac{3^2}{5^2}\right) = -9.0 \text{ kNm}$$

Moment distribution

Joint	A	B		C		D
Member	AB	BA	BC	CB	CD	DC
DF		0.5	0.5	0.45	0.55	
FEM	−10.0	−10.0	7.5	7.5	−9.0	−9.0
D.M.		1.3	1.3	0.7	0.8	
C.O.	0.7		0.3	0.7		0.4
D.M.		−0.2	−0.2	−0.3	−0.4	
C.O.	−0.1					−0.2
Final moment	−9.4	−8.9	8.9	8.6	−8.6	−8.8

Determine the equivalent applied force: The magnitude of equivalent applied force, R′ can be determined considering the shear equilibrium of the members following the same procedure as discussed above.

$$H_a = \frac{-(9.4+8.9)}{3}\,\text{kN} = -6.1\,\text{kN}(\leftarrow)$$

$$H_d = \frac{-8.6-8.8}{5}\,\text{kN} = -3.5\,\text{kN}(\leftarrow)$$

Taking moment about O,

$$R'(4) - 6.1(7) - 3.5(10) + 9.4 + 8.8 = 0 \quad \therefore\; R' = 14.9\,\text{kN}\;(\rightarrow)$$

(c) Final end moments

Since the applied force (R') to cause the joint translation should be equal and opposite to the reaction (R) obtained at the assumed restraint in the non-sway frame, the sway moments must be modified by applying a correction factor of (32.3/14.9).

$$M_{ab} = 3.8 + \left(\frac{32.3}{14.9}\right)(-9.4) = -16.6\,\text{kNm}$$

$$M_{ba} = -M_{bc} = 7.6 + \left(\frac{32.3}{14.9}\right)(-8.9) = -11.7\,\text{kNm}$$

$$M_{cb} = -M_{cd} = 8.6 + \left(\frac{32.3}{14.9}\right)(8.6) = 27.2\,\text{kNm}$$

$$M_{dc} = -1.4 + \left(\frac{0.9}{5.6}\right)(-7.9) = -23.4\,\text{kNm}$$

7.9   ANALYSIS OF MULTI-STORY FRAMES

In the previous sections, the moment-distribution analysis is carried out for the structures having one joint translation only. In a multi-story frame, the more than one joint translation may occur depending on the support conditions. The moment-distribution analysis of such frames is carried out by splitting the given structure into a series of independent frames in which only one joint translation is allowed to occur. The final end moments of the multi-story frames are obtained by summing of all independent cases using the principle of superposition.

This procedure is explained by considering a two-story frame subjected to the lateral forces, P_1 and P_2, as shown in Figure 7.11(a). This frame can undergo joint translations at each floor level. Thus, there are two joint translations which are required to consider in the moment-distribution analysis. Accordingly, two independent analysis cases would be necessary to consider the joint translation effects in the calculation of end moments.

In first case, the joint translation is allowed only at B keeping the joint D restrained against the translation as shown in Figure 7.11(b). The moment-distribution analysis of this frame is carried out only applying a lateral force F_1 at joint B following the procedure discussed earlier. Let the support reaction at the joint D corresponding to this loading and support conditions be mF_1, where m is the proportionality constant with respect to the applied force F_1.

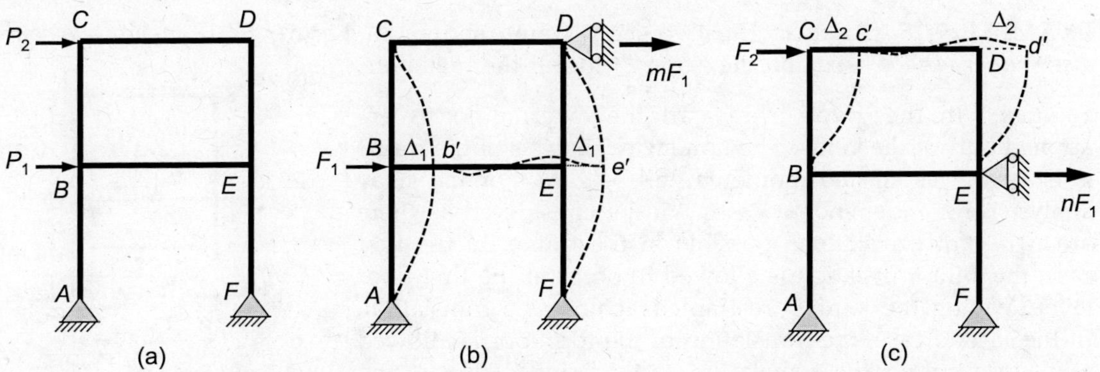

FIGURE 7.11 (a) Multi-story frame with sway loading, (b) Frame with joint translation prevented at top story, (c) Frame with joint translation prevented at bottom story.

In the second case, the translation at the joint E is restrained by using a lateral support. The moment-distribution analysis of the frame is carried out for a lateral force F_2 is applied at the joint C. Let the support reaction at E be nF_2, where the proportionality constant n is calculated with respect to F_2.

The superposition of two analysis cases shown in Figure 7.11(b) and (c) would result in the final case shown in Figure 7.12 in which the internal forces and reactions are the linear combinations of F_1 and F_2.

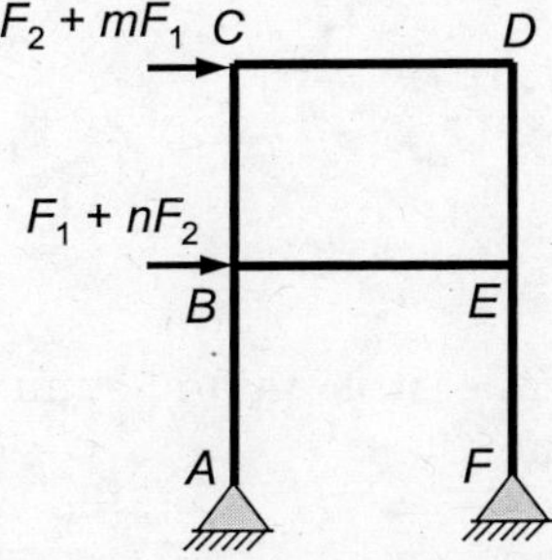

FIGURE 7.12 Combination of sway forces at each story level.

Now comparing Figure 7.11(a) and Figure 7.12, the actual values of F_1 and F_2 can be solved using the following expressions.

$$F_1 + nF_2 = P_1 \tag{7.16a}$$

$$mF_1 + F_2 = P_2 \tag{7.16b}$$

The similar procedure can be adopted for multi-story frames with any number of joint translations. In case, the frame members of a structures are subjected to the lateral loads acting away from the joints, a non-sway analysis has to be carried out additionally. For the non-sway analysis, all joint translations are required to be restrained simultaneously by using the lateral supports. The following examples illustrate the application of moment-distribution method in the multi-story frames.

EXAMPLE 7.18 Analyze the two-story frame shown in Figure E7.18 using *moment-distribution method*. Assume the same *EI* for all the members.

Solution: In the given frame, all the external loads are applied only at the joints. The non-sway analysis of the frame is irrelevant as all end moments will be zero. For the sway analysis, two independent cases will be considered as there are two joint translations possible in the frame. In the first case, the joint translation is allowed to occur at the first-floor level keeping the joint D restrained against the translation. In the second case, the translation of the top floor is allowed while preventing the translation at the joint E. These two cases are shown in following figure.

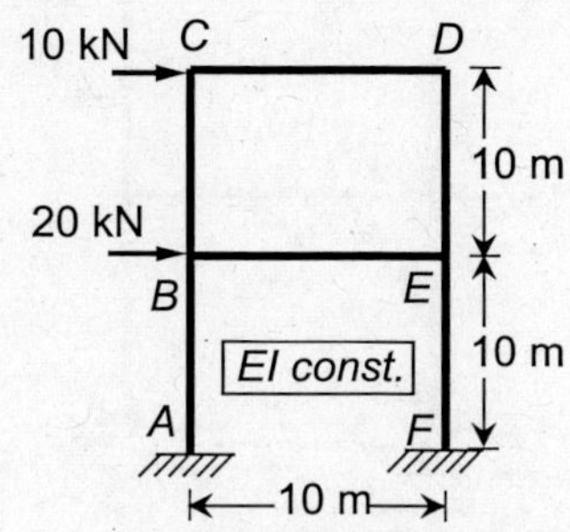

FIGURE E7.18

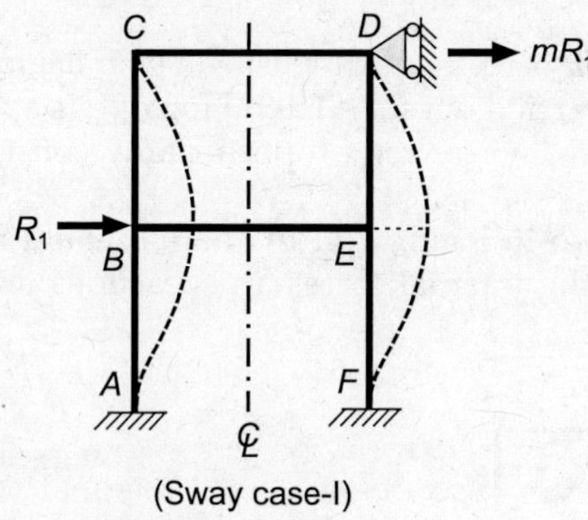

(Sway case-I)

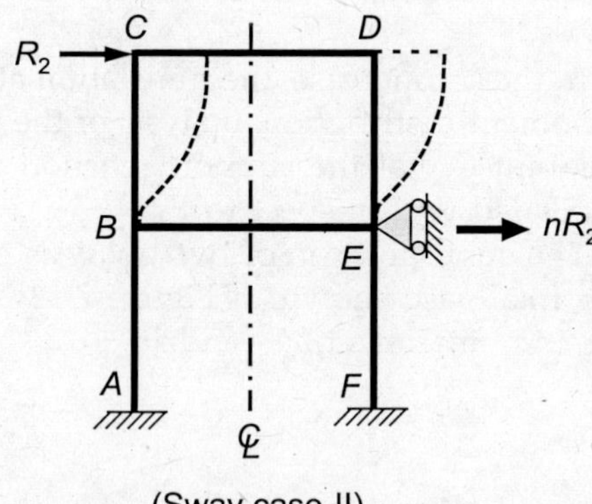

(Sway case-II)

(a) Analysis of Case-I

As evident from the deflected shape, it is an anti-symmetric frame. Hence, it is required to solve only one-half of the frame.

Distribution factors

$$S_{ba} = S_{bc} = S_{cb} = \frac{4EI}{10} = 0.4EI$$

$$S_{be} = S_{cd} = \frac{6EI}{10} = 0.6EI$$

$$DF_{ba} = DF_{bc} = \frac{0.4}{0.4 + 0.4 + 0.6} = 0.29$$

$$DF_{be} = \frac{0.6}{0.4 + 0.4 + 0.6} = 0.42$$

$$DF_{cb} = \frac{0.4}{0.4 + 0.6} = 0.4$$

$$DF_{cd} = \frac{0.6}{0.4 + 0.6} = 0.6$$

Fixed end moments

$$M_{ab}^f = M_{ba}^f = -\frac{6EI\Delta_1}{10^2} = -10 \text{ kNm (say)}$$

$$M_{bc}^f = -M_{cb}^f = \frac{6EI\Delta_1}{10^2} = 10 \text{ kNm}$$

$$M_{be}^f = -M_{cd}^f = 0$$

Moment distribution

Joint	A		B			C	
Member	AB	BA	BC	BE	CB	CD	
DF		0.29	0.29	0.42	0.4	0.6	
FEM	−10.0	−10.0	10.0		10.0		
D.M.					−4.0	−6.0	
C.O.			−2.0				
D.M.		0.6	0.6	0.8			
C.O.	0.3						
Final moment	−9.7	−9.4	8.6	0.8	6.0	−6.0	

Due to the anti-symmetry case, the end moments of the member in the other half of the frame would be exactly same and of same sign.

Determine the applied force and horizontal reaction at support D: Consider the free body diagrams of the members *AB* and *BC*. The shear forces normal to the longitudinal axes of members are determined as follows:

Considering the equilibrium member *AB*, $H_a = \dfrac{-(9.7 + 9.4)}{10} \text{kN} = -1.9 \text{ kN} (\leftarrow)$

Considering the equilibrium member *BC*, $H_b = \dfrac{8.6 + 6.0}{10} \text{ kN} = 1.5 \text{ kN} (\rightarrow)$

Shear forces in the member *DE* and *EF* would same as obtained above. The frame with the end moments and shear forces is shown below:

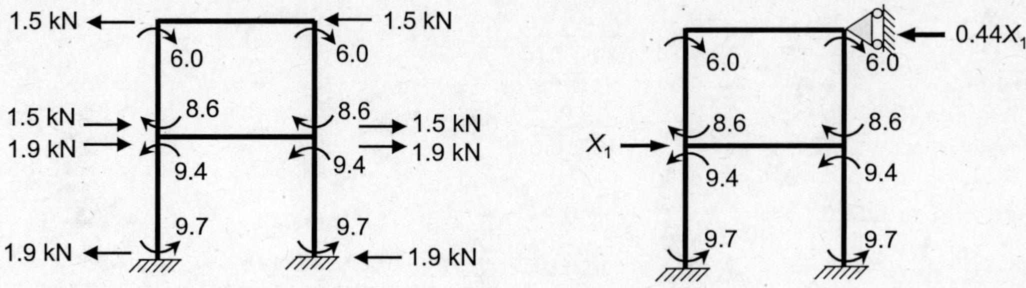

For the assumed deflection, the equivalent applied force, $X_1 = 2 \times (1.9 + 1.5)$ kN = 6.8 kN ($\rightarrow$)

The reaction at the support at D, $mX_1 = (1.5 + 1.5)$ kN = 30 kN ($\leftarrow$)

$$\therefore \quad m = \frac{3.0}{6.8} = 0.44$$

(b) Analysis of Case-II

As evident from the deflected shape, it is an anti-symmetric frame. Hence, it is required to solve only one-half of the frame. Similar procedure is followed to analysis the Case-II.

Fixed end moments

$$M_{ab}^f = -M_{ba}^f = 0; \quad M_{bc}^f = M_{cb}^f = -\frac{6EI\Delta_2}{10^2} = -10 \text{ kNm (say)}$$

$$M_{be}^f = M_{cd}^f = 0$$

Moment distribution

Joint	A		B		C	
Member	AB	BA	BC	BE	CB	CD
DF		0.29	0.29	0.42	0.4	0.6
FEM			−10.0		−10.0	
D.M.		2.9	2.9	4.2	4.0	6.0
C.O.	1.5		2.0		1.5	
D.M.		−0.6	−0.6	−0.8	−0.6	−0.9
C.O.	−0.3					
Final moment	1.2	2.3	−5.7	3.4	−5.1	5.1

Because of the anti-symmetry case, the end moments of the member in the other half of the frame would be exactly same and of same sign.

Determine the applied force and horizontal reaction at support E

Considering the equilibrium member AB, $H_a = \dfrac{1.2 + 2.3}{10}$ kN $= 0.3$ kN $(\rightarrow)$

Considering the equilibrium member BC,

$$H_b = \frac{-(5.7 + 5.1)}{10} \text{ kN} = -1.1 \text{ kN} (\leftarrow)$$

Shear forces in the member DE and EF would same as obtained above. The frame with the end moments and shear forces is shown below.

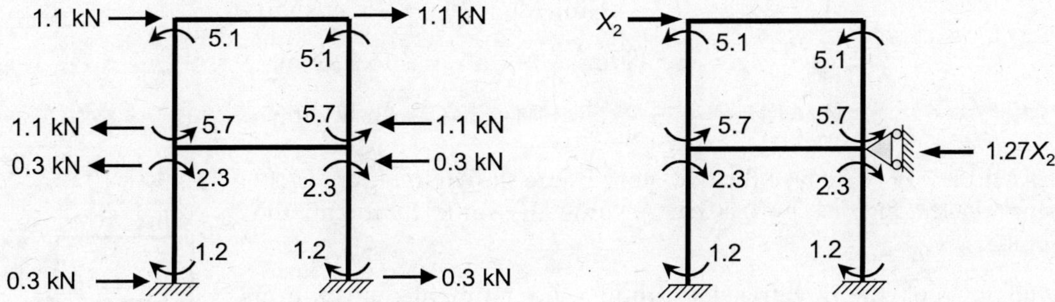

For the assumed deflection, the equivalent applied force, $X_2 = (1.1 + 1.1)$ kN $= 2.2$ kN $(\rightarrow)$

The reaction at the support at E,

$$nX_2 = 2 \times (1.1 + 0.3) \text{ kN} = 2.8 \text{ kN} (\leftarrow)$$

$$\therefore \qquad n = \frac{2.8}{2.2} = 1.27$$

(c) Superposition

The actual values of R_1 and R_2 can be obtained by superimposing the lateral forces and reactions at each floor level and equating them to the applied lateral forces. Thus,

$$R_1 + nR_2 = 20 \quad \text{and} \quad mR_1 + R_2 = 10$$

Putting the values of m and n,

$$R_1 - 1.27 R_2 = 20$$

$$-0.44 R_1 + R_2 = 10$$

Solving, $R_1 = 74.1$ kN; $R_2 = 42.6$ kN.

(d) Final end moments

The final end moments of the frame can be obtained by using the correction factors based on actual values of R_1 and R_2.

$$M_{ab} = (-9.7)\left(\frac{74.1}{6.8}\right) + (1.2)\left(\frac{42.6}{2.2}\right) = -82.5 \text{ kNm}$$

$$M_{ba} = (-9.4)\left(\frac{74.1}{6.8}\right) + (2.3)\left(\frac{42.6}{2.2}\right) = -57.9 \text{ kNm}$$

$$M_{bc} = (8.6)\left(\frac{74.1}{6.8}\right) + (-5.7)\left(\frac{42.6}{2.2}\right) = -16.6 \text{ kNm}$$

$$M_{be} = (0.8)\left(\frac{74.1}{6.8}\right) + (3.4)\left(\frac{42.6}{2.2}\right) = 74.5 \text{ kNm}$$

$$M_{cb} = -M_{cd} = (6)\left(\frac{74.1}{6.8}\right) + (-5.1)\left(\frac{42.6}{2.2}\right) = -33.4 \text{ kNm}$$

$$M_{fe} = M_{ab} = -82.5 \text{ kNm}; \quad M_{ef} = M_{ba} = -57.9 \text{ kNm}$$

$$M_{eb} = M_{be} = 74.5 \text{ kNm}; \quad M_{ed} = M_{bc} = -16.6 \text{ kNm}$$

$$M_{de} = M_{cb} = -33.4 \text{ kNm}; \quad M_{dc} = M_{cd} = 33.4 \text{ kNm}$$

EXAMPLE 7.19   Analyze the two-story frame shown in Figure E7.19 using *moment-distribution method*. Assume the same *EI* for all the members.

Solution: For the given sway frame, the moment-distribution analysis should be carried out in two stages, namely, non-sway analysis and sway analysis. Since there are two joint translations possible in the frame, two independent analysis cases with one joint translation allowed at any time.

(a) Non-sway analysis

For non-sway analysis, all joint translations must be restrained by using the lateral supports.

Distribution factors

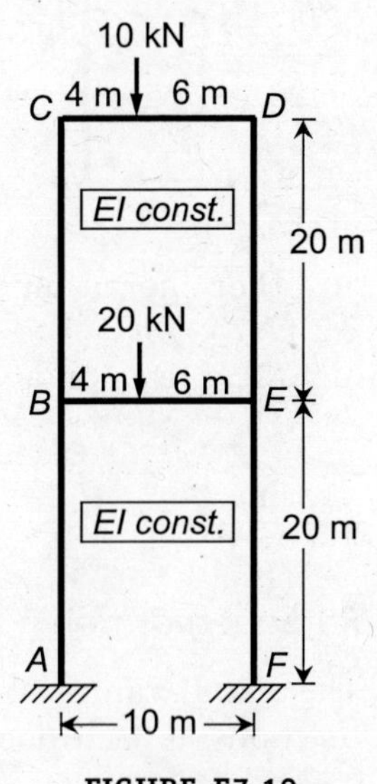

$$S_{ba} = S_{bc} = S_{ef} = S_{ed} = \frac{4EI}{20} = 0.2EI \; ; \; S_{be} = S_{cd} = \frac{4EI}{10} = 0.4EI$$

$$DF_{ba} = DF_{bc} = \frac{0.2}{0.2 + 0.2 + 0.4} = 0.25 \; ; \; DF_{be} = \frac{0.4}{0.4 + 0.4 + 0.6} = 0.5$$

$$DF_{ef} = DF_{ed} = 0.25; \quad DF_{eb} = 0.5$$

$$DF_{cb} = DF_{de} = \frac{0.2}{0.2 + 0.4} = 0.33 \; ; \; DF_{cd} = DF_{dc} = \frac{0.4}{0.2 + 0.4} = 0.67$$

Fixed end moments

$$M_{be}^{f} = -\frac{(20)(4)(6)^2}{10^2} = -28.8 \text{ kNm}$$

$$M_{eb}^{f} = \frac{(20)(6)(4)^2}{10^2} = 19.2 \text{ kNm}$$

$$M_{cd}^{f} = -\frac{(10)(4)(6)^2}{10^2} = -14.4 \text{ kNm}$$

$$M_{dc}^{f} = \frac{(10)(6)(4)^2}{10^2} = 9.6 \text{ kNm}$$

All other fixed end moments are zero. The moment distribution analysis is carried out as presented below:

Joint	A	B			E			C		D		F
Member	AB	BA	BC	BE	EB	EF	ED	CB	CD	DC	DE	FE
D.F.		0.25	0.25	0.50	0.50	0.25	0.25	0.33	0.67	0.67	0.33	
FEM	0	0	0	−28.8	19.2	0	0	0	−14.4	9.6	0	0
D.M.	0	7.2	7.2	14.4	−9.6	−4.8	−4.8	4.8	9.6	−6.4	−3.2	0
C.O.	3.6	0	1.8	−4.8	7.2	0	−1.6	3.6	−3.2	4.8	−2.4	−2.4
D.M.	0	0.7	0.8	1.5	−2.8	−1.4	−1.4	−0.1	−0.3	−1.6	−0.8	0
C.O.	0.3	0	0	−1.4	0.8	0	−0.4	0.4	−0.8	−0.2	−0.7	−0.7
D.M.	0	0.4	0.3	0.7	−0.2	−0.1	−0.1	0.1	0.3	0.6	0.3	0
C.O.	0.2											0
Final	4.1	8.3	10.1	−18.4	14.6	−6.3	−8.3	8.8	−8.8	6.8	−6.8	−3.1

The horizontal reactions at the lateral supports at *D* and *E* are determined using the end moments as summarized in the following figure.

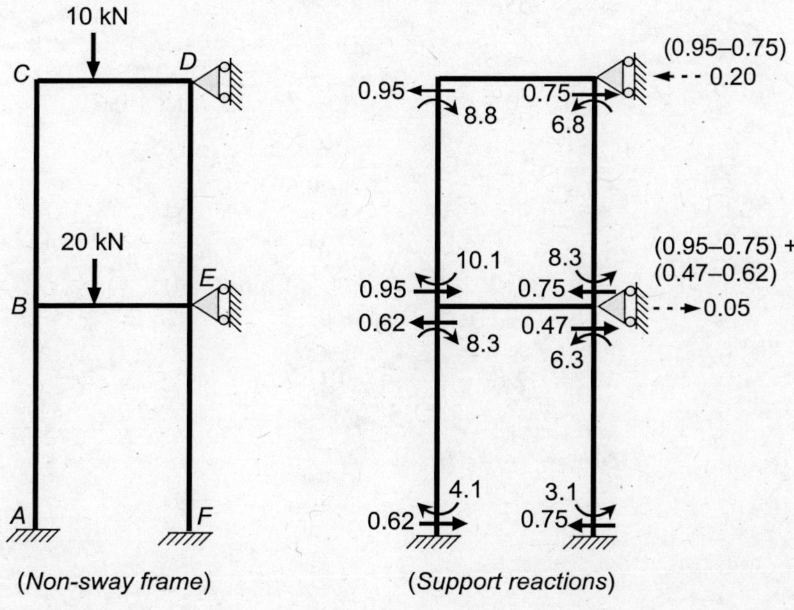

(b) Analysis of Sway Case-I

The sway frames with the applicable lateral supports in two cases are shown in the following figure. As evident from the deflected shape, the frames are anti-symmetric. Hence, it is required to solve only one-half of the frame.

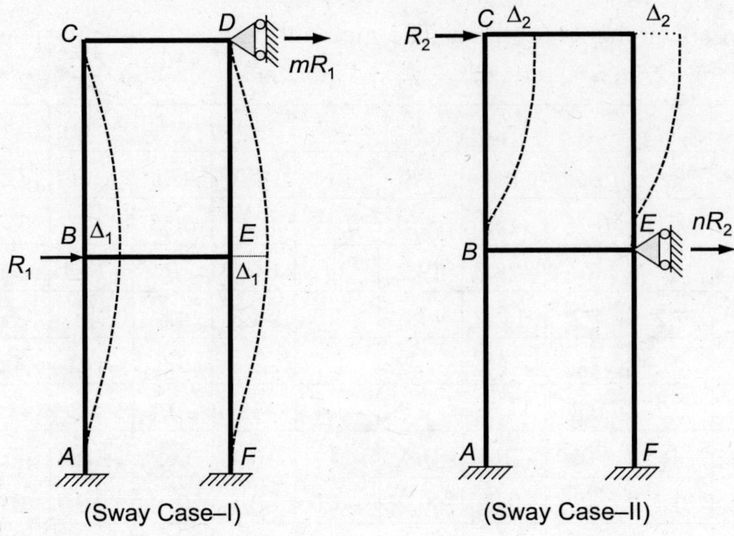

Distribution factors

$$S_{ba} = S_{bc} = S_{cb} = \frac{4EI}{20} = 0.2EI$$

$$S_{be} = S_{cd} = \frac{6EI}{10} = 0.6EI$$

$$DF_{ba} = DF_{bc} = \frac{0.2}{0.2+0.2+0.6} = 0.2 \,; \; DF_{be} = \frac{0.6}{0.2+0.2+0.6} = 0.6$$

$$DF_{cb} = \frac{0.2}{0.2+0.6} = 0.25 \,; \; DF_{cd} = \frac{0.6}{0.2+0.6} = 0.75$$

Fixed end moments

$$M_{ab}^{f} = M_{ba}^{f} = -\frac{6EI\Delta_1}{20^2} = -10 \text{ kNm (say)}$$

$$M_{bc}^{f} = -M_{cb}^{f} = \frac{6EI\Delta_1}{20^2} = 10 \text{ kNm}$$

$$M_{be}^{f} = -M_{cd}^{f} = 0$$

Moment distribution

Joint	A		B		C	
Member	AB	BA	BC	BE	CB	CD
DF		0.2	0.2	0.6	0.25	0.75
FEM	−10.0	−10.0	10.0		10.0	
D.M.					−2.5	−7.5
C.O.			−1.3			
D.M.		0.3	0.2	0.8		
C.O.	0.2					
Final moment	−9.8	−9.7	8.9	0.8	7.5	−7.5

Due to the anti-symmetry case, the end moments of the member in the other half of the frame would be exactly same and of same sign. The ratio of reaction force at the lateral support to the applied force is computed from the end moments.

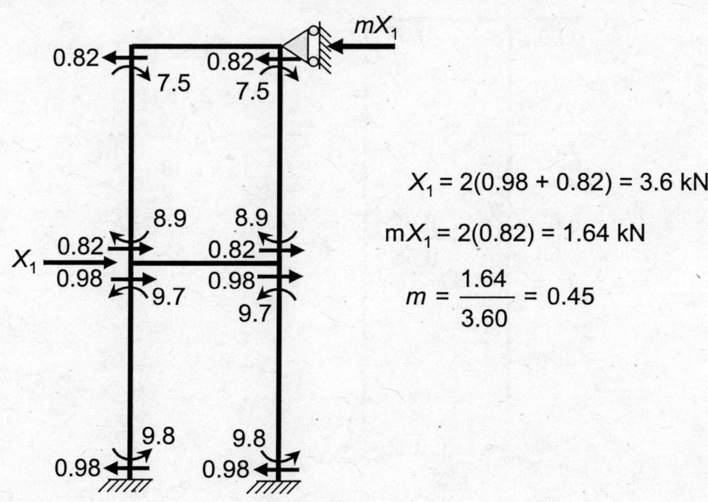

(c) Analysis of Sway Case-II

Similar procedure as discussed above is also followed to analysis the Case-II.

Fixed end moments

$$M_{ab}^{f} = -M_{ba}^{f} = 0$$

$$M_{bc}^{f} = M_{cb}^{f} = -\frac{6EI\Delta_2}{20^2} = -10 \text{ kNm (say)}$$

$$M_{be}^{f} = M_{cd}^{f} = 0$$

Moment distribution

Joint	A	B			C	
Member	AB	BA	BC	BE	CB	CD
DF		0.2	0.2	0.6	0.25	0.75
FEM			−10.0		−10.0	
D.M.		2.0	2.0	6.0	2.5	7.5
C.O.	1.0		1.2		1.0	
D.M.		−0.2	−0.2	−0.8	−0.2	−0.8
C.O.	−0.1					
Final moment	9.1	1.8	−7.0	5.2	−6.7	6.7

Due to the anti-symmetry case, the end moments of the member in the other half of the frame would be exactly same and of same sign. The ratio of reaction force at the lateral support to the applied force is computed from the end moments.

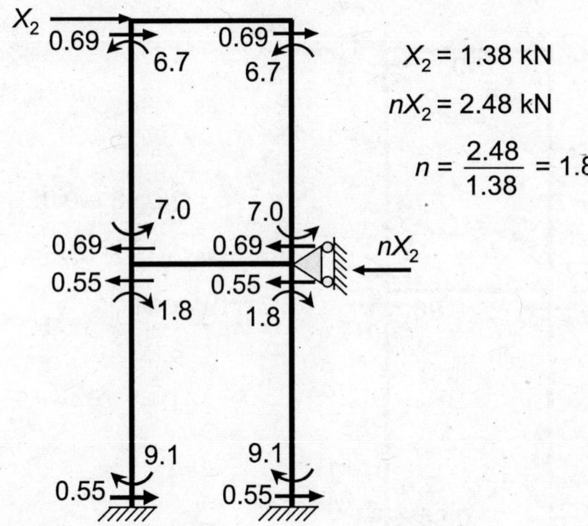

(d) Superposition

The actual values of R_1 and R_2 can be obtained by superimposing the lateral forces and reactions at each floor level and equating them to the applied lateral forces. Thus,

$$R_1 + nR_2 = -0.05 \quad \text{and} \quad mR_1 + R_2 = 0.2$$

Putting the values of m and n,

$$R_1 - 1.8R_2 = -0.05$$

$$-0.45R_1 + R_2 = 0.2$$

Solving, $R_1 = 1.63$ kN; $R_2 = 0.93$ kN

(e) Final end moments

The final end moments of the frame can be obtained by using the correction factors based on actual values of R_1 and R_2.

$$M_{ab} = 4.1 + (-9.8)\left(\frac{1.63}{3.6}\right) + (9.1)\left(\frac{0.93}{1.38}\right) = 5.8 \text{ kNm}$$

$$M_{ba} = 8.3 + (-9.7)\left(\frac{1.63}{3.6}\right) + (1.8)\left(\frac{0.93}{1.38}\right) = 5.1 \text{ kNm}$$

$$M_{bc} = 10.1 + (8.9)\left(\frac{1.63}{3.6}\right) + (-7.0)\left(\frac{0.93}{1.38}\right) = 9.4 \text{ kNm}$$

$$M_{be} = -18.4 + (0.8)\left(\frac{1.63}{3.6}\right) + (5.2)\left(\frac{0.93}{1.38}\right) = -14.5 \text{ kNm}$$

$$M_{cb} = -M_{cd} = 8.8 + (7.5)\left(\frac{1.63}{3.6}\right) + (-6.7)\left(\frac{0.93}{1.38}\right) = 7.7 \text{ kNm}$$

$$M_{fe} = -3.1 + (-9.8)\left(\frac{1.63}{3.6}\right) + (9.1)\left(\frac{0.93}{1.38}\right) = -1.4 \text{ kNm}$$

$$M_{ef} = -6.3 + (-9.7)\left(\frac{1.63}{3.6}\right) + (1.8)\left(\frac{0.93}{1.38}\right) = -9.5 \text{ kNm}$$

$$M_{eb} = 14.6 + (0.8)\left(\frac{1.63}{3.6}\right) + (5.2)\left(\frac{0.93}{1.38}\right) = 18.5 \text{ kNm}$$

$$M_{ed} = -8.3 + (8.9)\left(\frac{1.63}{3.6}\right) + (-7.0)\left(\frac{0.93}{1.38}\right) = -9.0 \text{ kNm}$$

$$M_{de} = -M_{dc} = -6.8 + (7.5)\left(\frac{1.63}{3.6}\right) + (-6.7)\left(\frac{0.93}{1.38}\right) = -7.9 \text{ kNm}$$

7.10   PROBLEMS

7.1   Determine the reactions and bending moments at the supports of the beam shown below using the Moment-Distribution method. Draw the shear force and bending moment diagrams.

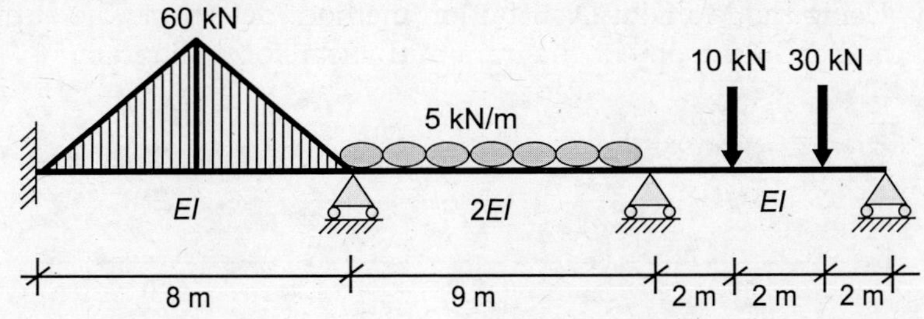

7.2 Determine the reactions and bending moments at the supports of the beam shown below using the Moment-Distribution method. Draw the shear force and bending moment diagrams.

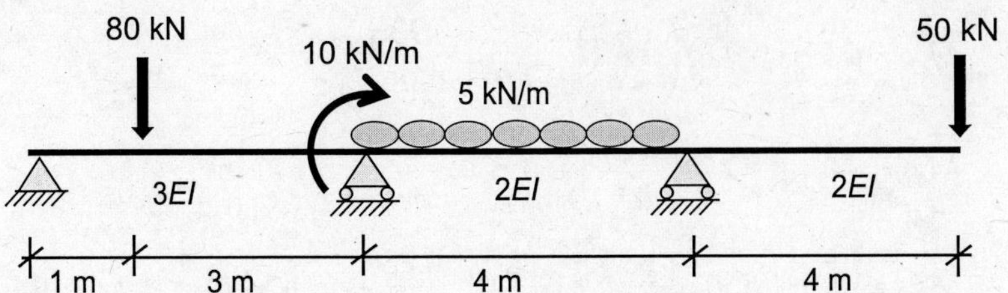

7.3 Determine the reactions and the bending moments at the supports of the beam supported using the Moment-Distribution method. The support B is settled down by 20 mm. Assume the value of $EI =1000$ kNm^2.

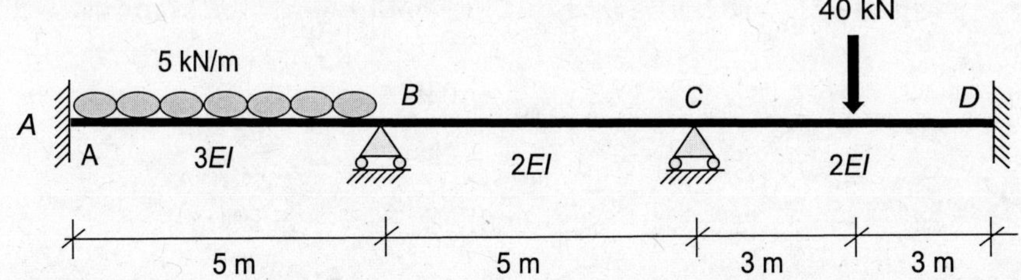

7.4 Determine the reactions and the bending moments at the supports of the beam as shown below using the Moment-Distribution method. The support B is settled down by 50 mm and the support C is rotated clockwise by 0.02 radian. Assume the value of $EI = 1000$ kNm^2.

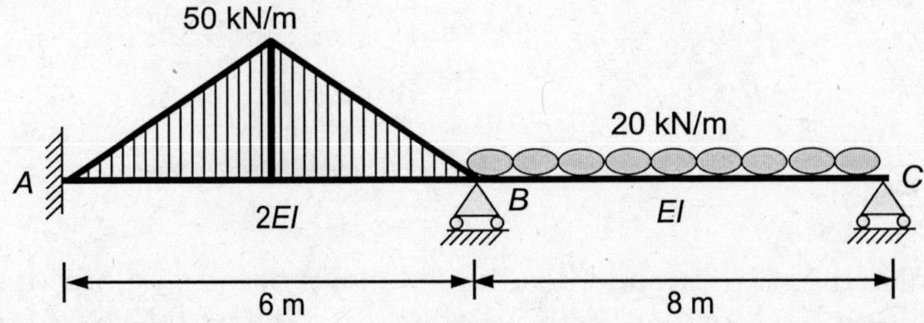

7.5 A beam is subjected to the uniformly distributed loads as shown in the figure below. Using the Moment-Distribution method, determine the support reactions. Draw the bending moment diagram and shear force diagram.

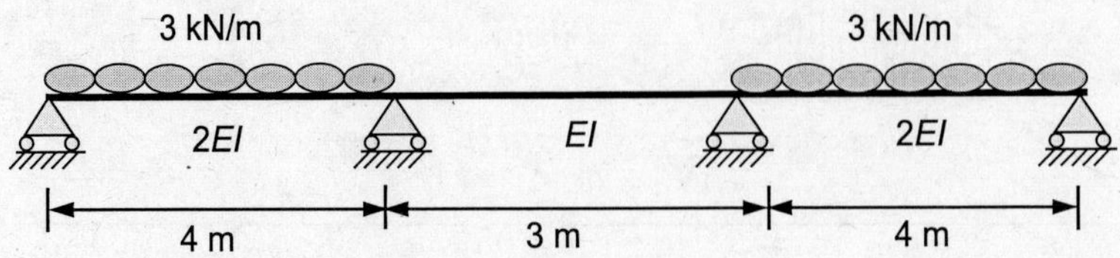

7.6 Determine the support reactions and bending moments at the joints of the beam supported on a spring and carrying a uniformly distributed load w throughout its length as shown in the following figure. Use the Moment-Distribution method. Take $w = 10$ kN/m; $L = 4$ m; $k = 10$ kN/mm; $EI = 78125$ kNm^2.

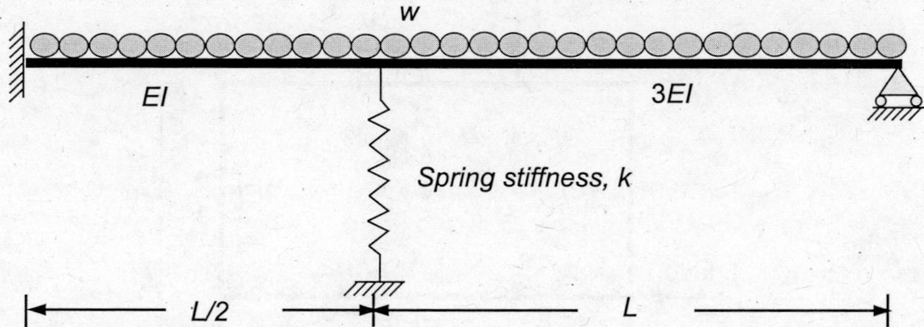

7.7 Determine the bending moments at the joints and the support reactions of the frame shown below using the Moment-Distribution method. Draw the bending moment and shear force diagrams.

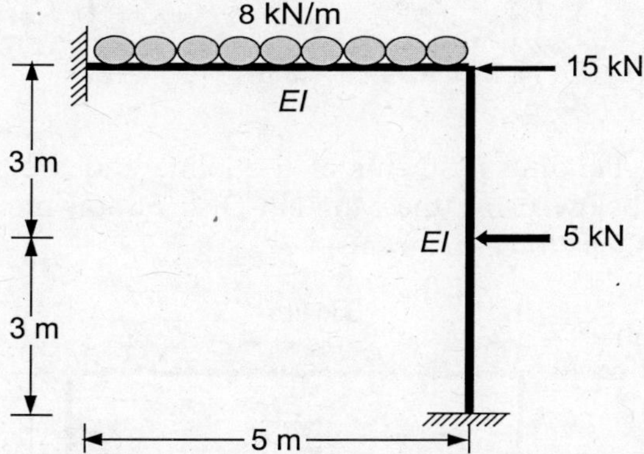

7.8 Determine the bending moments at the joints and the support reactions of the frame shown below using the Moment-Distribution method. Draw the bending moment and shear force diagrams.

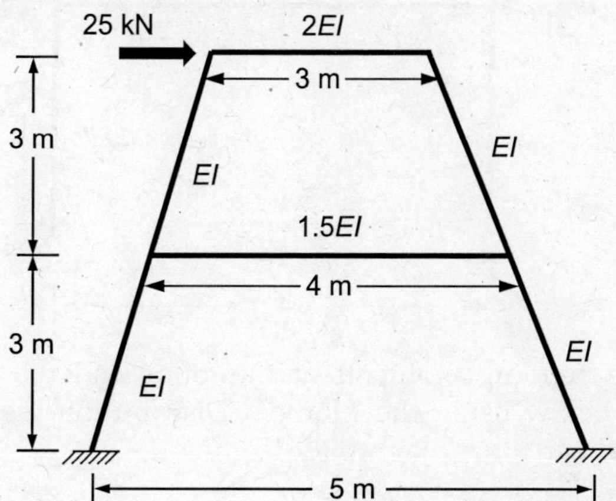

7.9   Determine the bending moments at the joints and the support reactions of the frame shown below using the Moment-Distribution method. Draw the bending moment and shear force diagrams.

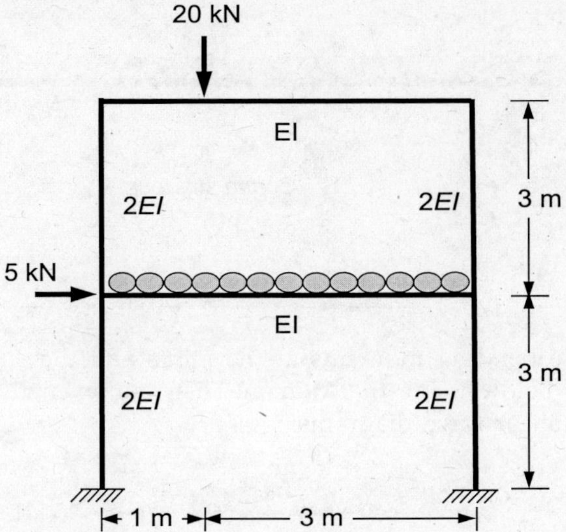

7.10   Determine the bending moments at the joints and the support reactions of the frame shown below using the Moment-Distribution method. Draw the bending moment and shear force diagrams.

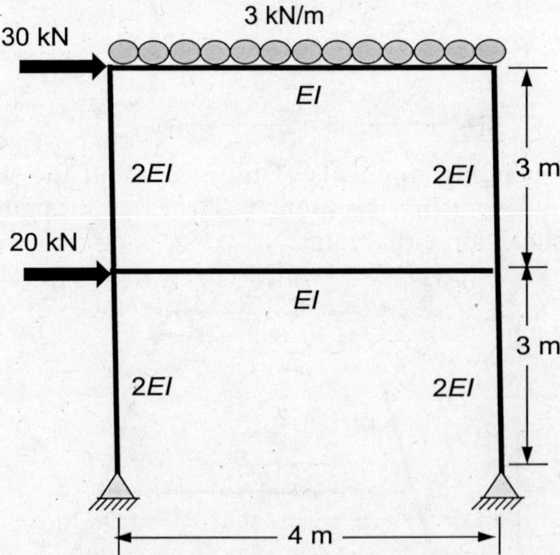

7.11   Determine the bending moments at the joints and the support reactions of the frame shown below using the Moment-Distribution method. Draw the bending moment and shear force diagrams.

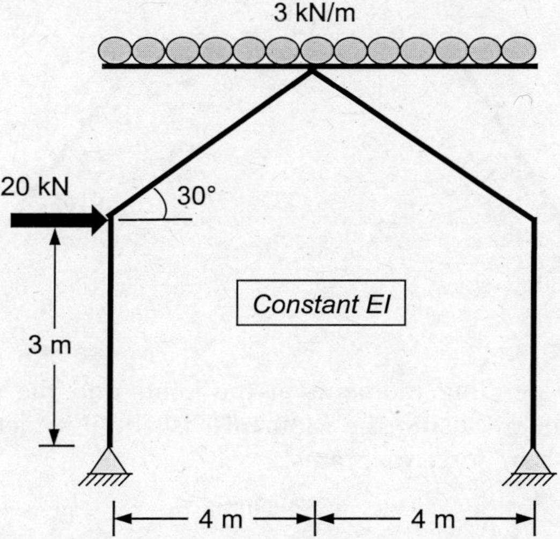

7.12 Determine the bending moments at the joints and the support reactions of the frame shown below using the Moment-Distribution method. Draw the bending moment and shear force diagrams.

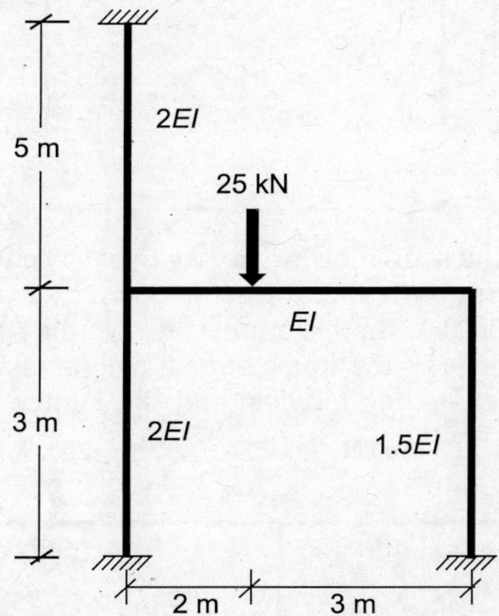

7.13 Determine the bending moments at the joints of the frame shown below using the Moment-Distribution method. Draw the bending moment and shear force diagrams.

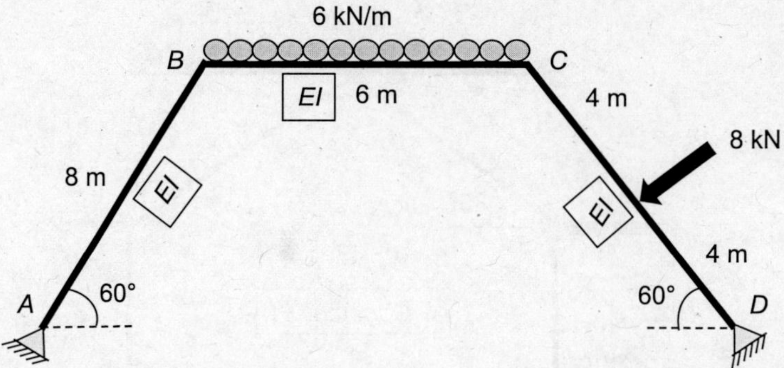

7.14 Determine the bending moments at the joints and the support reactions of the frame shown below using the Moment-Distribution method. Draw the bending moment and shear force diagrams.

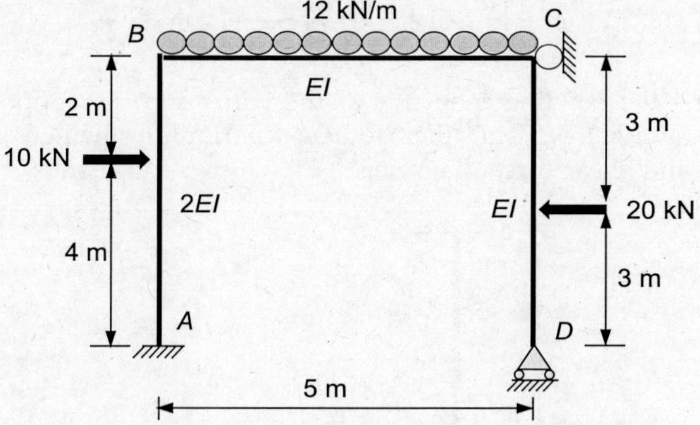

7.15 A rectangular frame shown below carries two concentrated load and a uniformly distributed load. Supports D and E settle down by 3 mm and 2 mm, respectively. Assume $EI = 100000$ kNm^2. Determine the bending moments at the joints and the support reactions of the frame shown below using the Moment-Distribution method. Draw the bending moment and shear force diagrams.

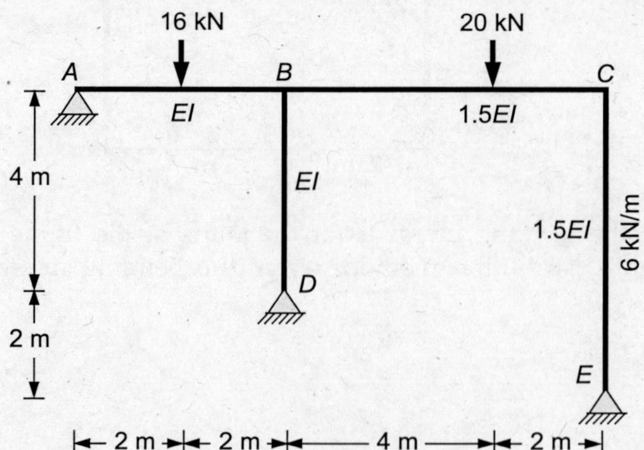

Matrix Force Method

8.1 GENERAL

The classical methods of structural analysis, such as the method of consistent deformations and the slope-deflection method are very effective in solving simple problems. Though these methods provide a fundamental basis of analysis, their use had been limited in the past primarily due to the difficulty in solving the simultaneous equations. The growing demand for a better method of analysis of complex structures led to the development of matrix analysis methods. The rapid development of high-speed digital computers facilitated the better implementation of the matrix methods of analysis. The matrix methods have their unique theoretical basis and particular procedures. These methods are based on the finite-element concept in which the force-displacement relationship of a structure is derived from the element levels.

The matrix methods of analysis of structures can be broadly classified into two categories, namely, the matrix force (flexibility) method and the matrix displacement (stiffness) method. The matrix force method considers the member forces as the basic unknowns and relates the forces to the corresponding displacements by flexibility matrices. On the other hand, the matrix displacement method treats the nodal displacements as the basic unknowns and relates the displacements with the corresponding forces by stiffness matrices. The matrix force method has been discussed in this chapter, whereas the matrix displacement method is presented in the next chapter. It is worth mentioning that a duality exists between both these methods.

8.2 BASIC CONCEPTS AND MATRIX NOTATIONS

The fundamental concept of matrix analysis is that the structures, such as beams, trusses, and rigid frames are represented as a combination of structural elements connected together at a finite number of discrete points. These points are called *nodes* or *nodal points*. It is considered that all external loads are applied at these discrete points. The term *"nodes"* is slightly different from the term *"joints"* that primarily represent the discrete points either at the member ends or at the supports. On the other hand, the nodes include these joints as well as any discrete points between the supports or member ends where concentrated forces act in a structure. For example, in the structure shown in Figure 8.1, A, *B* and *C* are the joints, whereas *i*, *j*, *k*, *l*, ... are the nodes or nodal points.

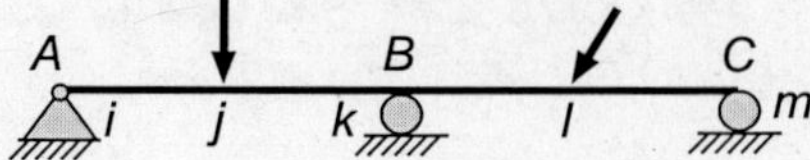

FIGURE 8.1   Definition nodes.

If a member is subjected to a distributed load as shown in Figure 8.2, the equivalent forces and bending moments should be computed assuming both ends of the member as the fixed ends. The support reactions and moments at the fixed ends are required to be transferred to the nodal points at the supports. The applied forces and moments should be of equal magnitude and opposite directions. The analysis of the structure with all loads acting at the nodes is carried out. The fixed end moments and reactions are then added to the corresponding values obtained from the matrix analysis to get the final reactions and bending moments of the structure.

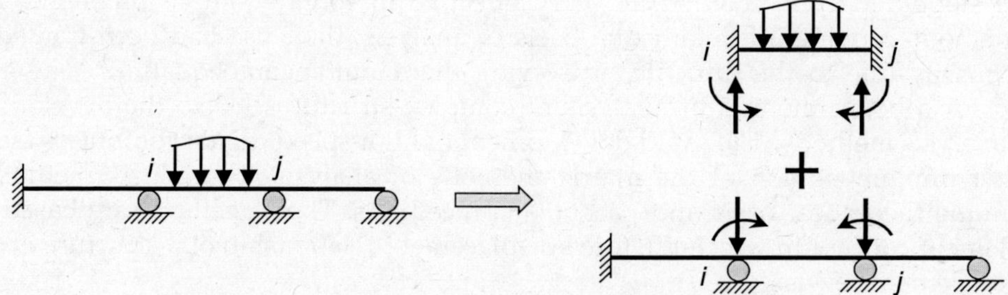

FIGURE 8.2   Nodal load equivalence for distributed loads.

Any structure can be considered as a combination of several members. For example, a truss is composed of many two-force members with pinned ends. A rigid frame consists of several three-force members. All elements and the structure as a whole must satisfy three basic conditions as follows:

- Equilibrium equations
- Displacement compatibility conditions
- Force-displacement relationships

As stated earlier, the principle of superposition is also valid as the behaviour of all members is considered to be linearly elastic.

8.2.1 Force and Displacement Matrices

In matrix methods of analysis, it is important to define the notations for force and deformation quantities of a structure. Further, these quantities for the whole structure must be differentiated from those of members. The notations adopted in the matrix methods have been discussed in the following sections.

Consider a rigid frame shown in Figure 8.3(a). This frame is composed of three members. These members are denoted by a, b and c. The frame is subjected to external loads (i.e., forces and moments) R_1, R_2, and R_3 at the joints. Thus, there are three nodes in the frame and accordingly, the forces R_1, R_2, and R_3 represent the *nodal forces*. The deformations r_1, r_2, and r_3 at the nodes are called the *nodal displacements*. The nodal forces and nodal displacements represent the force and deformation quantities of the structure at the global level. Both nodal force and nodal displacement matrices are represented by column matrices as shown below:

$$\text{Nodal force matrix, } \{R\} = \begin{Bmatrix} R_1 \\ R_2 \\ \dots \\ \dots \\ R_n \end{Bmatrix}$$

$$\text{Nodal displacement matrix, } \{r\} = \begin{Bmatrix} r_1 \\ r_2 \\ \dots \\ \dots \\ r_n \end{Bmatrix}$$

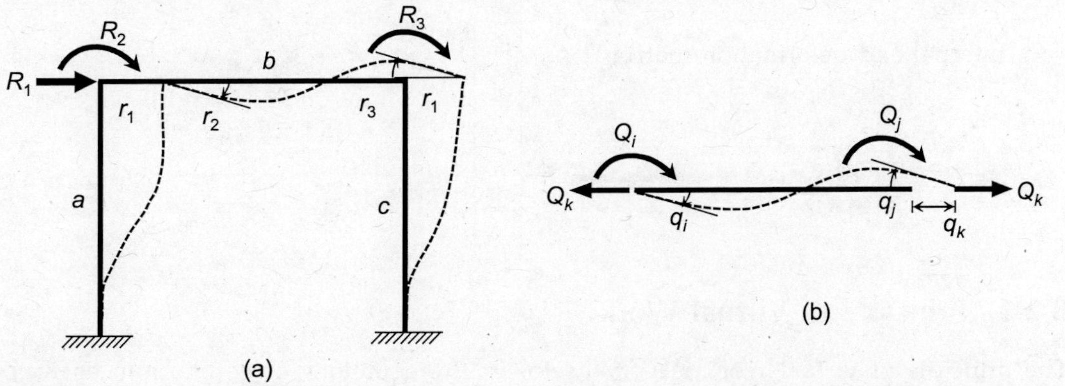

FIGURE 8.3 Definition of (a) nodal forces and nodal displacements, (b) element forces and element deformations.

Figure 8.3(b) shows the internal force and deformation quantities of members at the local level. A typical frame element is subjected to an end moment, a shear force, and an axial force at each node. Accordingly, a member is subjected to a total of two end moments, two end shear forces, and two axial forces. Since all forces/moments act at the nodes of an element, the end shear forces can be expressed in terms of the end moments. Therefore, the end shear forces are not considered as independent forces. Further, since the frame members are regarded as axially inextensible, the magnitude of axial force would be the same at both ends of an element. Thus, each frame element is subjected to three independent internal forces, Q_i, Q_i and Q_k representing two end moments and one axial force, respectively. The corresponding deformations are q_i, q_j and q_k representing two end rotations and one axial end deformation. Each member has three end forces and three end deformations. The total number of internal end forces and end deformation matrices of all members are represented in the column matrix format.

$$\text{Internal end force matrix, } \{Q\} = \begin{Bmatrix} Q_i^a \\ Q_j^a \\ Q_k^a \\ Q_i^b \\ Q_j^b \\ \dots \\ \dots \\ Q_k^m \end{Bmatrix}$$

$$\text{Internal end deformation matrix, } \{q\} = \begin{Bmatrix} q_i^a \\ q_j^a \\ q_k^a \\ q_i^b \\ q_j^b \\ \dots \\ \dots \\ q_k^m \end{Bmatrix}$$

8.2.2   Principle of Virtual Work

The principle of virtual work is very useful in the equations of equilibrium as well as displacement compatibility conditions. This principle states that the external virtual work of an elastic structural system in equilibrium must be equal to the internal virtual work. The virtual work may be considered as a result of either a virtual force or a virtual displacement.

If the virtual displacements are used in the calculation of virtual work, the principle of virtual work in the matrix form can be written as follows:

$$\{\delta r\}^T \{R\} = \{\delta q\}^T \{Q\} \tag{8.1}$$

Similarly, if the virtual forces are used in the calculation of virtual work, the principle of virtual work can be expressed as follows:

$$\{\delta R\}^T \{r\} = \{\delta Q\}^T \{q\} \tag{8.2}$$

8.3 FORCE-TRANSFORMATION (EQUILIBRIUM) MATRIX

In a statically determinate structure, the member internal forces can be expressed in terms of the external nodal forces. For example, the member internal moment at any point is the sum of the product of external forces and the lever arm. Similarly, the member internal moment is the sum of all external end moments. Thus, the equilibrium equation between the internal element forces and external nodal forces can be written as follows:

$$Q_1 = b_{11}R_1 + b_{12}R_2 + b_{13}R_3 + \dots + b_{1n}R_n$$

$$Q_2 = b_{21}R_1 + b_{22}R_2 + b_{23}R_3 + \dots + b_{2n}R_n$$

$$\dots \qquad \dots \qquad \dots \qquad \dots$$

$$\dots \qquad \dots \qquad \dots \qquad \dots \tag{8.3}$$

$$Q_m = b_{m1}R_1 + b_{m2}R_2 + b_{m3}R_3 + \dots + b_{mn}R_n$$

where, Q_1 may be Q_i^a; Q_2 may be Q_j^a; and so on … Eqn. (8.3) can be written in the matrix form as follows:

$$\begin{Bmatrix} Q_1 \\ Q_2 \\ \dots \\ Q_m \end{Bmatrix} = \begin{bmatrix} b_{11} & b_{12} & \dots & b_{1n} \\ b_{21} & b_{22} & \dots & b_{2n} \\ \dots & \dots & \dots & \dots \\ b_{m1} & b_{m2} & \dots & b_{mn} \end{bmatrix} \begin{Bmatrix} R_1 \\ R_2 \\ \dots \\ R_n \end{Bmatrix} \tag{8.4}$$

Simply,
$$\{Q\} = [b]\{R\} \tag{8.5}$$

where,
$$[b] = \begin{bmatrix} b_{11} & b_{12} & \dots & b_{1n} \\ b_{21} & b_{22} & \dots & b_{2n} \\ \dots & \dots & \dots & \dots \\ b_{m1} & b_{m2} & \dots & b_{mn} \end{bmatrix} \tag{8.6}$$

The matrix $[b]$ relating the internal end forces to the external nodal forces is called the *force transformation (equilibrium)* matrix. The matrix $[b]$ is a rectangular matrix in which the element b_{ij} represents the value of internal force, Q_i caused by a unit value of external nodal force, R_j. Depending on the nature of internal force Q and external nodal force R, the element b_{ij} may represent the length (lever arm) unit or a constant value.

For a statically indeterminate structure, the member internal forces cannot be determined using the external forces in the equilibrium equations alone. As discussed in the Method of Consistent Deformation, the statically indeterminate structure can be transformed into a statically determinate (primary) structure by removing the redundant forces. The final internal end forces are the sum of the end forces of the primary structure and the structure with redundant forces only. If the external nodal force matrix is $\{R\}$ and the redundant force matrix is $\{X\}$, the internal end force matrix $\{Q\}$ can be expressed as follows:

$$\{Q\} = [b_R]\{R\} + [b_X]\{X\} \tag{8.7}$$

or
$$\{Q\} = \begin{bmatrix} b_R & b_X \end{bmatrix} \begin{Bmatrix} R \\ X \end{Bmatrix} \tag{8.8}$$

where, $[b_R]$ and $[b_X]$ are the force-transformation matrices representing the influence of $\{R\}$ and $\{X\}$ considered independently, respectively.

8.4 DISPLACEMENT COMPATIBILITY CONDITION

Displacement compatibility presents the displacement continuity of the structure under the application of external loads. For statically indeterminate structures, the displacement corresponding to the redundant forces must be zero if the supports are unyielding. If the nodal displacement matrix corresponding to the redundant forces is represented by $\{r_X\}$, the displacement compatibility condition can be defined as follows:

$$\{r_X\} = \{0\} \tag{8.9}$$

These displacement compatibility conditions must be included in the analysis procedure as the equilibrium equations are not sufficient to solve the statically indeterminate structures.

8.5 ELEMENT FLEXIBILITY MATRIX

Since the nodal displacements of a structure are strongly dependent on the nodal forces, a relationship between these displacements and forces can be established at the global level. Similarly, the member end deformations can be related to the member internal forces. In this section, the force-displacement relationships are derived at the member (local) level as well as the structure (global) level.

First, the relationship between the member displacements and the member internal forces is established in the following section. Consider a typical frame member with the assumed deformed shape shown in Figure 8.4. The clockwise end moments and rotations are considered positive. The tensile axial forces and deformations are also taken as positive.

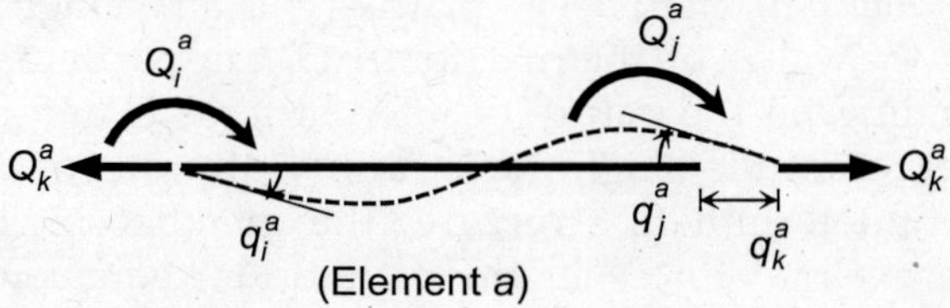

FIGURE 8.4 Element forces and deformations of a frame member.

Internal forces and displacements matrices of element a are given by

$$\{Q^a\} = \begin{Bmatrix} Q_i^a \\ Q_j^a \\ Q_k^a \end{Bmatrix} \quad \text{and} \quad \{q^a\} = \begin{Bmatrix} q_i^a \\ q_j^a \\ q_k^a \end{Bmatrix}$$

Flexibility coefficients relate the member deformations to the element internal forces. A *flexibility coefficient* f_{ij} is defined as the displacement at the point i due to a unit action (i.e., force/moment) at point j, all other points being unloaded. Using the flexibility coefficients for member a, the element deformations can be expressed in terms of the member forces. Accordingly,

$$q_i^a = f_{ii}^a Q_i^a + f_{ij}^a Q_j^a + f_{ik}^a Q_k^a$$

$$q_j^a = f_{ji}^a Q_i^a + f_{jj}^a Q_j^a + f_{jk}^a Q_k^a \qquad (8.10)$$

$$q_k^a = f_{ki}^a Q_i^a + f_{kj}^a Q_j^a + f_{kk}^a Q_k^a$$

In matrix form, the above equations can be written as

$$\{q^a\} = [f^a]\{Q^a\} \qquad (8.11)$$

where, flexibility coefficient matrix, $[f^a] = \begin{bmatrix} f_{ii}^a & f_{ij}^a & f_{ik}^a \\ f_{ji}^a & f_{jj}^a & f_{jk}^a \\ f_{ki}^a & f_{kj}^a & f_{kk}^a \end{bmatrix}$ \qquad (8.12)

Using Eqn. (8.10), one can notice that

the flexibility coefficient, $f_{ii}^a = q_i^a$ for $Q_i^a = 1$ and $Q_j^a = Q_k^a = 0$,

the flexibility coefficient, $f_{ij}^a = q_i^a$ for $Q_j^a = 1$ and $Q_i^a = Q_k^a = 0$,

the flexibility coefficient, $f_{kj}^a = q_k^a$ for $Q_j^a = 1$ and $Q_i^a = Q_k^a = 0$, and so on...

The above force-displacement relationship has been derived for the element a. A similar procedure can be followed for all the elements of a structure. Accordingly,

$$\{q^b\} = [f^b]\{Q^b\}; \{q^c\} = [f^c]\{Q^c\}; \text{ and so on...}$$

If a structure is composed of several elements (a, b, c), the element force and deformation matrices can be written as the combinations of these forces and deformations of all elements.

If $\{Q\} = \begin{Bmatrix} Q^a \\ Q^b \\ ... \\ ... \end{Bmatrix}$; and $\{q\} = \begin{Bmatrix} q^a \\ q^b \\ ... \\ ... \end{Bmatrix}$, the force-deformation relationships of all elements

can be written as follows:

$$\begin{Bmatrix} q^a \\ q^b \\ ... \\ ... \end{Bmatrix} = \begin{bmatrix} f^a & & \\ & f^b & \\ & & ... \\ & & & ... \end{bmatrix} \begin{Bmatrix} Q^a \\ Q^b \\ ... \\ ... \end{Bmatrix} \qquad (8.13)$$

or
$$\{q\} = [f]\{Q\} \tag{8.14}$$

The matrix $[f]$ is a diagonal matrix in which each component represents the flexibility matrix of an element.

Element flexibility matrices for different types of members are derived below. Broadly, three types of members are used in the structures. These are truss members, frame members, and beams.

(a) Truss member

Consider a truss element a of length L, area of cross-section A, and Modulus of elasticity E as shown in Figure 8.5. A truss member carries only one axial force. Consider one end of the member as hinged, whereas the other end is allowed to displace axially. The axial displacement (q_a) of the member a carrying an axial force Q_a is determined as follows:

$$q_a = \frac{Q_a L}{AE} \tag{8.15}$$

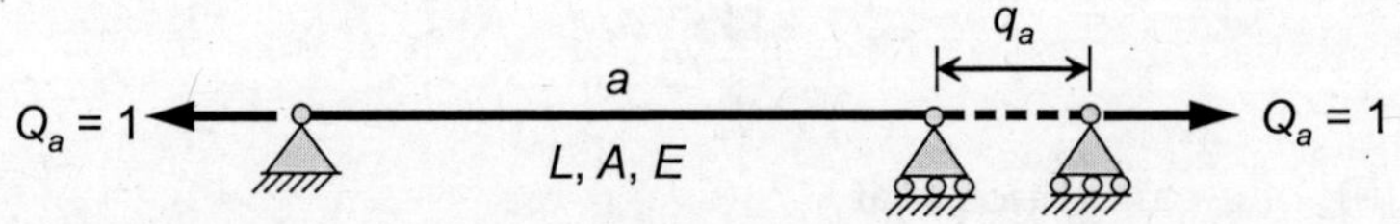

q_a

a

$Q_a = 1$ $\qquad L, A, E \qquad$ $Q_a = 1$

FIGURE 8.5 Element force and deformation of truss member.

Flexibility coefficient (f_a) is the axial displacement q_a at one end of the truss member when the value of $Q_a = 1$ and keeping other end at the restrained condition. Using Eqn. (8.15), the flexibility coefficient of a truss member can be determined as follows:

$$[f^a] = \left[\frac{L}{AE}\right] \tag{8.16}$$

(b) Frame member

Consider a frame element a of length L, flexural rigidity EI, and area of cross-section A. As shown in Figure 8.4, a frame member has three degrees of freedom, i.e., two end rotations $(q_i$ and $q_j)$ and one axial displacement (q_k). The flexibility coefficient matrix $[f^a]$ for member a is given by

$$[f^a] = \begin{bmatrix} f_{ii}^a & f_{ij}^a & f_{ik}^a \\ f_{ji}^a & f_{jj}^a & f_{jk}^a \\ f_{ki}^a & f_{kj}^a & f_{kk}^a \end{bmatrix}$$

$$
\begin{array}{c|c|c}
Q_i = 1 & Q_j = 1 & Q_k = 1 \\
Q_j = 0 & Q_i = 0 & Q_i = 0 \\
Q_k = 0 & Q_k = 0 & Q_j = 0
\end{array}
$$

By the definition, the flexibility coefficient f_{ij} is equal to the end deformation, q_i when the element is subjected to the force $Q_j = 1$ ensuring all other forces are zero, i.e., $Q_i = Q_k = 0$. Thus, each column of the flexibility matrix $[f^a]$ represents all element deformations for a particular unit force. For example, the components of the first column represent the member end deformations when $Q_i = 1$ and $Q_j = Q_k = 0$. To determine the member end deformations, the boundary conditions of the element must satisfy these conditions.

Consider the first column of the flexibility matrix in which $Q_i = 1$ and $Q_j = Q_k = 0$. Since Q_i and Q_j are the moments and the Q_k is the axial force, $Q_j = 0$ represent the pinned boundary condition. The structure with the applicable boundary conditions and subjected to a unit moment (i.e., $Q_i = 1$) is shown in Figure 8.6(a).

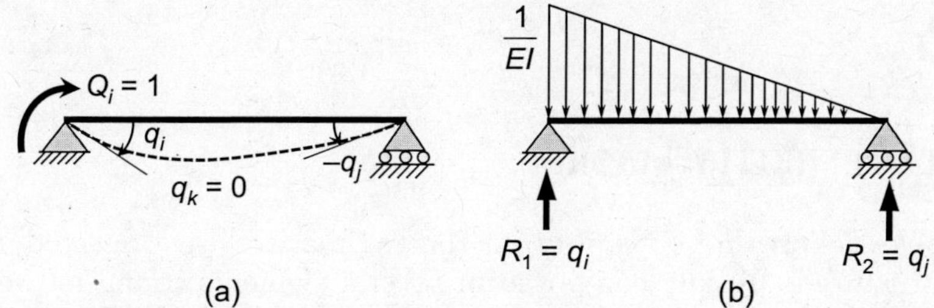

FIGURE 8.6 Procedure for deriving the flexibility coefficients of a frame member.

The end rotations (q_i and q_j) of the structure due to the applied unit moment can be determined using the conjugate-beam method. Note that the axial displacement, $q_k = 0$. Figure 8.6(b) shows the conjugate beam with $(1/EI)$ as the loading. The reactions at the supports would be equal to the end rotations. Considering the equilibrium of the conjugate beam,

$$R_1 = \frac{L}{3EI} \quad \text{and} \quad R_2 = \frac{L}{6EI} \tag{8.17}$$

$$f_{ii} = q_i = \frac{L}{3EI}; \; f_{ji} = q_j = -\frac{L}{6EI}; \; f_{ki} = q_k = 0 \tag{8.18}$$

A similar procedure can be adopted to determine the flexibility coefficients corresponding to the middle column.

$$f_{ij} = -\frac{L}{6EI}; \; f_{jj} = \frac{L}{3EI}; \; f_{kj} = 0 \tag{8.19}$$

For the third column, only a unit axial force is applied. One can find that

$$f_{ik} = f_{jk} = 0; \; f_{kk} = \frac{L}{AE} \tag{8.20}$$

Accordingly, the flexibility coefficient matrix for a frame is given by

$$[f^a] = \begin{bmatrix} \dfrac{L}{3EI} & -\dfrac{L}{6EI} & 0 \\[2ex] -\dfrac{L}{6EI} & \dfrac{L}{3EI} & 0 \\[2ex] 0 & 0 & \dfrac{L}{AE} \end{bmatrix} \tag{8.21}$$

(c) Beam

The flexibility coefficient matrix for a beam can be determined from the flexibility coefficient matrix for a frame member by neglecting the effect of axial force. Accordingly, the flexibility coefficient matrix of a beam can be obtained as follows:

$$[f^a] = \begin{bmatrix} \dfrac{L}{3EI} & -\dfrac{L}{6EI} \\ -\dfrac{L}{6EI} & \dfrac{L}{3EI} \end{bmatrix} = \dfrac{L}{6EI}\begin{bmatrix} 2 & -1 \\ -1 & 2 \end{bmatrix} \tag{8.22}$$

8.6 TOTAL FLEXIBILITY MATRIX

In the previous section, the flexibility coefficient matrices are derived for individual elements. A structure is usually composed of several elements connected to each other at many common nodes. It is therefore required to derive the *total flexibility matrix* or *flexibility matrix of the structure*. While the *element flexibility matrices* relate the element internal forces (Q) to the end deformations (q) of the elements (at the *local level*), the *total flexibility matrix* establishes the relationship between the nodal forces (R) and the nodal deformations (r) (at the *global level*). The relationship between the internal forces and deformations with the nodal forces and deformations can be established using the principle of virtual work.

8.6.1 Statically Determinate Structures

Equation (8.2) is reproduced as follows:

$$\{\delta R\}^T \{r\} = \{\delta Q\}^T \{q\}$$

For the statically determinate structures,

$$\{Q\} = [b]\{R\} \quad \text{and} \quad \{q\} = [f]\{Q\}$$

Thus, $\{\delta Q\} = [b]\{\delta R\}$

$$\{\delta Q\}^T = \{\delta R\}^T \{b\}^T \tag{8.23}$$

Using Eqn. (8.5) in Eqn. (8.14),

$$\{q\} = [f][b]\{R\} \tag{8.24}$$

Substituting Eqns. (8.23) and (8.24) in Eqn. (8.2),

$$\{\delta R\}^T \{r\} = \{\delta R\}^T [b]^T [f][b]\{R\}$$

or

$$\{r\} = [b]^T [f][b]\{R\} \tag{8.25}$$

or

$$\{r\} = [F]\{R\} \tag{8.26}$$

where,

$$[F] = [b]^T [f][b] \tag{8.27}$$

[F] is known as *total flexibility matrix* or *flexibility matrix of statically determinate structure.* Eqn. (8.26) provides the relationship between the nodal deformations and the nodal external forces in the statically determinate structures.

8.6.2 Statically Indeterminate Structures

For statically indeterminate structures, the internal element forces $\{Q\}$ are the function of external forces $\{R\}$ and the redundant forces $\{X\}$.

$$\{Q\} = \begin{bmatrix} b_R & b_X \end{bmatrix} \begin{Bmatrix} R \\ X \end{Bmatrix} = [b_R]\{R\} + [b_X]\{X\} \tag{8.28}$$

$$\{\delta Q\} = [b_R]\{\delta R\} + [b_X]\{\delta X\} \tag{8.29}$$

$$\{\delta Q\}^T = \{\delta R\}^T [b_R]^T + \{\delta X\}^T [b_X]^T \tag{8.30}$$

The nodal deformation matrix of a statically have two components. The nodal deformation matrix corresponding to the external nodal force matrix $\{R\}$ and the redundant force matrix $\{X\}$ are $\{r_R\}$ and $\{r_X\}$, respectively. Accordingly, the principle of virtual work for the statically indeterminate structures can be written as

$$\{\delta R\}^T \{r_R\} + \{\delta X\}^T \{r_X\} = \{\delta Q\}^T \{q\} \tag{8.31}$$

Using Eqn. (8.28), the element end deformation matrix $\{q\}$ for a statically indeterminate structure is given by

$$\{q\} = [f]\{Q\}$$

$$\{q\} = [f][b_R]\{R\} + [f][b_X]\{X\} \tag{8.32}$$

Using Eqn. (8.32) in Eqn. (8.31), one would get

$$\{\delta R\}^T \{r_R\} + \{\delta X\}^T \{r_X\} = \{\delta Q\}^T [f][b_R]\{R\} + \{\delta Q\}^T [f][b_X]\{X\}$$

Using Eqn. (8.30) in the above equation,

$$\{\delta R\}^T \{r_R\} + \{\delta X\}^T \{r_X\} = \{\delta R\}^T \left([b_R]^T [f][b_R]\{R\} + [b_R]^T [f][b_X]\{X\}\right)$$

$$+ \{\delta X\}^T \left([b_X]^T [f][b_R]\{R\} + [b_X]^T [f][b_X]\{X\}\right)$$

Comparing the virtual forces on left and right sides of the above equation,

$$\{r_R\} = [b_R]^T [f][b_R]\{R\} + [b_R]^T [f][b_X]\{X\} \tag{8.33}$$

$$\{r_X\} = [b_X]^T [f][b_R]\{R\} + [b_X]^T [f][b_X]\{X\} \tag{8.34}$$

These equations may be rearranged as

$$\{r_R\} = [F_{RR}]\{R\} + [F_{RX}]\{X\} \tag{8.35}$$

$$\{r_X\} = [F_{XR}]\{R\} + [F_{XX}]\{X\} \tag{8.36}$$

where,

$$[F_{RR}] = [b_R]^T [f][b_R]; \ [F_{XR}] = [b_X]^T [f][b_R] \tag{8.37}$$

$$[F_{RX}] = [b_R]^T [f][b_X]; \ [F_{XX}] = [b_X]^T [f][b_X]$$

Eqns. (8.5) and (8.37) can be arranged as:

$$\begin{Bmatrix} r_R \\ r_X \end{Bmatrix} = \begin{bmatrix} F_{RR} & F_{RX} \\ F_{XR} & F_{XX} \end{bmatrix} \begin{Bmatrix} R \\ X \end{Bmatrix} \tag{8.38}$$

It is worth mentioning that the nodal deformation matrix $\{r_R\}$ cannot be determined unless all components of the redundant force matrix $\{X\}$ are computed. The matrix $\{X\}$ can be determined using the compatibility condition in the nodal displacement matrix $\{r_X\}$ corresponding to the redundant forces.

For structures with unyielding supports, the compatibility condition is $\{r_X\} = \{0\}$. Accordingly, Eqn. (8.38) can be written as follows:

$$\begin{Bmatrix} r_R \\ 0 \end{Bmatrix} = \begin{bmatrix} F_{RR} & F_{RX} \\ F_{XR} & F_{XX} \end{bmatrix} \begin{Bmatrix} R \\ X \end{Bmatrix} \tag{8.39}$$

Therefore,

$$\{r_R\} = [F_{RR}]\{R\} + [F_{RX}]\{X\} \tag{8.40}$$

and

$$[F_{XR}]\{R\} + [F_{XX}]\{X\} = \{0\}$$

$$\therefore \quad \{X\} = - [F_{XX}]^{-1}[F_{XR}]\{R\} \tag{8.41}$$

Substituting Eqn. (8.40) in Eqn. (8.39), the nodal deformation matrix can be determined as follows:

$$\{r_R\} = \left([F_{RR}] - [F_{RX}][F_{XX}]^{-1}[F_{XR}]\right)\{R\} \tag{8.42}$$

Simply,

$$\{r\} = [F']\{R\} \tag{8.43}$$

where,

$$[F'] = [F_{RR}] - [F_{RX}][F_{XX}]^{-1}[F_{XR}] \tag{8.44}$$

Putting Eqn. (8.40) in Eqn. (8.28), the element force matrix $\{Q\}$ can be determined as follows:

$$\{Q\} = \left([b_R] - [b_X][F_{XX}]^{-1}[F_{XR}]\right)\{R\} \tag{8.45}$$

or

$$\{Q\} = [b']\{R\} \tag{8.46}$$

in which

$$[b'] = [b_R] - [b_X][F_{XX}]^{-1}[F_{XR}] \tag{8.47}$$

$[b']$ is the force-transformation matrix of the statically indeterminate structures. Using equations (8.43) and (8.46), the following identities can be derived.

$$[F'] = [b_R]^T [f][b'] \tag{8.48}$$

$$[b_X]^T [f][b'] = \{0\} \tag{8.49}$$

8.7 STEP-BY-STEP PROCEDURE OF MATRIX FORCE METHOD

The analysis of structures using the Matrix force method primarily aimed at determining the element internal force matrix $\{Q\}$ and the nodal displacement matrix $\{r\}$. As discussed in the preceding sections, the computation of these matrices requires a slightly different procedure for the statically determinate structures and indeterminate structures. The step-by-step procedure of analysis of these structures is discussed in the following sections.

8.7.1 Analysis of Statically Determinate Structures

The procedure to analyze a statically determinate structure by the matrix force method is presented below:

1. Define the nodes and elements based on the positions of externally applied loads.
2. Define the external nodal force matrix, $\{R\}$ and the nodal deformation matrix, $\{r\}$.
3. Define the internal element force matrix, $\{Q\}$ and the element end deformation, $\{q\}$.
4. Determine the force-transformation matrix, $[b]$. All components of $[b]$ are computed using the equilibrium conditions by setting only one nodal force equal to unity and keeping all other nodal forces as zero at a particular instance. For example, $R_1 = 1$, $R_2 = R_3 = ... = 0$ and so on.
5. Determine the internal element force matrix, $\{Q\} = [b]\{R\}$.
6. Determine the element flexibility matrices ($[f^a]$, $[f^b]$, $[f^c]$, ...) depending on the type of members and compute the assembled element flexibility matrix $[f]$ as follows:

$$[f] = \begin{bmatrix} f^a & & & \\ & f^b & & \\ & & \cdots & \\ & & & \cdots \end{bmatrix}$$

7. Compute the total flexibility matrix of structure, $[F] = [b]^T [f][b]$.
8. Determine the nodal deformation matrix, $\{r\} = [F]\{R\}$

8.7.2 Analysis of Statically Indeterminate Structures

The procedure to analyze a statically indeterminate structure by the matrix force method is presented below:

1. Define the nodes and elements based on the positions of externally applied loads.
2. Identify the redundant forces in the structure.
3. Define the external nodal force matrix, $\{R\}$ and redundant force matrix, $\{X\}$
4. Define the nodal deformation matrix corresponding to the externally applied load and the redundant forces, $\{r_R\}$ and $\{r_X\}$, respectively.
5. Define the internal element force matrix, $\{Q\}$ and element end deformation, $\{q\}$.
6. Determine the force-transformation matrices, $[b_R]$ and $[b_X]$ corresponding to the nodal forces and redundant forces matrices, respectively. All components of a

particular column of $[b_R]$ and $[b_X]$ are computed by setting only one nodal force equal to one and keeping all other nodal forces to be zero. For example, $R_1 = 1$, $R_2 = R_3 = \dots = 0$; $X_1 = 1$, $X_2 = X_3 = \dots = 0$ and so on.

7. Determine the element flexibility matrices ($[f^a]$, $[f^b]$, $[f^c]$, ...) depending on the type of members and compute the assembled element flexibility matrix $[f]$ as follows:

$$[f] = \begin{bmatrix} f^a & & & \\ & f^b & & \\ & & \dots & \\ & & & \dots \end{bmatrix}$$

8. Compute the components of the total flexibility matrix of structure,

$$[F_{RR}] = [b_R]^T [f][b_R]; \quad [F_{XR}] = [b_X]^T [f][b_R]$$

$$[F_{RX}] = [b_R]^T [f][b_X]; \quad [F_{XX}] = [b_X]^T [f][b_X]$$

9. Determine the redundant force matrix, $\{X\} = -[F_{XX}]^{-1}[F_{XR}]\{R\}$.

10. Determine the nodal deformations matrix corresponding to external loads,

$$\{r_R\} = \left([F_{RR}] - [F_{RX}][F_{XX}]^{-1}[F_{XR}]\right)\{R\}$$

11. Determine the internal element force matrix,

$$\{Q\} = \left([b_R] - [b_X][F_{XX}]^{-1}[F_{XR}]\right)\{R\}.$$

The application of the Matrix force method in the statically determinate and indeterminate structures has been presented below:

EXAMPLE 8.1 Determine the bar forces and the joint displacements corresponding to the applied loads of the truss shown in Figure E8.1 using the Matrix force method. Assume the cross-sectional area (A) of each bar is 10 sq. cm. and the modulus elasticity (E) of material is 200 GPa.

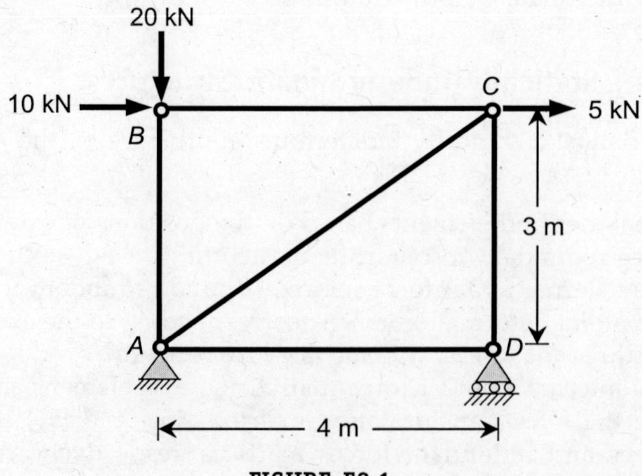

FIGURE E8.1

Solution: Since the degree of static determinacy is zero, the given truss is statically determinate. The nodal forces matrix and the corresponding nodal displacement matrix are as follows:

$$\{R\} = \begin{Bmatrix} R_1 \\ R_2 \\ R_3 \end{Bmatrix} = \begin{Bmatrix} 10 \\ 20 \\ 5 \end{Bmatrix} \text{kN}; \quad \{r\} = \begin{Bmatrix} r_1 \\ r_2 \\ r_3 \end{Bmatrix}$$

There are five elements as shown in the figure, namely, a, b, c, d, and e. The element internal force and end deformation matrices are as follows:

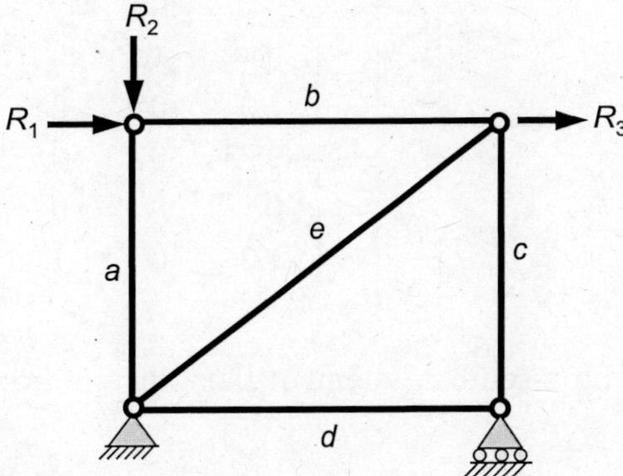

$$\{Q\} = \begin{Bmatrix} Q^a \\ Q^b \\ Q^c \\ Q^d \\ Q^e \end{Bmatrix}; \quad \{q\} = \begin{Bmatrix} q^a \\ q^b \\ q^c \\ q^d \\ q^e \end{Bmatrix}$$

Since it is a statically determinate structure and the matrix $\{R\}$ is known, the internal member forces (or bar forces) can be directly determined using the following expression:

$$\{Q\} = [b]\{R\}$$

The force-transformation matrix $[b]$ can be determined by computing all bar forces setting only one component of $\{R\}$ equal to one while keeping all other components of the matrix to be zero at a time. These are shown in the figures below:

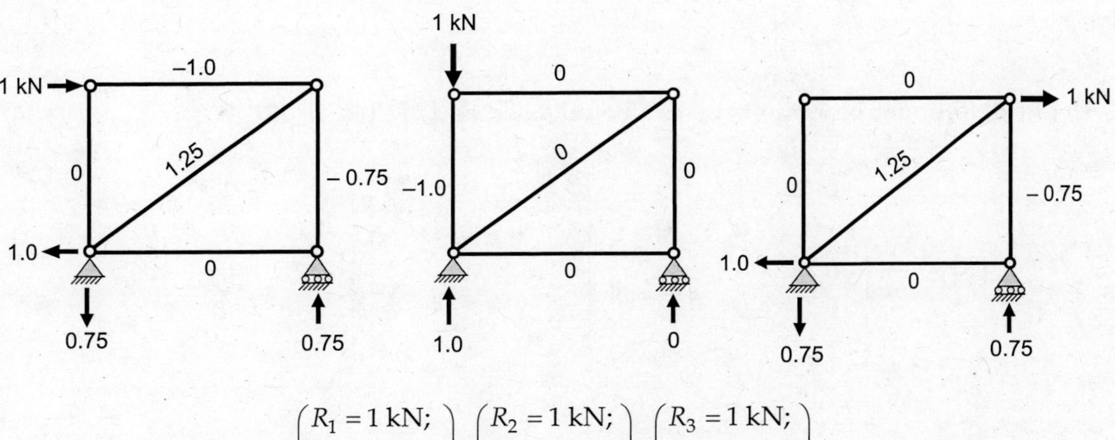

$$\left(\begin{matrix} R_1 = 1 \text{ kN}; \\ R_2 = R_3 = 0 \end{matrix} \right) \quad \left(\begin{matrix} R_2 = 1 \text{ kN}; \\ R_1 = R_3 = 0 \end{matrix} \right) \quad \left(\begin{matrix} R_3 = 1 \text{ kN}; \\ R_1 = R_2 = 0 \end{matrix} \right)$$

$$[b] = \begin{bmatrix} 0 & -1.0 & 0 \\ -1.0 & 0 & 0 \\ -0.75 & 0 & -0.75 \\ 0 & 0 & 0 \\ 1.25 & 0 & 1.25 \end{bmatrix}$$

Internal element force matrix,

$$\begin{Bmatrix} Q^a \\ Q^b \\ Q^c \\ Q^d \\ Q^e \end{Bmatrix} = \begin{bmatrix} 0 & -1.0 & 0 \\ -1.0 & 0 & 0 \\ -0.75 & 0 & -0.75 \\ 0 & 0 & 0 \\ 1.25 & 0 & 1.25 \end{bmatrix} = \begin{Bmatrix} 10 \\ 20 \\ 5 \end{Bmatrix} \text{ kN} \qquad \therefore \quad \begin{Bmatrix} Q^a \\ Q^b \\ Q^c \\ Q^d \\ Q^e \end{Bmatrix} = \begin{Bmatrix} -20.0 \\ -10.0 \\ -11.3 \\ 0 \\ 18.8 \end{Bmatrix} \text{ kN}$$

The assembled element flexibility matrix $[f]$ is given by

$$[f] = \begin{bmatrix} f^a & & & & \\ & f^b & & & \\ & & f^c & & \\ & & & f^d & \\ & & & & f^e \end{bmatrix} = \frac{1}{AE} \begin{bmatrix} L^a & & & & \\ & L^b & & & \\ & & L^c & & \\ & & & L^d & \\ & & & & L^e \end{bmatrix}$$

$$[f] = \frac{1}{AE} \begin{bmatrix} 3 & & & & \\ & 4 & & & \\ & & 3 & & \\ & & & 4 & \\ & & & & 5 \end{bmatrix}$$

The flexibility matrix of the truss can be obtained as $[F] = [b]^T [f][b]$

$$[F] = \frac{1}{AE} \begin{bmatrix} 0 & -1.0 & 0 \\ -1.0 & 0 & 0 \\ -0.75 & 0 & -0.75 \\ 0 & 0 & 0 \\ 1.25 & 0 & 1.25 \end{bmatrix}^T \begin{bmatrix} 3 & & & & \\ & 4 & & & \\ & & 3 & & \\ & & & 4 & \\ & & & & 5 \end{bmatrix} \begin{bmatrix} 0 & -1.0 & 0 \\ -1.0 & 0 & 0 \\ -0.75 & 0 & -0.75 \\ 0 & 0 & 0 \\ 1.25 & 0 & 1.25 \end{bmatrix}$$

$$[F] = \frac{1}{AE} \begin{bmatrix} 13.5 & 0 & 9.5 \\ 0 & 3 & 0 \\ 9.5 & 0 & 9.5 \end{bmatrix}$$

The nodal displacement matrix is computed as follows:

$$\{r\} = [F]\{R\}$$

$$\begin{Bmatrix} r_1 \\ r_2 \\ r_3 \end{Bmatrix} = \frac{1}{AE}\begin{bmatrix} 13.5 & 0 & 9.5 \\ 0 & 3 & 0 \\ 9.5 & 0 & 9.5 \end{bmatrix}\begin{Bmatrix} 10 \\ 20 \\ 5 \end{Bmatrix} = \frac{1}{AE}\begin{Bmatrix} 182.5 \\ 60 \\ 142.5 \end{Bmatrix}$$

$$\therefore \quad \begin{Bmatrix} r_1 \\ r_2 \\ r_3 \end{Bmatrix} = \begin{Bmatrix} 0.91 \\ 0.30 \\ 0.71 \end{Bmatrix} \text{ mm}$$

Note: In the above example, the number of components in the nodal displacement matrix is taken as three, which is the same as the number of external loads. As a result, the vertical displacement of joint C has not been computed. If it is required to determine the vertical displacement at joint C, the size of matrix $\{R\}$ and $\{r\}$ would be 4×1. In this case, the value of R_4 would be zero as there are no external loads present in this direction.

EXAMPLE 8.2 Determine the slope and deflection at points B and C of the beam shown in Figure E8.2 using the Matrix force method. Assume constant flexural rigidity (*EI*).

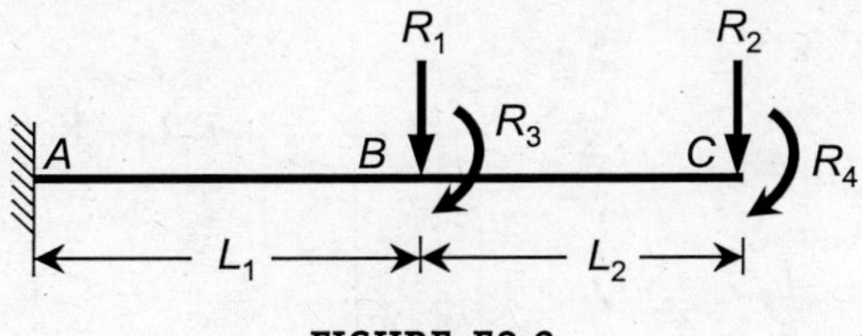

FIGURE E8.2

Solution: The beam is a statically determinate structure. The nodal forces matrix and the corresponding nodal displacement matrix are as follows:

$$\{R\} = \begin{Bmatrix} R_1 \\ R_2 \\ R_3 \\ R_4 \end{Bmatrix}; \quad \{r\} = \begin{Bmatrix} r_1 \\ r_2 \\ r_3 \\ r_4 \end{Bmatrix}$$

Two elements (a, b) are taken in the beam based on the position of external loads. Each beam element has two internal forces (i.e., moments) since the axial force is zero. Accordingly, the element internal force and deformation matrices are as follows:

$$\{Q\} = \begin{Bmatrix} Q_i^a \\ Q_j^a \\ Q_i^b \\ Q_j^b \end{Bmatrix}; \quad \{q\} = \begin{Bmatrix} q_i^a \\ q_j^a \\ q_i^b \\ q_j^b \end{Bmatrix}$$

$$\{Q\} = [b]\{R\}$$

The force-transformation matrix $[b]$ is given by:

$$[b] = \begin{bmatrix} -L_1 & -(L_1+L_2) & -1 & -1 \\ 0 & L_2 & 1 & 1 \\ 0 & -L_2 & 0 & -1 \\ 0 & 0 & 0 & 1 \end{bmatrix}$$

$$(R_1 = 1)\ (R_2 = 1)\ (R_3 = 1)\ (R_4 = 1)$$

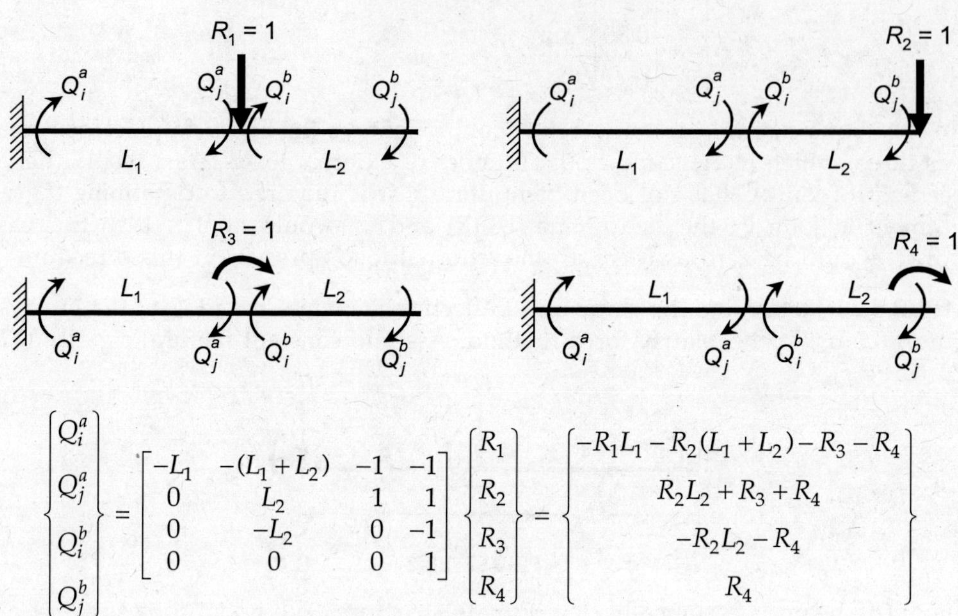

$$\begin{Bmatrix} Q_i^a \\ Q_j^a \\ Q_i^b \\ Q_j^b \end{Bmatrix} = \begin{bmatrix} -L_1 & -(L_1+L_2) & -1 & -1 \\ 0 & L_2 & 1 & 1 \\ 0 & -L_2 & 0 & -1 \\ 0 & 0 & 0 & 1 \end{bmatrix} \begin{Bmatrix} R_1 \\ R_2 \\ R_3 \\ R_4 \end{Bmatrix} = \begin{Bmatrix} -R_1 L_1 - R_2(L_1+L_2) - R_3 - R_4 \\ R_2 L_2 + R_3 + R_4 \\ -R_2 L_2 - R_4 \\ R_4 \end{Bmatrix}$$

Internal element force matrices are given by

$$[f^a] = \frac{L_1}{3EI} \begin{bmatrix} 2 & -1 \\ -1 & 2 \end{bmatrix}; \quad [f^b] = \frac{L_2}{3EI} \begin{bmatrix} 2 & -1 \\ -1 & 2 \end{bmatrix}$$

The assembled element flexibility matrix $[f]$ is given by

$$[f] = \begin{bmatrix} f^a & \\ & f^b \end{bmatrix} = \frac{1}{6EI} \begin{bmatrix} 2L_1 & -L_1 & & \\ -L_1 & 2L_1 & & \\ & & 2L_2 & -L_2 \\ & & -L_2 & 2L_2 \end{bmatrix}$$

The flexibility matrix of the truss can be obtained as $[F] = [b]^T [f][b]$

$$[F] = \begin{bmatrix} -L_1 & -(L_1+L_2) & -1 & -1 \\ 0 & L_2 & 1 & 1 \\ 0 & -L_2 & 0 & -1 \\ 0 & 0 & 0 & 1 \end{bmatrix}^T \times$$

$$\left(\frac{1}{6EI}\right) \begin{bmatrix} 2L_1 & -L_1 & & \\ -L_1 & 2L_1 & & \\ & & 2L_2 & -L_2 \\ & & -L_2 & 2L_2 \end{bmatrix} \begin{bmatrix} -L_1 & -(L_1+L_2) & -1 & -1 \\ 0 & L_2 & 1 & 1 \\ 0 & -L_2 & 0 & -1 \\ 0 & 0 & 0 & 1 \end{bmatrix}$$

$$\therefore \quad [F] = \left(\frac{1}{6EI}\right) \begin{bmatrix} 2L_1^3 & (2L_1^3+3L_1^2L_2) & 3L_1^2 & 3L_1^2 \\ & \begin{matrix}(2L_1^3+6L_1^2L_2)\\ +(2L_2^3+6L_1L_2^2)\end{matrix} & (3L_1^2+6L_1L_2) & (3L_1^2+6L_1L_2+3L_2^2) \\ & & 6L_1 & 6L_1 \\ Sym & & & (6L_1+6L_2) \end{bmatrix}$$

The nodal displacement matrix can be computed using the following equation:

$$\{r\} = [F]\{R\}$$

As a special case, if $L_1 = L_2 = L$, then

$$\begin{Bmatrix} Q_i^a \\ Q_j^a \\ Q_i^b \\ Q_j^b \end{Bmatrix} = \begin{Bmatrix} -R_1L-2R_2L-R_3-R_4 \\ R_2L+R_3+R_4 \\ -R_2L-R_4 \\ R_4 \end{Bmatrix}$$

$$\therefore \quad [F] = \left(\frac{1}{6EI}\right) \begin{bmatrix} 2L^3 & 5L^3 & 3L^2 & 3L^2 \\ & 16L^3 & 9L^2 & 12L^2 \\ & & 6L & 6L \\ Sym & & & 12L \end{bmatrix}$$

$$\therefore \quad \begin{Bmatrix} r_1 \\ r_2 \\ r_3 \\ r_4 \end{Bmatrix} = \begin{Bmatrix} 2R_1L^3+5R_2L^3+3R_3L^2+3R_4L^2 \\ 5R_1L^3+16R_2L^3+9R_3L^2+12R_4L^2 \\ 3R_1L^2+9R_2L^2+6R_3L+6R_4L \\ 3R_1L^2+12R_2L^2+6R_3L+12R_4L \end{Bmatrix}$$

EXAMPLE 8.3 Determine the member forces and the joint displacements corresponding to the applied loads of the truss shown in Figure E8.3 using the matrix force method. Assume the cross-sectional area (A) of each bar is 15 cm^2. and the modulus elasticity (E) of material is 200 GPa.

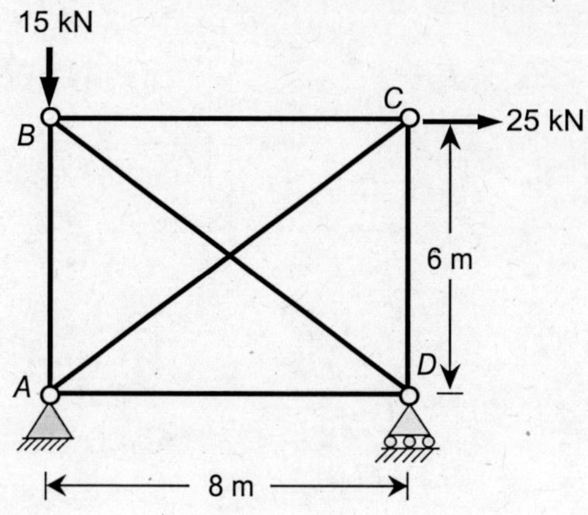

FIGURE E8.3

Solution: Since the degree of static determinacy is one, the given truss is statically indeterminate. It is necessary to choose one redundant force. Since the truss is indeterminate internally, it is assumed that the force in element e of redundant and equal to X as shown in the following figure.

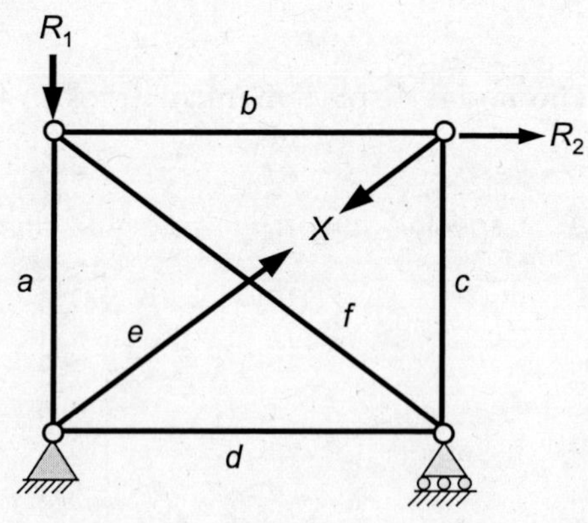

The nodal forces matrix and the corresponding nodal displacement matrix are as follows:

$$\{R\} = \begin{Bmatrix} R_1 \\ R_2 \end{Bmatrix} = \begin{Bmatrix} 15 \\ 25 \end{Bmatrix} \text{kN}; \quad \{r_R\} = \begin{Bmatrix} r_1 \\ r_2 \end{Bmatrix}$$

The redundant force and corresponding displacement matrices are $\{X\}$ and $\{r_X\}$, respectively. The element internal force and end deformation matrices are as follows:

$$\{Q\} = \begin{Bmatrix} Q^a \\ Q^b \\ Q^c \\ Q^d \\ Q^e \end{Bmatrix}; \quad \{q\} = \begin{Bmatrix} q^a \\ q^b \\ q^c \\ q^d \\ q^e \end{Bmatrix}$$

For statically indeterminate structure, $\{Q\} = [b_R]\{R\} + [b_X]\{X\}$.

The force-transformation matrices are determined as follows:

The matrix $[b_R]$ can be determined by computing all bar forces setting only one component of $\{R\}$ as unity while keeping all other components of the matrix to be zero at a time. Similarly, the matrix $[b_X]$ can be determined by considering the value of $X = 1$ while other forces are considered to be zero. These are shown in the figures below:

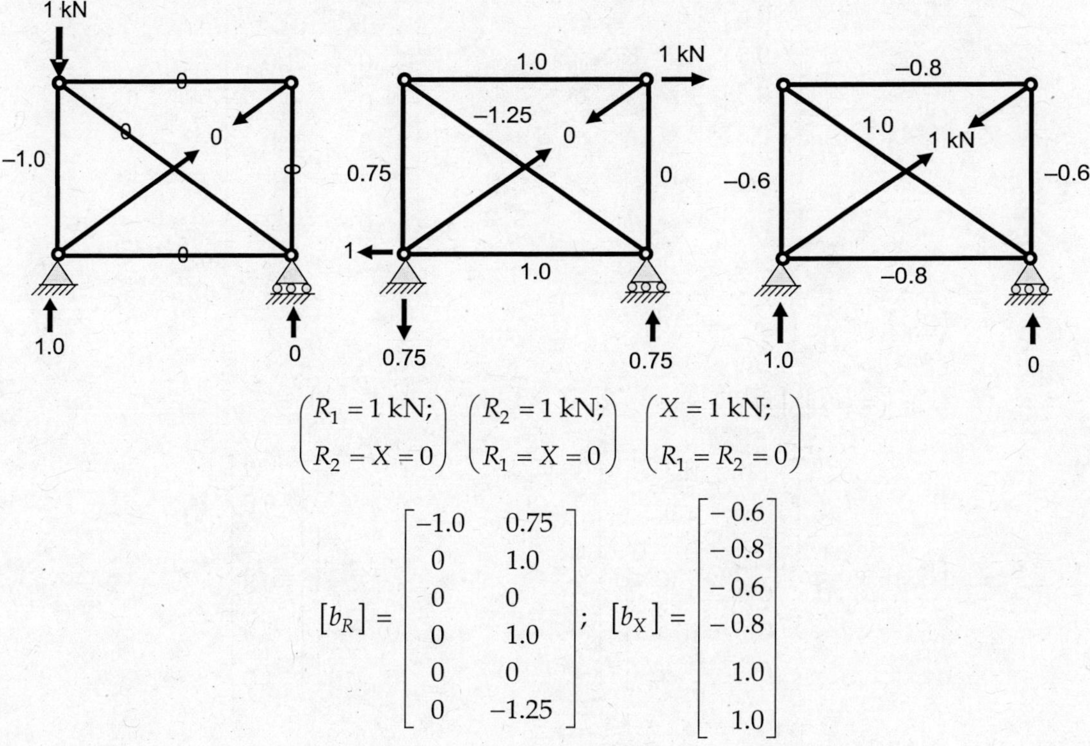

$$\begin{pmatrix} R_1 = 1\ \text{kN}; \\ R_2 = X = 0 \end{pmatrix} \quad \begin{pmatrix} R_2 = 1\ \text{kN}; \\ R_1 = X = 0 \end{pmatrix} \quad \begin{pmatrix} X = 1\ \text{kN}; \\ R_1 = R_2 = 0 \end{pmatrix}$$

$$[b_R] = \begin{bmatrix} -1.0 & 0.75 \\ 0 & 1.0 \\ 0 & 0 \\ 0 & 1.0 \\ 0 & 0 \\ 0 & -1.25 \end{bmatrix}; \quad [b_X] = \begin{bmatrix} -0.6 \\ -0.8 \\ -0.6 \\ -0.8 \\ 1.0 \\ 1.0 \end{bmatrix}$$

The assembled element flexibility matrix $[f]$ is given by

$$[f] = \begin{bmatrix} f^a & & & & & \\ & f^b & & & & \\ & & f^c & & & \\ & & & f^d & & \\ & & & & f^e & \\ & & & & & f^f \end{bmatrix} = \frac{1}{AE} \begin{bmatrix} 6 & & & & & \\ & 8 & & & & \\ & & 6 & & & \\ & & & 8 & & \\ & & & & 10 & \\ & & & & & 10 \end{bmatrix}$$

$$[F_{RR}] = [b_R]^T [f][b_R]$$

$$[F_{RR}] = \frac{1}{AE} \begin{bmatrix} -1.0 & 0.75 \\ 0 & 1.0 \\ 0 & 0 \\ 0 & 1.0 \\ 0 & 0 \\ 0 & -1.25 \end{bmatrix}^T \begin{bmatrix} 6 & & & & & \\ & 8 & & & & \\ & & 6 & & & \\ & & & 8 & & \\ & & & & 10 & \\ & & & & & 10 \end{bmatrix} \begin{bmatrix} -1.0 & 0.75 \\ 0 & 1.0 \\ 0 & 0 \\ 0 & 1.0 \\ 0 & 0 \\ 0 & -1.25 \end{bmatrix}$$

$$\therefore \quad [F_{RR}] = \frac{1}{AE}\begin{bmatrix} 6.0 & -4.5 \\ -4.5 & 35.0 \end{bmatrix}$$

$$[F_{XR}] = [b_X]^T [f][b_R]$$

$$[F_{XR}] = \frac{1}{AE}\begin{bmatrix} -0.6 \\ -0.8 \\ -0.6 \\ -0.8 \\ 1.0 \\ 1.0 \end{bmatrix}^T \begin{bmatrix} 6 & & & & & \\ & 8 & & & & \\ & & 6 & & & \\ & & & 8 & & \\ & & & & 10 & \\ & & & & & 10 \end{bmatrix} \begin{bmatrix} -1.0 & 0.75 \\ 0 & 1.0 \\ 0 & 0 \\ 0 & 1.0 \\ 0 & 0 \\ 0 & -1.25 \end{bmatrix}$$

$$\therefore \quad [F_{XR}] = \frac{1}{AE}\begin{bmatrix} 3.6 & -28.0 \end{bmatrix}$$

$$[F_{RX}] = [b_R]^T [f][b_X]$$

$$[F_{RX}] = \frac{1}{AE}\begin{bmatrix} -1.0 & 0.75 \\ 0 & 1.0 \\ 0 & 0 \\ 0 & 1.0 \\ 0 & 0 \\ 0 & -1.25 \end{bmatrix}^T \begin{bmatrix} 6 & & & & & \\ & 8 & & & & \\ & & 6 & & & \\ & & & 8 & & \\ & & & & 10 & \\ & & & & & 10 \end{bmatrix} \begin{bmatrix} -0.6 \\ -0.8 \\ -0.6 \\ -0.8 \\ 1.0 \\ 1.0 \end{bmatrix}$$

$$\therefore \quad [F_{RX}] = \frac{1}{AE}\begin{bmatrix} 3.6 \\ -28 \end{bmatrix}$$

Note that $[F_{RX}] = [F_{XR}]^T$

$$[F_{XX}] = [b_X]^T [f][b_X]$$

$$[F_{XX}] = \frac{1}{AE}\begin{bmatrix} -0.6 \\ -0.8 \\ -0.6 \\ -0.8 \\ 1.0 \\ 1.0 \end{bmatrix}^T \begin{bmatrix} 6 & & & & & \\ & 8 & & & & \\ & & 6 & & & \\ & & & 8 & & \\ & & & & 10 & \\ & & & & & 10 \end{bmatrix} \begin{bmatrix} -0.6 \\ -0.8 \\ -0.6 \\ -0.8 \\ 1.0 \\ 1.0 \end{bmatrix} = \frac{1}{AE}\begin{bmatrix} 34.56 \end{bmatrix}$$

The redundant force can be determined as follows:

$$\{X\} = -[F_{XX}]^{-1}[F_{XR}]\{R\}$$

$$\{X\} = -\left(\frac{AE}{34.56}\right)\left(\frac{1}{AE}\right)[3.6 \quad -28.0]\begin{Bmatrix}15\\25\end{Bmatrix} \text{ kN} = 18.7 \text{ kN}$$

The nodal displacement matrix is computed as follows:

$$\{r_R\} = \left([F_{RR}] - [F_{RX}][F_{XX}]^{-1}[F_{XR}]\right)\{R\}$$

or

$$\{r_R\} = [F_{RR}]\{R\} + [F_{RX}]\{X\}$$

$$\begin{Bmatrix}r_1\\r_2\end{Bmatrix} = \frac{1}{AE}\begin{bmatrix}6.0 & -4.5\\-4.5 & 35.0\end{bmatrix}\begin{Bmatrix}15\\25\end{Bmatrix} + \frac{1}{AE}\begin{bmatrix}3.6\\-28\end{bmatrix}(18.7) = \frac{1}{AE}\begin{Bmatrix}44.8\\284.0\end{Bmatrix}$$

$$\therefore \quad \begin{Bmatrix}r_1\\r_2\end{Bmatrix} = \begin{Bmatrix}0.15\\0.95\end{Bmatrix} \text{ mm}$$

The internal element force matrix, $\{Q\} = \left([b_R] - [b_X][F_{XX}]^{-1}[F_{XR}]\right)\{R\}$.

or

$$\{Q\} = [b_R]\{R\} + [b_X]\{X\}$$

$$\begin{Bmatrix}Q^a\\Q^b\\Q^c\\Q^d\\Q^e\end{Bmatrix} = \begin{bmatrix}-1.0 & 0.75\\0 & 1.0\\0 & 0\\0 & 1.0\\0 & 0\\0 & -1.25\end{bmatrix}\begin{Bmatrix}15\\25\end{Bmatrix} + \begin{bmatrix}-0.6\\-0.8\\-0.6\\-0.8\\1.0\\1.0\end{bmatrix}(18.7) = \begin{Bmatrix}-7.5\\10.0\\-11.2\\10.0\\18.7\\-12.5\end{Bmatrix} \text{ kN}$$

EXAMPLE 8.4   Determine the support reactions and the joint displacements of the beam *AB* supported on a spring as shown in Figure E8.4 using the Matrix force method. Assume that the stiffness of the spring is *K*, and the flexural rigidity of the beam is *EI*.

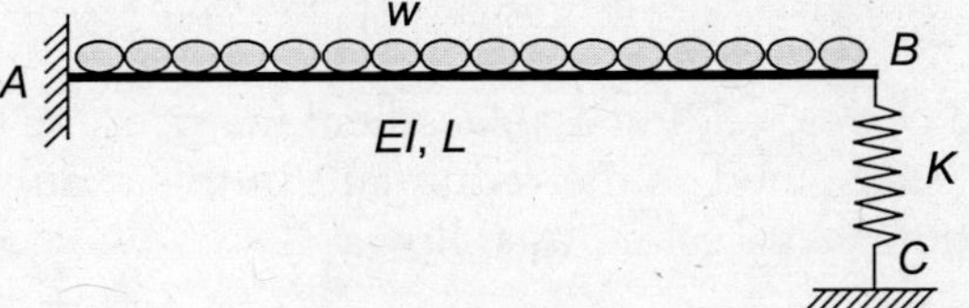

FIGURE E8.4

Solution: The degree of static indeterminacy of the structure is one. Hence, it is required to assume one redundant force in the analysis. Since the external applied load is uniformly distributed, it is required to transfer the loads as the nodal forces. For this purpose, the given structure can be considered as the superposition of the fixed-ended beam AB with the distributed load and the structure without loading. The end moments and reactions are shown in the following figure.

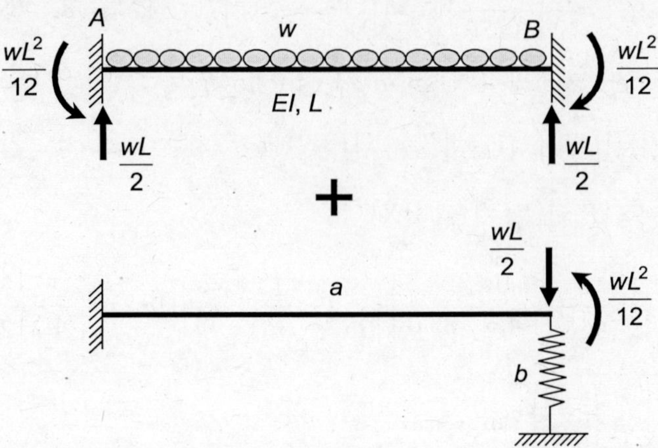

Since the left support A is fixed in both cases, there is no need to transfer any force from the fixed-ended beam. However, the equal and opposite support reaction and end moment are applied to the joint B as external loads of the structure. For the matrix force method of analysis, the structure with equivalent loads applied at the joint would be considered as shown below.

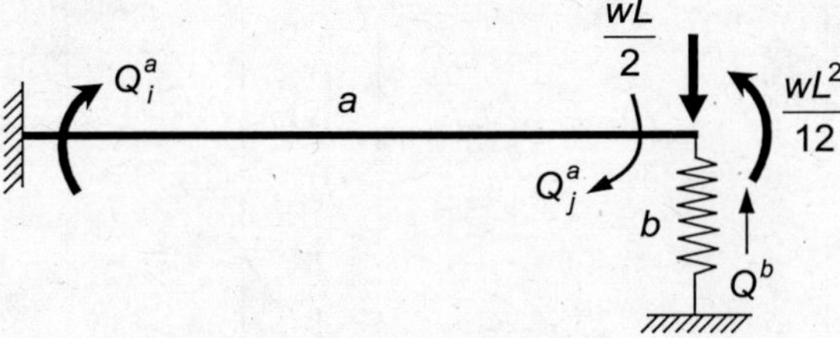

There are two elements, namely, a and b in the structure. The element a is a beam element, whereas the element b is a truss element as the spring can only carry an axial force. The nodal force and deformation matrices are as follows:

$$\{R\} = \begin{Bmatrix} R_1 \\ R_2 \end{Bmatrix} = \begin{Bmatrix} wL/2 \\ wL^2/12 \end{Bmatrix}; \quad \{r_R\} = \begin{Bmatrix} r_1 \\ r_2 \end{Bmatrix}$$

The redundant force and corresponding displacement matrices are $\{X\}$ and $\{r_X\}$, respectively. The force in the spring is assumed as the redundant in this example. The element internal force and end deformation matrices are as follows:

$$\{Q\} = \begin{Bmatrix} Q_i^a \\ Q_j^a \\ Q^b \end{Bmatrix}; \quad \{q\} = \begin{Bmatrix} q_i^a \\ q_j^a \\ q^b \end{Bmatrix}$$

$$\{Q\} = [b_R]\{R\} + [b_X]\{X\}$$

The force-transformation matrices are as follows:

$$[b_R] = \begin{bmatrix} -L & 1 \\ 0 & -1 \\ 0 & 0 \end{bmatrix}; \quad [b_X] = \begin{bmatrix} L \\ 0 \\ 1 \end{bmatrix}$$

The assembled element flexibility matrix $[f]$ is given by

$$[f] = \begin{bmatrix} f^a & \\ & f^b \end{bmatrix} = \begin{bmatrix} \dfrac{L}{3EI} & -\dfrac{L}{6EI} & \\ -\dfrac{L}{6EI} & \dfrac{L}{3EI} & \\ & & \dfrac{1}{K} \end{bmatrix}$$

Note that the flexibility coefficient of the spring is the inverse of its stiffness (K).

$$[F_{RR}] = [b_R]^T [f][b_R]$$

$$[F_{RR}] = \begin{bmatrix} -L & 1 \\ 0 & -1 \\ 0 & 0 \end{bmatrix}^T \begin{bmatrix} \dfrac{L}{3EI} & -\dfrac{L}{6EI} & 0 \\ -\dfrac{L}{6EI} & \dfrac{L}{3EI} & 0 \\ 0 & 0 & \dfrac{1}{K} \end{bmatrix} \begin{bmatrix} -L & 1 \\ 0 & -1 \\ 0 & 0 \end{bmatrix}$$

$$\therefore \quad [F_{RR}] = \begin{bmatrix} \dfrac{L^3}{3EI} & -\dfrac{L^2}{2EI} \\ -\dfrac{L^2}{2EI} & \dfrac{L}{EI} \end{bmatrix}$$

$$[F_{XR}] = [b_X]^T [f][b_R]$$

$$[F_{XR}] = \begin{bmatrix} L \\ 0 \\ 1 \end{bmatrix}^T \begin{bmatrix} \dfrac{L}{3EI} & -\dfrac{L}{6EI} & 0 \\ -\dfrac{L}{6EI} & \dfrac{L}{3EI} & 0 \\ 0 & 0 & \dfrac{1}{K} \end{bmatrix} \begin{bmatrix} -L & 1 \\ 0 & -1 \\ 0 & 0 \end{bmatrix}$$

$$\therefore \quad [F_{XR}] = \begin{bmatrix} -\dfrac{L^3}{3EI} & \dfrac{L^2}{2EI} \end{bmatrix}$$

$$[F_{RX}] = [F_{XR}]^T = \begin{bmatrix} -\dfrac{L^3}{3EI} \\[2ex] \dfrac{L^2}{2EI} \end{bmatrix}$$

$$[F_{XX}] = [b_X]^T [f] [b_X]$$

$$[F_{XX}] = \begin{bmatrix} L \\ 0 \\ 1 \end{bmatrix}^T \begin{bmatrix} \dfrac{L}{3EI} & -\dfrac{L}{6EI} & 0 \\[2ex] -\dfrac{L}{6EI} & \dfrac{L}{3EI} & 0 \\[2ex] 0 & 0 & \dfrac{1}{K} \end{bmatrix} \begin{bmatrix} L \\ 0 \\ 1 \end{bmatrix} = \begin{bmatrix} \dfrac{L^3}{3EI} + \dfrac{1}{K} \end{bmatrix}$$

The redundant force can be determined as follows:

$$\{X\} = -[F_{XX}]^{-1} [F_{XR}]\{R\}$$

$$\{X\} = -\begin{bmatrix} \dfrac{L^3}{3EI} + \dfrac{1}{K} \end{bmatrix}^{-1} \begin{bmatrix} -\dfrac{L^3}{3EI} & \dfrac{L^2}{2EI} \end{bmatrix} \begin{Bmatrix} wL/2 \\ wL^2/12 \end{Bmatrix}$$

$$\therefore \quad \{X\} = \left(\dfrac{3wL}{8} \right) \left(\dfrac{1}{1 + \dfrac{3EI}{KL^3}} \right)$$

The nodal displacement matrix is computed as follows:

$$\{r_R\} = \left([F_{RR}] - [F_{RX}][F_{XX}]^{-1}[F_{XR}] \right) \{R\}$$

or

$$\{r_R\} = [F_{RR}]\{R\} + [F_{RX}]\{X\}$$

$$\begin{Bmatrix} r_1 \\ r_2 \end{Bmatrix} = \begin{bmatrix} \dfrac{L^3}{3EI} & -\dfrac{L^2}{2EI} \\[2ex] -\dfrac{L^2}{2EI} & \dfrac{L}{EI} \end{bmatrix} \begin{Bmatrix} wL/2 \\ wL^2/12 \end{Bmatrix} + \begin{bmatrix} -\dfrac{L^3}{3EI} \\[2ex] \dfrac{L^2}{2EI} \end{bmatrix} \left(\dfrac{3wL}{8} \right) \left(\dfrac{1}{1 + \dfrac{3EI}{KL^3}} \right)$$

$$\therefore \quad r_1 = \left(\dfrac{3wL}{8} \right) \left(\dfrac{1}{K + \dfrac{3EI}{L^3}} \right) \quad \text{and} \quad r_2 = \left(\dfrac{wL^3 K}{48EI} - \dfrac{w}{2} \right) \left(\dfrac{1}{K + \dfrac{3EI}{L^3}} \right)$$

It is worth mentioning that the deflection (r_1) and slope (r_2) at the joint B are in the same direction as the nodal forces considered.

The internal element force matrix, $\{Q\} = [b_R]\{R\} + [b_X]\{X\}$

$$\begin{Bmatrix} Q_i^a \\ Q_j^a \\ Q^b \end{Bmatrix} = \begin{bmatrix} -L & 1 \\ 0 & -1 \\ 0 & 0 \end{bmatrix} \begin{Bmatrix} wL/2 \\ wL^2/12 \end{Bmatrix} + \begin{bmatrix} L \\ 0 \\ 1 \end{bmatrix} \left(\frac{3wL}{8} \right) \left(\frac{1}{1 + \dfrac{3EI}{KL^3}} \right)$$

$$\begin{Bmatrix} Q_i^a \\ Q_j^a \\ Q^b \end{Bmatrix} = \begin{Bmatrix} -\dfrac{5wL^2}{12} + \dfrac{3}{8}\left(\dfrac{wKL^5}{KL^3 + 3EI} \right) \\[2ex] -\dfrac{wL^2}{12} \\[2ex] \dfrac{3}{8}\left(\dfrac{wKL^4}{KL^3 + 3EI} \right) \end{Bmatrix}$$

The element internal force matrix shown above corresponds to the nodal forces, not for the actual uniformly distributed force. In order to obtain the final member forces, the support reactions/end moments of the fixed-ended bema must be added to the corresponding components. In this example, the fixed end moment corresponding to Q_j^a has to be added. The final element force matrix is computed as follows:

$$\begin{Bmatrix} Q_i^{a'} \\ Q_j^{a'} \\ Q^b \end{Bmatrix} = \begin{Bmatrix} -\dfrac{5wL^2}{12} + \dfrac{3}{8}\left(\dfrac{wKL^5}{KL^3 + 3EI} \right) \\[2ex] -\dfrac{wL^2}{12} \\[2ex] \dfrac{3}{8}\left(\dfrac{wKL^4}{KL^3 + 3EI} \right) \end{Bmatrix} + \begin{Bmatrix} -\dfrac{wL^2}{12} \\[2ex] \dfrac{wL^2}{12} \\[2ex] 0 \end{Bmatrix} = \begin{Bmatrix} -\dfrac{wL^2}{2} + \dfrac{3}{8}\left(\dfrac{wKL^5}{KL^3 + 3EI} \right) \\[2ex] 0 \\[2ex] \dfrac{3}{8}\left(\dfrac{wKL^4}{KL^3 + 3EI} \right) \end{Bmatrix}$$

Similarly, the fixed end moments and reactions should be added to obtain the support reactions as computed below:

$$M_a = Q_i^a = -\frac{wL^2}{2} + \frac{3}{8}\left(\frac{wKL^5}{KL^3 + 3EI} \right) \quad (\textit{Clockwise})$$

$$R_a = wL - \left(\frac{3wL}{8} \right) \left(\frac{1}{1 + \dfrac{3EI}{KL^3}} \right) \quad (\uparrow)$$

$$R_c = \left(\frac{3wL}{8} \right) \left(\frac{1}{1 + \dfrac{3EI}{KL^3}} \right) \quad (\uparrow)$$

EXAMPLE 8.5 Determine the end moments and reactions at supports B and C of the continuous beam shown in Figure E8.5 using the Matrix force method. Assume the flexural rigidity of the beam is EI.

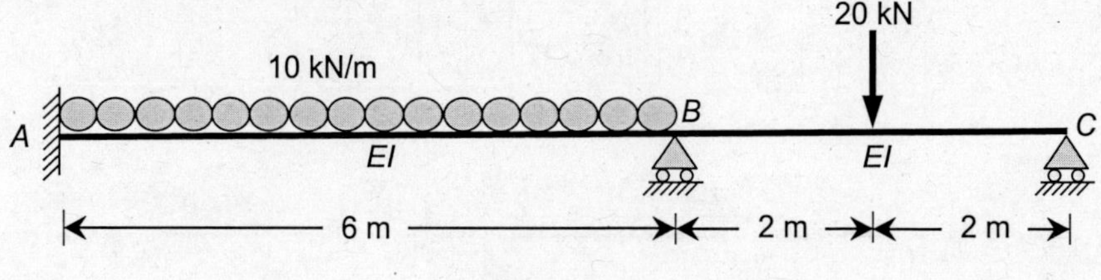

FIGURE E8.5

Solution: The degree of static indeterminacy of the beam is two. The given structure can be considered as the superposition of two fixed-ended beams so that all loads are transferred to the joints. The end moments and reactions for the spans AB and BC, assuming both ends are fixed, are shown in the following figure.

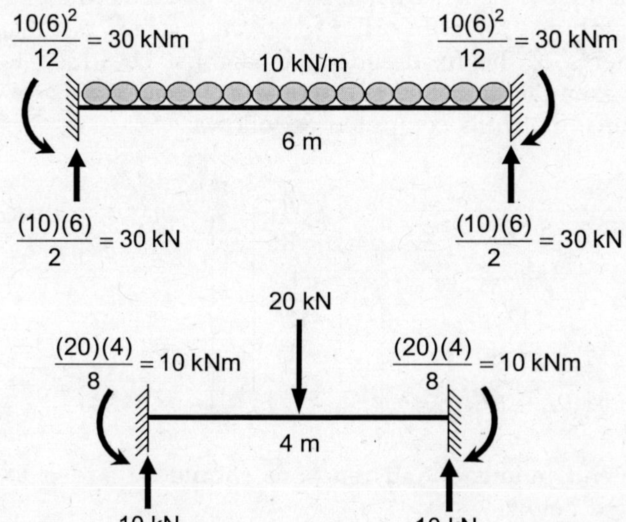

The net moments and forces are applied at the joints B and C of the beam in the opposite directions. The beam with all nodal forces is shown in the figure below.

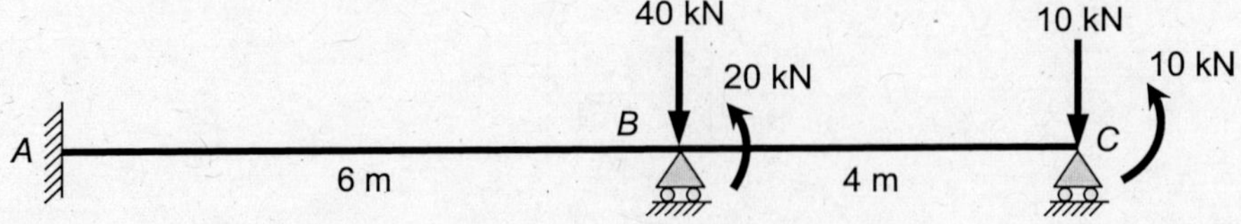

For the matrix formulation, the joint forces/moments are considered as the nodal forces. The reactions at the supports B and C are considered as the redundant forces. Thus, there are two elements, namely, a and b.

Accordingly,

$$\{R\} = \begin{Bmatrix} R_1 \\ R_2 \\ R_3 \\ R_4 \end{Bmatrix} = \begin{Bmatrix} 40 \\ 10 \\ 20 \\ 10 \end{Bmatrix}; \quad \{Q\} = \begin{Bmatrix} Q_i^a \\ Q_j^a \\ Q_i^b \\ Q_j^b \end{Bmatrix}; \quad \{X\} = \begin{Bmatrix} X_1 \\ X_2 \end{Bmatrix}$$

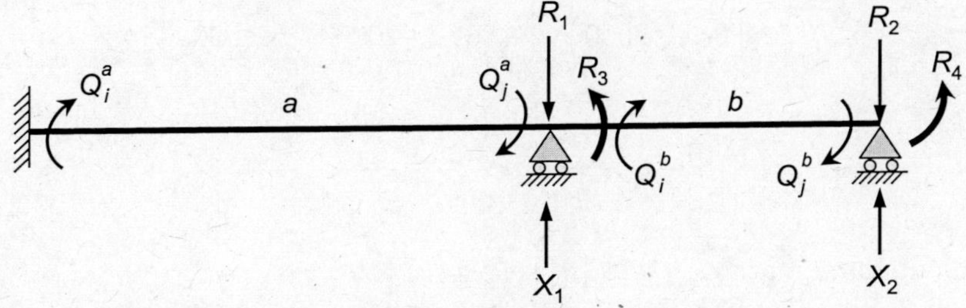

The force-transformation matrices are as follows:

$$[b_R] = \begin{bmatrix} -6 & -10 & 1 & 1 \\ 0 & 4 & -1 & -1 \\ 0 & -4 & 0 & 1 \\ 0 & 0 & 0 & -1 \end{bmatrix}; \quad [b_X] = \begin{bmatrix} 6 & 10 \\ 0 & -4 \\ 0 & 4 \\ 0 & 0 \end{bmatrix}$$

The assembled element flexibility matrix $[f]$ is given by

$$[f] = \begin{bmatrix} f^a & \\ & f^b \end{bmatrix} = \frac{1}{6EI} \begin{bmatrix} 12 & -6 & & \\ -6 & 12 & & \\ & & 8 & -4 \\ & & -4 & 8 \end{bmatrix}$$

$$[F_{RR}] = [b_R]^T [f][b_R]$$

$$[F_{RR}] = \frac{1}{6EI} \begin{bmatrix} 432 & 864 & -108 & -108 \\ & 2000 & -252 & -300 \\ & & 36 & 36 \\ Sym & & & 60 \end{bmatrix}$$

$$[F_{XR}] = [b_X]^T [f][b_R]$$

$$[F_{XR}] = \frac{1}{6EI} \begin{bmatrix} -432 & -864 & 108 & 108 \\ -864 & -2000 & 252 & 300 \end{bmatrix}$$

$$[F_{XX}] = [b_X]^T [f][b_X]$$

$$[F_{XX}] = \frac{1}{6EI}\begin{bmatrix} 832 & 864 \\ 864 & 2000 \end{bmatrix}$$

The redundant force can be determined as follows:

$$\{X\} = -[F_{XX}]^{-1}[F_{XR}]\{R\}$$

$$\{X\} = -\begin{bmatrix} 832 & 864 \\ 864 & 2000 \end{bmatrix}^{-1}\begin{bmatrix} -432 & -864 & 108 & 108 \\ -864 & -2000 & 252 & 300 \end{bmatrix}\begin{Bmatrix} 40 \\ 10 \\ 20 \\ 10 \end{Bmatrix}$$

$$\therefore \qquad \{X\} = \begin{Bmatrix} 44.0 \\ 4.3 \end{Bmatrix} \text{ kN}$$

Internal element force matrix, $\{Q\} = [b_R]\{R\} + [b_X]\{X\}$

$$\begin{Bmatrix} Q_i^a \\ Q_j^a \\ Q_i^b \\ Q_j^b \end{Bmatrix} = \begin{bmatrix} -6 & -10 & 1 & 1 \\ 0 & 4 & -1 & -1 \\ 0 & -4 & 0 & 1 \\ 0 & 0 & 0 & -1 \end{bmatrix}\begin{Bmatrix} 40 \\ 10 \\ 20 \\ 10 \end{Bmatrix} + \begin{bmatrix} 6 & 10 \\ 0 & -4 \\ 0 & 4 \\ 0 & 0 \end{bmatrix}\begin{Bmatrix} 44.0 \\ 4.3 \end{Bmatrix} = \begin{Bmatrix} -3.0 \\ -7.2 \\ -12.8 \\ -10.0 \end{Bmatrix}$$

The final end moments of the elements can be obtained by adding the corresponding fixed end moments.

Thus,

$$\begin{Bmatrix} Q_i^a \\ Q_j^a \\ Q_i^b \\ Q_j^b \end{Bmatrix} = \begin{Bmatrix} -3.0 \\ -7.2 \\ -12.8 \\ -10.0 \end{Bmatrix} + \begin{Bmatrix} -30.0 \\ 30.0 \\ -10.0 \\ 10.0 \end{Bmatrix} = \begin{Bmatrix} -33.0 \\ 22.8 \\ -22.8 \\ 0 \end{Bmatrix}$$

EXAMPLE 8.6 Determine the end moments of members of the rigid frame shown in Figure E8.6 using the matrix-force method.

Solution: The elements and internal forces of the rigid frame are shown below. The degree of static indeterminacy of the frame is three. Hence, it is required to identify three redundant forces. In this example, it is assumed that the end moments at the joints B, C and D are the redundant forces. Thus, the released structure would be a frame with hinges at these joints as shown in the figure.

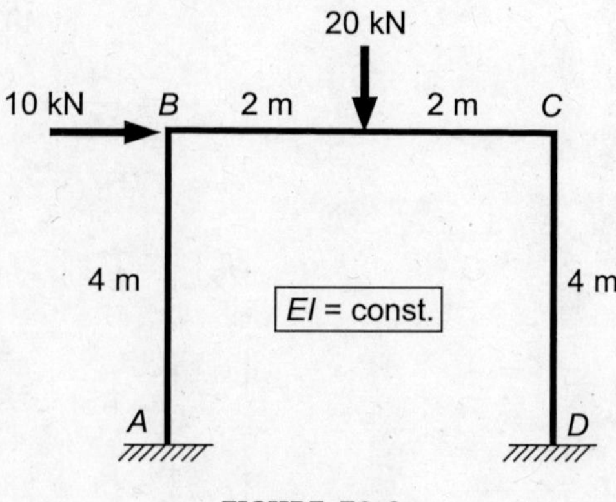

FIGURE E8.6

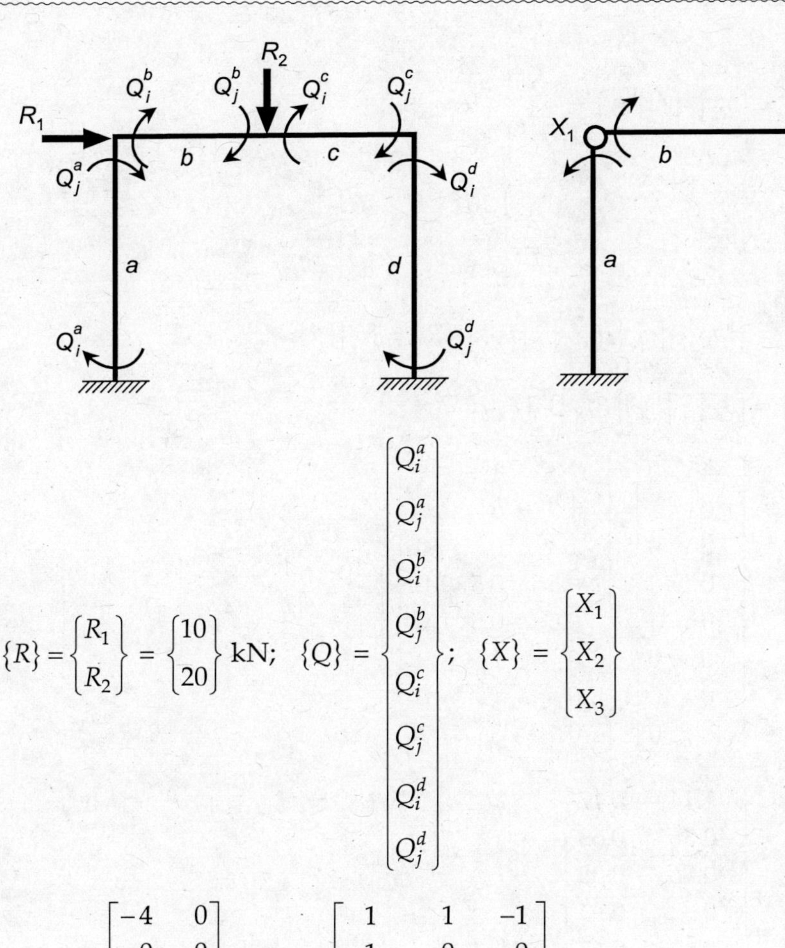

$$\{R\} = \begin{Bmatrix} R_1 \\ R_2 \end{Bmatrix} = \begin{Bmatrix} 10 \\ 20 \end{Bmatrix} \text{ kN}; \quad \{Q\} = \begin{Bmatrix} Q_i^a \\ Q_j^a \\ Q_i^b \\ Q_j^b \\ Q_i^c \\ Q_j^c \\ Q_i^d \\ Q_j^d \end{Bmatrix}; \quad \{X\} = \begin{Bmatrix} X_1 \\ X_2 \\ X_3 \end{Bmatrix}$$

$$[b_R] = \begin{bmatrix} -4 & 0 \\ 0 & 0 \\ 0 & 0 \\ 0 & -1 \\ 0 & 1 \\ 0 & 0 \\ 0 & 0 \\ 0 & 0 \end{bmatrix}; \quad [b_X] = \begin{bmatrix} 1 & 1 & -1 \\ -1 & 0 & 0 \\ 1 & 0 & 0 \\ -0.5 & 0.5 & 0 \\ 0.5 & -0.5 & 0 \\ 0 & 1 & 0 \\ 0 & -1 & 0 \\ 0 & 0 & 1 \end{bmatrix}$$

$$R_1 = 1 \quad R_2 = 1 \qquad X_1 = 1 \quad X_2 = 1 \quad X_2 = 1$$

$$[f] = \frac{2}{6EI} \begin{bmatrix} 4 & -2 & & & & & & \\ -2 & 4 & & & & & & \\ & & 2 & -1 & & & & \\ & & -1 & 2 & & & & \\ & & & & 2 & -1 & & \\ & & & & -1 & 2 & & \\ & & & & & & 4 & -2 \\ & & & & & & -2 & 4 \end{bmatrix}$$

$$[F_{XR}] = [b_X]^T [f][b_R] = \frac{2}{6EI}\begin{bmatrix} -24 & 3 \\ -16 & -3 \\ 16 & 0 \end{bmatrix}$$

$$[F_{XX}] = [b_X]^T [f][b_X] = \frac{2}{6EI}\begin{bmatrix} 16 & 4 & -6 \\ 4 & 12 & -2 \\ -6 & -2 & 8 \end{bmatrix}$$

$$\{Q\} = \left([b_R] - [b_X][F_{XX}]^{-1}[F_{XR}]\right)\{R\} = [b']\{R\}$$

$$[b'] = [b_R] - [b_X][F_{XX}]^{-1}[F_{XR}]$$

$$[b'] = \begin{bmatrix} -4 & 0 \\ 0 & 0 \\ 0 & 0 \\ 0 & -1 \\ 0 & 1 \\ 0 & 0 \\ 0 & 0 \\ 0 & 0 \end{bmatrix} - \begin{bmatrix} 1 & 1 & -1 \\ -1 & 0 & 0 \\ 1 & 0 & 0 \\ -0.5 & 0.5 & 0 \\ 0.5 & -0.5 & 0 \\ 0 & 1 & 0 \\ 0 & -1 & 0 \\ 0 & 0 & 1 \end{bmatrix} \begin{bmatrix} 16 & 4 & -6 \\ 4 & 12 & -2 \\ -6 & -2 & 8 \end{bmatrix}^{-1} \begin{bmatrix} -24 & 3 \\ -16 & -3 \\ 16 & 0 \end{bmatrix}$$

$$\therefore \quad [b'] = \begin{bmatrix} -1.14 & 0.17 \\ -0.86 & 0.33 \\ 0.86 & -0.33 \\ 0 & -0.67 \\ 0 & 0.67 \\ 0.86 & -0.33 \\ -0.86 & 0.33 \\ -1.14 & -0.17 \end{bmatrix}$$

$$\{Q\} = \begin{Bmatrix} Q_i^a \\ Q_j^a \\ Q_i^b \\ Q_j^b \\ Q_i^c \\ Q_j^c \\ Q_i^d \\ Q_j^d \end{Bmatrix} = \begin{bmatrix} -1.14 & 0.17 \\ -0.86 & 0.33 \\ 0.86 & -0.33 \\ 0 & -0.67 \\ 0 & 0.67 \\ 0.86 & -0.33 \\ -0.86 & 0.33 \\ -1.14 & -0.17 \end{bmatrix} \begin{Bmatrix} 10 \\ 20 \end{Bmatrix} = \begin{Bmatrix} -8.0 \\ -2.0 \\ 2.0 \\ -13.4 \\ 13.4 \\ 2.0 \\ -2.0 \\ -14.8 \end{Bmatrix} \text{kNm}$$

EXAMPLE 8.7 Determine the end moments of members of the rigid frame shown in Figure E8.7 using the matrix-force method.

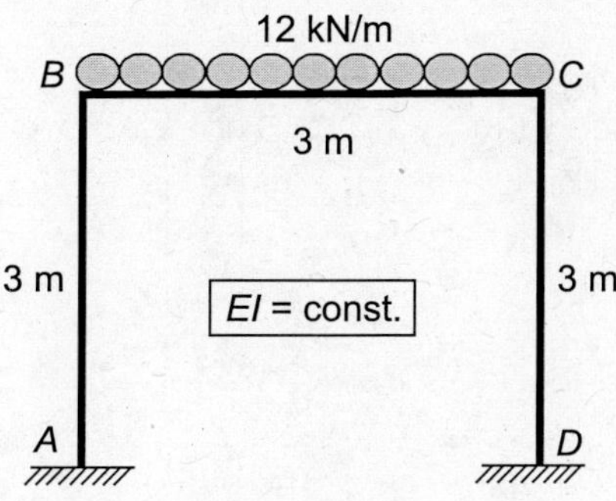

FIGURE E8.7

Solution: Since the loading on the frame is distributed over the beam, the equivalent nodal forces can be obtained by considering the fixed ended member BC. The frame with nodal forces is shown in the following figure.

The degree of static indeterminacy of the frame is three. However, the symmetrical loading would induce the axial forces in the columns. These axial forces would not cause any effect on the end moments of the frame. Hence, it is safe to neglect the axial force effects. As a result, the number of unknown redundant forces is two for the frame. Accordingly, the shear and moment at the support A are considered as the redundant forces in this example.

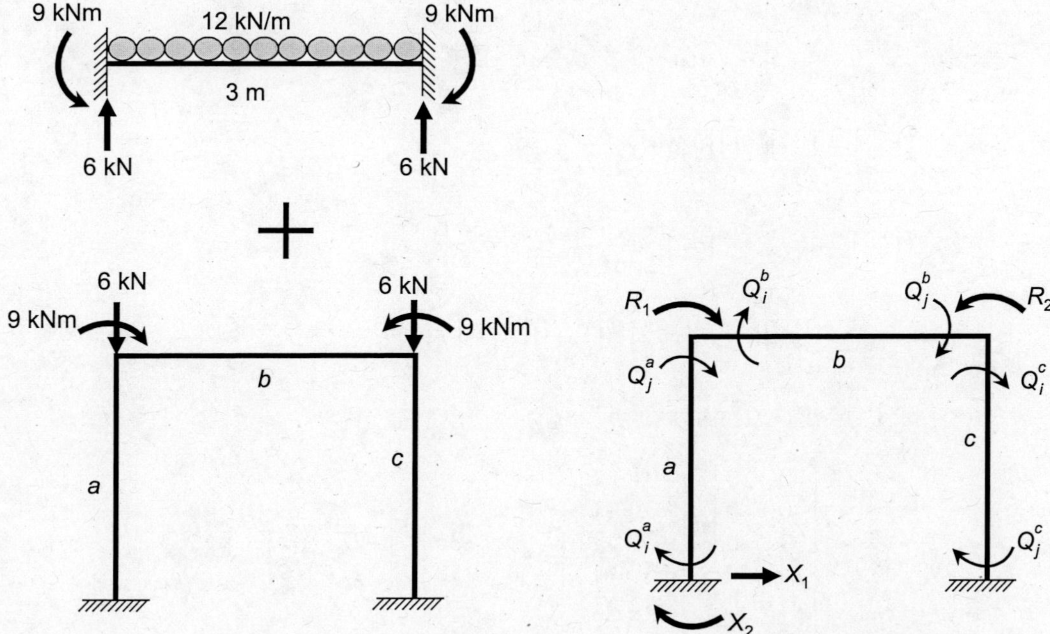

$$\{R\} = \begin{Bmatrix} R_1 \\ R_2 \end{Bmatrix} = \begin{Bmatrix} 9 \\ 9 \end{Bmatrix} \text{kNm}; \quad \{Q\} = \begin{Bmatrix} Q_i^a \\ Q_j^a \\ Q_i^b \\ Q_j^b \\ Q_i^c \\ Q_j^c \end{Bmatrix}; \quad \{X\} = \begin{Bmatrix} X_1 \\ X_2 \end{Bmatrix}$$

$$[b_R] = \begin{bmatrix} 0 & 0 \\ 0 & 0 \\ 1 & 0 \\ -1 & 0 \\ 1 & -1 \\ -1 & 1 \end{bmatrix} \quad ; \quad [b_X] = \begin{bmatrix} 0 & 1 \\ 3 & -1 \\ -3 & 1 \\ 3 & -1 \\ -3 & 1 \\ 0 & -1 \end{bmatrix}$$

$$R_1 = 1 \quad R_2 = 1 \qquad\qquad X_1 = 1 \quad X_2 = 1$$

$$[f] = \frac{3}{6EI} \begin{bmatrix} 2 & -1 & & & & \\ -1 & 2 & & & & \\ & & 2 & -1 & & \\ & & -1 & 2 & & \\ & & & & 2 & -1 \\ & & & & -1 & 2 \end{bmatrix}$$

$$[F_{XR}] = [b_X]^T [f][b_R] = \frac{3}{6EI} \begin{bmatrix} -27 & 9 \\ 12 & -6 \end{bmatrix}$$

$$[F_{XX}] = [b_X]^T [f][b_X] = \frac{3}{6EI} \begin{bmatrix} 90 & -36 \\ -36 & 18 \end{bmatrix}$$

$$\{Q\} = \Big([b_R] - [b_X][F_{XX}]^{-1}[F_{XR}] \Big)\{R\} = [b']\{R\}$$

$$[b'] = [b_R] - [b_X][F_{XX}]^{-1}[F_{XR}]$$

$$[b'] = \begin{bmatrix} 0 & 0 \\ 0 & 0 \\ 1 & 0 \\ -1 & 0 \\ 1 & -1 \\ -1 & 1 \end{bmatrix} - \begin{bmatrix} 0 & 1 \\ 3 & -1 \\ -3 & 1 \\ 3 & -1 \\ -3 & 1 \\ 0 & -1 \end{bmatrix} \begin{bmatrix} 90 & -36 \\ -36 & 18 \end{bmatrix}^{-1} \begin{bmatrix} -27 & 9 \\ 12 & -6 \end{bmatrix}$$

$$\therefore \quad [b'] = \begin{bmatrix} 0.33 & -0.67 \\ 0.83 & -0.17 \\ 0.17 & 0.17 \\ -0.17 & -0.17 \\ 0.17 & -0.83 \\ -0.67 & 0.33 \end{bmatrix}$$

$$\{Q\} = \begin{Bmatrix} Q_i^a \\ Q_j^a \\ Q_i^b \\ Q_j^b \\ Q_i^c \\ Q_j^c \end{Bmatrix} = \begin{bmatrix} 0.33 & -0.67 \\ 0.83 & -0.17 \\ 0.17 & 0.17 \\ -0.17 & -0.17 \\ 0.17 & -0.83 \\ -0.67 & 0.33 \end{bmatrix} \begin{Bmatrix} 9 \\ 9 \end{Bmatrix} = \begin{Bmatrix} 3.06 \\ 5.94 \\ 3.06 \\ -3.06 \\ -5.94 \\ -3.06 \end{Bmatrix} \text{kNm}$$

The final end moments can be obtained by adding the corresponding end moments as follows:

$$\begin{Bmatrix} Q_i^a \\ Q_j^a \\ Q_i^b \\ Q_j^b \\ Q_i^c \\ Q_j^c \end{Bmatrix} = \begin{Bmatrix} 3.06 \\ 5.94 \\ 3.06 \\ -3.06 \\ -5.94 \\ -3.06 \end{Bmatrix} + \begin{Bmatrix} 0 \\ 0 \\ -9 \\ 9 \\ 0 \\ 0 \end{Bmatrix} = \begin{Bmatrix} 3.06 \\ 5.94 \\ -5.94 \\ 5.94 \\ -5.94 \\ -3.06 \end{Bmatrix} \text{kNm}$$

8.8 PROBLEMS

8.1 Determine the rotations at the supports and the deflections at the loading points of the beam shown below using the Matrix force method.

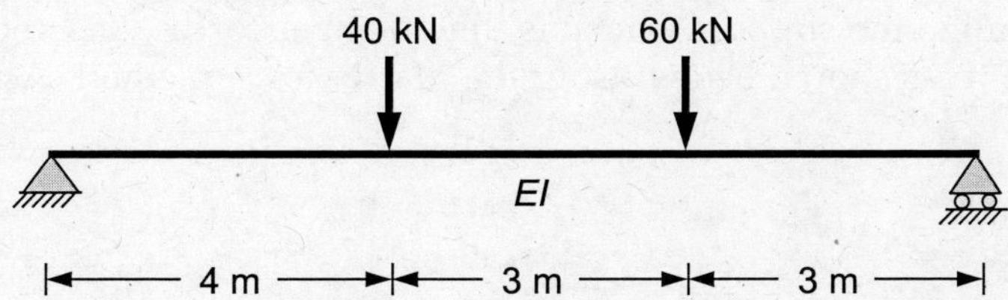

8.2 Determine the reactions and moments at the supports of the beam shown below using the Matrix force method. Spring flexibility is f_s.

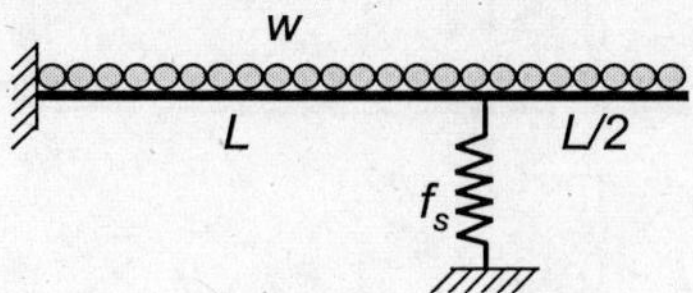

8.3 Determine the support reactions and the displacement at the loading point, and the rotation at the supports of the beam shown below using the Matrix force method.

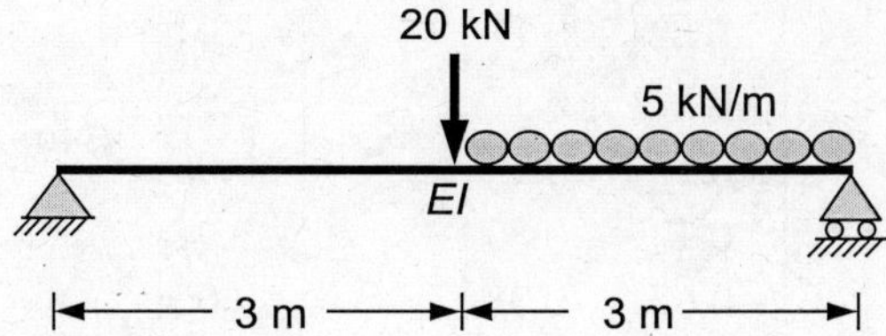

8.4 Determine the reactions and moments at the supports of the beam shown below using the Matrix force method.

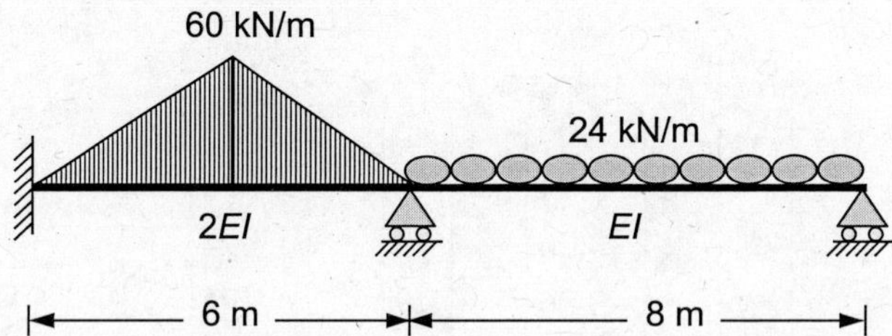

8.5 Determine the support reactions, support moments, the displacement at point C, and the rotation at B of the beam shown below using the Matrix force method. Assume constant flexural rigidity

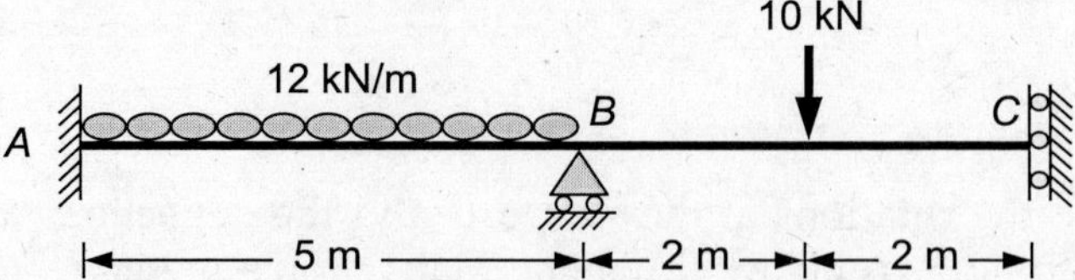

8.6 Determine the support reactions, the member forces, and the joint displacements of the truss shown below using the Matrix force method. Assume $AE = 8000$ kN.

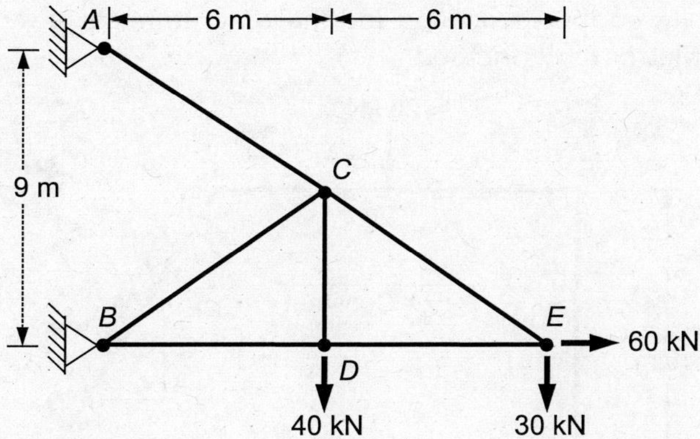

8.7 Determine the support reactions and the member forces of the truss shown below using the Matrix force method. Assume the value of AE for all members is 8000 kN.

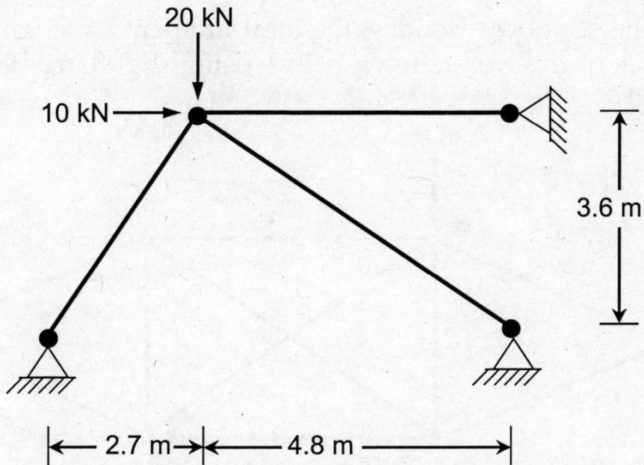

8.8 Determine the support reactions, the joint moments, and the rotation at B of the frame shown below using the Matrix force method.

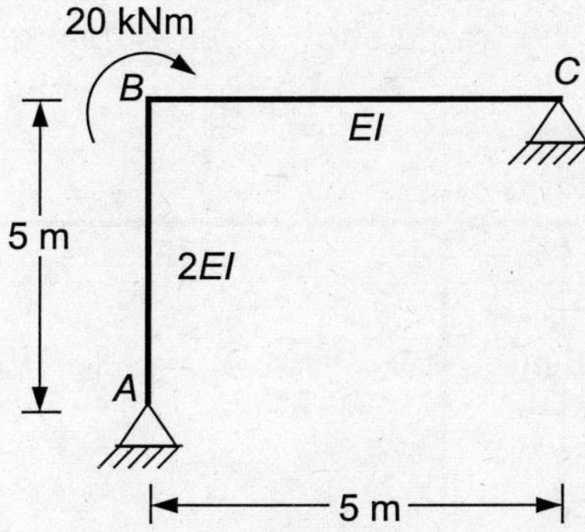

8.9 Determine the support reactions and the joint moments of the frame shown below using the Matrix force method.

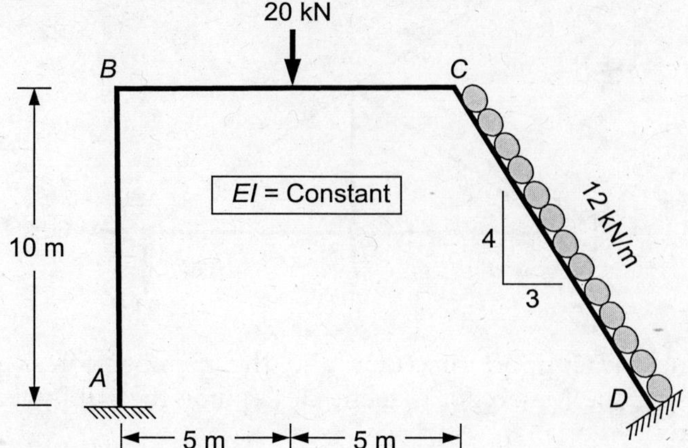

8.10 Determine the support reactions, the joint moments, the joint displacements, and joint rotations of the truss shown below using the Matrix force method. Assume $AE = 5000$ kN.

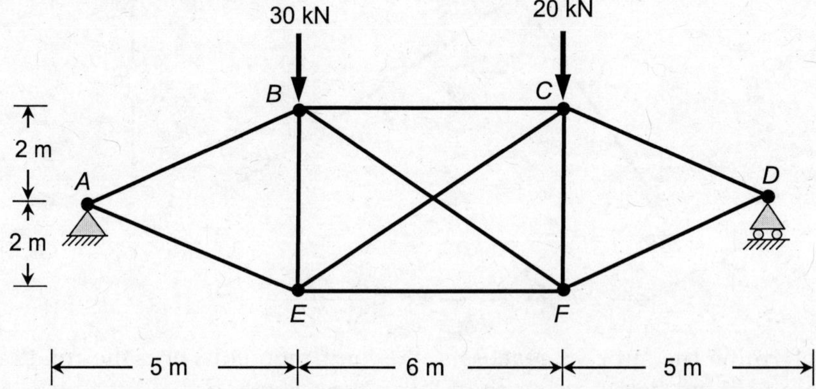

8.11 Determine the support reactions, the joint moments, the joint displacements, and joint rotations of the frame shown below using the Matrix force method.

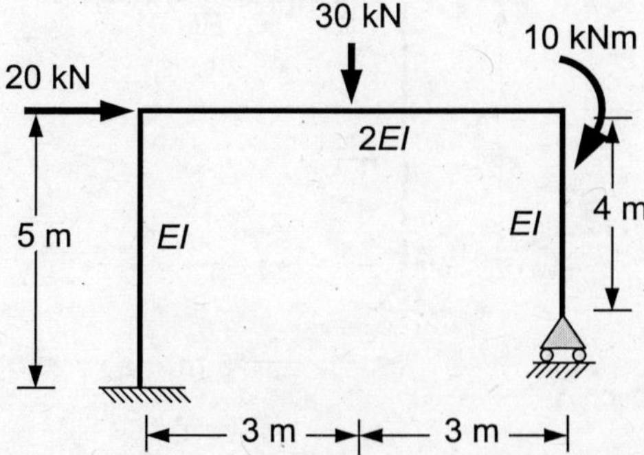

8.12 Determine the support reactions, the member forces, and the joint displacements of the truss shown below using the Matrix force method. Assume $AE = 5000$ kN.

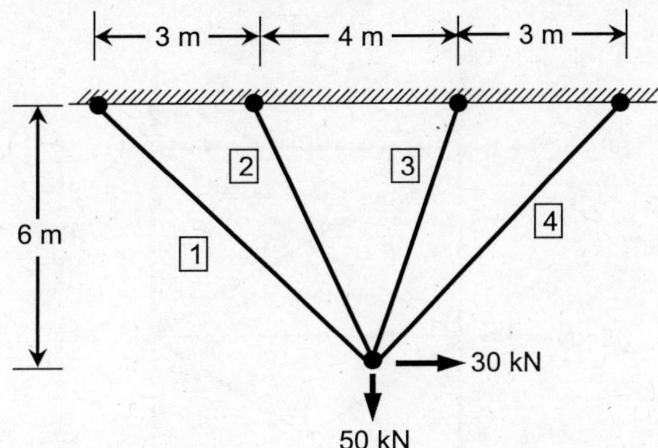

8.13 Determine the support reactions, the member forces, and the joint displacements of the truss shown below using the Matrix force method. Take $E = 200$ GPa.

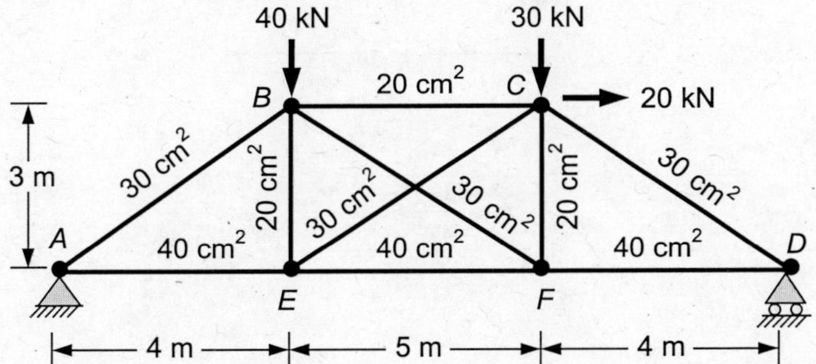

8.14 Determine the support reactions, the member forces, and the joint displacements of the truss shown below using the Matrix force method. Take $E = 200$ GPa.

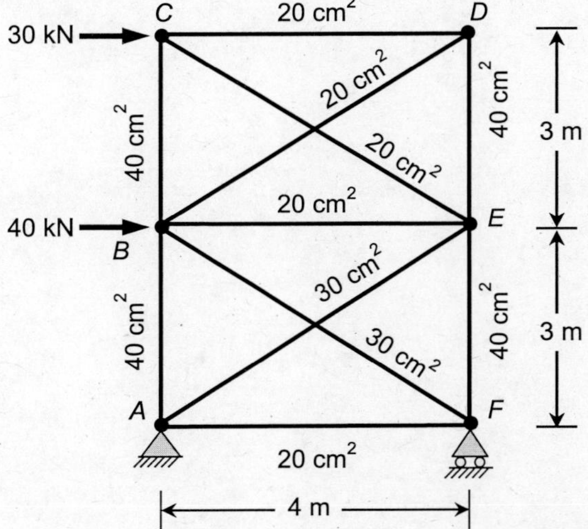

8.15 Determine the support reactions, the member forces, and the joint displacements of the truss shown below using the Matrix force method. Assume AE = 5000 kN.

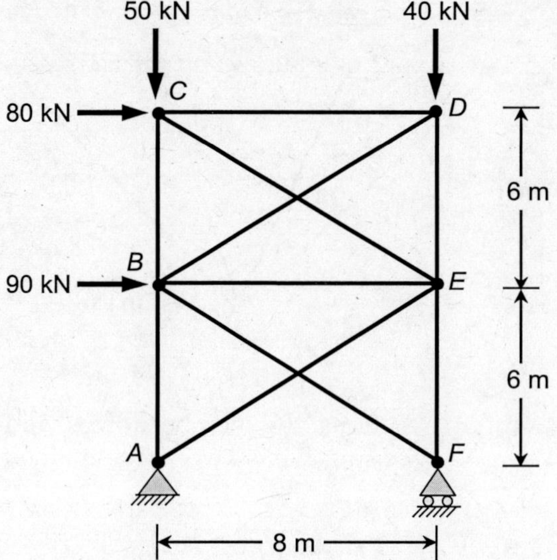

Matrix Displacement Method

9.1 GENERAL

The matrix-displacement method is also known as the stiffness method. The basic concepts of the matrix displacement method are similar to those of the matrix force method. In the matrix displacement method, the deformations are considered as the unknowns instead of forces. Thus, the nodal displacements are taken as the basic unknowns, which are solved using the three fundamental concepts of the structural analysis as follows:

- Displacement compatibility conditions
- Force-displacement relationships
- Equilibrium equations

In this method, the compatibility conditions are first satisfied at the nodes by correlating the external nodal displacements with the end deformations of the elements. The force-displacement relationships are then established for each element between the internal forces and the end deformations of elements (local level) as well as between the external nodal forces and the nodal displacements (global level). Finally, the internal forces and end deformations as well as the nodal displacements are determined using the equations of equilibrium at the nodes of the elements.

9.2 BASIC CONCEPTS AND MATRIX NOTATIONS

The same concepts as discussed in the matrix force method to define the elements and nodes are also adopted in the matrix displacement method. The points where the concentrated loads are applied or supports are considered as the nodes. The member between the consecutive nodes is defined as the elements. In the case of the distributed loads, the equivalent nodal forces are computed and are applied at the nodes. The same matrix notations used for elements and nodes as defined earlier are also used in this method.

9.2.1 Displacement Transformation (Compatibility) Matrix

Displacement compatibility condition in the matrix displacement method requires that the element deformations between two points in a structure must be consistent with the corresponding nodal displacements. For example, the displacements between two nodes may induce the displacements and/or rotations of the members. Thus, the end deformation of an element is the sum of the effects caused by all nodal displacements. Let a_{ij} represents the element deformation q_i induced by a nodal displacement r_j. Accordingly, each element end deformation can be written as follows:

$$q_1 = a_{11}r_1 + a_{12}r_2 + a_{13}r_3 + \ldots + a_{1n}r_n$$

$$q_2 = a_{21}r_1 + a_{22}r_2 + a_{23}r_3 + \ldots + a_{2n}r_n$$

$$\ldots \qquad \ldots \qquad \ldots \qquad \ldots$$

$$\ldots \qquad \ldots \qquad \ldots \qquad \ldots \tag{9.1}$$

$$q_m = a_{m1}r_1 + a_{m2}r_2 + a_{m3}r_3 + \ldots + a_{mn}r_n$$

where, $q_1 = q_i^a$, $q_2 = q_j^a$, $\ldots$ are the end deformations of elements and r_1, r_2, $\ldots$ are the nodal displacements. In the matrix form,

$$
\begin{Bmatrix} q_1 \\ q_2 \\ \ldots \\ q_m \end{Bmatrix} =
\begin{bmatrix}
a_{11} & a_{12} & \cdots & a_{1n} \\
a_{21} & a_{22} & \cdots & a_{2n} \\
\cdots & \cdots & \cdots & \cdots \\
a_{m1} & a_{m2} & \cdots & a_{mn}
\end{bmatrix}
\begin{Bmatrix} r_1 \\ r_2 \\ \ldots \\ r_n \end{Bmatrix}
\tag{9.2}
$$

Simply,
$$\{q\} = [a]\{r\} \tag{9.3}$$

in which $\{q\}$ = the element end deformation matrix, $\{r\}$ = the nodal displacement matrix and $[a]$ = displacement transformation matrix.

$$
[a] =
\begin{bmatrix}
a_{11} & a_{12} & \cdots & a_{1n} \\
a_{21} & a_{22} & \cdots & a_{2n} \\
\cdots & \cdots & \cdots & \cdots \\
a_{m1} & a_{m2} & \cdots & a_{mn}
\end{bmatrix}
\tag{9.4}
$$

Displacement transformation matrix $[a]$ relates the internal element deformations to the external nodal displacements. This matrix is essentially a coordinate transformation matrix representing the displacement compatibility of the structural system.

If the same forces and displacements are involved, the relationship between the displacement transformation matrix $[a]$ and the force transformation matrix $[b]$ can be established using the principle of virtual work.

$$\{\delta R\}^T \{r\} = \{\delta Q\}^T \{q\} \tag{9.5}$$

$$\{Q\} = [b]\{R\} \tag{9.6}$$

$$\{\delta Q\}^T = \{\delta R\}^T [b]^T \tag{9.7}$$

Using Eqns. (9.3) and (9.6) in Eqn. (9.5),

$$\{\delta R\}^T \{r\} = \{\delta R\}^T [b]^T [a]\{r\}$$

$$\therefore \qquad [a]^T [b] = [a][b]^T = [I] \tag{9.8}$$

9.2.2   Element Force-displacement (Stiffness) Matrix

Like the flexibility coefficient, the stiffness coefficient establishes the relationship between the force and displacement of an element. Accordingly, a stiffness coefficient k_{ij} is defined as the force developed at the point i due to a unit displacement at point j, keeping all other points (nodes) fixed. Using the principle of superposition, the internal force components of an element can be expressed in terms of all end deformations.

Consider a frame element shown in Figure 9.1. Internal forces and displacements matrices of element a are given by

$$\{Q^a\} = \begin{Bmatrix} Q_i^a \\ Q_j^a \\ Q_k^a \end{Bmatrix}; \quad \text{and} \quad \{q^a\} = \begin{Bmatrix} q_i^a \\ q_j^a \\ q_k^a \end{Bmatrix}$$

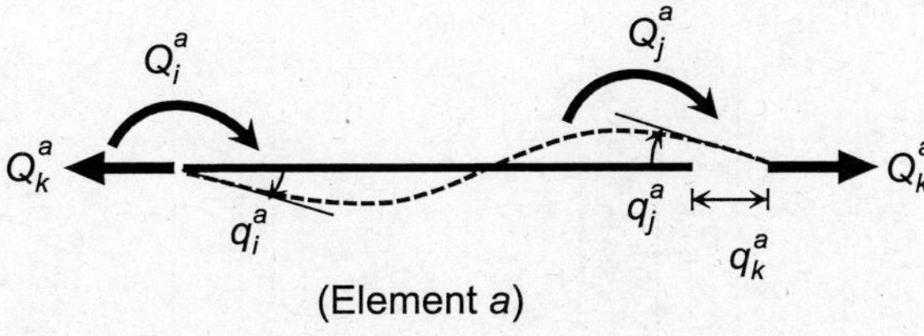

FIGURE 9.1   Element forces and deformations of a frame member.

The end moments/forces of the element can be expressed as a function of the end deformations. Accordingly,

$$Q_i^a = k_{ii}^a q_i^a + k_{ij}^a q_j^a + k_{ik}^a q_k^a$$

$$Q_j^a = k_{ji}^a q_i^a + k_{jj}^a q_j^a + k_{jk}^a q_k^a \qquad (9.9)$$

$$Q_k^a = k_{ki}^a q_i^a + k_{kj}^a q_j^a + k_{kk}^a q_k^a$$

In matrix form,

$$\begin{Bmatrix} Q_i^a \\ Q_j^a \\ Q_k^a \end{Bmatrix} = \begin{bmatrix} k_{ii}^a & k_{ij}^a & k_{ik}^a \\ k_{ji}^a & k_{jj}^a & k_{jk}^a \\ k_{ki}^a & k_{kj}^a & k_{kk}^a \end{bmatrix} \begin{Bmatrix} q_i^a \\ q_j^a \\ q_k^a \end{Bmatrix} \qquad (9.10)$$

$$\{Q^a\} = [k^a]\{q^a\} \qquad (9.11)$$

where, $[k^a] = \begin{bmatrix} k_{ii}^a & k_{ij}^a & k_{ik}^a \\ k_{ji}^a & k_{jj}^a & k_{jk}^a \\ k_{ki}^a & k_{kj}^a & k_{kk}^a \end{bmatrix}$ is the element stiffness matrix.

From Eqn. (9.10), the stiffness coefficient, $k_{ii}^a = Q_i^a$ for $q_i^a = 1$ and $q_j^a = q_k^a = 0$. Similarly, the stiffness coefficient, $k_{ij}^a = Q_i^a$ for $q_j^a = 1$ and $q_i^a = q_k^a = 0$, and so on...

A similar procedure can be followed for all the elements of a structure. Accordingly, $\{Q^b\} = [k^b]\{q^b\}$, $\{Q^c\} = [k^c]\{q^c\}$, and so on...

If a structure is composed of several elements (a, b, c), the element force and deformation matrices can be written as the combinations of these forces and deformations of all elements.

$$\text{If} \quad \{Q\} = \begin{Bmatrix} Q^a \\ Q^b \\ \cdots \\ \cdots \end{Bmatrix} ; \quad \text{and} \quad \{q\} = \begin{Bmatrix} q^a \\ q^b \\ \cdots \\ \cdots \end{Bmatrix} \quad \text{the force-deformation relationships of all elements}$$

can be obtained by assembling all the element stiffness matrices as follows:

$$\begin{Bmatrix} Q^a \\ Q^b \\ \cdots \\ \cdots \end{Bmatrix} = \begin{bmatrix} k^a & & & \\ & k^b & & \\ & & \cdots & \\ & & & \cdots \end{bmatrix} \begin{Bmatrix} q^a \\ q^b \\ \cdots \\ \cdots \end{Bmatrix} \tag{9.12}$$

or
$$\{Q\} = [k]\{q\} \tag{9.13}$$

The assembled element stiffness matrix $[k]$ is a diagonal matrix in which each component represents the individual stiffness matrices of all elements. Element stiffness matrices for truss members, frame members, and beams are derived below.

(a) Truss member

Consider a truss element a of length L, are of cross-section A, and Modulus of elasticity E as shown in Figure 9.2. By the definition, the element stiffness coefficient is the axial force required at a joint to induce the unit axial deformation at the same joint while restraining the displacement of other joints. The axial displacement (q_a) of the member carrying an axial force Q_a is determined as follows:

$$q_a = \frac{Q_a L}{AE} \tag{9.14}$$

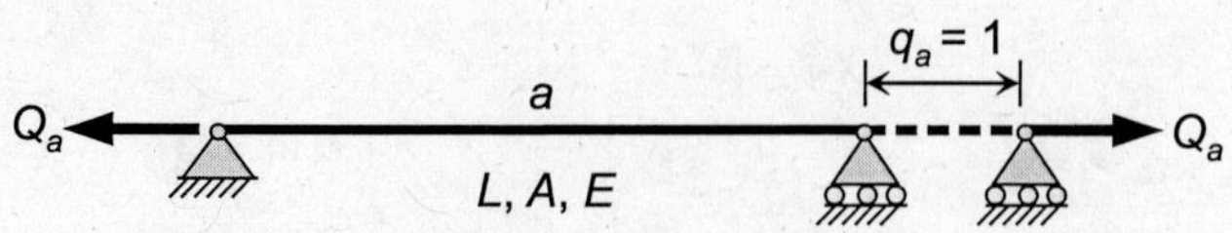

FIGURE 9.2

Stiffness coefficient (k^a) is the ratio of axial force, Q_a to the axial displacement, q_a. Thus, the stiffness coefficient for a truss member is given by

$$[k^a] = \left[\frac{AE}{L}\right] \tag{9.15}$$

(b) Frame member

Consider a frame element a of length L, flexural rigidity EI, and area of cross-section A. As shown in Figure 9.3, the element can undergo two end rotations (q_i and q_j) and one axial displacement (q_k). The element stiffness matrix $[k^a]$ for the member a is given by

$$[k^a] = \begin{bmatrix} k_{ii}^a & k_{ij}^a & k_{ik}^a \\ k_{ji}^a & k_{jj}^a & k_{jk}^a \\ k_{ki}^a & k_{kj}^a & k_{kk}^a \end{bmatrix}$$

$$q_i = 1 \quad q_j = 1 \quad q_k = 1$$

$$q_j = 0 \quad q_i = 0 \quad q_i = 0$$

$$q_k = 0 \quad q_k = 0 \quad q_j = 0$$

The stiffness coefficient k_{ij} is equal to the end moment, Q_i when the element is subjected to a unit deformation q_j keeping all other end deformations equal to zero, i.e., $q_i = q_k = 0$. Thus, each column of the element stiffness matrix represents all forces/reactions of the elements when subjected to a unit end deformation along one particular degree of freedom. For example, in order to determine the components of the first column of the element stiffness matrix, the end deformations $q_j = q_k = 0$. This results in the fixed support at the far end and a pinned support at the near end. A unit rotation is applied at the near end as shown in Figure 9.3(a).

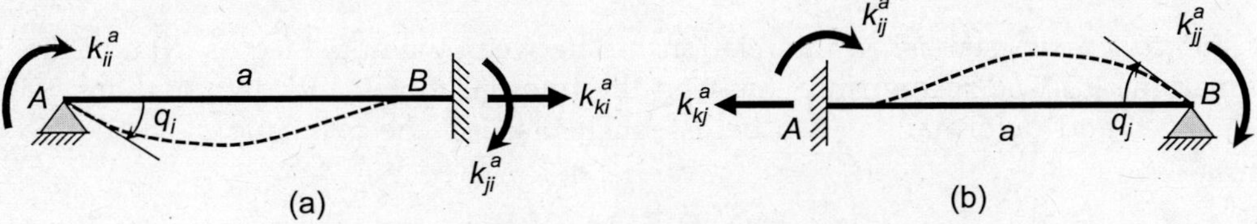

FIGURE 9.3   Stiffness coefficients of a frame member.

The end moment at A required to cause unit rotation at the same end can be obtained by using the slope-deflection equation as follows:

$$k_{ii}^a = 0 + \frac{2EI}{L}\left(2 \times 1 + 0 - \frac{3 \times 0}{L}\right) = \frac{4EI}{L} \tag{9.16a}$$

$$k_{ji}^a = \left(\frac{1}{2}\right)\left(\frac{4EI}{L}\right) = \frac{2EI}{L} \tag{9.16b}$$

$$k_{ki}^a = 0 \tag{9.16c}$$

The components of the second column of the matrix $[k^a]$ can be derived from the beam shown in Figure 9.3(b) as follows:

$$k_{ji}^a = \frac{2EI}{L}; \quad k_{jj}^a = \frac{4EI}{L}; \quad k_{kj}^a = 0 \tag{9.17}$$

Also,

$$k_{ki}^a = 0; \quad k_{kj}^a = 0; \quad k_{kk}^a = \frac{AE}{L} \tag{9.18}$$

Accordingly, the element stiffness matrix for a frame member is given by

$$[k^a] = \begin{bmatrix} \dfrac{4EI}{L} & \dfrac{2EI}{L} & 0 \\[2ex] \dfrac{2EI}{L} & \dfrac{4EI}{L} & 0 \\[2ex] 0 & 0 & \dfrac{AE}{L} \end{bmatrix} \tag{9.19}$$

(c) Beam

The stiffness coefficient matrix for a beam can be determined from the stiffness matrix for a frame member by neglecting the effect of axial force. Accordingly, the stiffness matrix of a beam can be obtained as follows:

$$[k^a] = \begin{bmatrix} \dfrac{4EI}{L} & \dfrac{2EI}{L} \\[2ex] \dfrac{2EI}{L} & \dfrac{4EI}{L} \end{bmatrix} = \frac{EI}{L} \begin{bmatrix} 4 & 2 \\ 2 & 4 \end{bmatrix} \tag{9.20}$$

9.2.3 Total Stiffness Matrix

Total stiffness matrix established the relationship between the nodal forces $\{R\}$ and the nodal deformations $\{r\}$. The relationship between the internal forces and deformations with the nodal forces and deformations can be established using the principle of virtual work as follows.

$$\{\delta r\}^T \{R\} = \{\delta q\}^T \{Q\} \tag{9.21}$$

Since, $\{q\} = [a]\{r\}$, $\{\delta q\} = [a]\{\delta r\}$

$$\{\delta q\}^T = \{\delta r\}^T [a]^T \tag{9.22}$$

$$\{Q\} = [k]\{q\}$$

$$\{Q\} = [k][a]\{r\} \tag{9.23}$$

Substituting Eqns. (9.22) and (9.22) in Eqn. (9.21),

$$\{\delta r\}^T \{R\} = \{\delta r\}^T [a]^T [k][a]\{r\}$$

or
$$\{R\} = [a]^T [k][a]\{r\} \tag{9.24}$$

$$\{R\} = [K]\{r\} \tag{9.25}$$

where,
$$[K] = [a]^T [k][a] \tag{9.26}$$

$[K]$ is known as the *total stiffness matrix* or *stiffness matrix* of the structure. Eqn. (9.26) provides the relationship between the nodal deformations and the nodal external forces. Since the nodal force matrix $\{R\}$ is unknown, the nodal displacement matrix $\{r\}$ can be determined as follows:

$$\{r\} = [K]^{-1}\{R\} \tag{9.27}$$

Comparing the nodal force and nodal displacement relationship obtained from the matric force method as discussed in the previous chapter,

$$\{r\} = [F]\{R\}$$

$$\therefore \qquad [F] = [K]^{-1} \tag{9.28}$$

The above relationship also valid at the element level, i.e.,

$$\left[f^a \right] = \left[k^a \right]^{-1} \tag{9.29}$$

9.2.4 Equilibrium Equations

The equilibrium condition at each node requires that the nodal forces corresponding to the unknown displacements must be equal to the applied loads. The internal element forces obtained from the end deformations represent the reaction components at the supports. Thus, both nodal forces and the sum of element internal forces obtained from the following equations must be in equilibrium.

$$\{R\} = [K]\{r\} \quad \text{and} \quad \{Q\} = [k][a]\{r\}$$

9.3 STEP-BY-STEP PROCEDURE OF MATRIX DISPLACEMENT METHOD

Unlike the matrix force method in which the analysis procedure is different for the statically determinate and indeterminate structures, a unique analysis procedure is adopted in the matrix displacement method for all structures. The step-by-step procedure of analysis of structures using the matrix displacement method is presented in the following sections.

1. Define the nodes and elements based on the positions of externally applied loads.
2. Identify all possible nodal unknown displacements (and rotations) in a structure.
3. Define the nodal force matrix $\{R\}$, the nodal deformation matrix $\{r\}$, the internal element force matrix $\{Q\}$, and the element end deformation, $\{q\}$.
4. Establish the displacement-transformation matrix $[a]$. All components of $[a]$ are determined by relating the element end deformations with the nodal displacements

using the deformed geometry of the structure. At a particular time, only one nodal displacement equal to the unit is considered and keeping all other nodal displacements equal to zero. For example, $r_1 = 1$, $r_2 = r_3 = \ldots = 0$, and so on.

5. Determine the element end deformation matrix, $\{q\} = [a]\{r\}$.

6. Determine the element stiffness matrices ($[k^a]$, $[k^b]$, $[k^c]$, ...) depending on the type of members and compute the assembled element stiffness matrix $[k]$ as follows:

$$[k] = \begin{bmatrix} k^a & & & \\ & k^b & & \\ & & \ddots & \\ & & & \ddots \end{bmatrix}$$

7. Compute the stiffness matrix of the structure, $[K] = [a]^T [k][a]$.

8. Determine the nodal deformation matrix, $\{r\} = [K]^{-1}\{R\}$.

9. Finally, compute the element internal force matrix, $\{Q\} = [k][a]\{r\}$.

EXAMPLE 9.1   Determine the bar forces and the joint displacements of the truss shown in Figure E9.1 using the Matrix displacement method. Assume that $AE/L = 1$ for all members.

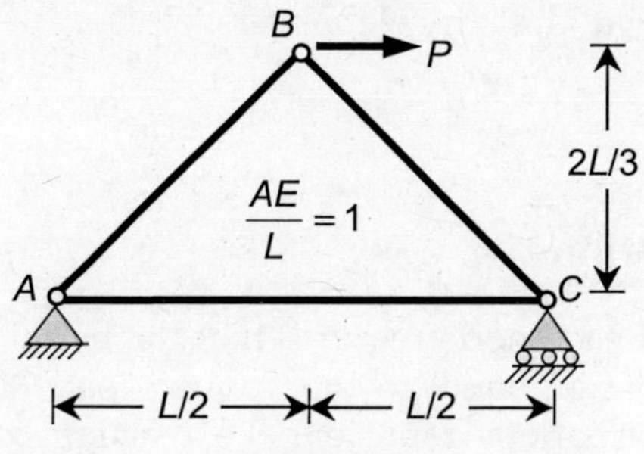

FIGURE E9.1

Solution: Three independent displacements are possible in the truss. These are one horizontal displacement at the roller support at joint C and two displacements at joint B. These displacements and corresponding forces are shown in the following figure. Since there are three members, the number of elements would be three, namely, a, b, and c.

The nodal displacement matrix and the nodal force matrix are as follows:

$$\{r\} = \begin{Bmatrix} r_1 \\ r_2 \\ r_3 \end{Bmatrix}; \quad \{R\} = \begin{Bmatrix} R_1 \\ R_2 \\ R_3 \end{Bmatrix} = \begin{Bmatrix} 0 \\ P \\ 0 \end{Bmatrix}$$

Similarly, the element internal force and end deformation matrices are as follows:

$$\{Q\} = \begin{Bmatrix} Q^a \\ Q^b \\ Q^c \end{Bmatrix} ; \quad \{q\} = \begin{Bmatrix} q^a \\ q^b \\ q^c \end{Bmatrix}$$

The element deformation matrix $\{q\}$ and the nodal displacement matrix $\{r\}$ is related as follows:

$$\{q\} = [a]\{r\}$$

The components of the displacement transformation matrix $[a]$ are the element deformations by applying a unit displacement along one degree of frame at a node and keeping all other nodal displacements as zero. These are shown in the figures below:

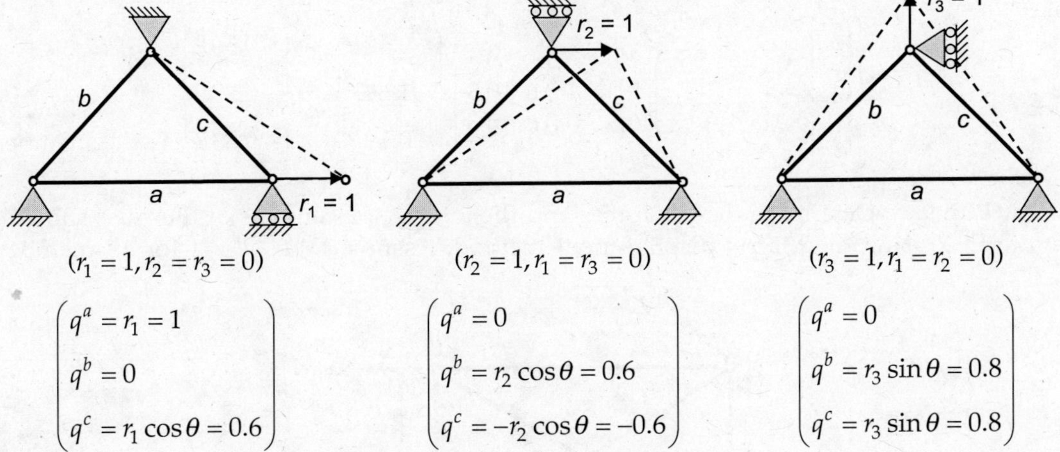

$$(r_1 = 1, r_2 = r_3 = 0) \qquad (r_2 = 1, r_1 = r_3 = 0) \qquad (r_3 = 1, r_1 = r_2 = 0)$$

$$\begin{pmatrix} q^a = r_1 = 1 \\ q^b = 0 \\ q^c = r_1 \cos\theta = 0.6 \end{pmatrix} \qquad \begin{pmatrix} q^a = 0 \\ q^b = r_2 \cos\theta = 0.6 \\ q^c = -r_2 \cos\theta = -0.6 \end{pmatrix} \qquad \begin{pmatrix} q^a = 0 \\ q^b = r_3 \sin\theta = 0.8 \\ q^c = r_3 \sin\theta = 0.8 \end{pmatrix}$$

Displacement transformation matrix is obtained as follows:

$$[a] = \begin{bmatrix} 1 & 0 & 0 \\ 0 & 0.6 & 0.8 \\ 0.6 & -0.6 & 0.8 \end{bmatrix}$$

The assembled element stiffness matrix $[k]$ is given by

$$[k] = \begin{bmatrix} k^a & & \\ & k^b & \\ & & k^c \end{bmatrix} = \begin{pmatrix} 1 & & \\ & 1 & \\ & & 1 \end{pmatrix} \quad \left(\because \frac{AE}{L} = 1 \right)$$

Stiffness matrix of the truss can be obtained as $[K] = [a]^T [k][a]$

$$[K] = \begin{bmatrix} 1 & 0 & 0 \\ 0 & 0.6 & 0.8 \\ 0.6 & -0.6 & 0.8 \end{bmatrix}^T \begin{pmatrix} 1 & & \\ & 1 & \\ & & 1 \end{pmatrix} \begin{bmatrix} 1 & 0 & 0 \\ 0 & 0.6 & 0.8 \\ 0.6 & -0.6 & 0.8 \end{bmatrix}$$

$$[K] = \begin{bmatrix} 1.36 & & Sym \\ -0.36 & 0.72 & \\ 0.48 & 0 & 1.28 \end{bmatrix}$$

Nodal displacements (or joint displacements) matrix, $\{r\} = [K]^{-1}\{R\}$

$$\begin{Bmatrix} r_1 \\ r_2 \\ r_3 \end{Bmatrix} = \begin{bmatrix} 1.36 & -0.36 & 0.48 \\ -0.36 & 0.72 & 0 \\ 0.48 & 0 & 1.28 \end{bmatrix}^{-1} \begin{Bmatrix} 0 \\ P \\ 0 \end{Bmatrix} = \begin{Bmatrix} 0.50P \\ 1.64P \\ -0.19P \end{Bmatrix}$$

Bar force (element forces) can be determined using the following expression:

$$\{Q\} = [k][a]\{r\}$$

$$\begin{Bmatrix} Q^a \\ Q^b \\ Q^c \end{Bmatrix} = \begin{pmatrix} 1 & & \\ & 1 & \\ & & 1 \end{pmatrix} \begin{bmatrix} 1 & 0 & 0 \\ 0 & 0.6 & 0.8 \\ 0.6 & -0.6 & 0.8 \end{bmatrix} \begin{Bmatrix} 0.50P \\ 1.64P \\ -0.19P \end{Bmatrix} = \begin{Bmatrix} 0.50P \\ 0.83P \\ -0.83P \end{Bmatrix}$$

EXAMPLE 9.2 Determine the bar forces and the joint displacements of the truss shown in Figure E9.2 using the Matrix displacement method. Assume that $A/L = 1$ for all members.

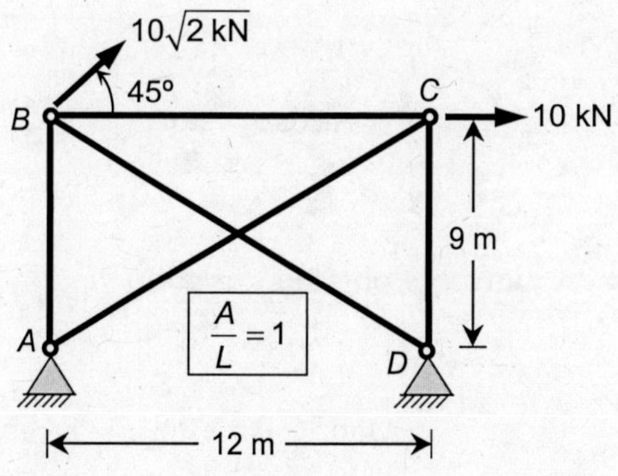

FIGURE E9.2

Solution: There are five elements (i.e., a, b, c, d, e) as identified in the following figure.

The nodal forces and the corresponding nodal displacement matrices are as follows:

$$\{R\} = \begin{Bmatrix} R_1 \\ R_2 \\ R_3 \\ R_4 \end{Bmatrix} = \begin{Bmatrix} 10 \\ 10 \\ 10 \\ 0 \end{Bmatrix} kN; \quad \{r\} = \begin{Bmatrix} r_1 \\ r_2 \\ r_3 \\ r_4 \end{Bmatrix}$$

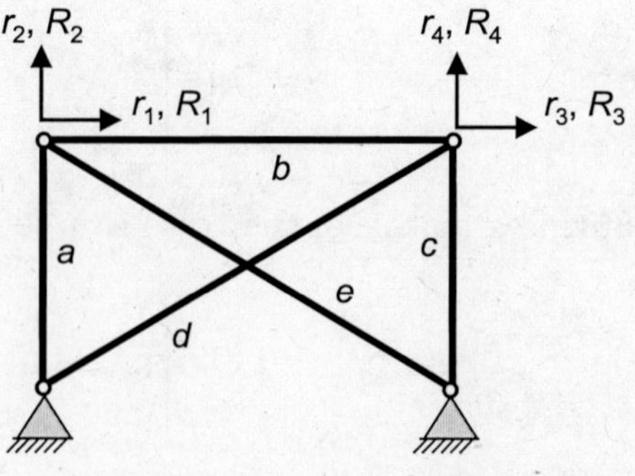

The element internal forces and deformations matrices are as follows:

$$\{Q\} = \begin{Bmatrix} Q^a \\ Q^b \\ Q^c \\ Q^d \\ Q^e \end{Bmatrix}; \quad \{q\} = \begin{Bmatrix} q^a \\ q^b \\ q^c \\ q^d \\ q^e \end{Bmatrix}$$

Displacement-transformation matrix $[a]$ is given by:

$$[a] = \begin{bmatrix} 0 & 1 & 0 & 0 \\ -1 & 0 & 1 & 0 \\ 0 & 0 & 0 & 1 \\ 0 & 0 & 0.8 & 0.6 \\ -0.8 & 0.6 & 0 & 0 \end{bmatrix}$$

$$(r_1 = 1) \quad (r_2 = 1) \quad (r_3 = 1) \quad (r_4 = 1)$$

The assembled element stiffness matrix $[k]$ is given by

$$[k] = \begin{bmatrix} k^a & & & & \\ & k^b & & & \\ & & k^c & & \\ & & & k^d & \\ & & & & k^e \end{bmatrix} = E \begin{pmatrix} 1 & & & & \\ & 1 & & & \\ & & 1 & & \\ & & & 1 & \\ & & & & 1 \end{pmatrix} \quad \left(\because \frac{A}{L} = 1 \right)$$

Stiffness matrix of the truss can be obtained as $[K] = [a]^T [k][a]$

$$[K] = E \begin{bmatrix} 0 & 1 & 0 & 0 \\ -1 & 0 & 1 & 0 \\ 0 & 0 & 0 & 1 \\ 0 & 0 & 0.8 & 0.6 \\ -0.8 & 0.6 & 0 & 0 \end{bmatrix}^T \begin{pmatrix} 1 & & & & \\ & 1 & & & \\ & & 1 & & \\ & & & 1 & \\ & & & & 1 \end{pmatrix} \begin{bmatrix} 0 & 1 & 0 & 0 \\ -1 & 0 & 1 & 0 \\ 0 & 0 & 0 & 1 \\ 0 & 0 & 0.8 & 0.6 \\ -0.8 & 0.6 & 0 & 0 \end{bmatrix}$$

$$[K] = E \begin{bmatrix} 1.64 & & & Sym \\ -0.48 & 1.36 & & \\ -1.0 & 0 & 1.64 & \\ 0 & 0 & 0.48 & 1.36 \end{bmatrix}$$

Nodal displacement matrix, $\{r\} = [K]^{-1}\{R\}$

$$\begin{Bmatrix} r_1 \\ r_2 \\ r_3 \\ r_4 \end{Bmatrix} = \frac{1}{E} \begin{bmatrix} 1.64 & & & Sym \\ -0.48 & 1.36 & & \\ -1.0 & 0 & 1.64 & \\ 0 & 0 & 0.48 & 1.36 \end{bmatrix}^{-1} \begin{Bmatrix} 10 \\ 10 \\ 10 \\ 0 \end{Bmatrix}$$

$$\therefore \qquad \begin{Bmatrix} r_1 \\ r_2 \\ r_3 \\ r_4 \end{Bmatrix} = \frac{1}{E} \begin{Bmatrix} 25.7 \\ 16.4 \\ 24.3 \\ -8.6 \end{Bmatrix}$$

Element forces can be determined using the following expression:

$$\{Q\} = [k][a]\{r\}$$

$$\begin{Bmatrix} Q^a \\ Q^b \\ Q^c \\ Q^d \\ Q^e \end{Bmatrix} = \begin{pmatrix} 1 & & & & \\ & 1 & & & \\ & & 1 & & \\ & & & 1 & \\ & & & & 1 \end{pmatrix} \begin{bmatrix} 0 & 1 & 0 & 0 \\ -1 & 0 & 1 & 0 \\ 0 & 0 & 0 & 1 \\ 0 & 0 & 0.8 & 0.6 \\ -0.8 & 0.6 & 0 & 0 \end{bmatrix} \begin{Bmatrix} 25.7 \\ 16.4 \\ 24.3 \\ -8.6 \end{Bmatrix}$$

$$\therefore \qquad \begin{Bmatrix} Q^a \\ Q^b \\ Q^c \\ Q^d \\ Q^e \end{Bmatrix} = \begin{Bmatrix} 16.4 \\ -1.4 \\ -8.6 \\ 14.3 \\ -10.7 \end{Bmatrix} kN$$

EXAMPLE 9.3 Determine the rotations at joints A and B and the bending moments at all joints of the rigid frame shown in Figure E9.3 using the Matrix displacement method. All members have the same flexural rigidity (EI).

Solution: For the given frame, all the joints and the point where the concentrated force is acting may be taken as the nodes. Accordingly, the number of elements would be four. Since it is required to determine the rotations at the joints only, the effect of applied concentrated load can be transferred to the joints. In such a case, only three nodes and two elements should be considered for the matrix displacement analysis. Thus, the given frame can be assumed as the sum of the fixed-ended beam BC and the frame with all forces and moments acting at the joints.

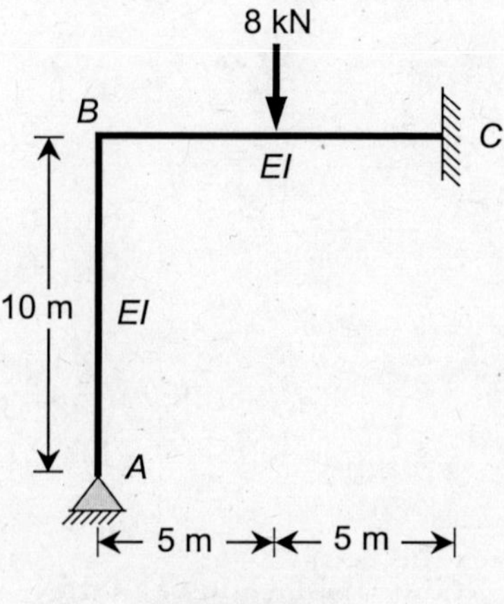

FIGURE E9.3

The end moments of the fixed ended beam are 10 kNm and the reactions at the fixed ends are 4 kN. The same reaction and end moment are applied at the joint B of the frame in the opposite directions as shown in the figure below.

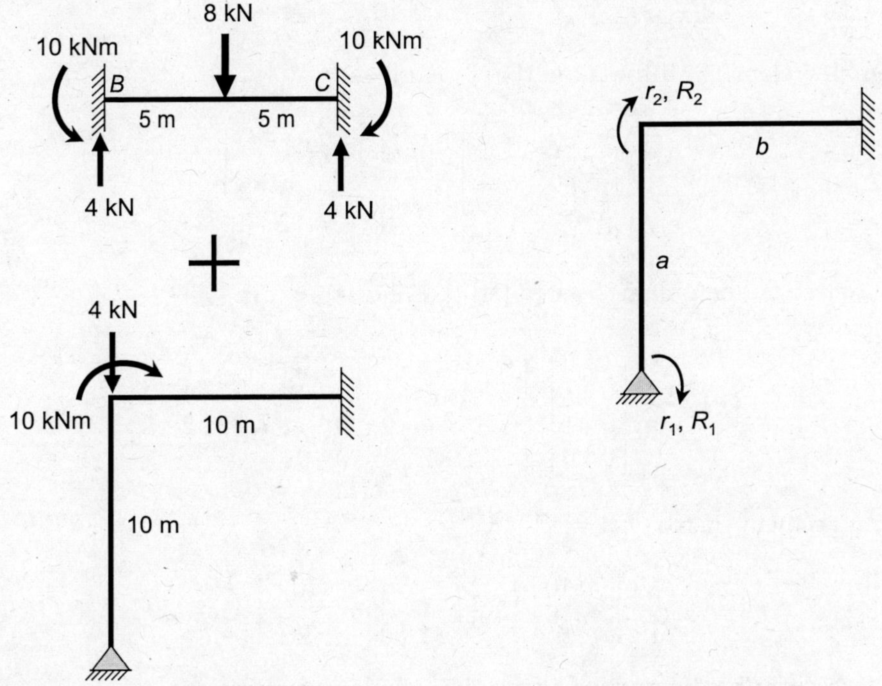

Two rotations are possible at joints A and B. The nodal displacement matrix and the nodal force matrix are as follows:

$$\{r\} = \begin{Bmatrix} r_1 \\ r_2 \end{Bmatrix}; \quad \{R\} = \begin{Bmatrix} R_1 \\ R_2 \end{Bmatrix} = \begin{Bmatrix} 0 \\ 10 \end{Bmatrix}$$

The element internal forces and deformations matrices are as follows:

$$\{Q\} = \begin{Bmatrix} Q_i^a \\ Q_j^a \\ Q_i^b \\ Q_j^b \end{Bmatrix}; \quad \{q\} = \begin{Bmatrix} q_i^a \\ q_j^a \\ q_i^b \\ q_j^b \end{Bmatrix}$$

Displacement-transformation matrix [a] can be obtained from the element deformation matrix by taking $r_1 = 1$ and $r_2 = 0$ in the first case and by considering $r_2 = 1$ and $r_1 = 0$ in the second case.

$$[a] = \begin{bmatrix} 1 & 0 \\ 0 & 1 \\ 0 & 1 \\ 0 & 0 \end{bmatrix}$$

The assembled element stiffness matrix $[k]$ is given by

$$[k] = \begin{bmatrix} k^a & \\ & k^b \end{bmatrix} = \frac{EI}{10} \begin{pmatrix} 4 & 2 & & \\ 2 & 4 & & \\ & & 4 & 2 \\ & & 2 & 4 \end{pmatrix}$$

Stiffness matrix can be obtained as $[K] = [a]^T [k][a]$

$$[K] = \frac{EI}{10} \begin{bmatrix} 1 & 0 \\ 0 & 1 \\ 0 & 1 \\ 0 & 0 \end{bmatrix}^T \begin{pmatrix} 4 & 2 & & \\ 2 & 4 & & \\ & & 4 & 2 \\ & & 2 & 4 \end{pmatrix} \begin{bmatrix} 1 & 0 \\ 0 & 1 \\ 0 & 1 \\ 0 & 0 \end{bmatrix} = \frac{EI}{10} \begin{bmatrix} 4 & 2 \\ 2 & 8 \end{bmatrix}$$

Nodal displacement matrix, $\{r\} = [K]^{-1}\{R\}$

$$\begin{Bmatrix} r_1 \\ r_2 \end{Bmatrix} = \frac{10}{EI} \begin{bmatrix} 4 & 2 \\ 2 & 8 \end{bmatrix}^{-1} \begin{Bmatrix} 0 \\ 10 \end{Bmatrix} = \frac{10}{1.4EI} \begin{Bmatrix} -1 \\ -2 \end{Bmatrix}$$

Element forces can be determined using the following expression:

$$\{Q\} = [k][a]\{r\}$$

$$\begin{Bmatrix} Q_i^a \\ Q_j^a \\ Q_i^b \\ Q_j^b \end{Bmatrix} = \frac{1}{1.4} \begin{pmatrix} 4 & 2 & 0 & 0 \\ 2 & 4 & 0 & 0 \\ 0 & 0 & 4 & 2 \\ 0 & 0 & 2 & 4 \end{pmatrix} \begin{bmatrix} 1 & 0 \\ 0 & 1 \\ 0 & 1 \\ 0 & 0 \end{bmatrix} \begin{Bmatrix} -1 \\ -2 \end{Bmatrix} = \begin{Bmatrix} 0 \\ 4.3 \\ 5.7 \\ 2.9 \end{Bmatrix}$$

The final member forces can be obtained by adding the forces/moments of the fixed beam at the respective nodes as follows:

$$\therefore \quad \begin{Bmatrix} Q_i^a \\ Q_j^a \\ Q_i^b \\ Q_j^b \end{Bmatrix} = \begin{Bmatrix} 0 \\ 4.3 \\ 5.7 \\ 2.9 \end{Bmatrix} + \begin{Bmatrix} 0 \\ 0 \\ -10 \\ 10 \end{Bmatrix} = \begin{Bmatrix} 0 \\ 4.3 \\ -4.3 \\ 12.9 \end{Bmatrix} \text{kNm}$$

EXAMPLE 9.4 Determine the end moments and the joint deformations of the rigid frame as shown in Figure E9.4 using the Matrix displacement method. The flexural rigidity of all members is EI.

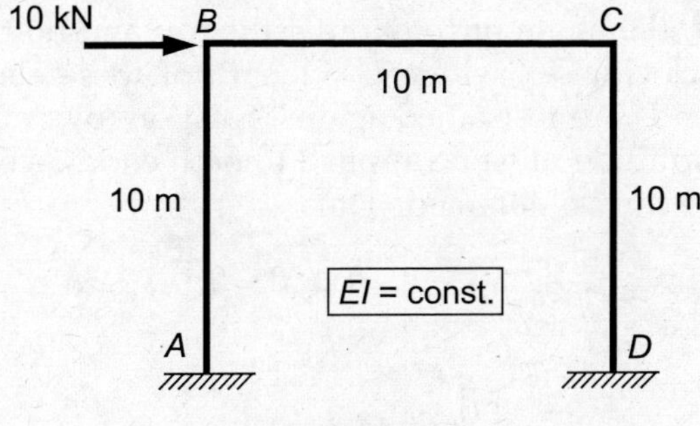

FIGURE E9.4

Solution: Three joint displacements (2 rotations and 1 displacement) are possible in the given frame. These nodal displacements are shown in the following figure.

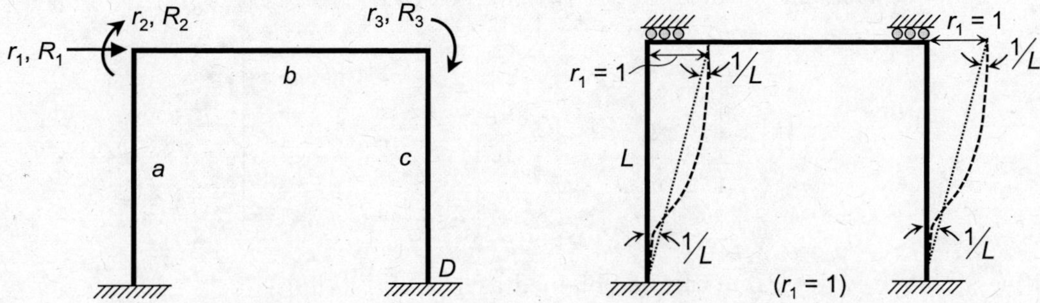

The nodal displacement matrix and the nodal force matrix are as follows:

$$\{r\} = \begin{Bmatrix} r_1 \\ r_2 \\ r_3 \end{Bmatrix}; \quad \{R\} = \begin{Bmatrix} R_1 \\ R_2 \\ R_3 \end{Bmatrix} = \begin{Bmatrix} 10 \\ 0 \\ 0 \end{Bmatrix}$$

The element internal forces and deformations matrices are as follows:

$$\{Q\} = \begin{Bmatrix} Q_i^a \\ Q_j^a \\ Q_i^b \\ Q_j^b \\ Q_i^c \\ Q_j^c \end{Bmatrix}; \quad \{q\} = \begin{Bmatrix} q_i^a \\ q_j^a \\ q_i^b \\ q_j^b \\ q_i^c \\ q_j^c \end{Bmatrix}$$

Displacement-transformation matrix [a] can be obtained from the element deformation matrix by setting one displacement to the unit and restraining all other displacements. In the first case, $r_1 = 1$ and $r_2 = r_3 = 0$. This means that no rotation is allowed at the joints B and C while allowing the horizontal unit displacement corresponding to r_1. Due to the unit horizontal displacement, the angle of rotation at the top and bottom ends of the elements a and c would be $1/L$ (clockwise), where L = length of these elements. Since the bottom ends are fixed and top ends are restrained against rotation (by definition), an anticlockwise rotation of the same magnitude must be applied to both ends. Accordingly, all components of element deformations can be obtained. Thus,

$$[a] = \begin{bmatrix} -0.1 & 0 & 0 \\ -0.1 & 1 & 0 \\ 0 & 1 & 0 \\ 0 & 0 & 1 \\ -0.1 & 0 & 1 \\ -0.1 & 0 & 0 \end{bmatrix}$$

The assembled element stiffness matrix $[k]$ is given by

$$[k] = \begin{bmatrix} k^a & & \\ & k^b & \\ & & k^c \end{bmatrix} = \frac{EI}{10} \begin{bmatrix} 4 & 2 & & & & \\ 2 & 4 & & & & \\ & & 4 & 2 & & \\ & & 2 & 4 & & \\ & & & & 4 & 2 \\ & & & & 2 & 4 \end{bmatrix}$$

Stiffness matrix of the frame can be obtained as $[K] = [a]^T [k][a]$

$$[K] = \frac{EI}{10} \begin{bmatrix} -0.1 & 0 & 0 \\ -0.1 & 1 & 0 \\ 0 & 1 & 0 \\ 0 & 0 & 1 \\ -0.1 & 0 & 1 \\ -0.1 & 0 & 0 \end{bmatrix}^T \begin{bmatrix} 4 & 2 & & & & \\ 2 & 4 & & & & \\ & & 4 & 2 & & \\ & & 2 & 4 & & \\ & & & & 4 & 2 \\ & & & & 2 & 4 \end{bmatrix} \begin{bmatrix} -0.1 & 0 & 0 \\ -0.1 & 1 & 0 \\ 0 & 1 & 0 \\ 0 & 0 & 1 \\ -0.1 & 0 & 1 \\ -0.1 & 0 & 0 \end{bmatrix}$$

$$\therefore \quad [K] = \frac{EI}{10} \begin{bmatrix} 0.24 & -0.6 & -0.6 \\ -0.6 & 8 & 2 \\ -0.6 & 2 & 8 \end{bmatrix}$$

Nodal displacement matrix, $\{r\} = [K]^{-1}\{R\}$

$$\begin{Bmatrix} r_1 \\ r_2 \\ r_3 \end{Bmatrix} = \frac{10}{EI} \begin{bmatrix} 0.24 & -0.6 & -0.6 \\ -0.6 & 8 & 2 \\ -0.6 & 2 & 8 \end{bmatrix}^{-1} \begin{Bmatrix} 10 \\ 0 \\ 0 \end{Bmatrix} = \frac{5.95}{EI} \begin{Bmatrix} 100 \\ 6 \\ 6 \end{Bmatrix}$$

Element forces can be determined using the following expression:

$$\{Q\} = [k][a]\{r\}$$

$$
\begin{Bmatrix} Q_i^a \\ Q_j^a \\ Q_i^b \\ Q_j^b \\ Q_i^c \\ Q_j^c \end{Bmatrix}
= \frac{5.95}{10}
\begin{bmatrix} 4 & 2 & & & & \\ 2 & 4 & & & & \\ & & 4 & 2 & & \\ & & 2 & 4 & & \\ & & & & 4 & 2 \\ & & & & 2 & 4 \end{bmatrix}
\begin{bmatrix} -0.1 & 0 & 0 \\ -0.1 & 1 & 0 \\ 0 & 1 & 0 \\ 0 & 0 & 1 \\ -0.1 & 0 & 1 \\ -0.1 & 0 & 0 \end{bmatrix}
\begin{Bmatrix} 100 \\ 6 \\ 6 \end{Bmatrix}
$$

$$
\therefore \quad
\begin{Bmatrix} Q_i^a \\ Q_j^a \\ Q_i^b \\ Q_j^b \\ Q_i^c \\ Q_j^c \end{Bmatrix}
= \begin{Bmatrix} -28.6 \\ -21.4 \\ 21.4 \\ 21.4 \\ -21.4 \\ 28.6 \end{Bmatrix}
$$

EXAMPLE 9.5 Determine the deflection and rotation at the joint B of the beam *AB* supported on a spring as shown in Figure E9.5 using the Matrix displacement method. Assume that the stiffness of the spring is K_s, and the flexural rigidity of the beam is *EI*.

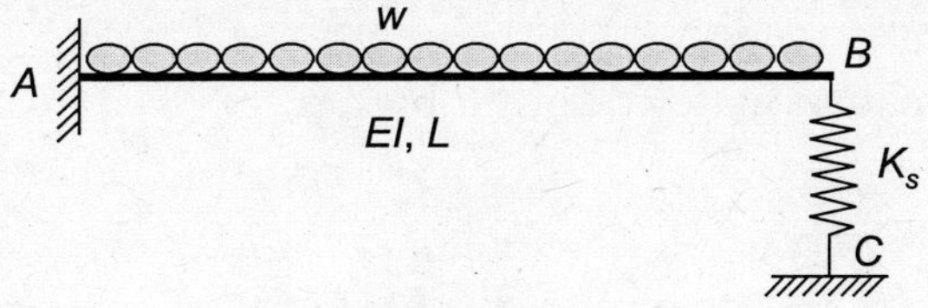

FIGURE E9.5

Solution: As discussed earlier, the given structure is considered as the superposition of the fixed-ended beam AB with the distributed load and the structure without loading. The end moments and reactions are shown in the following figure. The structure with equivalent loads applied at the joint is also shown below.

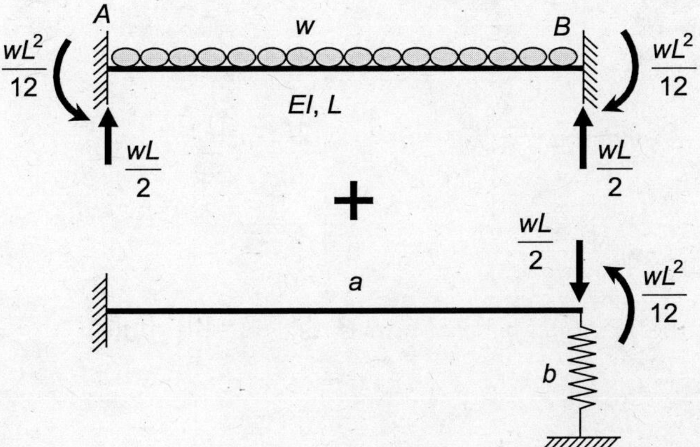

The nodal displacement and element end deformation are shown in the following figure.

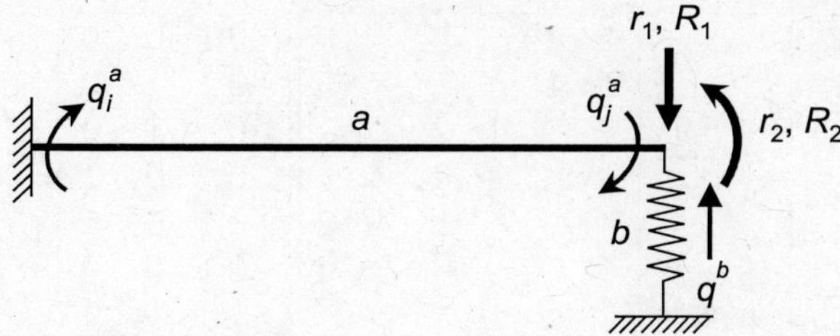

There are two elements, namely, a and b in the structure. Element a is a beam element, whereas the element b is a truss element as the spring can only carry an axial force. The nodal force and deformation matrices are as follows:

$$\{r\} = \begin{Bmatrix} r_1 \\ r_2 \end{Bmatrix}; \quad \{R\} = \begin{Bmatrix} R_1 \\ R_2 \end{Bmatrix} = \begin{Bmatrix} wL/2 \\ wL^2/12 \end{Bmatrix}$$

The element internal force and end deformation matrices are as follows:

$$\{Q\} = \begin{Bmatrix} Q_i^a \\ Q_j^a \\ Q^b \end{Bmatrix}; \quad \{q\} = \begin{Bmatrix} q_i^a \\ q_j^a \\ q^b \end{Bmatrix}$$

Displacement transformation matrix is given by

$$[a] = \begin{bmatrix} -\dfrac{1}{L} & 0 \\ -\dfrac{1}{L} & -1 \\ 1 & 0 \end{bmatrix}$$

The assembled element stiffness matrix $[k]$ is given by

$$[k] = \begin{bmatrix} k^a & \\ & k^b \end{bmatrix} = \begin{bmatrix} \dfrac{4EI}{L} & \dfrac{2EI}{L} & \\ \dfrac{2EI}{L} & \dfrac{4EI}{L} & \\ & & K_s \end{bmatrix}$$

Stiffness matrix of the frame can be obtained as $[K] = [a]^T [k][a]$

$$[K] = \begin{bmatrix} -\dfrac{1}{L} & 0 \\[2mm] -\dfrac{1}{L} & -1 \\[2mm] 1 & 0 \end{bmatrix}^{T} \begin{bmatrix} \dfrac{4EI}{L} & \dfrac{2EI}{L} \\[2mm] \dfrac{2EI}{L} & \dfrac{4EI}{L} \\[2mm] & & K_s \end{bmatrix} \begin{bmatrix} -\dfrac{1}{L} & 0 \\[2mm] -\dfrac{1}{L} & -1 \\[2mm] 1 & 0 \end{bmatrix}$$

$$\therefore \qquad [K] = \begin{bmatrix} \dfrac{12EI}{L^3} + K_s & \dfrac{6EI}{L^2} \\[4mm] \dfrac{6EI}{L^2} & \dfrac{4EI}{L} \end{bmatrix}$$

Nodal displacement matrix, $\{r\} = [K]^{-1}\{R\}$

$$\begin{Bmatrix} r_1 \\ r_2 \end{Bmatrix} = \begin{bmatrix} \dfrac{12EI}{L^3} + K_s & \dfrac{6EI}{L^2} \\[4mm] \dfrac{6EI}{L^2} & \dfrac{4EI}{L} \end{bmatrix}^{-1} \begin{Bmatrix} wL/2 \\[2mm] wL^2/12 \end{Bmatrix}$$

$$\therefore \qquad r_1 = \left(\frac{3wL}{8} \right) \left(\frac{1}{K_s + \dfrac{3EI}{L^3}} \right) \text{ and } r_2 = \left(\frac{wL^3 K_s}{48EI} - \frac{w}{2} \right) \left(\frac{1}{K_s + \dfrac{3EI}{L^3}} \right)$$

9.4 ANALYSIS USING MODIFIED STIFFNESS MATRIX

The element stiffness matrix of a beam given by Eqn. (9.20) is as follows:

$$[k^a] = \begin{bmatrix} \dfrac{4EI}{L} & \dfrac{2EI}{L} \\[2mm] \dfrac{2EI}{L} & \dfrac{4EI}{L} \end{bmatrix} = \frac{EI}{L}\begin{bmatrix} 4 & 2 \\ 2 & 4 \end{bmatrix}$$

In the above matrix, the components of the first column represent the element end moments obtained by applying a unit rotation at the joint i keeping the joint j as fixed. Whereas the first component of the matrix is the moment required to be applied at the joint i to cause a unit rotation at the same joint, the second component of the matrix is the carryover value of the applied moment which is one-half of the first component as shown in Figure 9.4.

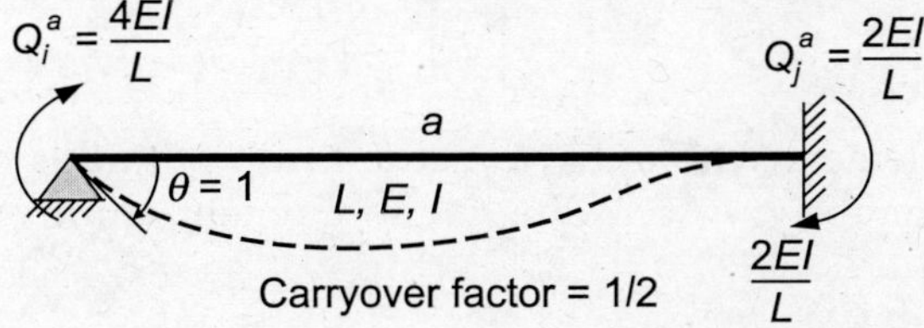

FIGURE 9.4 Modified stiffness coefficient for beam with far end fixed.

In a beam having one end pinned as shown in Figure 9.5, the final bending moment at the pinned support must be zero. This information can be considered in the formulation of the stiffness matrix at the beginning of the analysis. In the derivation of the modified stiffness matrix, it is required to compute the moment required to be applied at the end i to cause the unit rotation at the same joint keeping the far end is hinged (as it is). As discussed in the previous chapters, the value of applied moment end i, Q_i^a for $q_i^a = 1$ is given by

$$Q_i^a = \frac{3EI}{L} \tag{9.30}$$

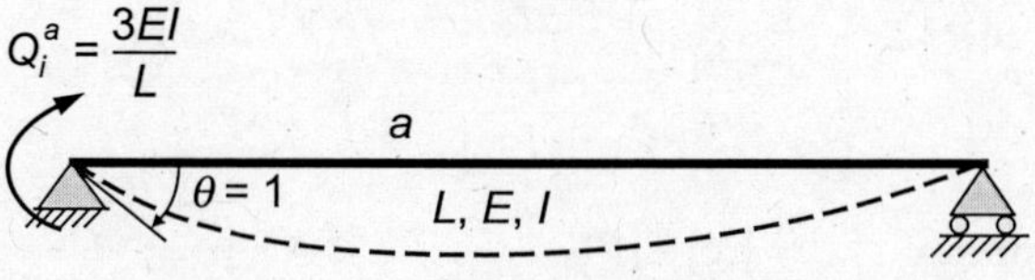

FIGURE 9.5   Modified stiffness coefficient for beam with far end hinged.

Since the moment at the end joint j, Q_j^a is zero, it is not necessary to consider the displacement and force components corresponding to joint j. Thus, the modified element stiffness matrix for the element having one end hinge would consist of one component only as given below:

$$\left[k^{a'} \right] = \left[\frac{3EI}{L} \right] \tag{9.31}$$

A similar approach may be followed for the elements with symmetric and anti-symmetric loadings (or deformed shapes). Figure 9.6 shows the element with a symmetric deformed shape. Since both end rotations are equal and opposite, the moment required to cause unit rotation at one end is given by

$$Q_i^a = \frac{2EI}{L} \tag{9.32}$$

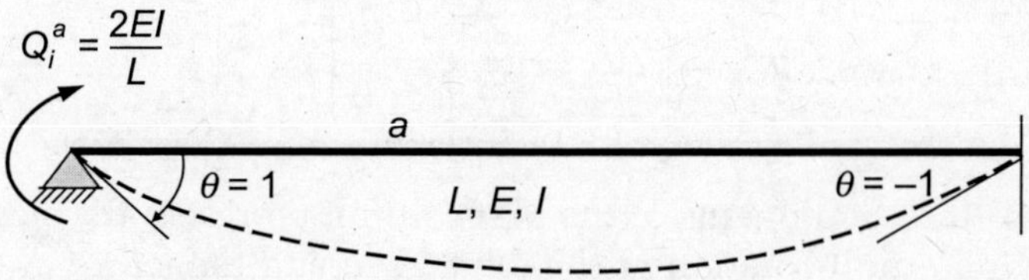

FIGURE 9.6   Modified stiffness coefficient for beam with symmetric loading.

The modified stiffness matrix of elements with the symmetric loadings is given by

$$\left[k^{a'} \right] = \left[\frac{2EI}{L} \right] \tag{9.33}$$

Figure 9.7 shows the element with antisymmetric loading. In this case, both end rotations are equal and of the same sign. Accordingly, the moment required to cause unit rotation at one end is given by

$$Q_i^a = \frac{6EI}{L} \tag{9.34}$$

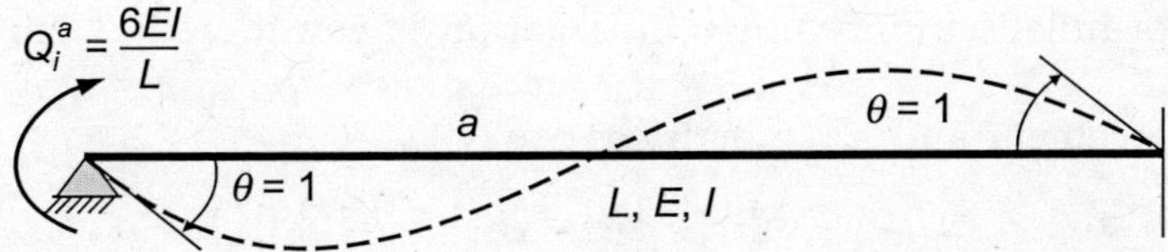

FIGURE 9.7 Modified stiffness coefficient for beam with anti-symmetric loading.

The modified stiffness matrix of elements with the symmetric loadings is given by

$$\left[k^{a'}\right] = \left[\frac{6EI}{L}\right] \tag{9.35}$$

The application of the modified stiffness matrix has been illustrated in the following examples. The main advantage of using the modified stiffness matrix is the analysis procedure is simplified and less time-consuming.

EXAMPLE 9.6 Determine the end moments of the continuous beam shown in Figure E9.6 using the Matrix displacement method. Assume the flexural rigidity of the beam is *EI*.

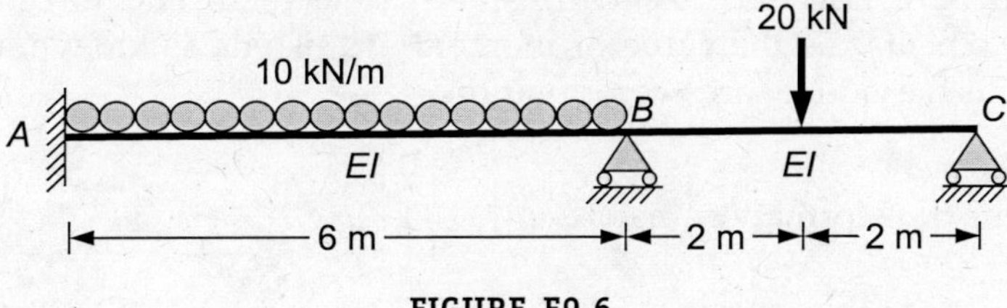

FIGURE E9.6

Solution: The given structure can be considered as the superposition of two fixed-ended beams so that all loads are transferred to the joints. The end moments and reactions for the spans AB and BC, assuming both ends are fixed, are shown in the following figure. Note that since the modified stiffness matrix would be used for the span BC having one hinged end, the modified fixed end moment should be computed at the joint B for the member BC.

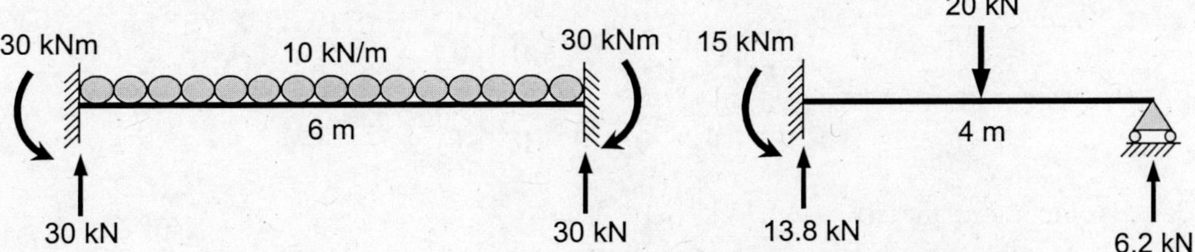

The net moments and forces are applied at the joints *B* and *C* of the beam in the opposite directions. The beam with all nodal forces is shown in the figure below.

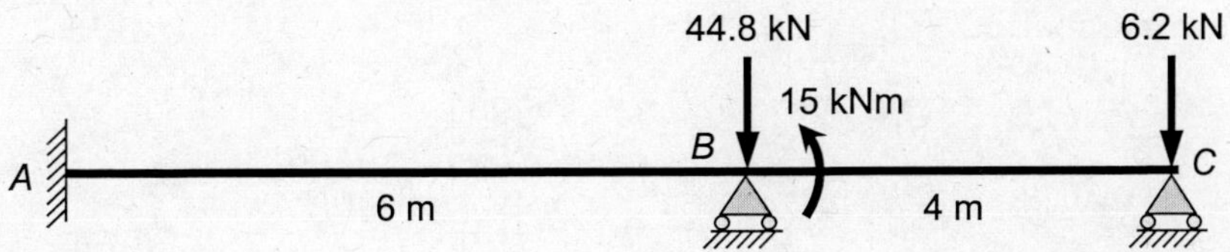

For the matrix formulation, only one joint rotation is considered as shown in the figure below.

Nodal displacement and force matrices are

$$\{r\} = \{r_1\}; \quad \{R\} = \{R_1\} = \{-15\}$$

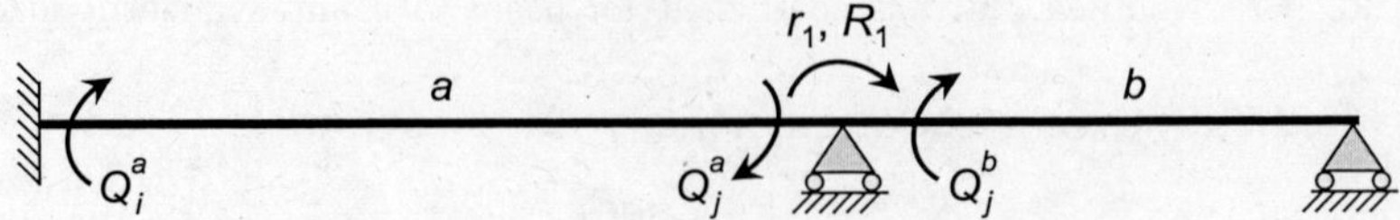

Element end force and deformation matrices are:

$$\{Q\} = \begin{Bmatrix} Q_i^a \\ Q_j^a \\ Q_i^b \end{Bmatrix}; \quad \{q\} = \begin{Bmatrix} q_i^a \\ q_j^a \\ q_i^b \end{Bmatrix}$$

In the above matrix, it not required to consider the element end force and deformation at joint j of the element b as the value of moment at this end is known and the same has been considered in the nodal force computation.

Displacement transformation matrix, $[a] = \begin{bmatrix} 0 \\ 1 \\ 1 \end{bmatrix}$

Element stiffness matrix in the assembled form is given by

$$[k] = \begin{bmatrix} k^a & \\ & k^{b'} \end{bmatrix} = EI \begin{bmatrix} 4/6 & 2/6 & \\ 2/6 & 4/6 & \\ & & 3/4 \end{bmatrix} = \frac{EI}{24} \begin{bmatrix} 16 & 8 & 0 \\ 8 & 16 & 0 \\ 0 & 0 & 18 \end{bmatrix}$$

Total stiffness matrix, $[K] = [a]^T [k][a]$

$$[K] = \frac{EI}{24} \begin{bmatrix} 0 \\ 1 \\ 1 \end{bmatrix}^T \begin{bmatrix} 16 & 8 & 0 \\ 8 & 16 & 0 \\ 0 & 0 & 18 \end{bmatrix} \begin{bmatrix} 0 \\ 1 \\ 1 \end{bmatrix} = \frac{34}{24} EI$$

Nodal displacement matrix, $\{r\} = [K]^{-1}\{R\}$

$$\{r_1\} = \begin{bmatrix} \dfrac{24}{34EI} \end{bmatrix} \{-15\} = \left\{ -\dfrac{10.59}{EI} \right\}$$

Element end force matrix, $\{Q\} = [k][a]\{r\}$

$$\begin{Bmatrix} Q_i^a \\ Q_j^a \\ Q_i^b \end{Bmatrix} = \frac{EI}{24} \begin{bmatrix} 16 & 8 & 0 \\ 8 & 16 & 0 \\ 0 & 0 & 18 \end{bmatrix} \begin{bmatrix} 0 \\ 1 \\ 1 \end{bmatrix} \left\{ -\dfrac{10.59}{EI} \right\}$$

$$\therefore \qquad \begin{Bmatrix} Q_i^a \\ Q_j^a \\ Q_i^b \end{Bmatrix} = \begin{Bmatrix} -3.5 \\ -7.1 \\ -7.9 \end{Bmatrix}$$

The final end moments of the elements can be obtained by adding the corresponding fixed end moments.

Thus,
$$\begin{Bmatrix} Q_i^a \\ Q_j^a \\ Q_i^b \\ Q_j^b \end{Bmatrix} = \begin{Bmatrix} -3.5 \\ -7.1 \\ -7.9 \\ 0 \end{Bmatrix} + \begin{Bmatrix} -30 \\ 30 \\ -15 \\ 0 \end{Bmatrix} = \begin{Bmatrix} -33.5 \\ 22.9 \\ -22.9 \\ 0 \end{Bmatrix}$$

EXAMPLE 9.7 Analyze the rigid frame shown in Figure E9.7 using the Matrix Displacement method. Assume the constant EI for all members.

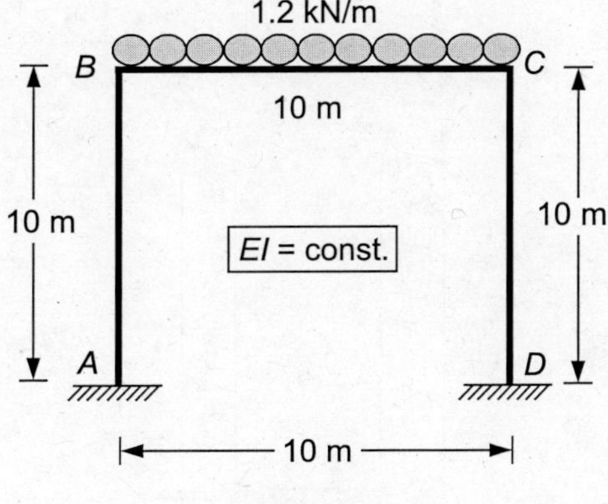

FIGURE E9.7

Solution: The distributed load on the member BC is transferred to both ends by assuming the fixed ended members. The given rigid frame is considered as the sum of the fixed end beam and the frame with all loads applied at the nodes. First, the frame with all nodal loads has been considered for the matrix displacement method. It can be seen that the frame is a non-sway frame since there is no horizontal force acting on the frame and the applied end moments are equal. The nodal forces, nodal displacements, and member end forces are shown in the figure below.

$$\{r\} = \begin{Bmatrix} r_1 \\ r_2 \end{Bmatrix}; \quad \{R\} = \begin{Bmatrix} R_1 \\ R_2 \end{Bmatrix} = \begin{Bmatrix} 10 \\ -10 \end{Bmatrix}$$

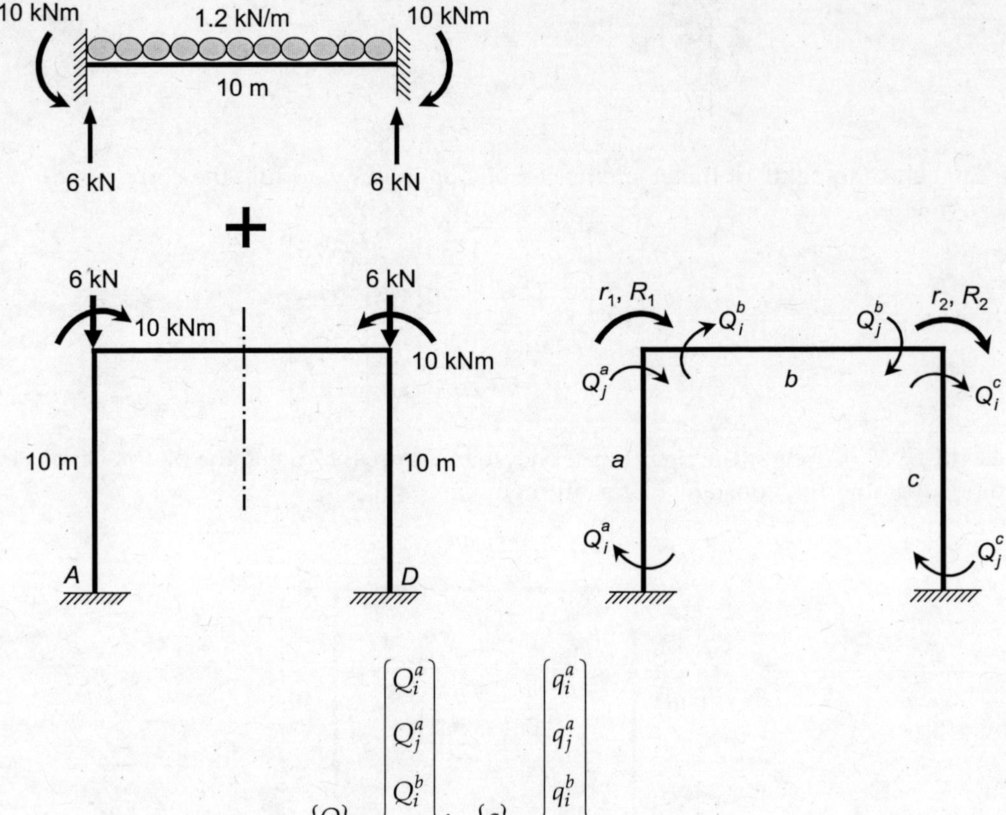

$$\{Q\} = \begin{Bmatrix} Q_i^a \\ Q_j^a \\ Q_i^b \\ Q_j^b \\ Q_i^c \\ Q_j^c \end{Bmatrix}; \quad \{q\} = \begin{Bmatrix} q_i^a \\ q_j^a \\ q_i^b \\ q_j^b \\ q_i^c \\ q_j^c \end{Bmatrix}$$

Displacement transformation matrix, $[a] = \begin{bmatrix} 0 & 0 \\ 1 & 0 \\ 1 & 0 \\ 0 & 1 \\ 0 & 1 \\ 0 & 0 \end{bmatrix}$

Element stiffness matrix is given by

$$[k] = \begin{bmatrix} k^a & & \\ & k^b & \\ & & k^c \end{bmatrix} = \frac{EI}{10} \begin{bmatrix} 4 & 2 & & & & \\ 2 & 4 & & & & \\ & & 4 & 2 & & \\ & & 2 & 4 & & \\ & & & & 4 & 2 \\ & & & & 2 & 4 \end{bmatrix}$$

Total stiffness matrix, $[K] = [a]^T [k][a]$

$$[K] = \frac{EI}{10} \begin{bmatrix} 0 & 0 \\ 1 & 0 \\ 1 & 0 \\ 0 & 1 \\ 0 & 1 \\ 0 & 0 \end{bmatrix}^T \begin{bmatrix} 4 & 2 & & & & \\ 2 & 4 & & & & \\ & & 4 & 2 & & \\ & & 2 & 4 & & \\ & & & & 4 & 2 \\ & & & & 2 & 4 \end{bmatrix} \begin{bmatrix} 0 & 0 \\ 1 & 0 \\ 1 & 0 \\ 0 & 1 \\ 0 & 1 \\ 0 & 0 \end{bmatrix} = \frac{EI}{10} \begin{bmatrix} 8 & 2 \\ 2 & 8 \end{bmatrix}$$

Nodal displacement matrix, $\{r\} = [K]^{-1}\{R\}$

$$\begin{Bmatrix} r_1 \\ r_2 \end{Bmatrix} = \left(\frac{10}{EI}\right)\begin{bmatrix} 8 & 2 \\ 2 & 8 \end{bmatrix}^{-1} \begin{Bmatrix} 10 \\ -10 \end{Bmatrix} = \left(\frac{10}{3EI}\right)\begin{Bmatrix} 5 \\ -5 \end{Bmatrix}$$

Element end force matrix, $\{Q\} = [k][a]\{r\}$

$$\begin{Bmatrix} Q_i^a \\ Q_j^a \\ Q_i^b \\ Q_j^b \\ Q_i^c \\ Q_j^c \end{Bmatrix} = \left(\frac{1}{3}\right)\begin{bmatrix} 4 & 2 & & & & \\ 2 & 4 & & & & \\ & & 4 & 2 & & \\ & & 2 & 4 & & \\ & & & & 4 & 2 \\ & & & & 2 & 4 \end{bmatrix} \begin{bmatrix} 0 & 0 \\ 1 & 0 \\ 1 & 0 \\ 0 & 1 \\ 0 & 1 \\ 0 & 0 \end{bmatrix} \begin{Bmatrix} 5 \\ -5 \end{Bmatrix} = \begin{Bmatrix} 3.33 \\ 6.67 \\ 3.33 \\ -3.33 \\ -6.67 \\ -3.33 \end{Bmatrix}$$

The final end moments are obtained by adding the corresponding fixed end moments.

$$\therefore \quad \begin{Bmatrix} Q_i^a \\ Q_j^a \\ Q_i^b \\ Q_j^b \\ Q_i^c \\ Q_j^c \end{Bmatrix} = \begin{Bmatrix} 3.33 \\ 6.67 \\ 3.33 \\ -3.33 \\ -6.67 \\ -3.33 \end{Bmatrix} + \begin{Bmatrix} 0 \\ 0 \\ -10 \\ 10 \\ 0 \\ 0 \end{Bmatrix} = \begin{Bmatrix} 3.33 \\ 6.67 \\ -6.67 \\ 6.67 \\ -6.67 \\ -3.33 \end{Bmatrix} \text{ kNm}$$

EXAMPLE 9.8 Re-analyze the rigid frame shown in Figure E9.7 using the modified stiffness method. Assume the constant EI for all members.

Solution: It can be noted that the loading on the frame is symmetric. As a result, only one-half of the frame can be considered in the matrix displacement method using the modified stiffness technique.

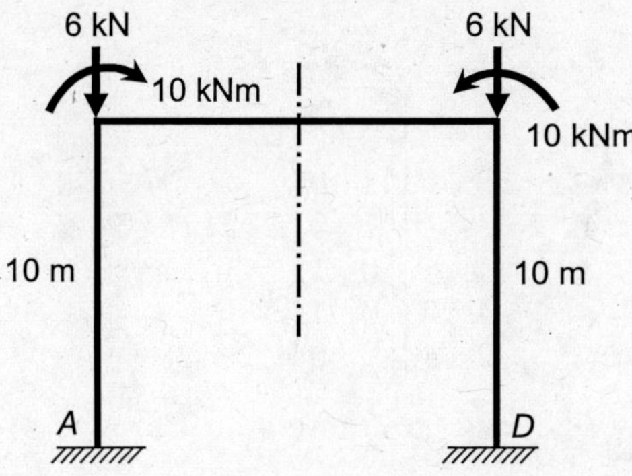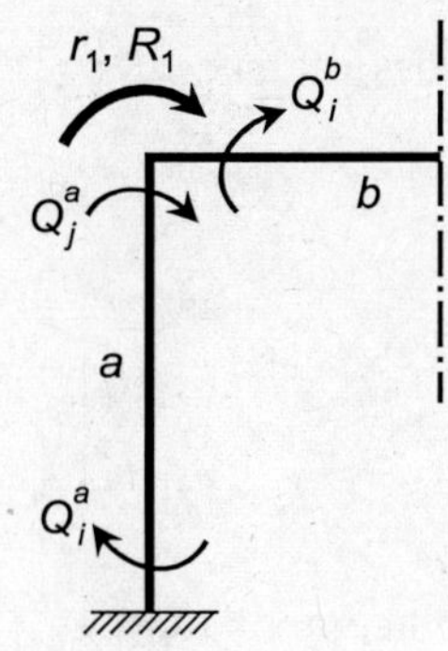

Nodal displacement and nodal force matrices are:

$$\{r\} = \{r_1\}; \quad \{R\} = \{R_1\} = \{10\}$$

Element end force and deformation matrices are:

$$\{Q\} = \begin{Bmatrix} Q_i^a \\ Q_j^a \\ Q_i^b \end{Bmatrix}; \quad \{q\} = \begin{Bmatrix} q_i^a \\ q_j^a \\ q_i^b \end{Bmatrix}$$

Displacement transformation matrix, $[a] = \begin{bmatrix} 0 \\ 1 \\ 1 \end{bmatrix}$

Element stiffness matrix in the assembled form is given by

$$[k] = \begin{bmatrix} k^a & \\ & k^{b'} \end{bmatrix} = \frac{EI}{10} \begin{bmatrix} 4 & 2 & \\ 2 & 4 & \\ & & 2 \end{bmatrix}$$

Total stiffness matrix, $[K] = [a]^T [k][a]$

$$[K] = \left(\frac{EI}{10}\right) \begin{bmatrix} 0 \\ 1 \\ 1 \end{bmatrix}^T \begin{bmatrix} 4 & 2 & 0 \\ 2 & 4 & 0 \\ 0 & 0 & 2 \end{bmatrix} \begin{bmatrix} 0 \\ 1 \\ 1 \end{bmatrix} = \frac{6EI}{10}$$

Nodal displacement matrix, $\{r\} = [K]^{-1}\{R\}$

$$\{r_1\} = \left[\frac{10}{6EI}\right]\{10\} = \left\{\frac{50}{3EI}\right\}$$

Element end force matrix, $\{Q\} = [k][a]\{r\}$

$$\begin{Bmatrix} Q_i^a \\ Q_j^a \\ Q_i^b \end{Bmatrix} = \frac{EI}{10} \begin{bmatrix} 4 & 2 & 0 \\ 2 & 4 & 0 \\ 0 & 0 & 2 \end{bmatrix} \begin{bmatrix} 0 \\ 1 \\ 1 \end{bmatrix} \left\{\frac{50}{3EI}\right\} = \begin{Bmatrix} 3.33 \\ 6.67 \\ 3.33 \end{Bmatrix}$$

The final end moments of the elements can be obtained by adding the corresponding fixed end moments.

Thus,
$$\begin{Bmatrix} Q_i^a \\ Q_j^a \\ Q_i^b \end{Bmatrix} = \begin{Bmatrix} 3.33 \\ 6.67 \\ 3.33 \end{Bmatrix} + \begin{Bmatrix} 0 \\ 0 \\ -10 \end{Bmatrix} = \begin{Bmatrix} 3.33 \\ 6.67 \\ -6.67 \end{Bmatrix} = \begin{Bmatrix} -Q_j^c \\ -Q_i^c \\ -Q_j^b \end{Bmatrix}$$

EXAMPLE 9.9 Determine the bending moment and rotation of the joint B of the continuous beam shown in Figure E9.9 using the matrix displacement method.

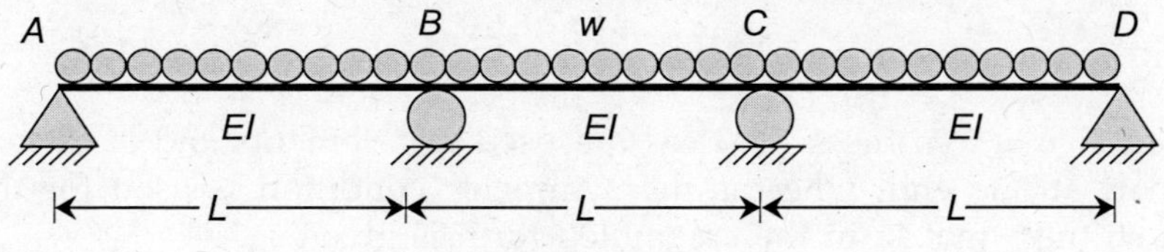

FIGURE E9.9

Solution: Since the loading and support conditions are symmetric, only one-half of the structure can be considered for the analysis. Further, ends A and D being supported on hinges, modified stiffness matrix may be used for spans the AB and CD.

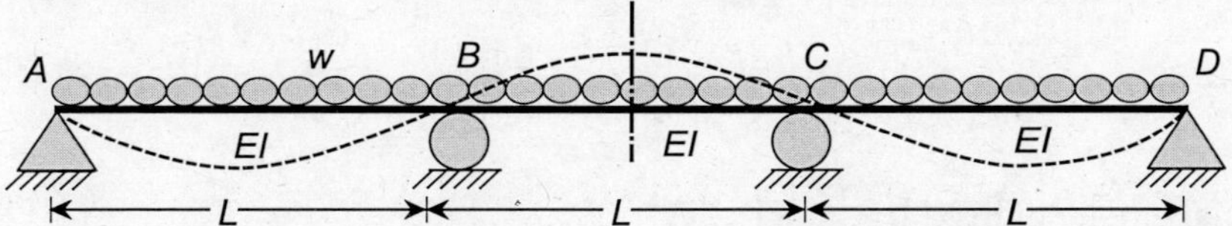

In this example, the left half of the beam has been considered for the matrix displacement method. The effect of distributed loads on the spans is transferred to the nodes by the equivalent moment and forces as shown below.

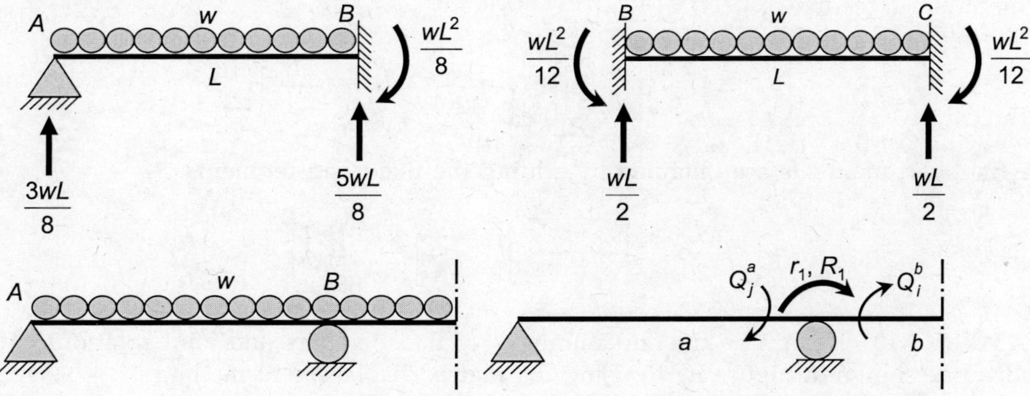

Nodal displacement and nodal force matrices are:

$$\{r\} = \{r_1\}; \quad \{R\} = \{R_1\} = \left\{ -\frac{wL^2}{8} + \frac{wL^2}{12} \right\} = \left\{ -\frac{wL^2}{24} \right\}$$

Element end force and deformation matrices are

$$\{Q\} = \left\{ \begin{matrix} Q_j^a \\ Q_i^b \end{matrix} \right\}; \quad \{q\} = \left\{ \begin{matrix} q_j^a \\ q_i^b \end{matrix} \right\}$$

Displacement transformation matrix, $[a] = \begin{bmatrix} 1 \\ 1 \end{bmatrix}$.

Modified element stiffness matrices are used for elements and b. Element a has the pinned support at one end, whereas the symmetry condition is used for the element b. The element stiffness matrix in the assembled form is given by

$$[k] = \begin{bmatrix} k^{a'} & \\ & k^{b'} \end{bmatrix} = \frac{EI}{L} \begin{bmatrix} 3 & \\ & 2 \end{bmatrix}$$

Total stiffness matrix, $[K] = [a]^T [k][a]$

$$[K] = \left(\frac{EI}{L} \right) \begin{bmatrix} 1 \\ 1 \end{bmatrix}^T \begin{bmatrix} 3 & 0 \\ 0 & 2 \end{bmatrix} \begin{bmatrix} 1 \\ 1 \end{bmatrix} = \frac{5EI}{L}$$

Nodal displacement matrix, $\{r\} = [K]^{-1}\{R\}$

$$\{r_1\} = \left[\frac{L}{5EI} \right] \left\{ -\frac{wL^2}{24} \right\} = \left\{ -\frac{wL^3}{120EI} \right\}$$

Element end force matrix, $\{Q\} = [k][a]\{r\}$

$$\left\{ \begin{matrix} Q_j^a \\ Q_i^b \end{matrix} \right\} = \left(\frac{EI}{L} \right) \begin{bmatrix} 3 & 0 \\ 0 & 2 \end{bmatrix} \begin{bmatrix} 1 \\ 1 \end{bmatrix} \left\{ -\frac{wL^3}{120EI} \right\} = \left(-\frac{wL^2}{120} \right) \left\{ \begin{matrix} 3 \\ 2 \end{matrix} \right\}$$

The final end moments are obtained by adding the fixed end moments.

Thus, $$\left\{ \begin{matrix} Q_j^a \\ Q_i^b \end{matrix} \right\} = \left(-\frac{wL^2}{120} \right) \left\{ \begin{matrix} 3 \\ 2 \end{matrix} \right\} + \left(\frac{wL^2}{24} \right) \left\{ \begin{matrix} 3 \\ -2 \end{matrix} \right\} = \left(\frac{wL^2}{10} \right) \left\{ \begin{matrix} 1 \\ -1 \end{matrix} \right\} = \left\{ \begin{matrix} -Q_i^c \\ -Q_j^b \end{matrix} \right\}$$

EXAMPLE 9.10 Determine the end moments of the members and joint rotations of the rigid frame shown in Figure E9.10 using the matrix displacement method.

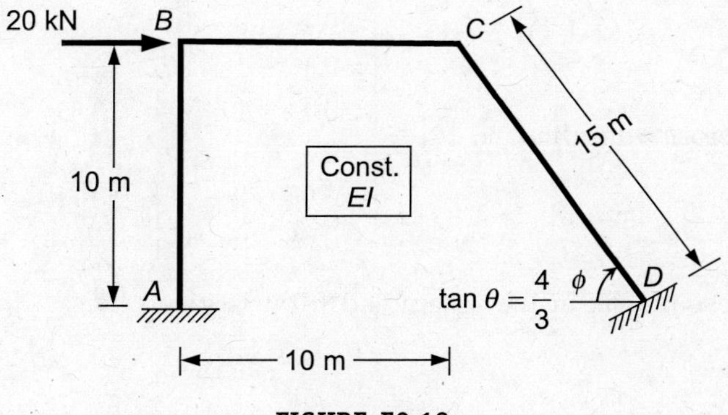

FIGURE E9.10

Solution: The elements, nodal forces, nodal displacements, element forces and element deformations are shown in the following figure. Using the displacement triangle, for a unit displacement normal to the element *a* would element *c* to displace by 1.25.

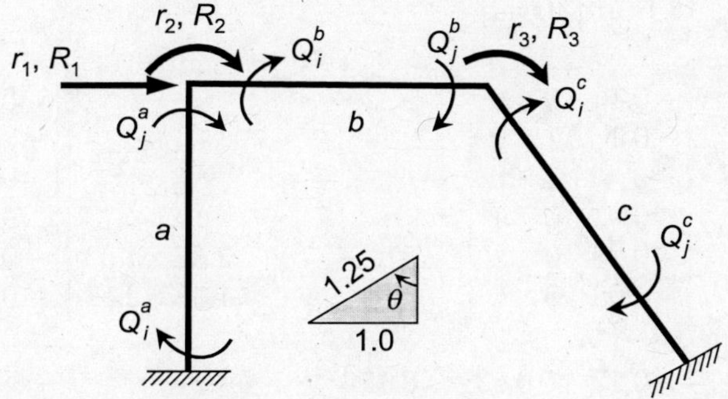

Nodal displacement and nodal force matrices are:

$$\{r\} = \begin{Bmatrix} r_1 \\ r_2 \\ r_3 \end{Bmatrix}; \quad \{R\} = \begin{Bmatrix} R_1 \\ R_2 \\ R_3 \end{Bmatrix} = \begin{Bmatrix} 20 \\ 0 \\ 0 \end{Bmatrix}$$

Element end force and deformation matrices are:

$$\{Q\} = \begin{Bmatrix} Q_i^a \\ Q_j^a \\ Q_i^b \\ Q_j^b \\ Q_i^c \\ Q_j^c \end{Bmatrix}; \quad \{q\} = \begin{Bmatrix} q_i^a \\ q_j^a \\ q_i^b \\ q_j^b \\ q_i^c \\ q_j^c \end{Bmatrix}$$

Displacement transformation matrix, $[a] = \begin{bmatrix} -0.1 & 0 & 0 \\ -0.1 & 1 & 0 \\ 0 & 1 & 0 \\ 0 & 0 & 1 \\ -0.08 & 0 & 1 \\ -0.08 & 0 & 0 \end{bmatrix}$

The element stiffness matrix in the assembled form is given by:

$$[k] = \begin{bmatrix} k^a & & \\ & k^b & \\ & & k^c \end{bmatrix} = \frac{EI}{10} \begin{bmatrix} 4 & 2 & & & & \\ 2 & 4 & & & & \\ & & 4 & 2 & & \\ & & 2 & 4 & & \\ & & & & 2.67 & 1.33 \\ & & & & 1.33 & 2.67 \end{bmatrix}$$

Total stiffness matrix, $[K] = [a]^T [k][a]$

$$[K] = \left(\frac{EI}{10}\right) \begin{bmatrix} -0.1 & 0 & 0 \\ -0.1 & 1 & 0 \\ 0 & 1 & 0 \\ 0 & 0 & 1 \\ -0.08 & 0 & 1 \\ -0.08 & 0 & 0 \end{bmatrix}^T \begin{bmatrix} 4 & 2 & & & & \\ 2 & 4 & & & & \\ & & 4 & 2 & & \\ & & 2 & 4 & & \\ & & & & 2.67 & 1.33 \\ & & & & 1.33 & 2.67 \end{bmatrix} \begin{bmatrix} -0.1 & 0 & 0 \\ -0.1 & 1 & 0 \\ 0 & 1 & 0 \\ 0 & 0 & 1 \\ -0.08 & 0 & 1 \\ -0.08 & 0 & 0 \end{bmatrix}$$

$$\therefore \quad [K] = \left(\frac{EI}{10}\right) \begin{bmatrix} 0.17 & -0.60 & -0.32 \\ -0.60 & 8.0 & 2.0 \\ -0.32 & 2.0 & 6.67 \end{bmatrix}$$

Nodal displacement matrix, $\{r\} = [K]^{-1}\{R\}$

$$\begin{Bmatrix} r_1 \\ r_2 \\ r_3 \end{Bmatrix} = \left(\frac{10}{EI}\right) \begin{bmatrix} 0.17 & -0.60 & -0.32 \\ -0.60 & 8.0 & 2.0 \\ -0.32 & 2.0 & 6.67 \end{bmatrix}^{-1} \begin{Bmatrix} 20 \\ 0 \\ 0 \end{Bmatrix} = \left(\frac{10}{EI}\right) \begin{Bmatrix} 164.59 \\ 11.21 \\ 4.53 \end{Bmatrix}$$

Element end force matrix, $\{Q\} = [k][a]\{r\}$

$$\begin{Bmatrix} Q_i^a \\ Q_j^a \\ Q_i^b \\ Q_j^b \\ Q_i^c \\ Q_j^c \end{Bmatrix} = \begin{Bmatrix} -76.3 \\ -53.9 \\ 53.9 \\ 40.6 \\ -40.6 \\ -46.6 \end{Bmatrix}$$

9.5   GENERAL FORMULATIONS OF MATRIX DISPLACEMENT METHOD

In the matrix displacement method discussed above, the nodal displacements and nodal forces are considered corresponding to the unknown joint deformations. This helps in determining the end deformations and end forces of elements. However, this procedure does not provide the reaction forces/moments at the supports directly. The matrix displacement method can be suitably modified to compute the unknown support reactions and moments directly.

Let the nodal displacement matrix corresponding to the unknown joint deformations is $\{r_R\}$ and the nodal displacement matrix corresponding to the joint deformations at the supports (assuming unrestrained) is $\{r_X\}$. Accordingly, $\{R\}$ and $\{X\}$ are the nodal external force and the nodal reaction force matrices, respectively. The element end deformation matrix $\{q\}$ is given by

$$\{q\} = \begin{bmatrix} a_R & a_X \end{bmatrix} \begin{Bmatrix} R \\ X \end{Bmatrix} \tag{9.36}$$

Thus, the displacement matrix $[a]$ is composed of two matrices, $[a_R]$ and $[a_X]$ corresponding to nodal force matrices, $\{R\}$ and $\{X\}$, respectively. Total stiffness matrix, $[K]$ can be written as follows:

$$[K] = [a]^T [k][a] = \begin{bmatrix} a_R & a_X \end{bmatrix}^T [k] \begin{bmatrix} a_R & a_X \end{bmatrix} \tag{9.37}$$

$$[K] = \begin{bmatrix} a_R{}^T k a_R & a_R{}^T k a_X \\ a_X{}^T k a_R & a_X{}^T k a_X \end{bmatrix} = \begin{bmatrix} K_{RR} & K_{RX} \\ K_{XR} & K_{XX} \end{bmatrix} \tag{9.38}$$

The nodal force and nodal displacement matrices can be related as follows:

$$\begin{Bmatrix} R \\ X \end{Bmatrix} = \begin{bmatrix} K_{RR} & K_{RX} \\ K_{XR} & K_{XX} \end{bmatrix} \begin{Bmatrix} r_R \\ r_X \end{Bmatrix} \tag{9.39}$$

Mostly, $\{r_X\} = \{0\}$, the nodal displacement matrix can be obtained as follows:

$$\{r_R\} = [K_{RR}]^{-1} \{R\} \tag{9.40}$$

Similarly, the nodal support reaction matrix can be determined as follows:

$$\{X\} = [K_{XR}]^{-1} \{R\} \tag{9.41}$$

9.6   PROBLEMS

9.1 Determine the rotations at the supports and the deflection at the mid-span of the beam shown below using the Matrix displacement method.

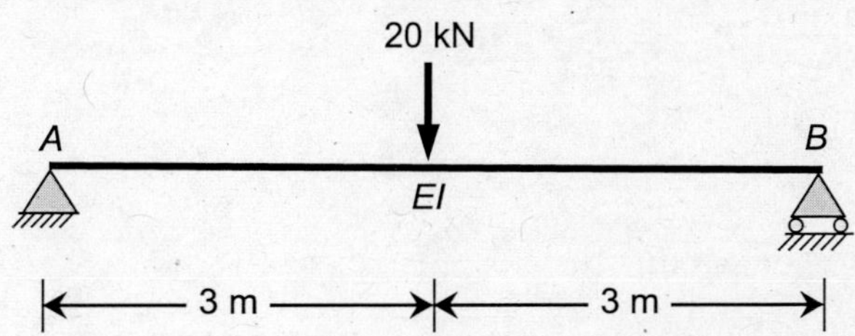

9.2 Determine the reactions and moments at the supports and the rotation at the supports B and C of the beam shown below using the Matrix displacement method.

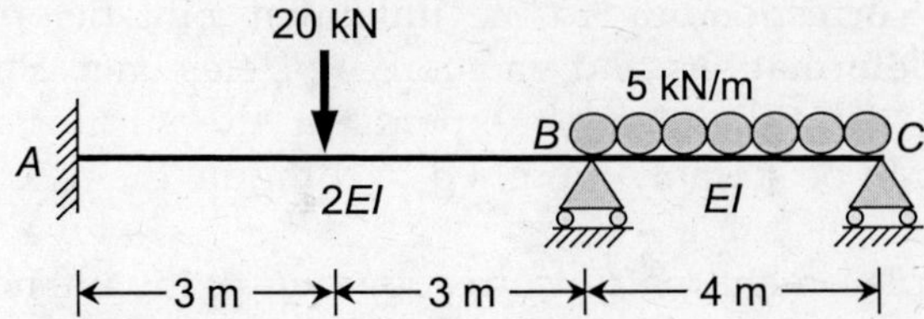

9.3 Analyze the continuous beam shown below using the Matrix displacement method. Determine the support reactions, the member end rotations, and the support moments.

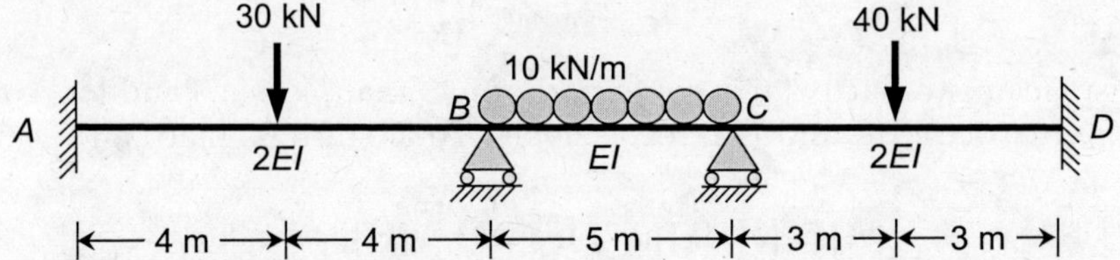

9.4 Determine the support reactions, the member forces and the joint displacements of the truss shown below using the Matrix displacement method. Take $E = 200$ GPa.

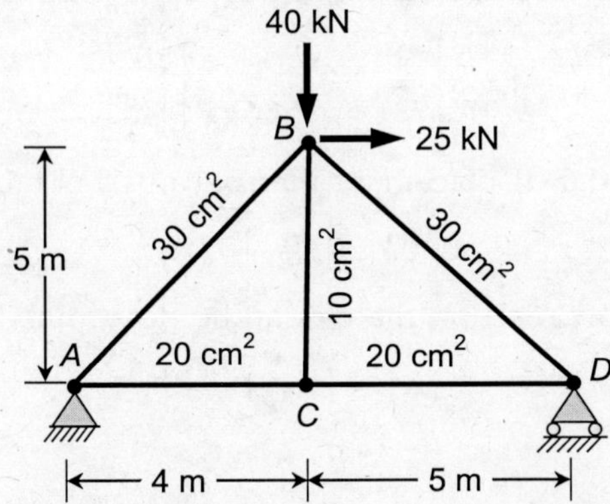

9.5 Determine the support reactions, the joint displacements, and the member forces of the truss shown below using the Matrix displacement method. Take $E = 200$ GPa.

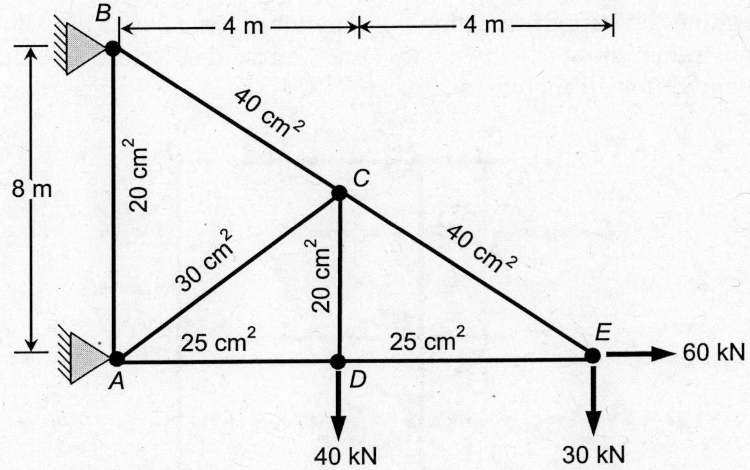

9.6 Determine the support reactions, the joint moments and the rotation at B of the beam shown below using the Matrix displacement method.

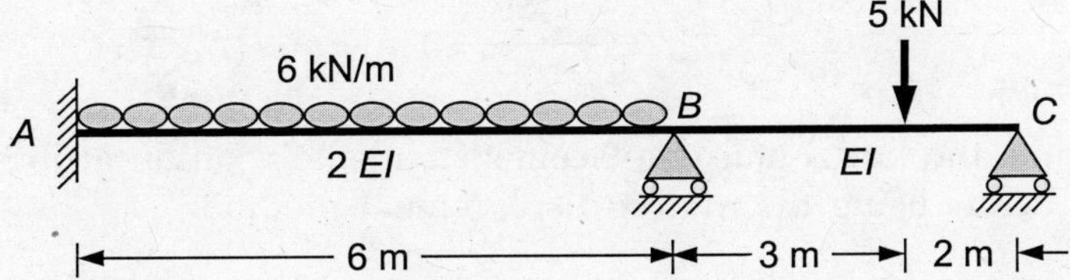

9.7 Determine the member stiffness factor at end *B* of the beam *AB* as shown below. Assume constant flexural rigidity (*EI*) of the beam.

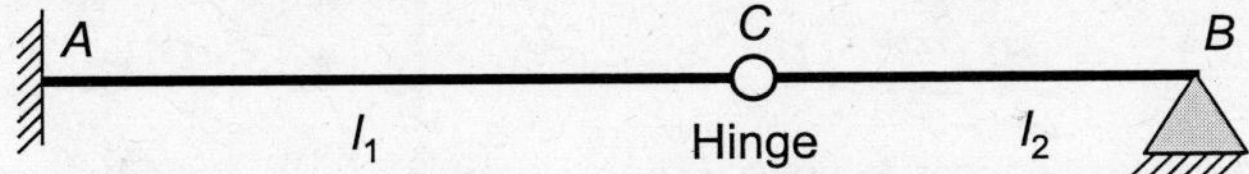

9.8 Determine the support reactions, the member forces and the joint displacement of the truss shown below using the Matrix force method. Assume $AE = 5000$ kN.

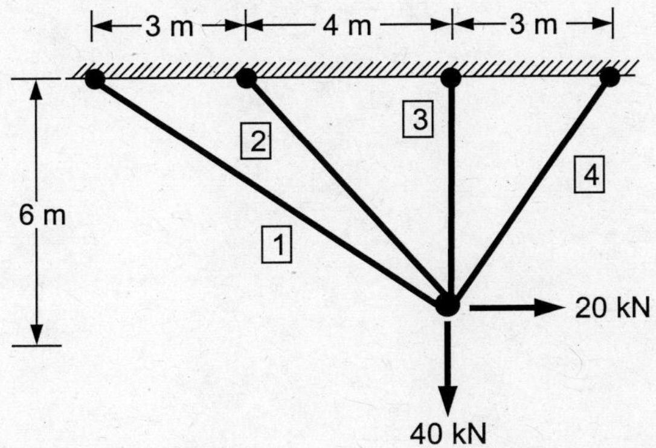

9.9 Determine the support reactions, the member forces, and the joint displacements of the frame shown below using the Matrix displacement method. Assume a constant *EI* for all members.

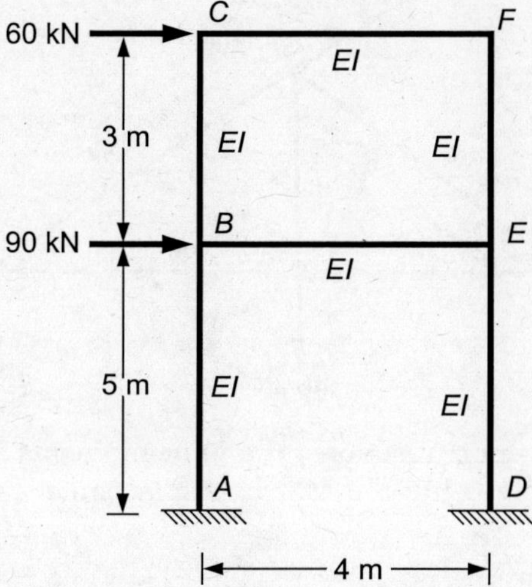

9.10 Determine the horizontal displacement and the rotation at joint *B* of the frame shown below using the Matrix displacement method.

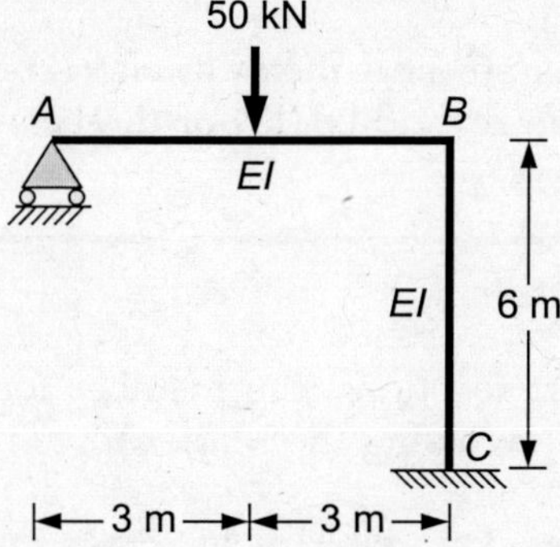

Direct Stiffness Matrix Method

10.1 GENERAL

Direct stiffness matrix method is similar to the *matrix-displacement method* (stiffness method). The basic difference between these two methods is the procedure adopted to derive the total stiffness matrix of the structure. In the *matrix displacement method* as discussed in the preceding chapter, the total stiffness matrix $[K]$ of a structure is derived from the assembled element stiffness matrix $[k]$ as follows:

$$[K] = [a]^T [k][a] \tag{10.1}$$

The *matrix displacement method* is well suited for small and simple structures. However, the size of the matrix $[k]$ would be very large when the structure to be analyzed is large and complex. In such a case, the computation of $[a]$ is time-consuming and tedious. Therefore, the matrix displacement method is practically not computationally efficient for large and complex structures.

In the *direct stiffness matrix* method, the total stiffness matrix $[K]$ is computed for each element and later, all these matrices are assembled to get the final stiffness matrix of the structure. Accordingly, each member (or element) in the direct stiffness matrix method is called as member-structure. Both the matrix displacement method and the direct stiffness matrix method are intended to establish the same total structure stiffness matrix $[K]$ relating the nodal forces and nodal displacements of a particular structure. As discussed in the preceding chapter, the nodal forces and nodal displacements are related as follows:

$$\begin{Bmatrix} R \\ X \end{Bmatrix} = \begin{bmatrix} K_{RR} & K_{RX} \\ K_{XR} & K_{XX} \end{bmatrix} \begin{Bmatrix} r_R \\ r_X \end{Bmatrix} \tag{10.2a}$$

$$\text{where, } [K] = \begin{bmatrix} K_{RR} & K_{RX} \\ K_{XR} & K_{XX} \end{bmatrix} \tag{10.2b}$$

In Eqn. (10.2b), the total stiffness matrix $[K]$ consider effects of unknown nodal displacements $\{r_X\}$ and the prescribed (known) nodal displacements at the supports $\{r_R\}$ separately.

There is a fundamental difference between the element stiffness matrix [k] and the total stiffness matrix, [K]. The element stiffness matrix [k] is computed at the local level, i.e., the matrix components are dependent on the orientation of elements. On the other hand, the total stiffness matrix [K] is computed at the global level. It is, therefore, required to establish a relationship between the local coordinate (element level) system and the global coordinate structure level) system to convert the element stiffness matrix [k] to the total stiffness matrix [K] of a particular member-structure.

10.2   COORDINATE TRANSFORMATION MATRIX

Consider a plane vector V as shown in Figure 10.1. Two orthogonal coordinate systems, namely, XY and xy are considered. Let the angle of rotation between X- and x-axes be θ. The components of vector V in the XY coordinate system are V_X and V_Y, whereas these components in the xy coordinate system are V_x and V_y as shown in the figure.

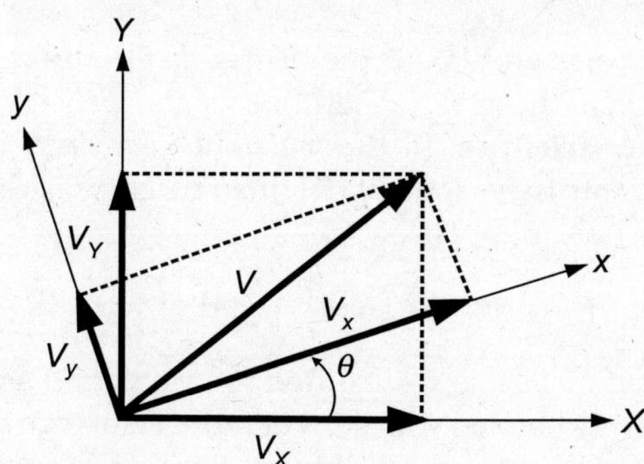

FIGURE 10.1   A plane vector in two orthogonal coordinate systems.

The vector components V_x and V_y can be expressed in terms of V_X and V_Y as follows:

$$\begin{Bmatrix} V_x \\ V_y \end{Bmatrix} = \begin{bmatrix} \cos\theta & \sin\theta \\ -\sin\theta & \cos\theta \end{bmatrix} \begin{Bmatrix} V_X \\ V_Y \end{Bmatrix} \tag{10.3}$$

or

$$\{V\} = [\theta]\{\overline{V}\} \tag{10.4}$$

in which $[\theta]$ = Rotational matrix between the XY system and the xy coordinate system.
Similarly, the components V_X and V_Y can be expressed in terms of V_x and V_y as follows:

$$\begin{Bmatrix} V_X \\ V_Y \end{Bmatrix} = \begin{bmatrix} \cos\theta & -\sin\theta \\ \sin\theta & \cos\theta \end{bmatrix} \begin{Bmatrix} V_x \\ V_y \end{Bmatrix} \tag{10.5}$$

or

$$\{\overline{V}\} = [\theta]^{-1}\{V\} \tag{10.6}$$

It can be seen that $[\theta]^T = [\theta]^{-1}$.

Hence, the rotational matrix $[\theta]$ is *orthogonal*. Thus, $\{\overline{V}\} = [\theta]^T \{V\}$ $\qquad$ (10.7)

Now consider an element *a*, shown in Figure 10.2, having two force (or displacement) components at both ends *i* and *j*. These end forces (or displacements) are shown in both local (*xy*) and global (*XY*) coordinate systems.

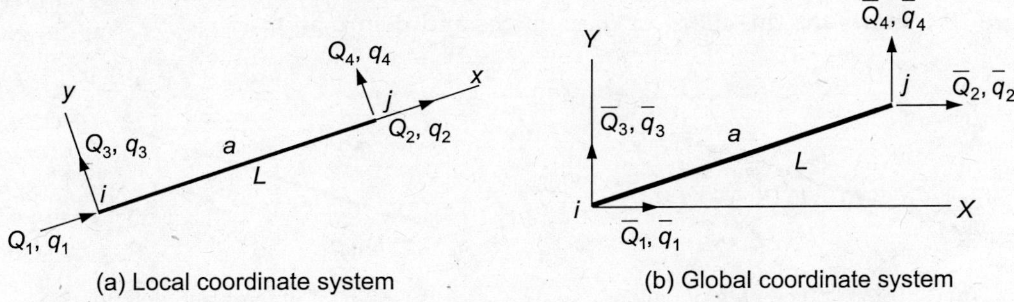

(a) Local coordinate system   (b) Global coordinate system

FIGURE 10.2   Element forces and deformations in local and global coordinate systems.

Assuming the angle of rotation between the global *XY* and the local *xy* systems is θ, the force (or displacement) components in the local coordinate system can be obtained as follows:

$$\begin{Bmatrix} Q_1 \\ Q_3 \end{Bmatrix} = \begin{bmatrix} \cos\theta & \sin\theta \\ -\sin\theta & \cos\theta \end{bmatrix} \begin{Bmatrix} \overline{Q}_1 \\ \overline{Q}_3 \end{Bmatrix} \tag{10.8}$$

and

$$\begin{Bmatrix} Q_2 \\ Q_4 \end{Bmatrix} = \begin{bmatrix} \cos\theta & \sin\theta \\ -\sin\theta & \cos\theta \end{bmatrix} \begin{Bmatrix} \overline{Q}_2 \\ \overline{Q}_4 \end{Bmatrix} \tag{10.9}$$

If the element has the end moments (or rotations) as shown in Figure 10.3, these quantities would be the same in both local and global coordinate systems, the axis about which the end moments (or rotations) remains the same in both systems. Mathematically,

$$\begin{Bmatrix} Q_5 \\ Q_6 \end{Bmatrix} = \begin{bmatrix} 1 & 0 \\ 0 & 1 \end{bmatrix} \begin{Bmatrix} \overline{Q}_5 \\ \overline{Q}_6 \end{Bmatrix} \tag{10.10}$$

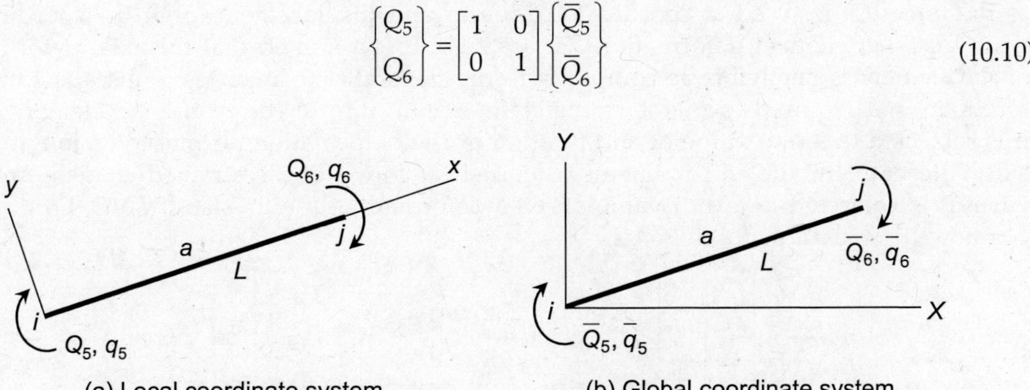

(a) Local coordinate system   (b) Global coordinate system

FIGURE 10.3   End moments and rotations in local and global coordinate systems.

10.3   ELEMENT STIFFNESS MATRIX (LOCAL COORDINATE SYSTEM)

Consider a beam element a of length, L, cross-section, A and constant flexural rigidity (EI). At each end, three forces/deformations (i.e., axial, shear, and moment) are possible. Figure 10.4 shows the directions of these forces and deformations.

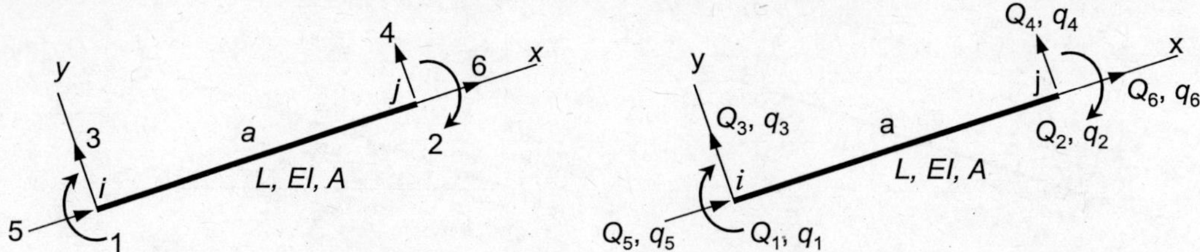

(a) Directions of force/deformations   (b) Element of force and deformations

FIGURE 10.4   Generalized elements forces.

The generalized force matrix $\{Q\}$ and deformation matrix $\{q\}$ can be related as follows:

$$\{Q\} = [k]\{q\}$$

or

$$
\begin{Bmatrix} Q_1 \\ Q_2 \\ Q_3 \\ Q_4 \\ Q_6 \\ Q_7 \end{Bmatrix} =
\begin{bmatrix}
k_{11} & k_{12} & k_{13} & k_{14} & k_{15} & k_{16} \\
k_{21} & k_{22} & k_{23} & k_{24} & k_{25} & k_{26} \\
k_{31} & k_{32} & k_{33} & k_{34} & k_{35} & k_{36} \\
k_{41} & k_{42} & k_{43} & k_{44} & k_{45} & k_{46} \\
k_{51} & k_{52} & k_{53} & k_{54} & k_{55} & k_{56} \\
k_{61} & k_{62} & k_{63} & k_{64} & k_{65} & k_{66}
\end{bmatrix}
\begin{Bmatrix} q_1 \\ q_2 \\ q_3 \\ q_4 \\ q_6 \\ q_7 \end{Bmatrix}
\tag{10.11}
$$

where, $\{q\}$ is the element stiffness matrix with reference to the local orthogonal coordinate x and y axes. x-axis is taken along the longitudinal centroidal axis of the element, y-axis is normal to the longitudinal axis of the element. The stiffness coefficient $k_{\alpha\beta}$ is defined as the force (Q) induced at coordinate α due to unit displacement (q) at the coordinate β while all other element deformations (q) are zero. It is assumed that the axial deformation of the element is negligible as compared to the flexural deformations. Thus, the axial end forces are not required to maintain the static equilibrium of the restrained element when it is subjected to a unit value of end rotation or the deformation normal to its longitudinal axis. Alternatively, the end moments or end shear forces of a restrained element are zero when it is subjected to a unit value of axial deformation at either end. Thus, Eqn. (10.11) can now be written as follows:

$$
\begin{Bmatrix} Q_1 \\ Q_2 \\ Q_3 \\ Q_4 \\ Q_5 \\ Q_6 \end{Bmatrix} =
\begin{bmatrix}
k_{11} & k_{12} & k_{13} & k_{14} & 0 & 0 \\
k_{21} & k_{22} & k_{23} & k_{24} & 0 & 0 \\
k_{31} & k_{32} & k_{33} & k_{34} & 0 & 0 \\
k_{41} & k_{42} & k_{43} & k_{44} & 0 & 0 \\
0 & 0 & 0 & 0 & k_{55} & k_{56} \\
0 & 0 & 0 & 0 & k_{65} & k_{66}
\end{bmatrix}
\begin{Bmatrix} q_1 \\ q_2 \\ q_3 \\ q_4 \\ q_5 \\ q_6 \end{Bmatrix}
\tag{10.12}
$$

The element stiffness matrix can be generated by applying a unit displacement at one end and determining the end forces induced in the restrained element. Using the procedure discussed in the previous chapter, all components of the element stiffness matrix can be derived as shown in Figure 10.5.

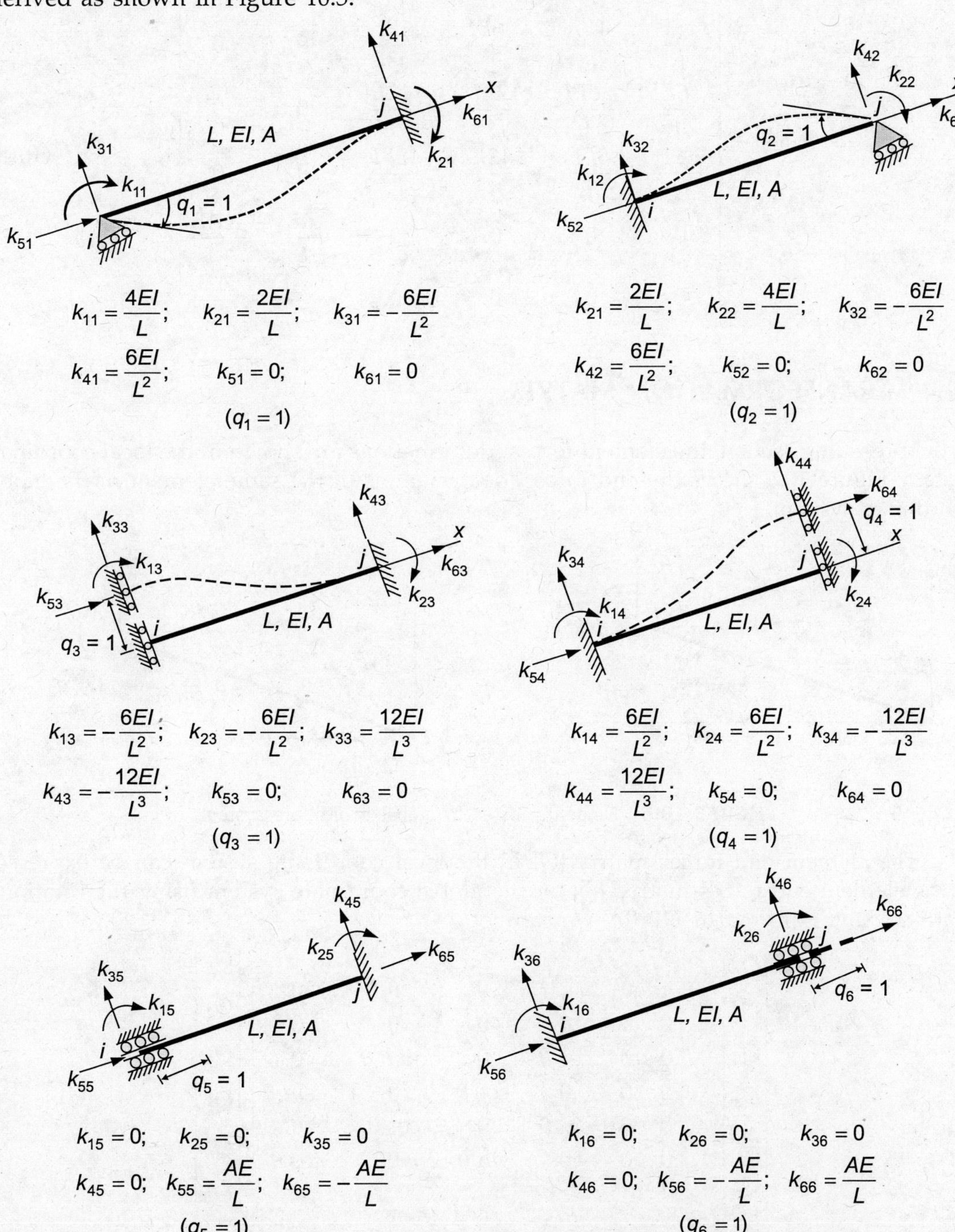

$$k_{11} = \frac{4EI}{L}; \quad k_{21} = \frac{2EI}{L}; \quad k_{31} = -\frac{6EI}{L^2}$$

$$k_{41} = \frac{6EI}{L^2}; \quad k_{51} = 0; \quad k_{61} = 0$$

$$(q_1 = 1)$$

$$k_{21} = \frac{2EI}{L}; \quad k_{22} = \frac{4EI}{L}; \quad k_{32} = -\frac{6EI}{L^2}$$

$$k_{42} = \frac{6EI}{L^2}; \quad k_{52} = 0; \quad k_{62} = 0$$

$$(q_2 = 1)$$

$$k_{13} = -\frac{6EI}{L^2}; \quad k_{23} = -\frac{6EI}{L^2}; \quad k_{33} = \frac{12EI}{L^3}$$

$$k_{43} = -\frac{12EI}{L^3}; \quad k_{53} = 0; \quad k_{63} = 0$$

$$(q_3 = 1)$$

$$k_{14} = \frac{6EI}{L^2}; \quad k_{24} = \frac{6EI}{L^2}; \quad k_{34} = -\frac{12EI}{L^3}$$

$$k_{44} = \frac{12EI}{L^3}; \quad k_{54} = 0; \quad k_{64} = 0$$

$$(q_4 = 1)$$

$$k_{15} = 0; \quad k_{25} = 0; \quad k_{35} = 0$$

$$k_{45} = 0; \quad k_{55} = \frac{AE}{L}; \quad k_{65} = -\frac{AE}{L}$$

$$(q_5 = 1)$$

$$k_{16} = 0; \quad k_{26} = 0; \quad k_{36} = 0$$

$$k_{46} = 0; \quad k_{56} = -\frac{AE}{L}; \quad k_{66} = \frac{AE}{L}$$

$$(q_6 = 1)$$

FIGURE 10.5 Stiffness coefficients of frame member.

The element stiffness matrix in the local coordinate system is given as follows:

$$[k] = \begin{bmatrix} \dfrac{4EI}{L} & \dfrac{2EI}{L} & -\dfrac{6EI}{L^2} & \dfrac{6EI}{L^2} & 0 & 0 \\[2ex] \dfrac{2EI}{L} & \dfrac{4EI}{L} & -\dfrac{6EI}{L^2} & \dfrac{6EI}{L^2} & 0 & 0 \\[2ex] -\dfrac{6EI}{L^2} & -\dfrac{6EI}{L^2} & \dfrac{12EI}{L^3} & -\dfrac{12EI}{L^3} & 0 & 0 \\[2ex] \dfrac{6EI}{L^2} & \dfrac{6EI}{L^2} & -\dfrac{12EI}{L^3} & \dfrac{12EI}{L^3} & 0 & 0 \\[2ex] 0 & 0 & 0 & 0 & \dfrac{AE}{L} & -\dfrac{AE}{L} \\[2ex] 0 & 0 & 0 & 0 & -\dfrac{AE}{L} & \dfrac{AE}{L} \end{bmatrix} \tag{10.13}$$

10.4 TRANSFORMATION MATRIX

In the preceding section, the element forces/deformations are shown in the local coordinate system. Figure 10.6 shows the end forces/deformations of the same element in the global coordinate system.

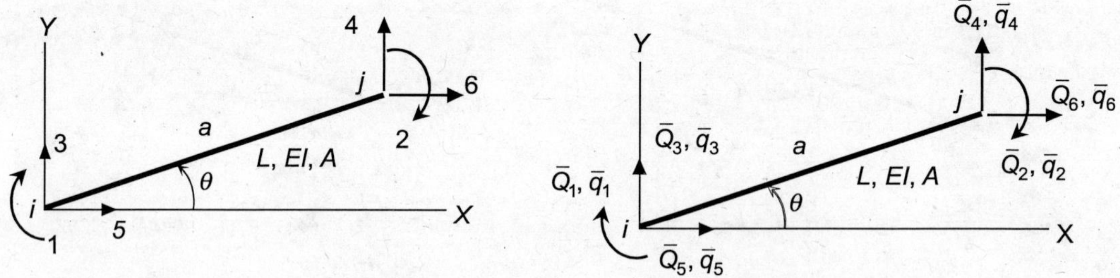

FIGURE 10.6 Elements forces in global coordinate system.

The element end forces matrix $\{Q\}$ in the local coordinate system can be expressed in the element end forces matrix $\{\bar{Q}\}$ in the global coordinate system using the rotational matrix. Using Eqns. (10.8)–(10.10), we can get

$$\begin{Bmatrix} Q_1 \\ Q_2 \\ Q_3 \\ Q_4 \\ Q_5 \\ Q_6 \end{Bmatrix} = \begin{bmatrix} 1 & 0 & 0 & 0 & 0 & 0 \\ 0 & 1 & 0 & 0 & 0 & 0 \\ 0 & 0 & \cos\theta & 0 & -\sin\theta & 0 \\ 0 & 0 & 0 & \cos\theta & 0 & -\sin\theta \\ 0 & 0 & \sin\theta & 0 & \cos\theta & 0 \\ 0 & 0 & 0 & \sin\theta & 0 & \cos\theta \end{bmatrix} \begin{Bmatrix} \bar{Q}_1 \\ \bar{Q}_2 \\ \bar{Q}_3 \\ \bar{Q}_4 \\ \bar{Q}_5 \\ \bar{Q}_6 \end{Bmatrix} \tag{10.14}$$

$$(\textit{Local}) \qquad\qquad\qquad\qquad\qquad\qquad (\textit{Global})$$

$${Q} = [a]{\bar{Q}} \tag{10.15}$$

where, $[a]$ is called as *transformation matrix* or *compatibility matrix* for a frame member, which transforms the element end forces/deformations in the local coordinates to the global coordinates.

$$[a] = \begin{bmatrix} 1 & 0 & 0 & 0 & 0 & 0 \\ 0 & 1 & 0 & 0 & 0 & 0 \\ 0 & 0 & \cos\theta & 0 & -\sin\theta & 0 \\ 0 & 0 & 0 & \cos\theta & 0 & -\sin\theta \\ 0 & 0 & \sin\theta & 0 & \cos\theta & 0 \\ 0 & 0 & 0 & \sin\theta & 0 & \cos\theta \end{bmatrix} \tag{10.16}$$

In terms of element end deformations, Eqn. (10.15) can be written as follows:

$${q} = [a]{\bar{q}} \tag{10.17}$$

It can be the matrix $[a]$ is a square and orthogonal matrix, i.e., $[a]^{-1} = [a]^{T}$. Hence, the element end forces/deformations can be transformed from the global to the local coordinate systems as follows:

$${\bar{Q}} = [a]^{T}{Q} \tag{10.18}$$

$${\bar{q}} = [a]^{T}{q} \tag{10.19}$$

where, the matrices ${q}$ and ${\bar{q}}$ are the element end deformations in the local and global coordinate systems, respectively.

It can be seen that the element end force matrix ${Q}$ in the local coordinate system is obtained by multiplying the transformation matrix $[a]$ to the element force matrix ${\bar{Q}}$ in the global coordinate system. On the other hand, the element end deformation matrix ${\bar{q}}$ in the global coordinate system is obtained by multiplying the transpose of the transformation matrix $[a]^{T}$ to the element end deformation matrix ${q}$ in the local coordinate system. This is known as *principle of contragredience*.

i.e.,
$${Q} = [a]{\bar{Q}} \; ; \; {\bar{q}} = [a]^{T}{q}$$

Thus, the *principle of contragredience* states that if two statically equivalent force systems (i.e., ${Q}$ and ${\bar{Q}}$) are connected by a transformation matrix ($[a]$), then their displacement systems (i.e., ${\bar{q}}$ and ${q}$) are related in the reverse order by the transformation matrix ($[a]^{T}$).

10.5 ELEMENT STIFFNESS MATRIX (GLOBAL COORDINATE SYSTEM)

The element stiffness matrix $[k]$ relating the element end forces and the end deformations in the local coordinate system is given by

$${Q} = [k]{q} \tag{10.20}$$

Similarly, the element stiffness matrix $[\bar{k}]$ relating the element end forces and the end deformations in the global coordinate system is written as

$$\{\bar{Q}\} = [\bar{k}]\{\bar{q}\} \tag{10.21}$$

Using Eqns. (10.15) and (10.17) in Eqn. (10.20),

$$[a]\{\bar{Q}\} = [k][a]\{\bar{q}\} \tag{10.22}$$

or

$$\{\bar{Q}\} = [a]^{-1}[k][a]\{\bar{q}\} \tag{10.23}$$

Since, $[a]^{-1} = [a]^T$

$$\{\bar{Q}\} = [a]^T[k][a]\{\bar{q}\} \tag{10.24}$$

Comparing Eqns. (10.21) and (10.23), the element stiffness matrix in the global coordinate system can be obtained as follows:

$$\left[\bar{k}\right] = [a]^T[k][a] \tag{10.25}$$

The global stiffness matrices for truss, beam, and frame members are presented in the following sections:

10.5.1 Global Stiffness Matrix of Truss Member

Figure 10.2 shows the end forces and deformations of a truss member in the local and global coordinate systems. Since both ends of a truss member are pinned, the end moments/rotations are irrelevant and should not be considered in the derivation of the stiffness matrix of the truss member. The element stiffness matrix in the global coordinate system is derived from the element stiffness matrix in the local coordinate system using the transformation matrix. We would consider two possible end conditions between the local coordinate and global coordinate systems.

Parallel supports

Figure 10.7 shows a truss member with all applicable nodes and element forces/deformations. Let's assume the member is supported on two supports whose line of actions are parallel. Thus, the same global coordinate systems are taken at both ends of the member in which the angle of rotation (θ) between the global and local coordinate axes are the same at both ends.

Element stiffness matrix in the local coordinate system of the truss member is given by

$$[k] = \begin{bmatrix} \dfrac{AE}{L} & -\dfrac{AE}{L} & 0 & 0 \\ -\dfrac{AE}{L} & \dfrac{AE}{L} & 0 & 0 \\ 0 & 0 & 0 & 0 \\ 0 & 0 & 0 & 0 \end{bmatrix} = \left(\dfrac{AE}{L}\right) \begin{array}{c} \begin{array}{cccc} 1 & 2 & 3 & 4 \end{array} \\ \begin{bmatrix} 1 & -1 & 0 & 0 \\ -1 & 1 & 0 & 0 \\ 0 & 0 & 0 & 0 \\ 0 & 0 & 0 & 0 \end{bmatrix} \begin{array}{c} 1 \\ 2 \\ 3 \\ 4 \end{array} \end{array} \tag{10.26}$$

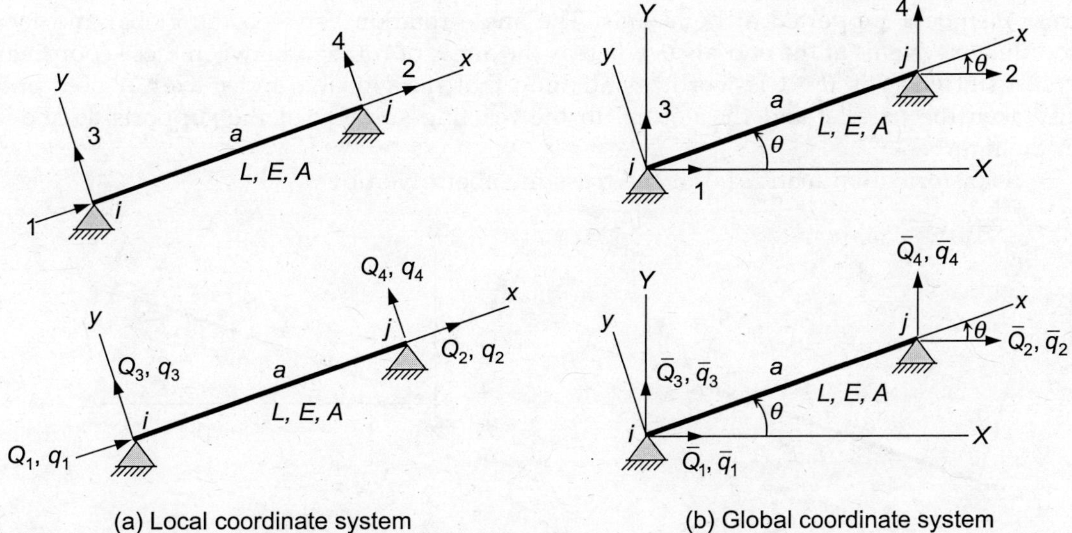

(a) Local coordinate system　　　　　(b) Global coordinate system

FIGURE 10.7　Element forces/deformations of truss member in local and global coordinate systems for parallel supports.

Using Eqns. (10.8) and (10.9), the transformation matrix $[a]$ of the truss member can be obtained as follows:

$$[a] = \begin{matrix} & 1 & 2 & 3 & 4 & \\ & \begin{bmatrix} \cos\theta & 0 & \sin\theta & 0 \\ 0 & \cos\theta & 0 & \sin\theta \\ -\sin\theta & 0 & \cos\theta & 0 \\ 0 & -\sin\theta & 0 & \cos\theta \end{bmatrix} & & & & \begin{matrix} 1 \\ 2 \\ 3 \\ 4 \end{matrix} \end{matrix} \qquad (10.27)$$

The global stiffness matrix $[\bar{k}]$ is given by $\left[\bar{k}\right] = [a]^T[k][a]$. Thus,

$$[\bar{k}] = \left(\frac{AE}{L}\right) \begin{matrix} & 1 & 2 & 3 & 4 & \\ & \begin{bmatrix} C_x^2 & -C_x^2 & C_xC_y & -C_xC_y \\ -C_x^2 & C_x^2 & -C_xC_y & C_xC_y \\ C_xC_y & -C_xC_y & C_y^2 & -C_y^2 \\ -C_xC_y & C_xC_y & -C_y^2 & C_y^2 \end{bmatrix} & & & & \begin{matrix} 1 \\ 2 \\ 3 \\ 4 \end{matrix} \end{matrix} \qquad (10.28)$$

where, $C_x = \cos\theta$ and $C_y = \sin\theta$. Similar to the local stiffness matrix, the global stiffness matrix is also symmetric.

Inclined supports

Now consider a truss member supported on two supports whose lines of actions are not parallel. This case arises when one of the supports is inclined. Figure 10.8 shows the truss members supported at both ends. The angle rotation between the global and local coordinate systems at the end i is θ, whereas the angle of rotation between these coordinate systems at end j is θ'. It is worth mentioning that the global x and y axes at both ends are taken the parallel and the normal to the reacting surfaces of the supports as shown in the figure.

Transformation matrix $[a]$ of the truss member given by

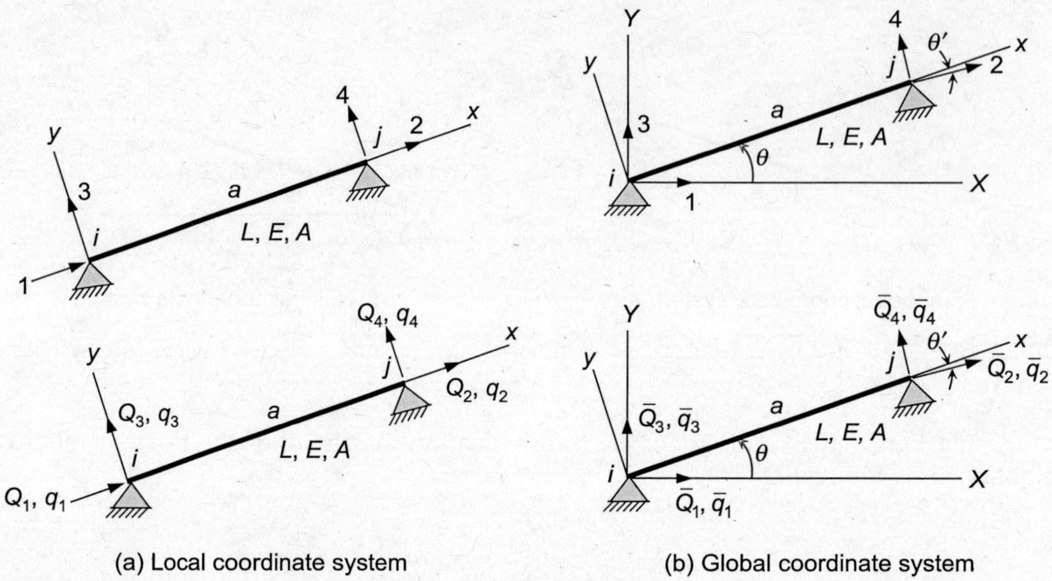

(a) Local coordinate system   (b) Global coordinate system

FIGURE 10.8   Element forces/deformations of truss member in local and global coordinate systems for inclined supports.

$$[a] = \begin{bmatrix} \cos\theta & 0 & \sin\theta & 0 \\ 0 & \cos\theta' & 0 & \sin\theta' \\ -\sin\theta & 0 & \cos\theta & 0 \\ 0 & -\sin\theta' & 0 & \cos\theta' \end{bmatrix} \begin{matrix} 1 \\ 2 \\ 3 \\ 4 \end{matrix} \tag{10.29}$$

Element stiffness matrix in the local coordinate system remains same as follows:

$$[k] = \begin{bmatrix} \dfrac{AE}{L} & -\dfrac{AE}{L} & 0 & 0 \\ -\dfrac{AE}{L} & \dfrac{AE}{L} & 0 & 0 \\ 0 & 0 & 0 & 0 \\ 0 & 0 & 0 & 0 \end{bmatrix} = \left(\dfrac{AE}{L}\right)\begin{bmatrix} 1 & -1 & 0 & 0 \\ -1 & 1 & 0 & 0 \\ 0 & 0 & 0 & 0 \\ 0 & 0 & 0 & 0 \end{bmatrix}\begin{matrix} 1 \\ 2 \\ 3 \\ 4 \end{matrix}$$

The global stiffness matrix $[\bar{k}]$ is given by $\left[\bar{k}\right] = [a]^T[k][a]$. Thus,

$$[\bar{k}] = \left(\frac{AE}{L}\right)\begin{matrix} & 1 & 2 & 3 & 4 \\ \begin{bmatrix} C_x^2 & -C_xC_x' & C_xC_y & -C_xC_y' \\ -C_xC_x' & C_x'^2 & -C_x'C_y & C_x'C_y' \\ C_xC_y & -C_x'C_y & C_y^2 & -C_yC_y' \\ -C_xC_y' & C_x'C_y' & -C_yC_y' & C_y'^2 \end{bmatrix} & \begin{matrix} 1 \\ 2 \\ 3 \\ 4 \end{matrix} \end{matrix} \qquad (10.30)$$

where, $C_x = \cos\theta$, $C_y = \sin\theta$, $C_x' = \cos\theta'$ and $C_y' = \sin\theta'$.

10.5.2 Global Stiffness Matrix of Frame Member

The element stiffness matrix of the frame in the local coordinate system is given by Eqn. (10.13) and the transformation matrix is given by Eqn. (10.16). Thus, $\left[\bar{k}\right] = [a]^T[k][a]$.

Accordingly, the global stiffness matrix for a frame member is given in Eqn. (10.31) on next page.

If the frame member is considered to be axially inextensible, then the global stiffness matrix of the frame member is obtained by setting the value of $AE/L = 0$.

$$[\bar{k}] = \begin{bmatrix} \dfrac{4EI}{L} & & & & & Sym. \\[2mm] \dfrac{2EI}{L} & \dfrac{4EI}{L} & & & & \\[2mm] -\dfrac{6EI}{L^2}C_x & -\dfrac{6EI}{L^2}C_x & \dfrac{12EI}{L^3}C_x^2 & & & \\[2mm] \dfrac{6EI}{L^2}C_x & \dfrac{6EI}{L^2}C_x & -\dfrac{12EI}{L^3}C_x^2 & \dfrac{12EI}{L^3}C_x^2 & & \\[2mm] \dfrac{6EI}{L^2}C_y & \dfrac{6EI}{L^2}C_y & -\dfrac{12EI}{L^3}C_xC_y & \dfrac{12EI}{L^3}C_xC_y & \dfrac{12EI}{L^3}C_y^2 & \\[2mm] -\dfrac{6EI}{L^2}C_y & -\dfrac{6EI}{L^2}C_y & \dfrac{12EI}{L^3}C_xC_y & -\dfrac{12EI}{L^3}C_xC_y & -\dfrac{12EI}{L^3}C_y^2 & \dfrac{12EI}{L^3}C_y^2 \end{bmatrix}$$

$$(10.32)$$

10.5.3 Global Stiffness Matrix of Beam Member

A beam member is a special case of the frame member in which the effect of axial force/deformation is considered to be negligible. Thus, the global stiffness matrix of a beam member is the same as the global stiffness matrix of a frame member considering the axial forces/deformations effects to be negligible. Hence, Eqn. (10.30) also represent the global stiffness matrix for a beam member.

$$[\bar{k}] = \begin{bmatrix} \dfrac{4EI}{L} & & & & & Sym. \\[2ex] \dfrac{2EI}{L} & \dfrac{4EI}{L} & & & & \\[2ex] -\dfrac{6EI}{L^2}C_x & -\dfrac{6EI}{L^2}C_x & \left(\dfrac{12EI}{L^3}C_x^{\,2}+\dfrac{AE}{L}C_y^{\,2}\right) & & & \\[2ex] \dfrac{6EI}{L^2}C_x & \dfrac{6EI}{L^2}C_x & -\left(\dfrac{12EI}{L^3}C_x^{\,2}+\dfrac{AE}{L}C_y^{\,2}\right) & \left(\dfrac{12EI}{L^3}C_x^{\,2}+\dfrac{AE}{L}C_y^{\,2}\right) & & \\[2ex] \dfrac{6EI}{L^2}C_y & \dfrac{6EI}{L^2}C_y & \left(-\dfrac{12EI}{L^3}+\dfrac{AE}{L}\right)C_xC_y & \left(\dfrac{12EI}{L^3}-\dfrac{AE}{L}\right)C_xC_y & \left(\dfrac{12EI}{L^3}C_y^{\,2}+\dfrac{AE}{L}C_x^{\,2}\right) & \\[2ex] -\dfrac{6EI}{L^2}C_y & -\dfrac{6EI}{L^2}C_y & \left(\dfrac{12EI}{L^3}-\dfrac{AE}{L}\right)C_xC_y & \left(-\dfrac{12EI}{L^3}+\dfrac{AE}{L}\right)C_xC_y & -\left(\dfrac{12EI}{L^3}C_y^{\,2}+\dfrac{AE}{L}C_x^{\,2}\right) & \left(\dfrac{12EI}{L^3}C_y^{\,2}+\dfrac{AE}{L}C_x^{\,2}\right) \end{bmatrix} \qquad (10.31)$$

10.6   STRUCTURE STIFFNESS MATRIX

The *structure stiffness matrix* relates the nodal forces to the corresponding nodal displacements of a structure. Since the nodal force and nodal displacement matrices are established in the global coordinate system, the *structure stiffness matrix* is formed by assembling the global stiffness matrices of all member-structures. The stiffness components of various member-structures meeting at a common node are taken into account in the assembling process. The relationship between the nodal forces (R) and the nodal deformations (r) is expressed as follows:

$$\{R\} = [K]\{r\} \tag{10.33}$$

$$\begin{Bmatrix} R_1 \\ R_2 \\ \dots \\ \dots \\ R_n \end{Bmatrix} = \begin{bmatrix} K_{11} & K_{12} & \dots & \dots & K_{1n} \\ K_{21} & K_{22} & \dots & \dots & K_{2n} \\ \dots & \dots & \dots & \dots & \dots \\ \dots & \dots & \dots & \dots & \dots \\ K_{n1} & K_{n2} & \dots & \dots & K_{nn} \end{bmatrix} \begin{Bmatrix} r_1 \\ r_2 \\ \dots \\ \dots \\ r_n \end{Bmatrix} \tag{10.34}$$

where $[K]$ = *Structure stiffness matrix* or *total stiffness matrix*. The stiffness coefficient $K_{\alpha\beta}$ is defined as the force-induced at coordinate α due to unit displacement at the coordinate β while all other element deformations are zero. During the assembling process, the stiffness coefficients of the same subscripts from the members connected at a common node are added together. Since all possible nodal forces are considered in the direct stiffness matrix method, the nodal force matrix $\{R\}$ consists of the applied external forces as well as the reaction forces at the supports in the global coordinate system. Similarly, the nodal displacement matrix $\{r\}$ contains the unknown member end displacements as well as the prescribed (known) end deformations at the supports. As a result, all unknown reaction forces/member deformations of structure can be obtained directly by solving the Eqn. (10.34). The member forces/deformations in the local coordinates can then be obtained by using the transformation matrix.

10.7   STEP-BY-STEP PROCEDURE OF DIRECT STIFFNESS MATRIX METHOD

The general step-by-step procedure of analysis of structures using the direct stiffness matrix method is presented in the following sections.

1. Define the nodes and elements of the structure. The interconnected points and the supporting points are considered as the nodes.
2. Identify and number all possible nodal forces/deformations with reference to a set of global coordinate axes. All nodes corresponding to the unknown displacements are numbered first sequentially and then, all nodes with prescribed (known) displacements are numbered. Establish the nodal force matrix $\{R\}$ and nodal displacement matrix $\{r\}$.
3. Identify the element end coordinates and determine the element stiffness matrix $[k]$ in the local coordinate system for each element.

4. Determine the transformation matrix $[a]$ based on the angle of rotation between the global coordinate axes and the local coordinate axes.

5. Establish the element stiffness matrix $\left[\overline{k}\right]$ in the global coordinate system for all elements using the expression $\left[\overline{k}\right] = [a]^T [k][a]$.

6. Obtain the structure stiffness matrix $[K]$ by assembling the element stiffness matrix $\left[\overline{k}\right]$ in the global coordinate system for all elements.

7. Write the relationship between the nodal force matrix $[R]$ and the nodal displacement matrix $[r]$ using the structure stiffness matrix $[K]$. Divide the nodal displacements into two groups: one is the unknown nodal displacement matrix $\{r_R\}$ corresponding to the known nodal forces $[R]$ and the other is the known nodal displacement matrix $[r_X]$ corresponding to the unknown nodal forces $[X]$.

$$\left\{\frac{R}{X}\right\} = \left[\begin{array}{c|c} K_{RR} & K_{RX} \\ \hline K_{XR} & K_{XX} \end{array}\right] \left\{\frac{r_R}{r_X}\right\}$$

8. Determine the unknown displacement matrix $\{r_R\}$ and the unknown nodal force matrix $\{X\}$ by solving the above matrix.

9. Obtain the member end forces in the global coordinate system as a linear function of the nodal displacements and then, these forces are transformed in the local coordinate system using the transformation matrix.

$$\{Q\} = [k][a]\{\overline{q}\}$$

10.8 ANALYSIS OF TRUSSES WITH MEMBER DEFORMATION DUE TO OTHER SOURCES

In trusses, members may undergo deformations due to change in temperature or fabrication errors (lack of fit) excluding the effects of externally applied loads. These deformations induce additional forces in the members. This effect can be computed separately and added to the member forces generated due to externally applied loads using the principle of superposition. Primary, the additional deformation may result due to thermal change or fabrication errors.

Effect of thermal changes

Now consider a portion of a truss shown in Figure 10.9. Assume that the diagonal member AB is subjected to an action that would increase the member length. If the member would be assumed to deform freely, the increase in the length of this member would be BB' which is represented by Δ. Since the member deformation is restrained at both ends, the internal compressive force will be induced in the member. This is similar to the force induced by applying equal and opposite deformation at the end of member AB. The node numbering of forces and deformations at both ends of the member in the local coordinator and the global coordinate systems are shown in the figure. Accordingly, $q_2 = \Delta$.

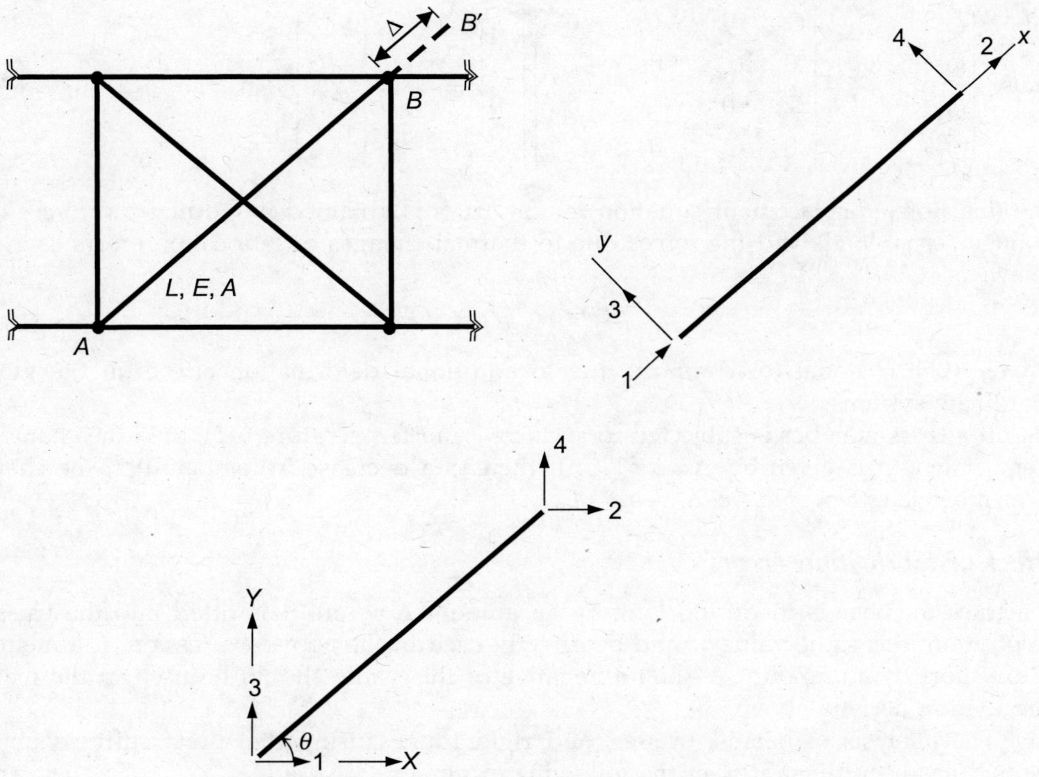

FIGURE 10.9 Truss member subjected to temperature change.

Since $\{Q\} = [k][\{q\}]$, the member forces in the local coordinate systems are given by

$$\begin{Bmatrix} Q_1 \\ Q_2 \\ Q_3 \\ Q_4 \end{Bmatrix} = \frac{AE}{L} \begin{bmatrix} 1 & -1 & 0 & 0 \\ -1 & 1 & 0 & 0 \\ 0 & 0 & 0 & 0 \\ 0 & 0 & 0 & 0 \end{bmatrix} \begin{Bmatrix} 0 \\ -\Delta \\ 0 \\ 0 \end{Bmatrix} = \frac{AE\Delta}{L} \begin{Bmatrix} 1 \\ -1 \\ 0 \\ 0 \end{Bmatrix}$$

We know that $\{\overline{Q}\} = [a]^T \{Q\}$, where

$$[a] = \begin{bmatrix} \cos\theta & 0 & \sin\theta & 0 \\ 0 & \cos\theta & 0 & \sin\theta \\ -\sin\theta & 0 & \cos\theta & 0 \\ 0 & -\sin\theta & 0 & \cos\theta \end{bmatrix} = \begin{bmatrix} C_x & 0 & C_y & 0 \\ 0 & C_x & 0 & C_y \\ -C_y & 0 & C_x & 0 \\ 0 & -C_y & 0 & C_x \end{bmatrix}$$

$$[a]^T = \begin{bmatrix} C_x & 0 & -C_y & 0 \\ 0 & C_x & 0 & -C_y \\ C_y & 0 & C_x & 0 \\ 0 & C_y & 0 & C_x \end{bmatrix}$$

Thus,
$$\left\{\begin{array}{c}\overline{Q}_1\\\overline{Q}_2\\\overline{Q}_3\\\overline{Q}_4\end{array}\right\}=\frac{AE\Delta}{L}\begin{bmatrix}C_x & 0 & -C_y & 0\\0 & C_x & 0 & -C_y\\C_y & 0 & C_x & 0\\0 & C_y & 0 & C_x\end{bmatrix}\left\{\begin{array}{c}1\\-1\\0\\0\end{array}\right\}=\frac{AE\Delta}{L}\left\{\begin{array}{c}C_x\\-C_x\\C_y\\-C_y\end{array}\right\}$$

The final force-displacement equation for the truss is obtained by adding the forces due to the external loads and the forces due to thermal changes or fabrication errors as given below:

$$\{R\}=[K]\{r\}+\{\overline{Q}'\} \tag{10.35}$$

where, $\{\overline{Q}'\}$ = nodal force matrix due to additional deformation effects in the global coordinate systems.

If a truss member is subjected to an increase in temperature (δT), then the change in member length is given by $\Delta=\alpha.\delta T.L$. If there is a decrease in temperature, the change in member length would be $\Delta=-\alpha.\delta T.L$.

Effect of fabrication errors

If a truss member is made too long by an amount Δ before it is fitted into the truss at its position, this same value would be directly used in the above expression. If a member is too short by an amount Δ, then a negative of this value should be used in the matrix formulation derived above.

The analysis of beams, trusses, and rigid frames using the direct stiffness matrix method has been illustrated in the following examples.

EXAMPLE 10.1   Determine the bar forces, the joint displacements, and the support reactions of the truss shown in Figure E10.1 using the direct stiffness matrix method. Assume that $AE/L = 1$ for all members.

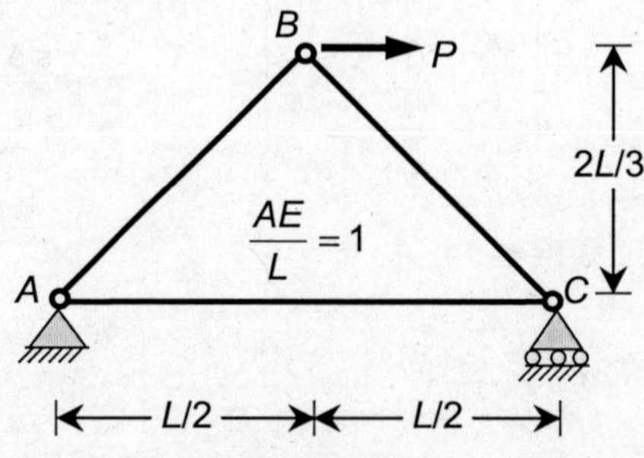

FIGURE E10.1

Solution: The first step is to identify and number the elements and nodes of the structure. Since it is required to determine the joint displacements as well as the support reactions, all nodes have to be considered. As shown in the following figure, all nodes have been numbered in the global ordinate system. The numbering of nodes has been carried out keeping all unknown displacements sequentially as one group followed by all nodes with the prescribed node displacements.

The nodal displacement matrix and the nodal force matrix are as follows:

$$\{r\} = \begin{Bmatrix} r_1 \\ r_2 \\ r_3 \\ r_4 \\ r_5 \\ r_6 \end{Bmatrix} = \begin{Bmatrix} r_1 \\ r_2 \\ r_3 \\ 0 \\ 0 \\ 0 \end{Bmatrix}; \quad \{R\} = \begin{Bmatrix} R_1 \\ R_2 \\ R_3 \\ R_4 \\ R_5 \\ R_6 \end{Bmatrix} = \begin{Bmatrix} 0 \\ P \\ 0 \\ R_4 \\ R_5 \\ R_6 \end{Bmatrix}$$

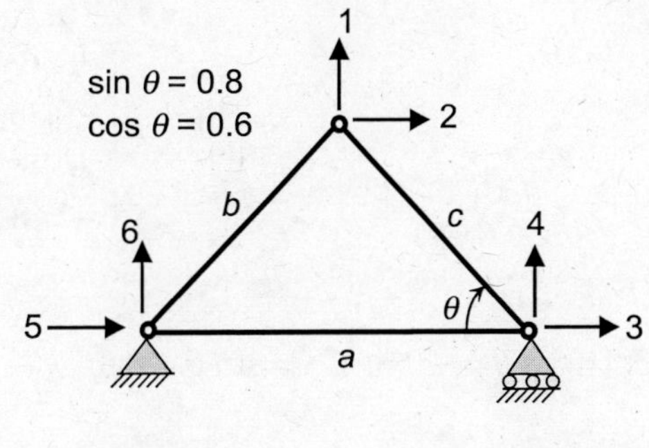

The stiffness matrix of elements in the global coordinate system is then computed using the following equation.

$$[\bar{k}] = \left(\frac{AE}{L}\right) \begin{bmatrix} C_x^2 & -C_x^2 & C_x C_y & -C_x C_y \\ -C_x^2 & C_x^2 & -C_x C_y & C_x C_y \\ C_x C_y & -C_x C_y & C_y^2 & -C_y^2 \\ -C_x C_y & C_x C_y & -C_y^2 & C_y^2 \end{bmatrix}$$

Element a

$$C_x = \cos \theta = 1; \quad C_y = \sin \theta = 0; \quad \frac{AE}{L} = 1$$

6   4

5 →   *a*   → 3

θ = 0

$$[\bar{k}^a] = \begin{matrix} & 5 & 3 & 6 & 4 \\ & \begin{bmatrix} 1 & -1 & 0 & 0 \\ -1 & 1 & 0 & 0 \\ 0 & 0 & 0 & 0 \\ 0 & 0 & 0 & 0 \end{bmatrix} & \begin{matrix} 5 \\ 3 \\ 6 \\ 4 \end{matrix} \end{matrix}$$

Element b

$$C_x = \cos \theta = 0.6; \quad C_y = \sin \theta = 0.8; \quad \frac{AE}{L} = 1$$

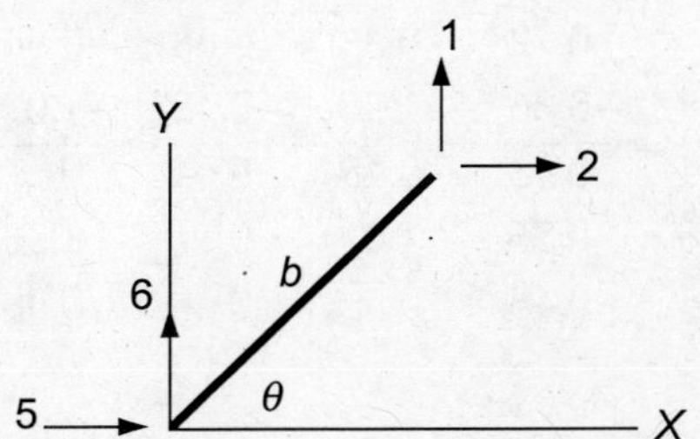

$$
\left[\overline{k}^{b}\right] =
\begin{array}{cccc}
5 & 2 & 6 & 1 \\
\end{array}
$$

$$
\left[\overline{k}^{b}\right] =
\begin{bmatrix}
0.36 & -0.36 & 0.48 & -0.48 \\
-0.36 & 0.36 & -0.48 & 0.48 \\
0.48 & -0.48 & 0.64 & -0.64 \\
-0.48 & 0.48 & -0.64 & 0.64
\end{bmatrix}
\begin{array}{c}5\\2\\6\\1\end{array}
$$

Element c

$$C_x = \cos\theta = -0.6; \quad C_y = \sin\theta = 0.8; \quad \frac{AE}{L} = 1$$

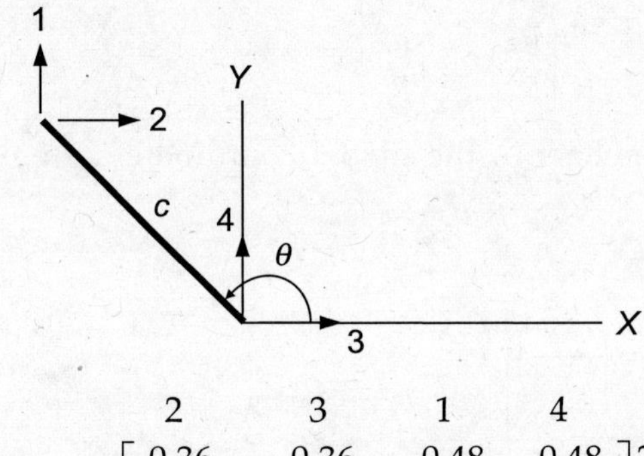

$$
\left[\overline{k}^{c}\right] =
\begin{array}{cccc}
2 & 3 & 1 & 4 \\
\end{array}
$$

$$
\left[\overline{k}^{c}\right] =
\begin{bmatrix}
0.36 & -0.36 & -0.48 & 0.48 \\
-0.36 & 0.36 & 0.48 & -0.48 \\
-0.48 & 0.48 & 0.64 & -0.64 \\
0.48 & -0.48 & -0.64 & 0.64
\end{bmatrix}
\begin{array}{c}2\\3\\1\\4\end{array}
$$

The next step is to determine the structure stiffness matrix by assembling all the element stiffness matrices.

$$
[K] =
\begin{array}{cccccc}
1 & 2 & 3 & 4 & 5 & 6 \\
\end{array}
$$

$$
[K] =
\begin{bmatrix}
1.28 & & & & & sym \\
0 & 0.72 & & & & \\
0.48 & -0.36 & 1.36 & & & \\
-0.64 & 0.48 & -0.48 & 0.64 & & \\
-0.48 & -0.36 & -1.0 & 0 & 1.36 & \\
-0.64 & -0.48 & 0 & 0 & 0.48 & 0.64
\end{bmatrix}
\begin{array}{c}1\\2\\3\\4\\5\\6\end{array}
$$

Since, $\{R\} = [K]\{r\}$

$$
\begin{Bmatrix}
0 \\ P \\ 0 \\ \hline R_4 \\ R_5 \\ R_6
\end{Bmatrix}
=
\begin{bmatrix}
1.28 & 0 & 0.48 & -0.64 & -0.48 & -0.64 \\
0 & 0.72 & -0.36 & 0.48 & -0.36 & -0.48 \\
0.48 & -0.36 & 1.36 & -0.48 & -1.0 & 0 \\
\hline
-0.64 & 0.48 & -0.48 & 0.64 & 0 & 0 \\
-0.48 & -0.36 & -1.0 & 0 & 1.36 & 0.48 \\
-0.64 & -0.48 & 0 & 0 & 0.48 & 0.64
\end{bmatrix}
\begin{Bmatrix}
r_1 \\ r_2 \\ r_3 \\ \hline 0 \\ 0 \\ 0
\end{Bmatrix}
$$

Unknown nodal displacements can be obtained as follows:

$$\begin{Bmatrix} r_1 \\ r_2 \\ r_3 \end{Bmatrix} = \begin{bmatrix} 1.28 & 0 & 0.48 \\ 0 & 0.72 & -0.36 \\ 0.48 & -0.36 & 1.36 \end{bmatrix}^{-1} \begin{Bmatrix} 0 \\ P \\ 0 \end{Bmatrix}$$

$$\therefore \quad \begin{Bmatrix} r_1 \\ r_2 \\ r_3 \end{Bmatrix} = \begin{Bmatrix} -0.19 \\ 1.64 \\ 0.50 \end{Bmatrix} P$$

Support reactions can be obtained as follows:

$$\begin{Bmatrix} R_4 \\ R_5 \\ R_6 \end{Bmatrix} = \begin{bmatrix} -0.64 & 0.48 & -0.48 \\ -0.48 & -0.36 & -1.0 \\ -0.64 & -0.48 & 0 \end{bmatrix} \begin{Bmatrix} -0.19 \\ 1.64 \\ 0.50 \end{Bmatrix} P$$

$$\therefore \quad \begin{Bmatrix} R_4 \\ R_5 \\ R_6 \end{Bmatrix} = \begin{Bmatrix} 0.67 \\ -1.0 \\ -0.67 \end{Bmatrix} P$$

Member forces can be determined using the nodal displacements and the support reactions obtained previously. First, the member forces in the global coordinate system are computed. Then, the member forces in the local coordinate system are determined by using the transformation matrix.

Member a

Member forces in the global coordinate systems are obtained as follows:

$$\{\overline{Q}^a\} = \left[\,\overline{k}^a\,\right]\{\overline{q}^a\}$$

$$\begin{Bmatrix} \overline{Q}_5^a \\ \overline{Q}_3^a \\ \overline{Q}_6^a \\ \overline{Q}_4^a \end{Bmatrix} = \begin{bmatrix} 1 & -1 & 0 & 0 \\ -1 & 1 & 0 & 0 \\ 0 & 0 & 0 & 0 \\ 0 & 0 & 0 & 0 \end{bmatrix} \begin{Bmatrix} \overline{q}_5^a \\ \overline{q}_3^a \\ \overline{q}_6^a \\ \overline{q}_4^a \end{Bmatrix} = \begin{bmatrix} 1 & -1 & 0 & 0 \\ -1 & 1 & 0 & 0 \\ 0 & 0 & 0 & 0 \\ 0 & 0 & 0 & 0 \end{bmatrix} \begin{Bmatrix} r_5 \\ r_3 \\ r_6 \\ r_4 \end{Bmatrix} = \begin{Bmatrix} -0.5 \\ 0.5 \\ 0 \\ 0 \end{Bmatrix} P$$

Note that $\overline{q}_i = r_i$

Member forces in the local coordinate system are obtained as follows:

$$\{Q^a\} = [a]\{\overline{Q}^a\}$$

where, $[a] = \begin{bmatrix} \cos\theta & 0 & \sin\theta & 0 \\ 0 & \cos\theta & 0 & \sin\theta \\ -\sin\theta & 0 & \cos\theta & 0 \\ 0 & -\sin\theta & 0 & \cos\theta \end{bmatrix}$

Thus,

$$\begin{Bmatrix} Q_5^a \\ Q_3^a \\ Q_6^a \\ Q_4^a \end{Bmatrix} = \begin{bmatrix} 1 & 0 & 0 & 0 \\ 0 & 1 & 0 & 0 \\ 0 & 0 & 1 & 0 \\ 0 & 0 & 0 & 1 \end{bmatrix} \begin{Bmatrix} -0.5 \\ 0.5 \\ 0 \\ 0 \end{Bmatrix} P = \begin{Bmatrix} -0.5 \\ 0.5 \\ 0 \\ 0 \end{Bmatrix} P$$

Hence, the force in element a is $0.5P$ (tensile).

Member b

Member forces in the global coordinate systems are obtained as follows:

$$\{\overline{Q}^b\} = [\overline{k}^b]\{\overline{q}^b\}$$

$$\begin{Bmatrix} \overline{Q}_5^b \\ \overline{Q}_2^b \\ \overline{Q}_6^b \\ \overline{Q}_1^b \end{Bmatrix} = \begin{bmatrix} 0.36 & -0.36 & 0.48 & -0.48 \\ -0.36 & 0.36 & -0.48 & 0.48 \\ 0.48 & -0.48 & 0.64 & -0.64 \\ -0.48 & 0.48 & -0.64 & 0.64 \end{bmatrix} \begin{Bmatrix} \overline{q}_5^b \\ \overline{q}_2^b \\ \overline{q}_6^b \\ \overline{q}_1^b \end{Bmatrix}$$

$$\begin{Bmatrix} \overline{Q}_5^b \\ \overline{Q}_2^b \\ \overline{Q}_6^b \\ \overline{Q}_1^b \end{Bmatrix} = \begin{bmatrix} 0.36 & -0.36 & 0.48 & -0.48 \\ -0.36 & 0.36 & -0.48 & 0.48 \\ 0.48 & -0.48 & 0.64 & -0.64 \\ -0.48 & 0.48 & -0.64 & 0.64 \end{bmatrix} \begin{Bmatrix} 0 \\ 1.64P \\ 0 \\ -0.19P \end{Bmatrix} = \begin{Bmatrix} -0.50 \\ 0.50 \\ 0.67 \\ -0.67 \end{Bmatrix} P$$

Note that $\overline{q}_i = r_i$

Member forces in the local coordinate system are obtained as follows:

$$\{Q^a\} = [a]\{\overline{Q}^a\}$$

Thus,

$$\begin{Bmatrix} Q_5^b \\ Q_2^b \\ Q_6^b \\ Q_1^b \end{Bmatrix} = \begin{bmatrix} 0.6 & 0 & 0.8 & 0 \\ 0 & 0.6 & 0 & 0.8 \\ -0.8 & 0 & 0.6 & 0 \\ 0 & -0.8 & 0 & 0.6 \end{bmatrix} \begin{Bmatrix} -0.50 \\ 0.50 \\ 0.67 \\ -0.67 \end{Bmatrix} P = \begin{Bmatrix} -0.83 \\ 0.83 \\ 0 \\ 0 \end{Bmatrix} P$$

Hence, the force in element a is $0.83P$ (tensile).

Member c

Member forces in the global coordinate systems are obtained as follows:

$$\{\overline{Q}^c\} = [\overline{k}^c]\{\overline{q}^c\}$$

$$\begin{Bmatrix} \overline{Q}_2^c \\ \overline{Q}_3^c \\ \overline{Q}_1^c \\ \overline{Q}_4^c \end{Bmatrix} = \begin{bmatrix} 0.36 & -0.36 & -0.48 & 0.48 \\ -0.36 & 0.36 & 0.48 & -0.48 \\ -0.48 & 0.48 & 0.64 & -0.64 \\ 0.48 & -0.48 & -0.64 & 0.64 \end{bmatrix} \begin{Bmatrix} \overline{q}_2^c \\ \overline{q}_3^c \\ \overline{q}_1^c \\ \overline{q}_4^c \end{Bmatrix}$$

$$\begin{Bmatrix} \overline{Q}_2^c \\ \overline{Q}_3^c \\ \overline{Q}_1^c \\ \overline{Q}_4^c \end{Bmatrix} = \begin{bmatrix} 0.36 & -0.36 & -0.48 & 0.48 \\ -0.36 & 0.36 & 0.48 & -0.48 \\ -0.48 & 0.48 & 0.64 & -0.64 \\ 0.48 & -0.48 & -0.64 & 0.64 \end{bmatrix} \begin{Bmatrix} 1.64P \\ 0.50P \\ -0.19P \\ 0 \end{Bmatrix} = \begin{Bmatrix} 0.50 \\ -0.50 \\ -0.67 \\ 0.67 \end{Bmatrix} P$$

Member forces in the local coordinate system are obtained as follows:

$$\{Q^a\} = [a]\{\overline{Q}^a\}$$

Thus,

$$\begin{Bmatrix} Q_2^c \\ Q_3^c \\ Q_1^c \\ Q_4^c \end{Bmatrix} = \begin{bmatrix} -0.6 & 0 & 0.8 & 0 \\ 0 & -0.6 & 0 & 0.8 \\ -0.8 & 0 & -0.6 & 0 \\ 0 & -0.8 & 0 & -0.6 \end{bmatrix} \begin{Bmatrix} 0.50 \\ -0.50 \\ -0.67 \\ 0.67 \end{Bmatrix} P = \begin{Bmatrix} -0.83 \\ 0.83 \\ 0 \\ 0 \end{Bmatrix} P$$

Hence, the force in element *a* is $0.83P$ (compression).

EXAMPLE 10.2 Determine the joint displacements and the support reactions of the truss shown in Figure E10.2 using the direct stiffness matrix method.

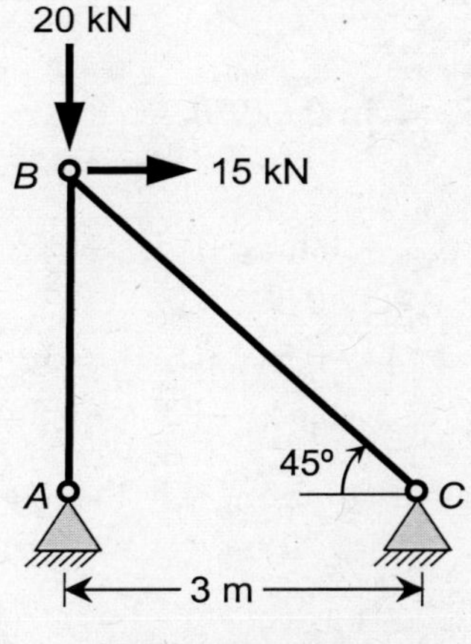

FIGURE E10.2

Solution: All nodes have been numbered in the global coordinate system as shown in the following figure. The numbering of nodes has been carried out keeping all unknown displacements sequentially as one group followed by all nodes with the prescribed node displacements.

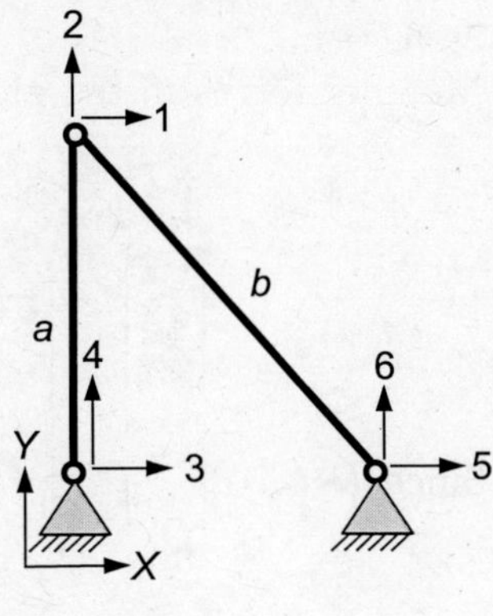

The nodal displacement matrix and the nodal force matrix are as follows:

$$\{r\} = \begin{Bmatrix} r_1 \\ r_2 \\ r_3 \\ r_4 \\ r_5 \\ r_6 \end{Bmatrix} = \begin{Bmatrix} r_1 \\ r_2 \\ 0 \\ 0 \\ 0 \\ 0 \end{Bmatrix}; \quad \{R\} = \begin{Bmatrix} R_1 \\ R_2 \\ R_3 \\ R_4 \\ R_5 \\ R_6 \end{Bmatrix} = \begin{Bmatrix} 15 \\ -20 \\ R_3 \\ R_4 \\ R_5 \\ R_6 \end{Bmatrix}$$

The stiffness matrix of elements in the global coordinate system is then computed using the following equation.

$$[\bar{k}] = \left(\frac{AE}{L}\right) \begin{bmatrix} C_x^2 & -C_x^2 & C_x C_y & -C_x C_y \\ -C_x^2 & C_x^2 & -C_x C_y & C_x C_y \\ C_x C_y & -C_x C_y & C_y^2 & -C_y^2 \\ -C_x C_y & C_x C_y & -C_y^2 & C_y^2 \end{bmatrix}$$

Element a

$\theta = 90°; \ C_x = \cos\theta = 0; \ C_y = \sin\theta = 1$

$$[\bar{k}^a] = \frac{AE}{3} \begin{array}{cccc} 3 & 1 & 4 & 2 \\ \begin{bmatrix} 0 & 0 & 0 & 0 \\ 0 & 0 & 0 & 0 \\ 0 & 0 & 1 & -1 \\ 0 & 0 & -1 & 1 \end{bmatrix} & \begin{matrix} 3 \\ 1 \\ 4 \\ 2 \end{matrix} \end{array}$$

Element b

$\theta = 135°; \ C_x = \cos\theta = -0.707; \ C_y = \sin\theta = 0.707$

$$[\bar{k}^b] = \frac{AE}{4.24} \begin{array}{cccc} 5 & 1 & 6 & 2 \\ \begin{bmatrix} 0.5 & -0.5 & -0.5 & 0.5 \\ -0.5 & 0.5 & 0.5 & -0.5 \\ -0.5 & 0.5 & 0.5 & -0.5 \\ 0.5 & -0.5 & -0.5 & 0.5 \end{bmatrix} & \begin{matrix} 5 \\ 1 \\ 6 \\ 2 \end{matrix} \end{array}$$

The structure stiffness matrix by assembling all the element stiffness matrices is given below:

$$[K] = \frac{AE}{4.24}\begin{bmatrix} & 1 & 2 & 3 & 4 & 5 & 6 & \\ 0.5 & & & & & sym \\ -0.5 & 1.91 & & & & \\ 0 & 0 & 0 & & & \\ 0 & -1.41 & 0 & 1.41 & & \\ -0.5 & 0.5 & 0 & 0 & 0.5 & \\ 0.5 & -0.5 & 0 & 0 & -0.5 & 0.5 \end{bmatrix}\begin{matrix} 1 \\ 2 \\ 3 \\ 4 \\ 5 \\ 6 \end{matrix}$$

Since $\{R\} = [K]\{r\}$

$$\begin{Bmatrix} 15 \\ -20 \\ \hline R_3 \\ R_4 \\ R_5 \\ R_6 \end{Bmatrix} = \frac{4.24}{AE}\begin{bmatrix} 0.5 & -0.5 & 0 & 0 & -0.5 & 0.5 \\ -0.5 & 1.91 & 0 & -1.41 & 0.5 & -0.5 \\ \hline 0 & 0 & 0 & 0 & 0 & 0 \\ 0 & -1.41 & 0 & 1.41 & 0 & 0 \\ -0.5 & 0.5 & 0 & 0 & 0.5 & -0.5 \\ 0.5 & -0.5 & 0 & 0 & -0.5 & 0.5 \end{bmatrix}\begin{Bmatrix} r_1 \\ r_2 \\ \hline 0 \\ 0 \\ 0 \\ 0 \end{Bmatrix}$$

Joint displacements can be obtained as follows:

$$\begin{Bmatrix} r_1 \\ r_2 \end{Bmatrix} = \frac{4.24}{AE}\begin{bmatrix} 0.5 & -0.5 \\ -0.5 & 1.91 \end{bmatrix}^{-1}\begin{Bmatrix} 15 \\ -20 \end{Bmatrix} \qquad \therefore \quad \begin{Bmatrix} r_1 \\ r_2 \end{Bmatrix} = \frac{4.24}{AE}\begin{Bmatrix} 26.45 \\ -3.55 \end{Bmatrix}$$

Support reactions can be obtained as follows:

$$\begin{Bmatrix} R_3 \\ R_4 \\ R_5 \\ R_6 \end{Bmatrix} = \begin{bmatrix} 0 & 0 \\ 0 & -1.41 \\ -0.5 & 0.5 \\ 0.5 & -0.5 \end{bmatrix}\begin{Bmatrix} 26.45 \\ -3.55 \end{Bmatrix} \qquad \therefore \quad \begin{Bmatrix} R_3 \\ R_4 \\ R_5 \\ R_6 \end{Bmatrix} = \begin{Bmatrix} 0 \\ 5 \\ -15 \\ 15 \end{Bmatrix} \text{kN}$$

EXAMPLE 10.3 Determine the support reactions and the joint displacements of the truss shown in Figure E10.3 using the Direct stiffness matrix method. Assume all members have the same area (A) of cross-section and modulus of elasticity (E).

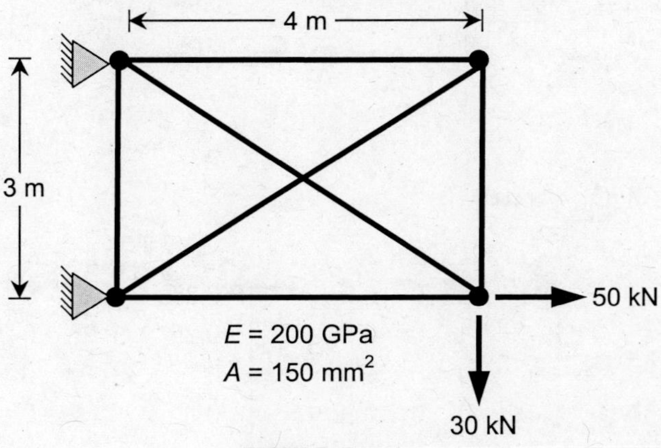

FIGURE E10.3

Solution: All nodes and elements of the truss are shown in the following figure.

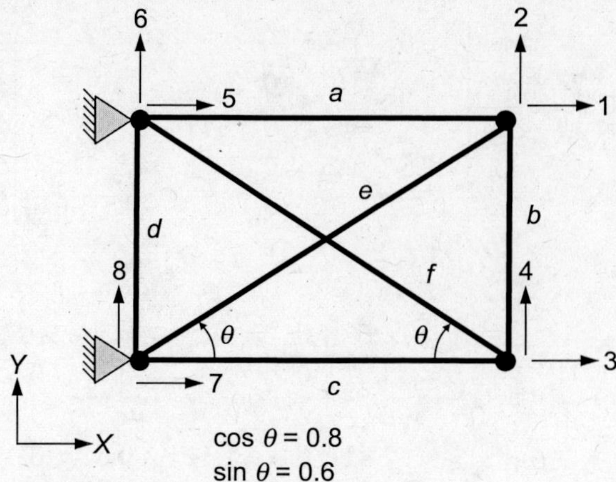

The nodal displacement matrix and the nodal force matrix are as follows:

$$\{r\} = \begin{Bmatrix} r_1 \\ r_2 \\ r_3 \\ r_4 \\ r_5 \\ r_6 \\ r_7 \\ r_8 \end{Bmatrix} = \begin{Bmatrix} r_1 \\ r_2 \\ r_3 \\ r_4 \\ 0 \\ 0 \\ 0 \\ 0 \end{Bmatrix}; \quad \{R\} = \begin{Bmatrix} R_1 \\ R_2 \\ R_3 \\ R_4 \\ R_5 \\ R_6 \\ R_7 \\ R_8 \end{Bmatrix} = \begin{Bmatrix} 0 \\ 0 \\ 50 \\ -30 \\ R_5 \\ R_6 \\ R_7 \\ R_8 \end{Bmatrix}$$

The stiffness matrix of elements in the global coordinate system is then computed using the following equation.

$$[\bar{k}] = \left(\frac{AE}{L}\right) \begin{bmatrix} C_x^2 & -C_x^2 & C_xC_y & -C_xC_y \\ -C_x^2 & C_x^2 & -C_xC_y & C_xC_y \\ C_xC_y & -C_xC_y & C_y^2 & -C_y^2 \\ -C_xC_y & C_xC_y & -C_y^2 & C_y^2 \end{bmatrix}$$

Element a
$\theta = 0°$; $C_x = \cos\theta = 1$; $C_y = \sin\theta = 0$

$$[\bar{k}^a] = \frac{AE}{4} \begin{array}{c} \begin{array}{cccc} 5 & 1 & 6 & 2 \end{array} \\ \begin{bmatrix} 1 & -1 & 0 & 0 \\ -1 & 1 & 0 & 0 \\ 0 & 0 & 0 & 0 \\ 0 & 0 & 0 & 0 \end{bmatrix} \begin{array}{c} 5 \\ 1 \\ 6 \\ 2 \end{array} \end{array} = AE \begin{array}{c} \begin{array}{cccc} 5 & 1 & 6 & 2 \end{array} \\ \begin{bmatrix} 0.25 & -0.25 & 0 & 0 \\ -0.25 & 0.25 & 0 & 0 \\ 0 & 0 & 0 & 0 \\ 0 & 0 & 0 & 0 \end{bmatrix} \begin{array}{c} 5 \\ 1 \\ 6 \\ 2 \end{array} \end{array}$$

Element b

$\theta = 90°;\ C_x = \cos\theta = 0;\ C_y = \sin\theta = 1$

$$
\left[\bar{k}^b\right] = \frac{AE}{3}
\begin{array}{cccc}
3 & 1 & 4 & 2
\end{array}
\begin{bmatrix}
0 & 0 & 0 & 0 \\
0 & 0 & 0 & 0 \\
0 & 0 & 1 & -1 \\
0 & 0 & -1 & 1
\end{bmatrix}
\begin{array}{c}
3 \\ 1 \\ 4 \\ 2
\end{array}
= AE
\begin{array}{cccc}
3 & 1 & 4 & 2
\end{array}
\begin{bmatrix}
0 & 0 & 0 & 0 \\
0 & 0 & 0 & 0 \\
0 & 0 & 0.33 & -0.33 \\
0 & 0 & -0.33 & 0.33
\end{bmatrix}
\begin{array}{c}
3 \\ 1 \\ 4 \\ 2
\end{array}
$$

Element c

$\theta = 0°;\ C_x = \cos\theta = 1;\ C_y = \sin\theta = 0$

$$
\left[\bar{k}^c\right] = \frac{AE}{4}
\begin{array}{cccc}
7 & 3 & 8 & 4
\end{array}
\begin{bmatrix}
1 & -1 & 0 & 0 \\
-1 & 1 & 0 & 0 \\
0 & 0 & 0 & 0 \\
0 & 0 & 0 & 0
\end{bmatrix}
\begin{array}{c}
7 \\ 3 \\ 8 \\ 4
\end{array}
= AE
\begin{array}{cccc}
7 & 3 & 8 & 4
\end{array}
\begin{bmatrix}
0.25 & -0.25 & 0 & 0 \\
-0.25 & 0.25 & 0 & 0 \\
0 & 0 & 0 & 0 \\
0 & 0 & 0 & 0
\end{bmatrix}
\begin{array}{c}
7 \\ 3 \\ 8 \\ 4
\end{array}
$$

Element d

$\theta = 90°;\ C_x = \cos\theta = 0;\ C_y = \sin\theta = 1$

$$
\left[\bar{k}^d\right] = \frac{AE}{3}
\begin{array}{cccc}
7 & 5 & 8 & 6
\end{array}
\begin{bmatrix}
0 & 0 & 0 & 0 \\
0 & 0 & 0 & 0 \\
0 & 0 & 1 & -1 \\
0 & 0 & -1 & 1
\end{bmatrix}
\begin{array}{c}
7 \\ 5 \\ 8 \\ 6
\end{array}
= AE
\begin{array}{cccc}
7 & 5 & 8 & 6
\end{array}
\begin{bmatrix}
0 & 0 & 0 & 0 \\
0 & 0 & 0 & 0 \\
0 & 0 & 0.33 & -0.33 \\
0 & 0 & -0.33 & 0.33
\end{bmatrix}
\begin{array}{c}
7 \\ 5 \\ 8 \\ 6
\end{array}
$$

Element e

$C_x = \cos\theta = 0.8;\ C_y = \sin\theta = 0.6$

$$
\left[\bar{k}^e\right] = \frac{AE}{5}
\begin{array}{cccc}
7 & 1 & 8 & 2
\end{array}
\begin{bmatrix}
0.64 & -0.64 & 0.48 & -0.48 \\
-0.64 & 0.64 & -0.48 & 0.48 \\
0.48 & -0.48 & 0.36 & -0.36 \\
-0.48 & 0.48 & -0.36 & 0.36
\end{bmatrix}
\begin{array}{c}
7 \\ 1 \\ 8 \\ 2
\end{array}
$$

or

$$
\left[\bar{k}^e\right] = AE
\begin{array}{cccc}
7 & 1 & 8 & 2
\end{array}
\begin{bmatrix}
0.13 & -0.13 & 0.10 & -0.10 \\
-0.13 & 0.13 & -0.10 & 0.10 \\
0.10 & -0.10 & 0.07 & -0.07 \\
-0.10 & 0.10 & -0.07 & 0.07
\end{bmatrix}
\begin{array}{c}
7 \\ 1 \\ 8 \\ 2
\end{array}
$$

Element f
$$C_x = \cos\theta = -0.8; \quad C_y = \sin\theta = 0.6$$

$$\left[\bar{k}^f\right] = \frac{AE}{5}\begin{array}{c}\\[6pt]\end{array}\begin{matrix}3 & 5 & 4 & 6\end{matrix}$$

$$\left[\bar{k}^f\right] = \frac{AE}{5}\begin{bmatrix} 0.64 & -0.64 & -0.48 & 0.48 \\ -0.64 & 0.64 & 0.48 & -0.48 \\ -0.48 & 0.48 & 0.36 & -0.36 \\ 0.48 & -0.48 & -0.36 & 0.36 \end{bmatrix}\begin{matrix}3\\5\\4\\6\end{matrix}$$

or

$$\left[\bar{k}^f\right] = AE\begin{bmatrix} 0.13 & -0.13 & -0.10 & 0.10 \\ -0.13 & 0.13 & 0.10 & -0.10 \\ -0.10 & 0.10 & 0.07 & -0.07 \\ 0.10 & -0.10 & -0.07 & 0.07 \end{bmatrix}\begin{matrix}3\\5\\4\\6\end{matrix}$$

with columns labelled $\begin{matrix}3 & 5 & 4 & 6\end{matrix}$.

The structure stiffness matrix is computed as follows:

$$[K] = AE\begin{bmatrix} 0.38 & & & & & & & \\ 0.10 & 0.40 & & & & & & \\ 0 & 0 & 0.38 & & & & & \\ 0 & -0.33 & -0.10 & 0.40 & & & & \\ -0.25 & 0 & -0.13 & 0.10 & 0.38 & & & \\ 0 & 0 & 0.10 & -0.07 & -0.10 & 0.40 & & \\ -0.13 & -0.10 & -0.25 & 0 & 0 & 0 & 0.38 & \\ -0.10 & -0.07 & 0 & 0 & 0 & -0.33 & 0.10 & 0.40 \end{bmatrix}\begin{matrix}1\\2\\3\\4\\5\\6\\7\\8\end{matrix}$$

with columns labelled $\begin{matrix}1 & 2 & 3 & 4 & 5 & 6 & 7 & 8\end{matrix}$ and *sym* in the upper right.

Joint displacements can be obtained as follows:

$$\begin{Bmatrix} r_1 \\ r_2 \\ r_3 \\ r_4 \end{Bmatrix} = \frac{1}{AE}\begin{bmatrix} 0.38 & 0.10 & 0 & 0 \\ 0.10 & 0.40 & 0 & -0.33 \\ 0 & 0 & 0.38 & -0.10 \\ 0 & -0.33 & -0.10 & 0.40 \end{bmatrix}^{-1}\begin{Bmatrix} 0 \\ 0 \\ 50 \\ -30 \end{Bmatrix} \quad \therefore \quad \begin{Bmatrix} r_1 \\ r_2 \\ r_3 \\ r_4 \end{Bmatrix} = \frac{1}{AE}\begin{Bmatrix} 47.58 \\ -180.80 \\ 77.70 \\ -204.74 \end{Bmatrix}$$

Support reactions can be obtained as follows:

$$\begin{Bmatrix} R_5 \\ R_6 \\ R_7 \\ R_8 \end{Bmatrix} = \begin{bmatrix} -0.25 & 0 & -0.13 & 0.10 \\ 0 & 0 & 0.10 & -0.07 \\ -0.13 & -0.10 & -0.25 & 0 \\ -0.10 & -0.07 & 0 & 0 \end{bmatrix}\begin{Bmatrix} 47.58 \\ -180.80 \\ 77.70 \\ -204.74 \end{Bmatrix} \quad \therefore \quad \begin{Bmatrix} R_5 \\ R_6 \\ R_7 \\ R_8 \end{Bmatrix} = \begin{Bmatrix} -42.5 \\ 22.1 \\ -7.5 \\ 7.9 \end{Bmatrix}\text{kN}$$

EXAMPLE 10.4 Determine the joint displacements, and the support reactions of the truss shown in Figure E10.4 using the direct stiffness matrix method. Assume the same area (A) of cross-section and modulus (E) of elasticity for all members.

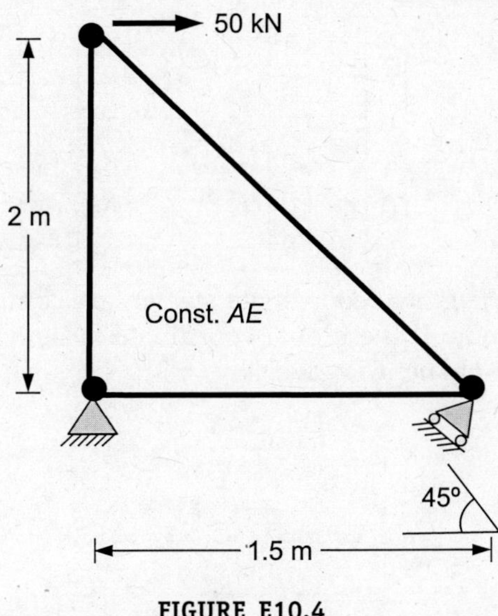

FIGURE E10.4

Solution: As shown in the following figure, two different global coordinate systems (XY and X′Y′) are considered for this truss, in which X′Y′ system is taken along the directions of displacement and reaction of the inclined support. The main goal of choosing two coordinate systems is to reduce the computational efforts.

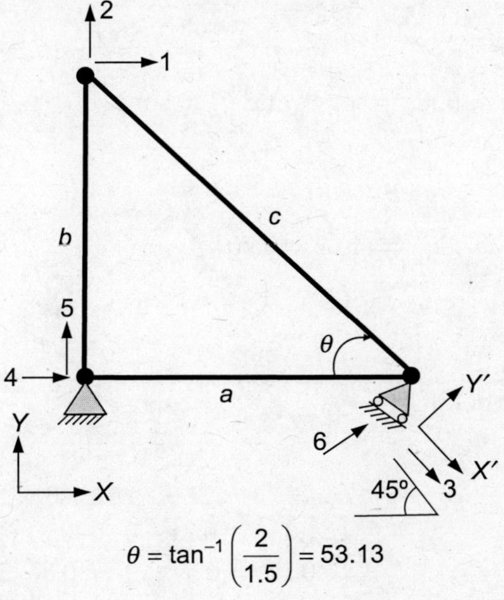

$$\theta = \tan^{-1}\left(\frac{2}{1.5}\right) = 53.13$$

The nodal displacement and the nodal force matrices are as follows:

$$\{r\} = \begin{Bmatrix} r_1 \\ r_2 \\ r_3 \\ r_4 \\ r_5 \\ r_6 \end{Bmatrix} = \begin{Bmatrix} r_1 \\ r_2 \\ r_3 \\ 0 \\ 0 \\ 0 \end{Bmatrix}; \quad \{R\} = \begin{Bmatrix} R_1 \\ R_2 \\ R_3 \\ R_4 \\ R_5 \\ R_6 \end{Bmatrix} = \begin{Bmatrix} 50 \\ 0 \\ 0 \\ R_4 \\ R_5 \\ R_6 \end{Bmatrix}$$

The stiffness matrices in the global coordinate system are computed considering the global coordinate axes at both ends. If the global coordinate systems are the same at both ends, the stiffness matrix of the element is given by

$$[\bar{k}] = \left(\frac{AE}{L}\right) \begin{bmatrix} C_x^2 & -C_x^2 & C_xC_y & -C_xC_y \\ -C_x^2 & C_x^2 & -C_xC_y & C_xC_y \\ C_xC_y & -C_xC_y & C_y^2 & -C_y^2 \\ -C_xC_y & C_xC_y & -C_y^2 & C_y^2 \end{bmatrix}$$

If two global coordinate systems exist at both ends of the element, the stiffness matrix of the element is computed as follows:

$$[\bar{k}] = \left(\frac{AE}{L}\right) \begin{bmatrix} C_x^2 & -C_xC_x' & C_xC_y & -C_xC_y' \\ -C_xC_x' & C_x'^2 & -C_x'C_y & C_x'C_y' \\ C_xC_y & -C_x'C_y & C_y^2 & -C_yC_y' \\ -C_xC_y' & C_x'C_y' & -C_yC_y' & C_y'^2 \end{bmatrix}$$

where, $C_x = \cos\theta$, $C_y = \sin\theta$, $C_x' = \cos\theta'$ and $C_y' = \sin\theta'$

Element a

$\theta = 0°$; $C_x = \cos\theta = 1$; $C_y = \sin\theta = 0$

$\theta' = 45°$; $C_x' = \cos\theta' = 0.707$; $C_y' = \sin\theta' = 0.707$

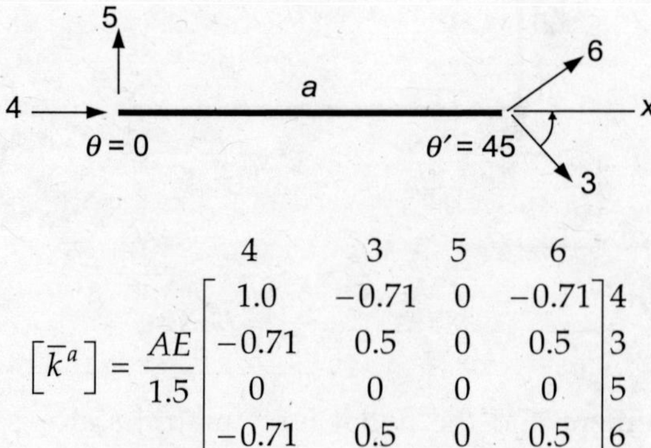

$$[\bar{k}^a] = \frac{AE}{1.5} \begin{bmatrix} & 4 & 3 & 5 & 6 & \\ 1.0 & -0.71 & 0 & -0.71 & 4 \\ -0.71 & 0.5 & 0 & 0.5 & 3 \\ 0 & 0 & 0 & 0 & 5 \\ -0.71 & 0.5 & 0 & 0.5 & 6 \end{bmatrix}$$

Element b

$\theta = 90°;\ C_x = \cos\theta = 0;\ C_y = \sin\theta = 1$

$$[\bar{k}^b] = \frac{AE}{2}\begin{array}{cccc} 4 & 1 & 5 & 2 \end{array}\begin{bmatrix} 0 & 0 & 0 & 0 \\ 0 & 0 & 0 & 0 \\ 0 & 0 & 1 & -1 \\ 0 & 0 & -1 & 1 \end{bmatrix}\begin{array}{c} 4 \\ 1 \\ 5 \\ 2 \end{array}$$

Element c

$\theta = -53.13°;\ C_x = \cos\theta = 0.6;\ C_y = \sin\theta = -0.8$

$\theta' = -8.13°;\ C'_x = \cos\theta' = 0.99;\ C'_y = \sin\theta' = -0.14$

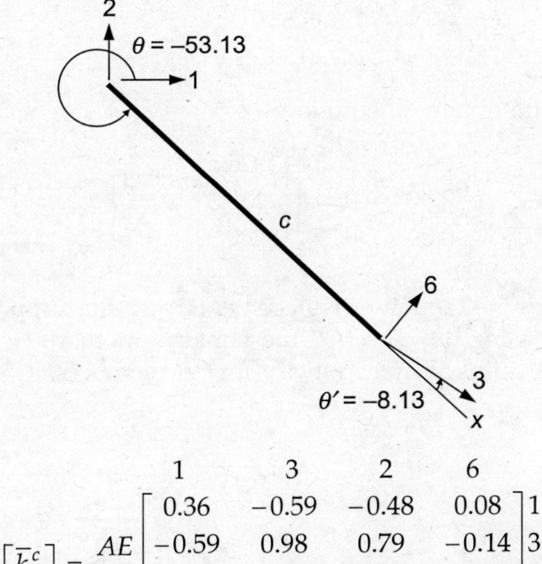

$$[\bar{k}^c] = \frac{AE}{2.5}\begin{array}{cccc} 1 & 3 & 2 & 6 \end{array}\begin{bmatrix} 0.36 & -0.59 & -0.48 & 0.08 \\ -0.59 & 0.98 & 0.79 & -0.14 \\ -0.48 & 0.79 & 0.64 & -0.11 \\ 0.08 & -0.14 & -0.11 & 0.02 \end{bmatrix}\begin{array}{c} 1 \\ 3 \\ 2 \\ 6 \end{array}$$

The structure stiffness matrix by assembling all the element stiffness matrices is given below:

$$[K] = AE\begin{array}{cccccc} 1 & 2 & 3 & 4 & 5 & 6 \end{array}\begin{bmatrix} 0.14 & & & & & sym \\ -0.19 & 0.76 & & & & \\ -0.24 & 0.32 & 0.72 & & & \\ 0 & 0 & -0.47 & 0.67 & & \\ 0 & -0.5 & 0 & 0 & 0.5 & \\ 0.03 & -0.04 & 0.27 & -0.47 & 0 & 0.34 \end{bmatrix}\begin{array}{c} 1 \\ 2 \\ 3 \\ 4 \\ 5 \\ 6 \end{array}$$

$$\{R\} = [K]\{r\}$$

$$
\left\{\begin{array}{c} 50 \\ 0 \\ 0 \\ \hline R_4 \\ R_5 \\ R_6 \end{array}\right\} = AE \left[\begin{array}{ccc|ccc} 0.14 & -0.19 & -0.24 & 0 & 0 & 0.03 \\ -0.19 & 0.76 & 0.32 & 0 & -0.50 & -0.04 \\ -0.24 & 0.32 & 0.72 & -0.47 & 0 & 0.27 \\ \hline 0 & 0 & -0.47 & 0.67 & 0 & -0.47 \\ 0 & -0.50 & 0 & 0 & 0.50 & 0 \\ 0.03 & -0.04 & 0.27 & -0.47 & 0 & 0.34 \end{array}\right] \left\{\begin{array}{c} r_1 \\ r_2 \\ r_3 \\ \hline 0 \\ 0 \\ 0 \end{array}\right\}
$$

Joint displacements can be obtained as follows:

$$
\left\{\begin{array}{c} r_1 \\ r_2 \\ r_3 \end{array}\right\} = \frac{1}{AE}\left[\begin{array}{ccc} 0.14 & -0.19 & -0.24 \\ -0.19 & 0.76 & 0.32 \\ -0.24 & 0.32 & 0.72 \end{array}\right]^{-1} \left\{\begin{array}{c} 50 \\ 0 \\ 0 \end{array}\right\}
$$

$$
\therefore \qquad \left\{\begin{array}{c} r_1 \\ r_2 \\ r_3 \end{array}\right\} = \frac{1}{AE}\left\{\begin{array}{c} 1025.5 \\ 138.3 \\ 280.3 \end{array}\right\}
$$

Support reactions can be obtained as follows:

$$
\left\{\begin{array}{c} R_4 \\ R_5 \\ R_6 \end{array}\right\} = \left[\begin{array}{ccc} 0 & 0 & -0.47 \\ 0 & -0.50 & 0 \\ 0.03 & -0.04 & 0.27 \end{array}\right]\left\{\begin{array}{c} 1025.5 \\ 138.3 \\ 280.3 \end{array}\right\} \qquad \therefore \quad \left\{\begin{array}{c} R_4 \\ R_5 \\ R_6 \end{array}\right\} = \left\{\begin{array}{c} -131.8 \\ -69.2 \\ 103.7 \end{array}\right\} \text{kN}
$$

EXAMPLE 10.5 Determine the joint displacements, and the support reactions of the truss shown in Figure E10.5 using the direct stiffness matrix method. Assume the same area (A) of cross-section and modulus (E) of elasticity for all members.

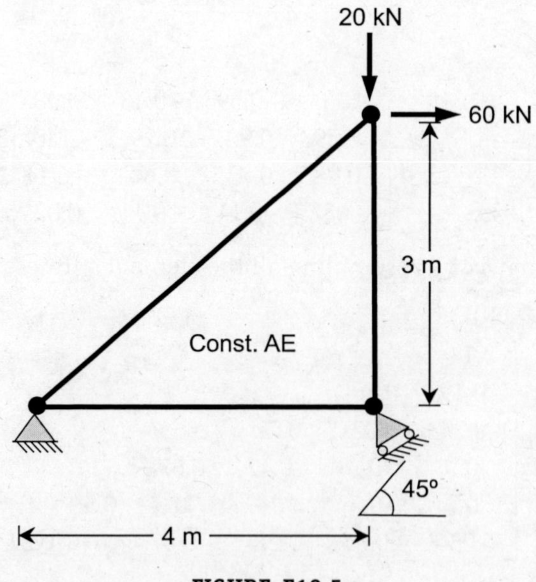

FIGURE E10.5

Solution: As shown in the following figure, two different global coordinate systems (XY and X′Y′) are considered for this truss, in which X′Y′ system is taken along the directions of displacement and reaction of the inclined support.

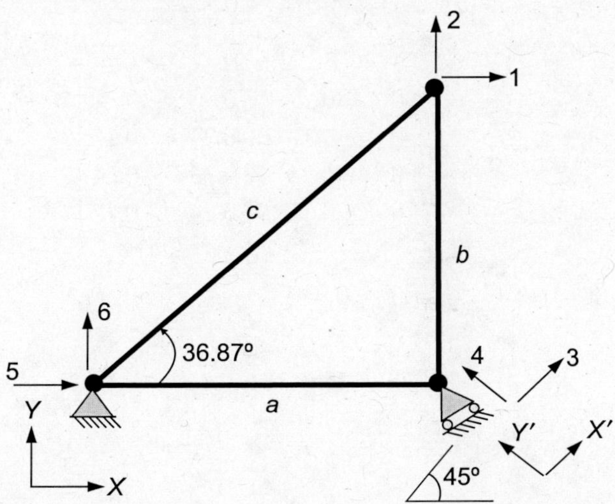

The nodal displacement and the nodal force matrices are as follows:

$$\{r\} = \begin{Bmatrix} r_1 \\ r_2 \\ r_3 \\ r_4 \\ r_5 \\ r_6 \end{Bmatrix} = \begin{Bmatrix} r_1 \\ r_2 \\ r_3 \\ 0 \\ 0 \\ 0 \end{Bmatrix}; \{R\} = \begin{Bmatrix} R_1 \\ R_2 \\ R_3 \\ R_4 \\ R_5 \\ R_6 \end{Bmatrix} = \begin{Bmatrix} 60 \\ -20 \\ 0 \\ R_4 \\ R_5 \\ R_6 \end{Bmatrix}$$

The stiffness matrices in the global coordinate system are computed considering the global coordinate axes at both ends.

Element a

$\theta = 0°$; $C_x = \cos \theta = 1$; $C_y = \sin \theta = 0$

$\theta' = -45°$; $C'_x = \cos \theta' = 0.707$; $C'_y = \sin \theta' = -0.707$

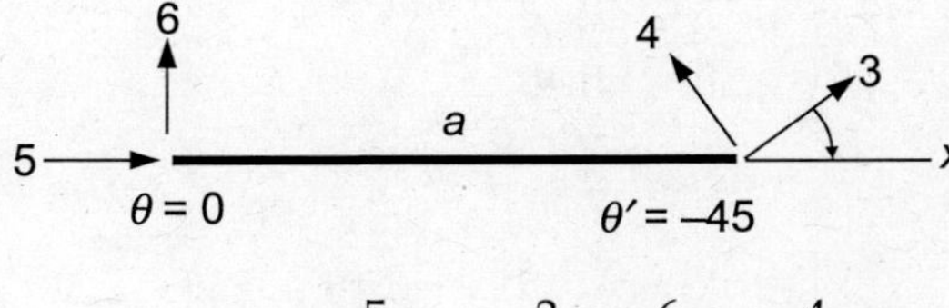

$$\begin{bmatrix} \bar{k}^a \end{bmatrix} = AE \begin{array}{cccc} 5 & 3 & 6 & 4 \\ \begin{bmatrix} 0.25 & -0.18 & 0 & 0.18 \\ -0.18 & 0.13 & 0 & -0.13 \\ 0 & 0 & 0 & 0 \\ 0.18 & -0.13 & 0 & 0.13 \end{bmatrix} & \begin{matrix} 5 \\ 3 \\ 6 \\ 4 \end{matrix} \end{array}$$

Element b

$\theta = -90°; \; C_x = \cos\theta = 0; \; C_y = \sin\theta = -1$

$\theta' = -135°; \; C_x' = \cos\theta' = -0.707; \; C_y' = \sin\theta' = -0.707$

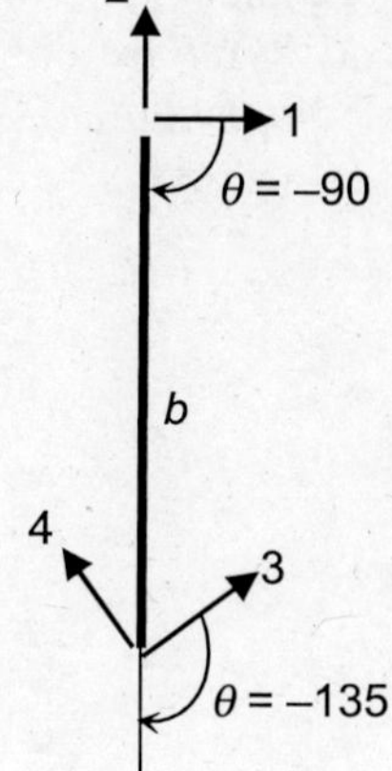

$$[\bar{k}^b] = AE \begin{array}{cccc} 1 & 3 & 2 & 4 \\ \begin{bmatrix} 0 & 0 & 0 & 0 \\ 0 & 0.17 & -0.24 & 0.17 \\ 0 & -0.24 & 0.33 & -0.24 \\ 0 & 0.17 & -0.24 & 0.17 \end{bmatrix} & \begin{matrix} 1 \\ 3 \\ 2 \\ 4 \end{matrix} \end{array}$$

Element c

$\theta = 36.87°; \; C_x = \cos\theta = 0.6; \; C_y = \sin\theta = 0.8$

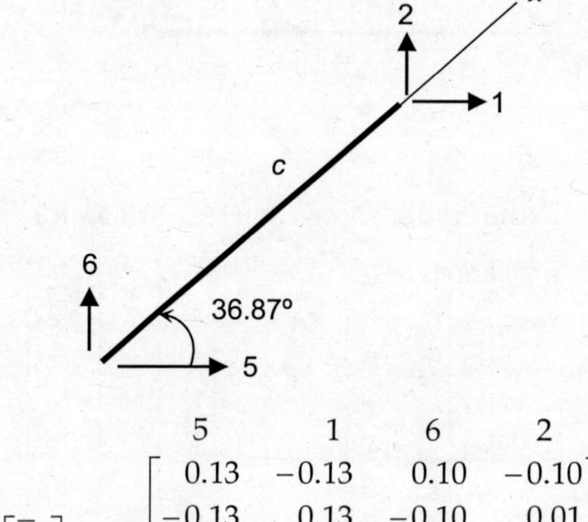

$$[\bar{k}^c] = AE \begin{array}{cccc} 5 & 1 & 6 & 2 \\ \begin{bmatrix} 0.13 & -0.13 & 0.10 & -0.10 \\ -0.13 & 0.13 & -0.10 & 0.01 \\ 0.10 & -0.10 & 0.07 & -0.07 \\ -0.10 & 0.10 & -0.07 & 0.07 \end{bmatrix} & \begin{matrix} 5 \\ 1 \\ 6 \\ 2 \end{matrix} \end{array}$$

The structure stiffness matrix by assembling all the element stiffness matrices is given below.

$$[K] = AE \begin{array}{cccccc} 1 & 2 & 3 & 4 & 5 & 6 \\ \begin{bmatrix} 0.13 & & & & & sym \\ 0.10 & 0.40 & & & & \\ 0 & -0.24 & 0.30 & & & \\ 0 & -0.24 & 0.04 & 0.30 & & \\ -0.13 & -0.10 & -0.18 & 0.18 & 0.38 & \\ -0.10 & -0.07 & 0 & 0 & 0.10 & 0.07 \end{bmatrix} & \begin{matrix} 1 \\ 2 \\ 3 \\ 4 \\ 5 \\ 6 \end{matrix} \end{array}$$

$$\{R\} = [K]\{r\}$$

$$\begin{Bmatrix} 60 \\ -20 \\ 0 \\ \hline R_4 \\ R_5 \\ R_6 \end{Bmatrix} = AE \left[\begin{array}{ccc:ccc} 0.13 & 0.10 & 0 & 0 & -0.13 & -0.10 \\ 0.10 & 0.40 & -0.24 & -0.24 & -0.10 & -0.07 \\ 0 & -0.24 & 0.30 & 0.04 & -0.18 & 0 \\ \hdashline 0 & -0.24 & 0.04 & 0.30 & 0.18 & 0 \\ -0.13 & -0.10 & -0.18 & 0.18 & 0.38 & 0.10 \\ -0.10 & -0.07 & 0 & 0 & 0.10 & 0.07 \end{array}\right] \begin{Bmatrix} r_1 \\ r_2 \\ r_3 \\ 0 \\ 0 \\ 0 \end{Bmatrix}$$

Joint displacements can be obtained as follows:

$$\begin{Bmatrix} r_1 \\ r_2 \\ r_3 \end{Bmatrix} = \frac{1}{AE} \begin{bmatrix} 0.13 & 0.10 & 0 \\ 0.10 & 0.40 & -0.24 \\ 0 & -0.24 & 0.30 \end{bmatrix}^{-1} \begin{Bmatrix} 60 \\ -20 \\ 0 \end{Bmatrix}$$

$$\therefore \quad \begin{Bmatrix} r_1 \\ r_2 \\ r_3 \end{Bmatrix} = \frac{1}{AE} \begin{Bmatrix} 849.8 \\ -504.7 \\ -403.7 \end{Bmatrix}$$

Support reactions can be obtained as follows:

$$\begin{Bmatrix} R_4 \\ R_5 \\ R_6 \end{Bmatrix} = \begin{bmatrix} 0 & -0.24 & 0.04 \\ -0.13 & -0.10 & -0.18 \\ -0.10 & -0.07 & 0 \end{bmatrix} \begin{Bmatrix} 849.8 \\ -504.7 \\ -403.7 \end{Bmatrix} \qquad \therefore \quad \begin{Bmatrix} R_4 \\ R_5 \\ R_6 \end{Bmatrix} = \begin{Bmatrix} 105.0 \\ 12.7 \\ -49.6 \end{Bmatrix} \text{ kN}$$

EXAMPLE 10.6 Determine the joint displacements, and the support reactions of the truss shown in Figure E10.6 using the direct stiffness matrix method. The member BC is too short by 20 mm. All members have the same area (A) of cross-section and modulus (E) of elasticity. The value of AE is 5000 kN.

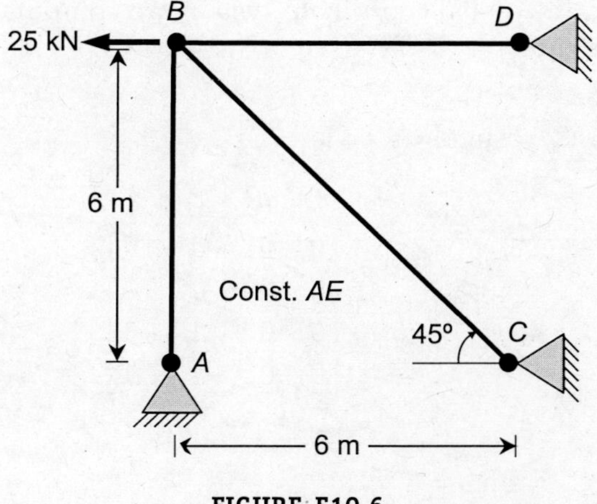

FIGURE E10.6

Solution: The nodes and elements of the truss are shown in the following figure.

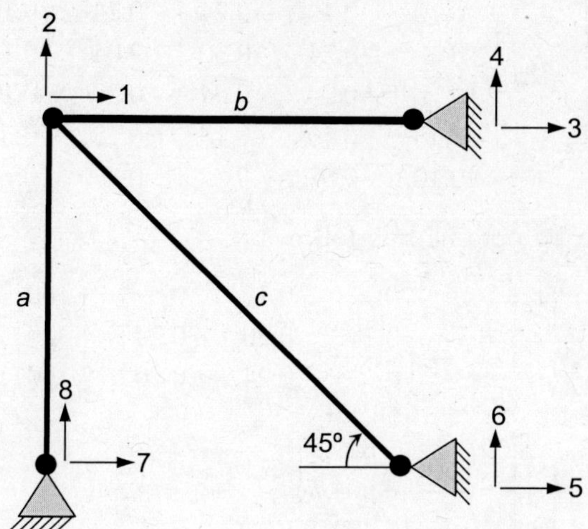

The nodal displacement and the nodal force matrices are as follows:

$$\{r\} = \begin{Bmatrix} r_1 \\ r_2 \\ r_3 \\ r_4 \\ r_5 \\ r_6 \\ r_7 \\ r_8 \end{Bmatrix} = \begin{Bmatrix} r_1 \\ r_2 \\ 0 \\ 0 \\ 0 \\ 0 \\ 0 \\ 0 \end{Bmatrix}; \quad \{R\} = \begin{Bmatrix} R_1 \\ R_2 \\ R_3 \\ R_4 \\ R_5 \\ R_6 \\ R_7 \\ R_8 \end{Bmatrix} = \begin{Bmatrix} -25 \\ 0 \\ R_3 \\ R_4 \\ R_5 \\ R_6 \\ R_7 \\ R_8 \end{Bmatrix}$$

The stiffness matrices in the global coordinate system are computed considering the global coordinate axes at both ends.

Element a

$\theta = 90°; \ C_x = \cos\theta = 0; \ C_y = \sin\theta = 1$

$$\begin{bmatrix} \bar{k}^a \end{bmatrix} = \frac{AE}{6} \begin{matrix} 7 & 1 & 8 & 2 \\ \begin{bmatrix} 0 & 0 & 0 & 0 \\ 0 & 0 & 0 & 0 \\ 0 & 0 & 1 & -1 \\ 0 & 0 & -1 & 1 \end{bmatrix} & \begin{matrix} 7 \\ 1 \\ 8 \\ 2 \end{matrix} \end{matrix}$$

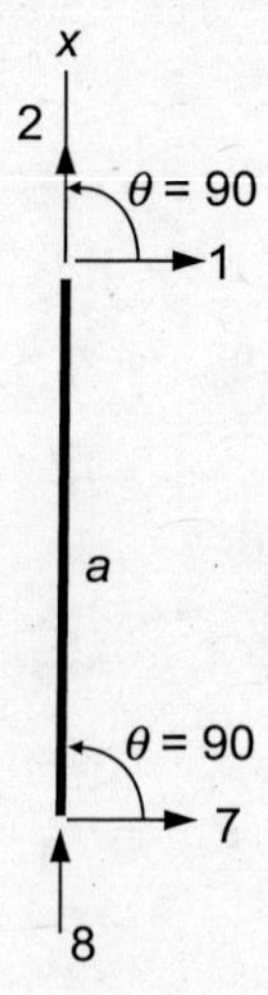

Element b

$\theta = 0°; \ C_x = \cos\theta = 1; \ C_y = \sin\theta = 0$

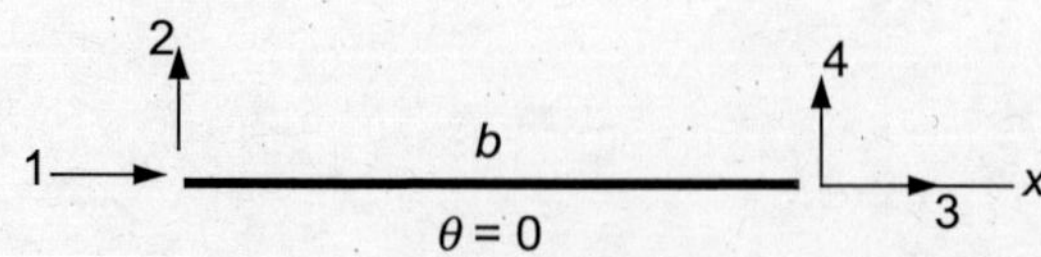

$$\left[\overline{k}^b\right] = \frac{AE}{6}\begin{array}{c}\\\begin{array}{cccc}1 & 3 & 2 & 4\end{array}\\\left[\begin{array}{cccc}1 & -1 & 0 & 0\\-1 & 1 & 0 & 0\\0 & 0 & 0 & 0\\0 & 0 & 0 & 0\end{array}\right]\begin{array}{c}1\\3\\2\\4\end{array}\end{array}$$

Element c

$\theta = 135°$; $C_x = \cos\theta = -0.707$; $C_y = \sin\theta = 0.707$

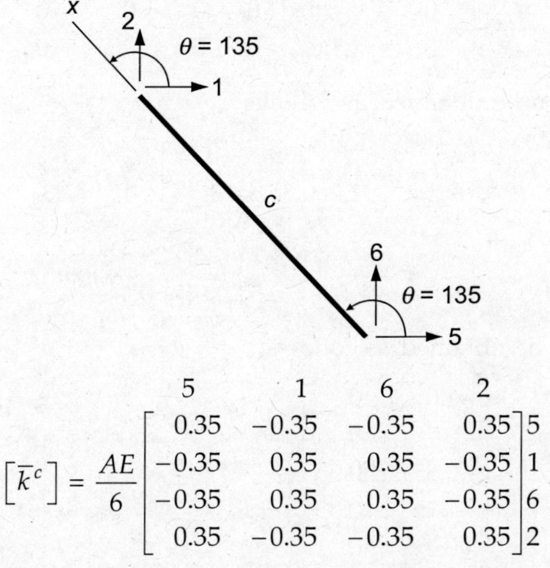

$$\left[\overline{k}^c\right] = \frac{AE}{6}\begin{array}{c}\begin{array}{cccc}5 & 1 & 6 & 2\end{array}\\\left[\begin{array}{cccc}0.35 & -0.35 & -0.35 & 0.35\\-0.35 & 0.35 & 0.35 & -0.35\\-0.35 & 0.35 & 0.35 & -0.35\\0.35 & -0.35 & -0.35 & 0.35\end{array}\right]\begin{array}{c}5\\1\\6\\2\end{array}\end{array}$$

Since element c is too short by 20 mm, the induced forces in this element in the global coordinate system are given by

$$\begin{Bmatrix}\overline{Q}_5'\\\overline{Q}_1'\\\overline{Q}_6'\\\overline{Q}_2'\end{Bmatrix} = \frac{AE\Delta}{L}\begin{Bmatrix}C_x\\-C_x\\C_y\\-C_y\end{Bmatrix} = \frac{(5000)(-0.02)}{6\sqrt{2}}\begin{Bmatrix}-0.707\\0.707\\0.707\\-0.707\end{Bmatrix} = \begin{Bmatrix}8.33\\-8.33\\-8.33\\8.33\end{Bmatrix}\text{kN}$$

Structure stiffness matrix is given by:

$$[K] = \frac{AE}{6}\begin{array}{c}\begin{array}{cccccccc}1 & 2 & 3 & 4 & 5 & 6 & 7 & 8\end{array}\\\left[\begin{array}{cccccccc}1.35 & & & & & & & \\-0.35 & 1.35 & & & & sym & & \\-1.0 & 0 & 1.0 & & & & & \\0 & 0 & 0 & 0 & & & & \\-0.35 & 0.35 & 0 & 0 & 0.35 & & & \\0.35 & -0.35 & 0 & 0 & -0.35 & 0.35 & & \\0 & 0 & 0 & 0 & 0 & 0 & 0 & \\0 & -1.0 & 0 & 0 & 0 & 0 & 0 & 1.0\end{array}\right]\begin{array}{c}1\\2\\3\\4\\5\\6\\7\\8\end{array}\end{array}$$

$$\{R\} = [K]\{r\} + \{\overline{Q}'\}$$

$$
\begin{Bmatrix} -25-8.33 \\ 8.33 \\ \hline R_3 \\ R_4 \\ R_5+8.33 \\ R_6-8.33 \\ R_7 \\ R_8 \end{Bmatrix}
= \frac{AE}{6}
\begin{bmatrix}
1.35 & -0.35 & & & & & & \\
-0.35 & 1.35 & & & sym & & & \\
\hline
-1.0 & 0 & 1.0 & & & & & \\
0 & 0 & 0 & 0 & & & & \\
-0.35 & 0.35 & 0 & 0 & 0.35 & & & \\
0.35 & -0.35 & 0 & 0 & -0.35 & 0.35 & & \\
0 & 0 & 0 & 0 & 0 & 0 & 0 & \\
0 & -1.0 & 0 & 0 & 0 & 0 & 0 & 1.0
\end{bmatrix}
\begin{Bmatrix} r_1 \\ r_2 \\ 0 \\ 0 \\ 0 \\ 0 \\ 0 \\ 0 \end{Bmatrix}
$$

Joint displacements can be obtained as follows:

$$
\begin{Bmatrix} r_1 \\ r_2 \end{Bmatrix}
= \frac{6}{AE}
\begin{bmatrix} 1.35 & -0.35 \\ -0.35 & 1.35 \end{bmatrix}^{-1}
\begin{Bmatrix} -33.33 \\ 8.33 \end{Bmatrix}
$$

$$
\therefore \quad
\begin{Bmatrix} r_1 \\ r_2 \end{Bmatrix}
= \frac{6}{5000}
\begin{Bmatrix} -24.7 \\ 0.3 \end{Bmatrix}
= \begin{Bmatrix} -29.6 \\ 0.4 \end{Bmatrix} mm
$$

Support reactions can be obtained as follows:

$$
\begin{Bmatrix} R_3 \\ R_4 \\ R_5+8.33 \\ R_6-8.33 \\ R_7 \\ R_8 \end{Bmatrix}
=
\begin{bmatrix}
-1.0 & 0 \\
0 & 0 \\
-0.35 & 0.35 \\
0.35 & -0.35 \\
0 & 0 \\
0 & -1.0
\end{bmatrix}
\begin{Bmatrix} -24.7 \\ 0.3 \end{Bmatrix}
\qquad \therefore \quad
\begin{Bmatrix} R_3 \\ R_4 \\ R_5 \\ R_6 \\ R_7 \\ R_8 \end{Bmatrix}
=
\begin{Bmatrix} 24.7 \\ 0 \\ 0.3 \\ -0.3 \\ 0 \\ -0.3 \end{Bmatrix} kN
$$

EXAMPLE 10.7   Determine the slope at the supports, and the deflection at the loading point of the beam shown in Figure E10.7 using the direct stiffness matrix method.

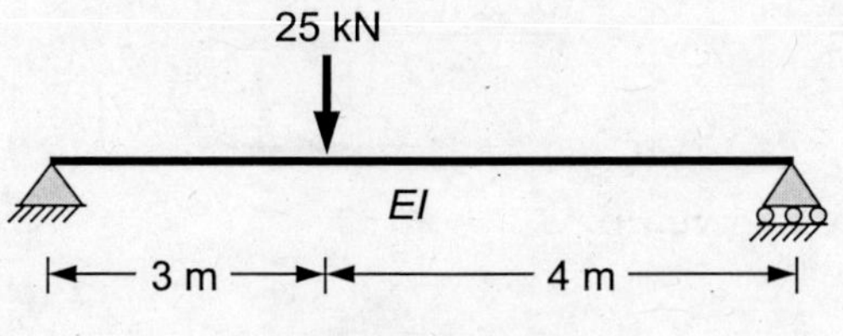

FIGURE E10.7

Solution: As per the direct stiffness matrix method, the total number of degrees of freedom should be six, i.e., two at each node (neglecting the effects of axial deformation of the beam). Since it is required to determine the joint displacements only, the degrees of freedom having zero displacements at the supports need not be considered. Accordingly, four degrees of freedom has been considered to reduce the computation efforts. All degrees of freedom in the beam is shown in the following figure.

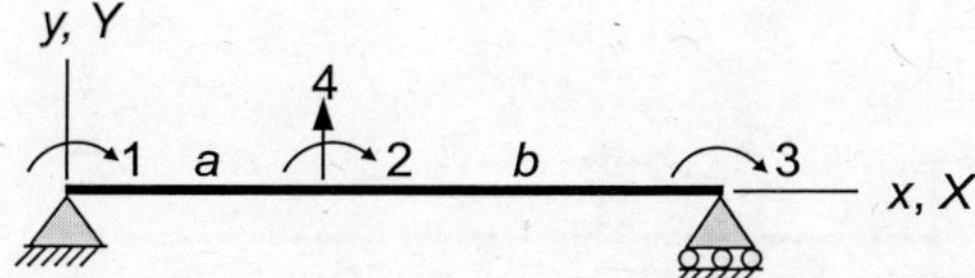

The nodal displacement and the nodal force matrices are as follows:

$$\{r\} = \begin{Bmatrix} r_1 \\ r_2 \\ r_3 \\ r_4 \end{Bmatrix} ; \quad \{R\} = \begin{Bmatrix} R_1 \\ R_2 \\ R_3 \\ R_4 \end{Bmatrix} = \begin{Bmatrix} 0 \\ 0 \\ 0 \\ -25 \end{Bmatrix}$$

The stiffness matrix in the global coordinate system can be computed from the stiffness matrix in the local coordinate system using the transformation matrix. Note that since the member is horizontal, the local and global coordinates coincide, i.e., $\theta = 0°$; $C_x = \cos \theta = 1$; $C_y = \sin \theta = 0$.

One can determine the element stiffness matrices for both the elements and assemble them to get the global stiffness matrix using the equations discussed above.

Alternatively, the global stiffness matrix can be directly determined using the same concept as discussed below. Only one displacement (or rotation) corresponding to a particular degree of freedom should be set to unity and the corresponding forces/reactions representing the stiffness coefficients can be obtained.

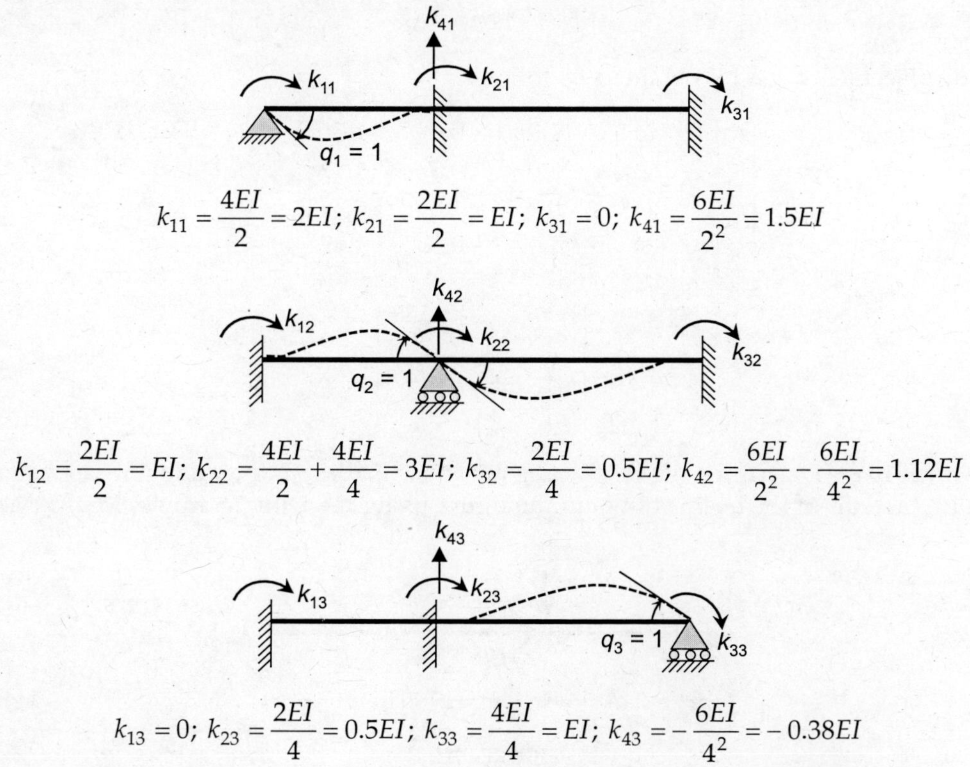

$$k_{11} = \frac{4EI}{2} = 2EI; \; k_{21} = \frac{2EI}{2} = EI; \; k_{31} = 0; \; k_{41} = \frac{6EI}{2^2} = 1.5EI$$

$$k_{12} = \frac{2EI}{2} = EI; \; k_{22} = \frac{4EI}{2} + \frac{4EI}{4} = 3EI; \; k_{32} = \frac{2EI}{4} = 0.5EI; \; k_{42} = \frac{6EI}{2^2} - \frac{6EI}{4^2} = 1.12EI$$

$$k_{13} = 0; \; k_{23} = \frac{2EI}{4} = 0.5EI; \; k_{33} = \frac{4EI}{4} = EI; \; k_{43} = -\frac{6EI}{4^2} = -0.38EI$$

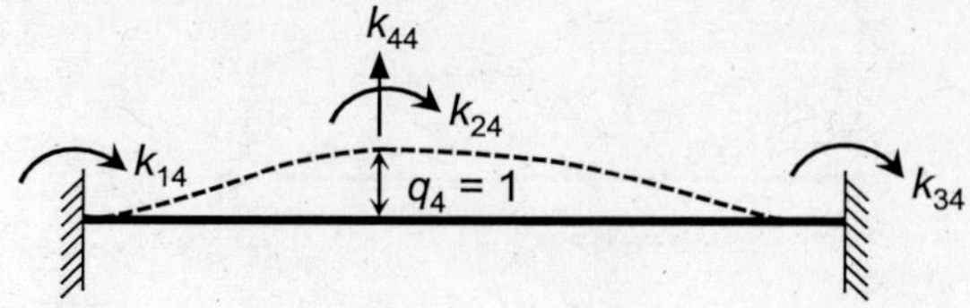

$$k_{14} = \frac{6EI}{2^2} = 1.5EI; \ k_{24} = \frac{6EI}{2^2} - \frac{6EI}{4^2} = 1.12EI; \ k_{34} = -\frac{6EI}{4^2} = -0.38EI; \ k_{44} = \frac{12EI}{2^3} + \frac{12EI}{4^3} = 1.69EI$$

Structure stiffness matrix is given by

$$[K] = EI \begin{array}{c} \\ \end{array} \begin{array}{cccc} 1 & 2 & 3 & 4 \\ \begin{bmatrix} 2.0 & 1.0 & 0 & 1.5 \\ 1.0 & 3.0 & 0.5 & 1.12 \\ 0 & 0.5 & 1.0 & -0.38 \\ 1.5 & 1.12 & -0.38 & 1.69 \end{bmatrix} \begin{array}{c} 1 \\ 2 \\ 3 \\ 4 \end{array} \end{array}$$

Force-displacement relationship is given by $\{R\} = [K]\{r\}$

$$\begin{Bmatrix} 0 \\ 0 \\ 0 \\ -25 \end{Bmatrix} = EI \begin{bmatrix} 2.0 & 1.0 & 0 & 1.5 \\ 1.0 & 3.0 & 0.5 & 1.12 \\ 0 & 0.5 & 1.0 & -0.38 \\ 1.5 & 1.12 & -0.38 & 1.69 \end{bmatrix} \begin{Bmatrix} r_1 \\ r_2 \\ r_3 \\ r_4 \end{Bmatrix}$$

Joint displacements can be obtained as follows:

$$\begin{Bmatrix} r_1 \\ r_2 \\ r_3 \\ r_4 \end{Bmatrix} = \frac{1}{EI} \begin{bmatrix} 2.0 & 1.0 & 0 & 1.5 \\ 1.0 & 3.0 & 0.5 & 1.12 \\ 0 & 0.5 & 1.0 & -0.38 \\ 1.5 & 1.12 & -0.38 & 1.69 \end{bmatrix}^{-1} \begin{Bmatrix} 0 \\ 0 \\ 0 \\ -25 \end{Bmatrix}$$

$$\therefore \quad \begin{Bmatrix} r_1 \\ r_2 \\ r_3 \\ r_4 \end{Bmatrix} = \frac{1}{EI} \begin{Bmatrix} 55.6 \\ 22.1 \\ -44.8 \\ -88.9 \end{Bmatrix}$$

EXAMPLE 10.8 Determine the displacements at the loading point and joints as well as the support reactions of the beam shown in the figure using the Direct Stiffness Matrix Method.

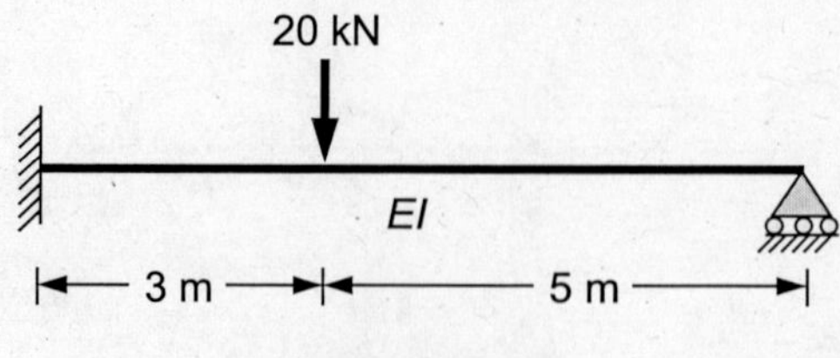

FIGURE E10.8

Solution: All degrees of freedom in the beam is shown in the following figure.

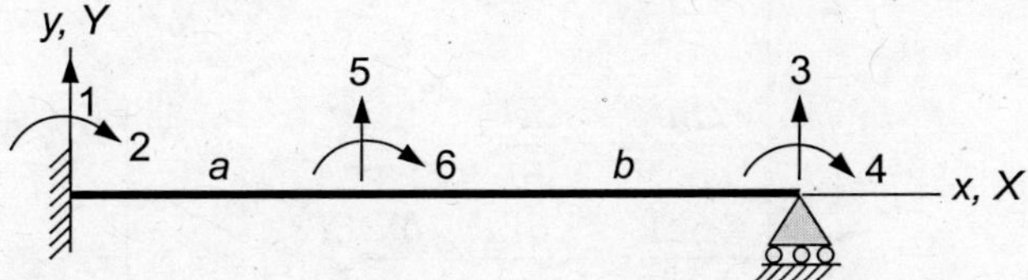

The nodal displacement and the nodal force matrices are as follows:

$$\{r\} = \begin{Bmatrix} r_1 \\ r_2 \\ r_3 \\ r_4 \\ r_5 \\ r_6 \end{Bmatrix} = \begin{Bmatrix} 0 \\ 0 \\ 0 \\ r_4 \\ r_5 \\ r_6 \end{Bmatrix}; \ \{R\} = \begin{Bmatrix} R_1 \\ R_2 \\ R_3 \\ R_4 \\ R_5 \\ R_6 \end{Bmatrix} = \begin{Bmatrix} R_1 \\ R_2 \\ R_3 \\ 0 \\ -20 \\ 0 \end{Bmatrix}$$

Note that since the member is horizontal, the local and global coordinates coincide, i.e.,
$\theta = 0°; C_x = \cos \theta = 1; C_y = \sin \theta = 0$

Element stiffness matrix for the element is given by

$$[k] = \begin{bmatrix} \dfrac{4EI}{L} & & & & sym \\ \dfrac{2EI}{L} & \dfrac{4EI}{L} & & & \\ -\dfrac{6EI}{L^2} & -\dfrac{6EI}{L^2} & \dfrac{12EI}{L^3} & & \\ \dfrac{6EI}{L^2} & \dfrac{6EI}{L^2} & -\dfrac{12EI}{L^3} & \dfrac{12EI}{L^3} \end{bmatrix}$$

Accordingly,

$$\begin{matrix} & \ \ 2 & \ \ \ 6 & \ \ \ 1 & \ \ \ 5 & \\ [\bar{k}^a] = & \begin{bmatrix} \dfrac{4EI}{3} & & & & sym \\ \dfrac{2EI}{3} & \dfrac{4EI}{3} & & & \\ -\dfrac{6EI}{3^2} & -\dfrac{6EI}{3^2} & \dfrac{12EI}{3^3} & & \\ \dfrac{6EI}{3^2} & \dfrac{6EI}{3^2} & -\dfrac{12EI}{3^3} & \dfrac{12EI}{3^3} \end{bmatrix} & \begin{matrix} 2 \\ \\ 6 \\ \\ 1 \\ \\ 5 \end{matrix} \end{matrix}$$

$$
\left[\overline{k}^b\right] =
\begin{matrix}
 & \;6 & \;\;4 & \;\;5 & \;\;3 & \\
\end{matrix}
\begin{bmatrix}
\dfrac{4EI}{5} & & & & \text{sym} \\[2ex]
\dfrac{2EI}{5} & \dfrac{4EI}{5} & & & \\[2ex]
-\dfrac{6EI}{5^2} & -\dfrac{6EI}{5^2} & \dfrac{12EI}{5^3} & & \\[2ex]
\dfrac{6EI}{5^2} & \dfrac{6EI}{5^2} & -\dfrac{12EI}{5^3} & \dfrac{12EI}{5^3} &
\end{bmatrix}
\begin{matrix}
6 \\[2ex] 4 \\[2ex] 5 \\[2ex] 3
\end{matrix}
$$

The structure stiffness matrix is given by

$$
[K] = EI
\begin{matrix}
1 & 2 & 3 & 4 & 5 & 6 \\
\end{matrix}
\begin{bmatrix}
0.44 & & & & & \text{sym} \\
-0.67 & 1.33 & & & & \\
0 & 0 & 0.10 & & & \\
0 & 0 & 0.24 & 0.80 & & \\
-4.48 & 0.67 & -0.10 & -0.24 & 0.54 & \\
-0.67 & 0.67 & 0.24 & 0.40 & 0.43 & 2.13
\end{bmatrix}
\begin{matrix}
1 \\ 2 \\ 3 \\ 4 \\ 5 \\ 6
\end{matrix}
$$

The force-displacement relationship is given by $\{R\} = [K]\{r\}$

$$
\begin{Bmatrix}
R_1 \\ R_2 \\ R_3 \\ \hline 0 \\ -20 \\ 0
\end{Bmatrix}
= EI
\left[
\begin{array}{ccc|ccc}
0.44 & -0.67 & 0 & 0 & -0.44 & -0.67 \\
-0.67 & 1.33 & 0 & 0 & 0.67 & 0.67 \\
0 & 0 & 0.10 & 0.24 & -0.10 & 0.24 \\
\hline
0 & 0 & 0.24 & 0.80 & -0.24 & 0.40 \\
-0.44 & 0.67 & -0.10 & -0.24 & 0.54 & 0.43 \\
-0.67 & 0.67 & 0.24 & 0.40 & 0.43 & 2.13
\end{array}
\right]
\begin{Bmatrix}
0 \\ 0 \\ 0 \\ \hline r_4 \\ r_5 \\ r_6
\end{Bmatrix}
$$

Joint displacements can be obtained as follows:

$$
\begin{Bmatrix} r_4 \\ r_5 \\ r_6 \end{Bmatrix}
= \frac{1}{EI}
\begin{bmatrix}
0.80 & -0.24 & 0.40 \\
-0.24 & 0.54 & 0.43 \\
0.40 & 0.43 & 2.13
\end{bmatrix}^{-1}
\begin{Bmatrix} 0 \\ -20 \\ 0 \end{Bmatrix}
$$

$$
\therefore \quad
\begin{Bmatrix} r_4 \\ r_5 \\ r_6 \end{Bmatrix}
= \frac{1}{EI}
\begin{Bmatrix} -28.4 \\ -64.2 \\ 18.3 \end{Bmatrix}
$$

Support reactions can be obtained as follows:

$$
\begin{Bmatrix} R_1 \\ R_2 \\ R_3 \end{Bmatrix}
=
\begin{bmatrix}
0 & -0.44 & -0.67 \\
0 & 0.67 & 0.67 \\
0.24 & -0.10 & 0.24
\end{bmatrix}
\begin{Bmatrix} -28.4 \\ -64.2 \\ 18.3 \end{Bmatrix}
\qquad
\therefore
\begin{Bmatrix} R_1 \\ R_2 \\ R_3 \end{Bmatrix}
=
\begin{Bmatrix} 16.0 \\ -30.8 \\ 4.0 \end{Bmatrix}
$$

EXAMPLE 10.9 Determine the joint displacements, the support reactions, and the member forces of the beam shown in Figure E10.9 using the Direct Stiffness Matrix Method.

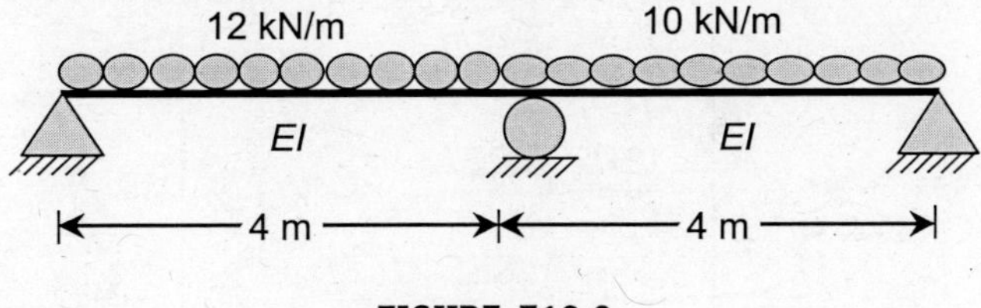

FIGURE E10.9

Solution: Since the loads are uniformly distributed over the span, it is required to transfer the load effects to the ends of members assuming the fixed ended beams. The end moments and reactions are now applied at the joints. The beam with nodal forces will be first considered for the analysis using the Direct stiffness method.

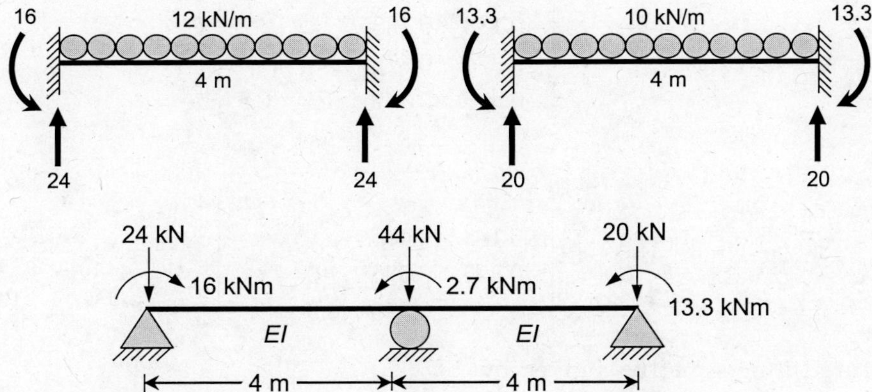

All degrees of freedom in the beam are shown in the following figure.

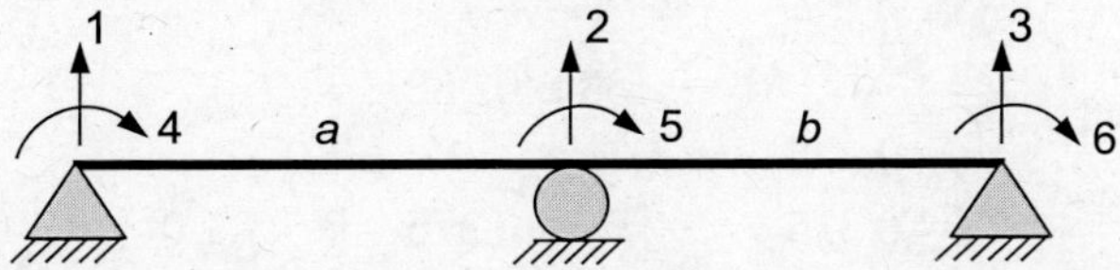

The nodal displacement and the nodal force matrices are as follows:

$$\{r\} = \begin{Bmatrix} r_1 \\ r_2 \\ r_3 \\ r_4 \\ r_5 \\ r_6 \end{Bmatrix} = \begin{Bmatrix} 0 \\ 0 \\ 0 \\ r_4 \\ r_5 \\ r_6 \end{Bmatrix}; \ \{R\} = \begin{Bmatrix} R_1 - 24 \\ R_2 - 44 \\ R_3 - 20 \\ 16 \\ -2.7 \\ -13.3 \end{Bmatrix}$$

The stiffness matrix for the element is given by

$$
[k] =
\begin{array}{cccc}
1 & 2 & 4 & 5
\end{array}
\begin{bmatrix}
\dfrac{12EI}{L^3} & & & sym \\[2ex]
-\dfrac{12EI}{L^3} & \dfrac{12EI}{L^3} & & \\[2ex]
-\dfrac{6EI}{L^2} & \dfrac{6EI}{L^2} & \dfrac{4EI}{L} & \\[2ex]
-\dfrac{6EI}{L^2} & \dfrac{6EI}{L^2} & \dfrac{2EI}{L} & \dfrac{4EI}{L}
\end{bmatrix}
\begin{array}{c}
1 \\[2ex] 2 \\[2ex] 4 \\[2ex] 5
\end{array}
$$

Accordingly,

$$
\left[\bar{k}^a\right] = EI
\begin{array}{cccc}
1 & 2 & 4 & 5
\end{array}
\begin{bmatrix}
0.188 & & & sym \\
-0.188 & 0.188 & & \\
-0.375 & 0.375 & 1.0 & \\
-0.375 & 0.375 & 0.5 & 0.5
\end{bmatrix}
\begin{array}{c}
1 \\ 2 \\ 4 \\ 5
\end{array}
$$

$$
\left[\bar{k}^b\right] = EI
\begin{array}{cccc}
2 & 3 & 5 & 6
\end{array}
\begin{bmatrix}
0.188 & & & sym \\
-0.188 & 0.188 & & \\
-0.375 & 0.375 & 1.0 & \\
-0.375 & 0.375 & 0.5 & 1.0
\end{bmatrix}
\begin{array}{c}
2 \\ 3 \\ 5 \\ 6
\end{array}
$$

The structure stiffness matrix is given by

$$
[K] = EI
\begin{array}{cccccc}
1 & 2 & 3 & 4 & 5 & 6
\end{array}
\begin{bmatrix}
0.188 & & & & & sym \\
-0.188 & 0.375 & & & & \\
0 & -0.188 & 0.188 & & & \\
-0.375 & 0.375 & 0 & 1.0 & & \\
-0.375 & 0 & 0.375 & 0.5 & 2.0 & \\
0 & -0.375 & 0.375 & 0 & 0.5 & 1.0
\end{bmatrix}
\begin{array}{c}
1 \\ 2 \\ 3 \\ 4 \\ 5 \\ 6
\end{array}
$$

The force-displacement relationship is given by $\{R\} = [K]\{r\}$

$$
\begin{Bmatrix}
R_1 - 24 \\
R_2 - 44 \\
R_3 - 20 \\
\hline
16.0 \\
-2.7 \\
-13.3
\end{Bmatrix}
= EI
\left[
\begin{array}{ccc|ccc}
0.188 & -0.188 & 0 & -0.375 & -0.375 & 0 \\
-0.188 & 0.375 & -0.188 & 0.375 & 0 & -0.375 \\
0 & -0.188 & 0.188 & 0 & 0.375 & 0.375 \\
\hline
-0.375 & 0.375 & 0 & 1.0 & 0.5 & 0 \\
-0.375 & 0 & 0.375 & 0.5 & 2.0 & 0.5 \\
0 & -0.375 & 0.375 & 0 & 0.5 & 1.0
\end{array}
\right]
\begin{Bmatrix}
0 \\ 0 \\ 0 \\ r_4 \\ r_5 \\ r_6
\end{Bmatrix}
$$

Joint displacements can be obtained as follows:

$$\begin{Bmatrix} r_4 \\ r_5 \\ r_6 \end{Bmatrix} = \frac{1}{EI} \begin{bmatrix} 1.0 & 0.5 & 0 \\ 0.5 & 2.0 & 0.5 \\ 0 & 0.5 & 1.0 \end{bmatrix}^{-1} \begin{Bmatrix} 16.0 \\ -2.7 \\ -13.3 \end{Bmatrix}$$

$$\therefore \quad \begin{Bmatrix} r_4 \\ r_5 \\ r_6 \end{Bmatrix} = \frac{1}{EI} \begin{Bmatrix} 17.35 \\ -2.70 \\ -11.95 \end{Bmatrix}$$

Support reactions can be obtained as follows:

$$\begin{Bmatrix} R_1 \\ R_2 \\ R_3 \end{Bmatrix} = \begin{bmatrix} -0.375 & -0.375 & 0 \\ 0.375 & 0 & -0.375 \\ 0 & 0.375 & 0.375 \end{bmatrix} \begin{Bmatrix} 17.5 \\ -2.70 \\ -11.95 \end{Bmatrix} + \begin{Bmatrix} 24 \\ 44 \\ 20 \end{Bmatrix} \qquad \therefore \quad \begin{Bmatrix} R_1 \\ R_2 \\ R_3 \end{Bmatrix} = \begin{Bmatrix} 18.5 \\ 55.0 \\ 14.5 \end{Bmatrix}$$

The member forces can be obtained by computing the element force matrix and then adding the effects of fixed end moments/reactions to the corresponding components.

Element a:

$$\{\bar{Q}^a\} = [\bar{k}^a]\{\bar{q}^a\}$$

$$\begin{Bmatrix} \bar{Q}_1^a \\ \bar{Q}_2^a \\ \bar{Q}_4^a \\ \bar{Q}_5^a \end{Bmatrix} = \begin{bmatrix} 0.188 & -0.188 & -0.375 & -0.375 \\ -0.188 & 0.188 & 0.375 & 0.375 \\ -0.375 & 0.375 & 1.0 & 0.5 \\ -0.375 & 0.375 & 0.5 & 1.0 \end{bmatrix} \begin{Bmatrix} 0 \\ 0 \\ 17.35 \\ -2.70 \end{Bmatrix} = \begin{Bmatrix} -5.5 \\ 5.5 \\ 16.0 \\ 6.0 \end{Bmatrix}$$

Element b:

$$\{\bar{Q}^b\} = [\bar{k}^b]\{\bar{q}^b\}$$

$$\begin{Bmatrix} \bar{Q}_2^b \\ \bar{Q}_3^b \\ \bar{Q}_5^b \\ \bar{Q}_6^b \end{Bmatrix} = \begin{bmatrix} 0.188 & -0.188 & -0.375 & -0.375 \\ -0.188 & 0.188 & 0.375 & 0.375 \\ -0.375 & 0.375 & 1.0 & 0.5 \\ -0.375 & 0.375 & 0.5 & 1.0 \end{bmatrix} \begin{Bmatrix} 0 \\ 0 \\ -2.70 \\ -11.95 \end{Bmatrix} = \begin{Bmatrix} 5.5 \\ -5.5 \\ -8.7 \\ -13.3 \end{Bmatrix}$$

The final end moments are as follows:

$$\begin{Bmatrix} \bar{Q}_4^a \\ \bar{Q}_5^a \\ \bar{Q}_5^b \\ \bar{Q}_6^b \end{Bmatrix} = \begin{Bmatrix} 16.0 \\ 6.0 \\ -8.7 \\ -13.3 \end{Bmatrix} + \begin{Bmatrix} -16.0 \\ 16.0 \\ -13.3 \\ 13.3 \end{Bmatrix} = \begin{Bmatrix} 0 \\ 22.0 \\ -22.0 \\ 0 \end{Bmatrix} \text{kNm}$$

EXAMPLE 10.10 Determine the joint displacements, the support reactions, and the member forces of the beam shown in Figure E10.10 using the Direct Stiffness Matrix Method.

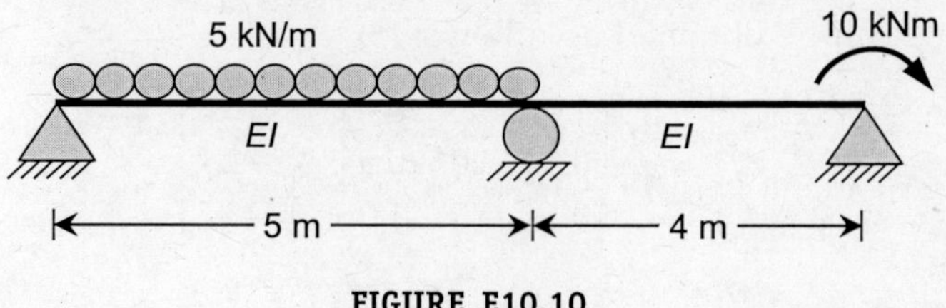

FIGURE E10.10

Solution: The beam with nodal forces derived from considering the fixed-ended beam for the uniformly distributed load will be first considered for the analysis as shown below.

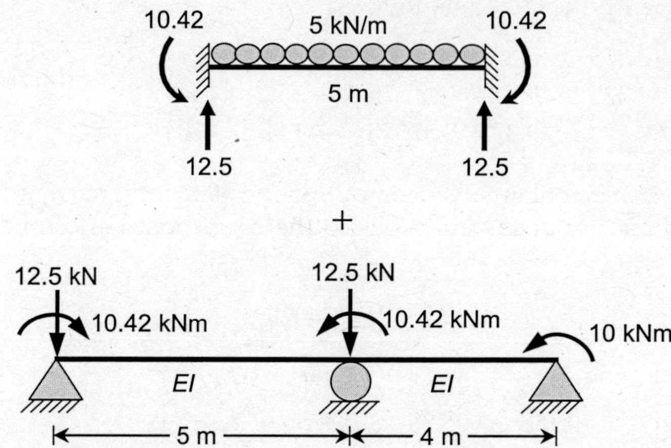

The beam with nodes and elements is shown in the following figure.

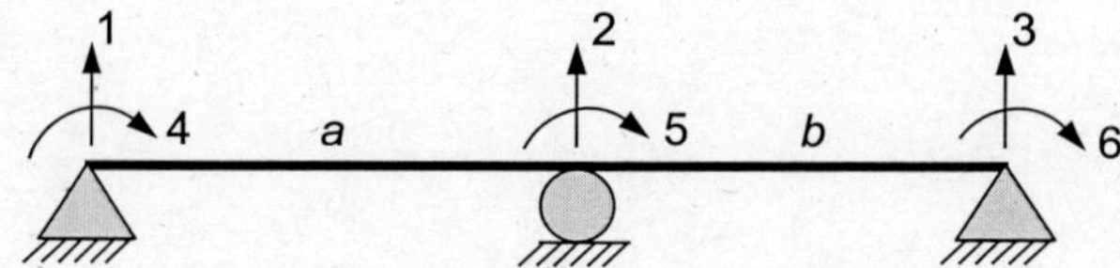

The nodal displacement and the nodal force matrices are as follows:

$$\{r\} = \begin{Bmatrix} r_1 \\ r_2 \\ r_3 \\ r_4 \\ r_5 \\ r_6 \end{Bmatrix} = \begin{Bmatrix} 0 \\ 0 \\ 0 \\ r_4 \\ r_5 \\ r_6 \end{Bmatrix}; \ \{R\} = \begin{Bmatrix} R_1 - 12.5 \\ R_2 - 12.5 \\ R_3 \\ 10.42 \\ -10.42 \\ -10.0 \end{Bmatrix}$$

The stiffness matrix for the element is given by

$$
[k] = \begin{bmatrix}
\dfrac{12EI}{L^3} & & & & sym \\[2ex]
-\dfrac{12EI}{L^3} & \dfrac{12EI}{L^3} & & \\[2ex]
-\dfrac{6EI}{L^2} & \dfrac{6EI}{L^2} & \dfrac{4EI}{L} & \\[2ex]
-\dfrac{6EI}{L^2} & \dfrac{6EI}{L^2} & \dfrac{2EI}{L} & \dfrac{4EI}{L}
\end{bmatrix}
\begin{matrix} 1 \\[2ex] 2 \\[2ex] 4 \\[2ex] 5 \end{matrix}
$$

$$
\begin{matrix} & 1 & 2 & 4 & 5 \end{matrix}
$$

Accordingly,

$$
\left[\overline{k}^a\right] = EI \begin{bmatrix}
0.096 & & & sym \\
-0.096 & 0.096 & & \\
-0.240 & 0.240 & 0.80 & \\
-0.240 & 0.240 & 0.40 & 0.80
\end{bmatrix}
\begin{matrix} 1 \\ 2 \\ 4 \\ 5 \end{matrix}
$$

$$
\begin{matrix} & 1 & 2 & 4 & 5 \end{matrix}
$$

$$
\left[\overline{k}^b\right] = EI \begin{bmatrix}
0.188 & & & sym \\
-0.188 & 0.188 & & \\
-0.375 & 0.375 & 1.0 & \\
-0.375 & 0.375 & 0.5 & 1.0
\end{bmatrix}
\begin{matrix} 2 \\ 3 \\ 5 \\ 6 \end{matrix}
$$

$$
\begin{matrix} & 2 & 3 & 5 & 6 \end{matrix}
$$

The structure stiffness matrix is given by

$$
[K] = EI \begin{bmatrix}
0.096 & & & & & sym \\
-0.096 & 0.284 & & & & \\
0 & -0.188 & 0.188 & & & \\
-0.240 & 0.240 & 0 & 0.8 & & \\
-0.240 & -0.135 & 0.375 & 0.4 & 1.8 & \\
0 & -0.375 & 0.375 & 0 & 0.5 & 1.0
\end{bmatrix}
\begin{matrix} 1 \\ 2 \\ 3 \\ 4 \\ 5 \\ 6 \end{matrix}
$$

$$
\begin{matrix} & 1 & 2 & 3 & 4 & 5 & 6 \end{matrix}
$$

The force-displacement relationship is given by $\{R\} = [K]\{r\}$

$$
\begin{Bmatrix}
R_1 - 12.5 \\
R_2 - 12.5 \\
R_3 \\
\hline
10.42 \\
-10.42 \\
-10.0
\end{Bmatrix} = EI
\left[\begin{array}{ccc|ccc}
0.096 & -0.096 & 0 & -0.240 & -0.240 & 0 \\
-0.096 & 0.284 & -0.188 & 0.240 & -0.135 & -0.375 \\
0 & -0.188 & 0.188 & 0 & 0.375 & 0.375 \\
\hline
-0.240 & 0.240 & 0 & 0.80 & 0.40 & 0 \\
-0.240 & -0.135 & 0.375 & 0.40 & 1.80 & 0.50 \\
0 & -0.375 & 0.375 & 0 & 0.50 & 1.00
\end{array}\right]
\begin{Bmatrix}
0 \\
0 \\
0 \\
r_4 \\
r_5 \\
r_6
\end{Bmatrix}
$$

Joint displacements can be obtained as follows:

$$\begin{Bmatrix} r_4 \\ r_5 \\ r_6 \end{Bmatrix} = \frac{1}{EI} \begin{bmatrix} 0.8 & 0.4 & 0 \\ 0.4 & 1.8 & 0.5 \\ 0 & 0.5 & 1.0 \end{bmatrix}^{-1} \begin{Bmatrix} 10.42 \\ -10.42 \\ -10.0 \end{Bmatrix}$$

$$\therefore \quad \begin{Bmatrix} r_4 \\ r_5 \\ r_6 \end{Bmatrix} = \frac{1}{EI} \begin{Bmatrix} 16.96 \\ -7.87 \\ -6.06 \end{Bmatrix}$$

Support reactions can be obtained as follows:

$$\begin{Bmatrix} R_1 \\ R_2 \\ R_3 \end{Bmatrix} = \begin{bmatrix} -0.240 & -0.240 & 0 \\ 0.240 & -0.135 & -0.375 \\ 0 & 0.375 & 0.375 \end{bmatrix} \begin{Bmatrix} 16.96 \\ -7.87 \\ -6.06 \end{Bmatrix} + \begin{Bmatrix} 12.5 \\ 12.5 \\ 0 \end{Bmatrix} \quad \therefore \quad \begin{Bmatrix} R_1 \\ R_2 \\ R_3 \end{Bmatrix} = \begin{Bmatrix} 10.3 \\ 20.0 \\ -5.3 \end{Bmatrix}$$

The member forces can be obtained by computing the element force matrix and then adding the effects of fixed end moments/reactions to the corresponding components.

Element a:

$$\{\bar{Q}^a\} = [\bar{k}^a]\{\bar{q}^a\}$$

$$\begin{Bmatrix} \bar{Q}_1^a \\ \bar{Q}_2^a \\ \bar{Q}_4^a \\ \bar{Q}_5^a \end{Bmatrix} = \begin{bmatrix} 0.096 & -0.096 & -0.240 & -0.240 \\ -0.096 & 0.096 & 0.240 & 0.240 \\ -0.240 & 0.240 & 0.80 & 0.40 \\ -0.240 & 0.240 & 0.40 & 0.80 \end{bmatrix} \begin{Bmatrix} 0 \\ 0 \\ 16.96 \\ -7.87 \end{Bmatrix} = \begin{Bmatrix} -2.18 \\ 2.18 \\ 10.42 \\ 0.49 \end{Bmatrix}$$

Element b:

$$\{\bar{Q}^b\} = [\bar{k}^b]\{\bar{q}^b\}$$

$$\begin{Bmatrix} \bar{Q}_2^b \\ \bar{Q}_3^b \\ \bar{Q}_5^b \\ \bar{Q}_6^b \end{Bmatrix} = \begin{bmatrix} 0.188 & -0.188 & -0.375 & -0.375 \\ -0.188 & 0.188 & 0.375 & 0.375 \\ -0.375 & 0.375 & 1.0 & 0.5 \\ -0.375 & 0.375 & 0.5 & 1.0 \end{bmatrix} \begin{Bmatrix} 0 \\ 0 \\ -7.87 \\ -6.06 \end{Bmatrix} = \begin{Bmatrix} 5.22 \\ -5.22 \\ -10.9 \\ -10.0 \end{Bmatrix}$$

The final end moments are as follows:

$$\begin{Bmatrix} \bar{Q}_4^a \\ \bar{Q}_5^a \\ \bar{Q}_5^b \\ \bar{Q}_6^b \end{Bmatrix} = \begin{Bmatrix} 10.42 \\ 0.49 \\ -10.9 \\ -10.0 \end{Bmatrix} + \begin{Bmatrix} -10.42 \\ 10.42 \\ 0 \\ 0 \end{Bmatrix} = \begin{Bmatrix} 0 \\ 10.9 \\ -10.9 \\ -10.0 \end{Bmatrix} \text{kNm}$$

EXAMPLE 10.11 Analyze the rigid frame shown in Figure E10.11 using the Direct Stiffness Matrix Method.

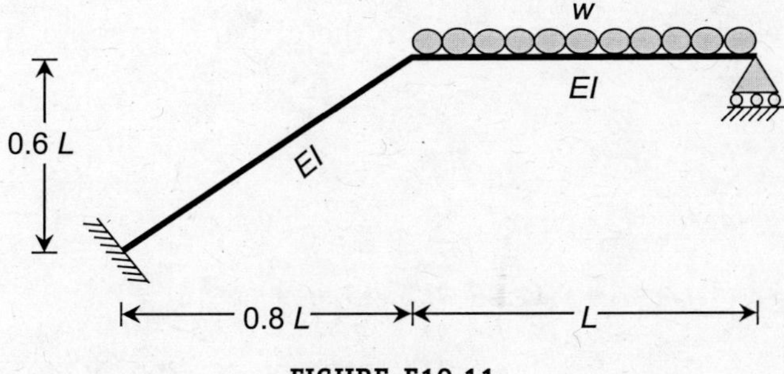

FIGURE E10.11

Solution: The equivalent forces and moments acting at the joints of the frame can be computed considering the fixed ended beam as shown below.

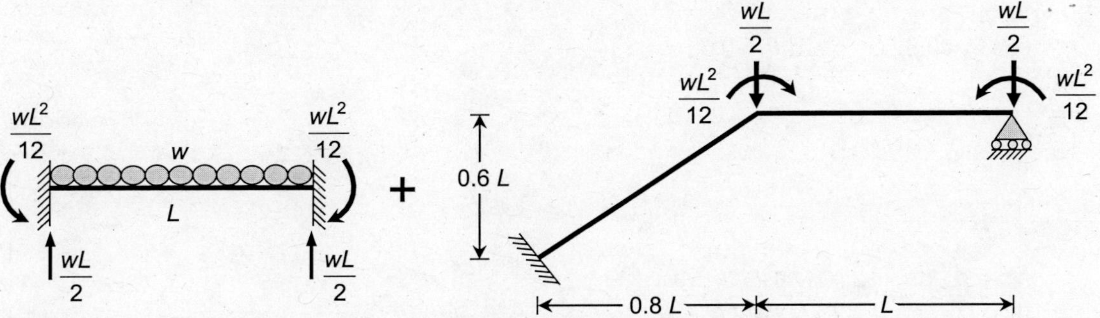

All the nodes and elements are shown in the following figure.

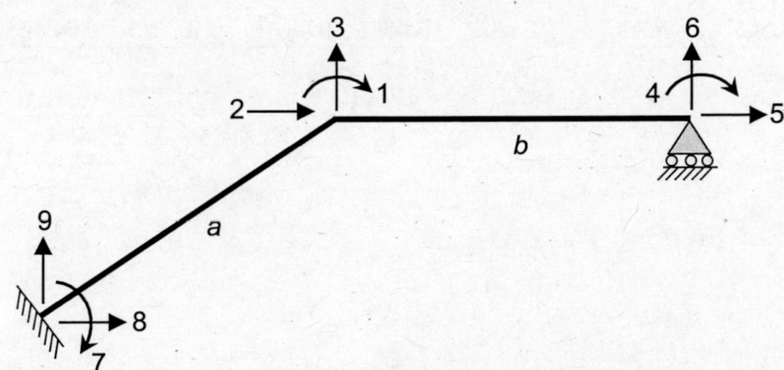

The nodal displacement and the nodal force matrices are as follows:

$$\{r\} = \begin{Bmatrix} r_1 \\ r_2 \\ r_3 \\ r_4 \\ r_5 \\ r_6 \\ r_7 \\ r_8 \\ r_9 \end{Bmatrix} = \begin{Bmatrix} r_1 \\ r_2 \\ r_3 \\ r_4 \\ r_5 \\ 0 \\ 0 \\ 0 \\ 0 \end{Bmatrix} ; \quad \{R\} = \begin{Bmatrix} R_1 \\ R_2 \\ R_3 \\ R_4 \\ R_5 \\ R_6 \\ R_7 \\ R_8 \\ R_9 \end{Bmatrix} = \begin{Bmatrix} wL^2/12 \\ 0 \\ -wL/2 \\ -wL^2/12 \\ 0 \\ R_6 - wL/2 \\ R_7 \\ R_8 \\ R_9 \end{Bmatrix}$$

Let's compute the element stiffness matrix in the global coordinate system.

$$[\overline{k}] = \begin{bmatrix}
\dfrac{4EI}{L} & & & & & \text{Sym.} \\[2ex]
\dfrac{2EI}{L} & \dfrac{4EI}{L} \\[2ex]
-\dfrac{6EI}{L^2}C_x & -\dfrac{6EI}{L^2}C_x & \left(\dfrac{12EI}{L^3}C_x{}^2 + \dfrac{AE}{L}C_y{}^2\right) \\[2ex]
\dfrac{6EI}{L^2}C_x & \dfrac{6EI}{L^2}C_x & -\left(\dfrac{12EI}{L^3}C_x{}^2 + \dfrac{AE}{L}C_y{}^2\right) & \left(\dfrac{12EI}{L^3}C_x{}^2 + \dfrac{AE}{L}C_y{}^2\right) \\[2ex]
\dfrac{6EI}{L^2}C_y & \dfrac{6EI}{L^2}C_y & \left(-\dfrac{12EI}{L^3} + \dfrac{AE}{L}\right)C_xC_y & \left(\dfrac{12EI}{L^3} - \dfrac{AE}{L}\right)C_xC_y & \left(\dfrac{12EI}{L^3}C_y{}^2 + \dfrac{AE}{L}C_x{}^2\right) \\[2ex]
-\dfrac{6EI}{L^2}C_y & -\dfrac{6EI}{L^2}C_y & \left(\dfrac{12EI}{L^3} - \dfrac{AE}{L}\right)C_xC_y & \left(-\dfrac{12EI}{L^3} + \dfrac{AE}{L}\right)C_xC_y & -\left(\dfrac{12EI}{L^3}C_y{}^2 + \dfrac{AE}{L}C_x{}^2\right) & \left(\dfrac{12EI}{L^3}C_y{}^2 + \dfrac{AE}{L}C_x{}^2\right)
\end{bmatrix}$$

Element a:
$$C_x = \cos\theta = 0.8; \quad C_y = \sin\theta = 0.6$$

$$\begin{array}{cccccc}
7 & 1 & 9 & 3 & 8 & 2
\end{array}$$

$$[\overline{k}^a] = \begin{bmatrix}
\dfrac{4EI}{L} & & & & & \text{Sym.} \\[2ex]
\dfrac{2EI}{L} & \dfrac{4EI}{L} \\[2ex]
-\dfrac{4.8EI}{L^2} & -\dfrac{4.8EI}{L^2} & \dfrac{7.68EI}{L^3} + \dfrac{0.36AE}{L} \\[2ex]
\dfrac{4.8EI}{L^2} & \dfrac{4.8EI}{L^2} & -\dfrac{7.68EI}{L^3} - \dfrac{0.36AE}{L} & \dfrac{7.68EI}{L^3} + \dfrac{0.48AE}{L} \\[2ex]
\dfrac{3.6EI}{L^2} & \dfrac{3.6EI}{L^2} & -\dfrac{5.76EI}{L^3} + \dfrac{0.48AE}{L} & \dfrac{5.76EI}{L^3} - \dfrac{0.48AE}{L} & \dfrac{4.32EI}{L^3} + \dfrac{0.64AE}{L} \\[2ex]
-\dfrac{3.6EI}{L^2} & -\dfrac{3.6EI}{L^2} & \dfrac{5.76EI}{L^3} - \dfrac{0.48AE}{L} & -\dfrac{5.76EI}{L^3} + \dfrac{0.48AE}{L} & -\dfrac{4.32EI}{L^3} - \dfrac{0.64AE}{L} & \dfrac{4.32EI}{L^3} + \dfrac{0.64AE}{L}
\end{bmatrix}\begin{matrix} 7 \\[2ex] 1 \\[2ex] 9 \\[2ex] 3 \\[2ex] 8 \\[2ex] 2 \end{matrix}$$

Element b:
$$C_x = \cos\theta = 1; \quad C_y = \sin\theta = 0$$

$$\begin{array}{cccccc}
1 & 4 & 3 & 6 & 2 & 5
\end{array}$$

$$[\overline{k}^b] = \begin{bmatrix}
\dfrac{4EI}{L} & & & & & \text{Sym.} \\[2ex]
\dfrac{2EI}{L} & \dfrac{4EI}{L} \\[2ex]
-\dfrac{6EI}{L^2} & -\dfrac{6EI}{L^2} & \dfrac{12EI}{L^3} \\[2ex]
\dfrac{6EI}{L^2} & \dfrac{6EI}{L^2} & -\dfrac{12EI}{L^3} & \dfrac{12EI}{L^3} \\[2ex]
0 & 0 & 0 & 0 & \dfrac{AE}{L} \\[2ex]
0 & 0 & 0 & 0 & -\dfrac{AE}{L} & \dfrac{AE}{L}
\end{bmatrix}\begin{matrix} 1 \\[2ex] 4 \\[2ex] 3 \\[2ex] 6 \\[2ex] 2 \\[2ex] 5 \end{matrix}$$

The structure stiffness matrix is given by

$$[K] = EI \begin{bmatrix} \dfrac{8EI}{L} & & & & & & & & \\[2ex] -\dfrac{3.6EI}{L^2} & \dfrac{4.32EI}{L^3} + \dfrac{1.64AE}{L} & & & & & & & \\[2ex] \dfrac{1.2EI}{L^2} & -\dfrac{5.76EI}{L^3} + \dfrac{0.48AE}{L} & \dfrac{19.68EI}{L^3} + \dfrac{0.36AE}{L} & & & & & & \\[2ex] \dfrac{2EI}{L} & 0 & -\dfrac{6EI}{L^2} & \dfrac{4EI}{L} & & & & & \\[2ex] 0 & -\dfrac{AE}{L} & 0 & 0 & \dfrac{AE}{L} & & & & \\[2ex] \dfrac{6EI}{L^2} & 0 & -\dfrac{12EI}{L^3} & \dfrac{6EI}{L^2} & 0 & \dfrac{12EI}{L^3} & & & \\[2ex] \dfrac{2EI}{L} & -\dfrac{3.6EI}{L^2} & \dfrac{4.8EI}{L^2} & 0 & 0 & 0 & \dfrac{4EI}{L} & & \\[2ex] \dfrac{3.6EI}{L^2} & -\dfrac{4.32EI}{L^3} - \dfrac{0.64AE}{L} & \dfrac{5.76EI}{L^3} - \dfrac{0.48AE}{L} & 0 & 0 & 0 & \dfrac{3.6EI}{L^2} & \dfrac{4.32EI}{L^3} + \dfrac{0.64AE}{L} & \\[2ex] -\dfrac{4.8EI}{L^2} & \dfrac{5.76EI}{L^3} - \dfrac{0.48AE}{L} & -\dfrac{7.68EI}{L^3} - \dfrac{0.36AE}{L} & 0 & 0 & 0 & -\dfrac{4.8EI}{L^2} & -\dfrac{5.76EI}{L^3} + \dfrac{0.48AE}{L} & \dfrac{7.68EI}{L^3} + \dfrac{0.36AE}{L} \end{bmatrix}$$

The force-displacement relationship is given by $\{R\} = [K]\{r\}$.

Joint displacements can be obtained by solving the part of the matrix equation as follows:

$$\begin{Bmatrix} r_1 \\ r_2 \\ r_3 \\ r_4 \\ r_5 \end{Bmatrix} = \frac{L}{EI} \begin{bmatrix} 8 & & & & \\ -3.6L & 3284.3L & & & \\ -1.2L & 954.2L & 739.7L & & \\ 2 & 0 & -6 & 4 & \\ 0 & -2000L & 0 & 0 & 2000L \end{bmatrix} \begin{Bmatrix} wL^2/12 \\ 0 \\ -wL/2 \\ -wL^2/12 \\ 0 \end{Bmatrix}$$

$$\therefore \quad \begin{Bmatrix} r_1 \\ r_2 \\ r_3 \\ r_4 \\ r_5 \end{Bmatrix} = \frac{wL^3}{EI} \begin{Bmatrix} 0.042 \\ 0.028L \\ -0.037L \\ -0.098 \\ 0.028L \end{Bmatrix}$$

Support reactions can be obtained as follows:

$$\begin{Bmatrix} R_6 \\ R_7 \\ R_8 \\ R_9 \end{Bmatrix} = \begin{bmatrix} 6 & 0 & -12 & 6 & 0 \\ 2 & -3.6 & 4.8 & 0 & 0 \\ 3.6 & -1284.3 & -954.2 & 0 & 0 \\ -4.8 & -954.2 & -727.7 & 0 & 0 \end{bmatrix} \begin{Bmatrix} 0.042 \\ 0.028L \\ -0.037L \\ -0.098 \\ 0.028L \end{Bmatrix} wL + \begin{Bmatrix} wL/2 \\ 0 \\ 0 \\ 0 \end{Bmatrix} \quad \therefore \quad \begin{Bmatrix} R_6 \\ R_7 \\ R_8 \\ R_9 \end{Bmatrix} = \begin{Bmatrix} 0.61 \\ -0.20L \\ 0 \\ 0.39 \end{Bmatrix} wL$$

10.9 PROBLEMS

10.1 Determine the rotation at the roller support and the deflection at the mid-span of the beam shown below using the Direct stiffness matrix method. Assume $EI = 1000$ kNm^2.

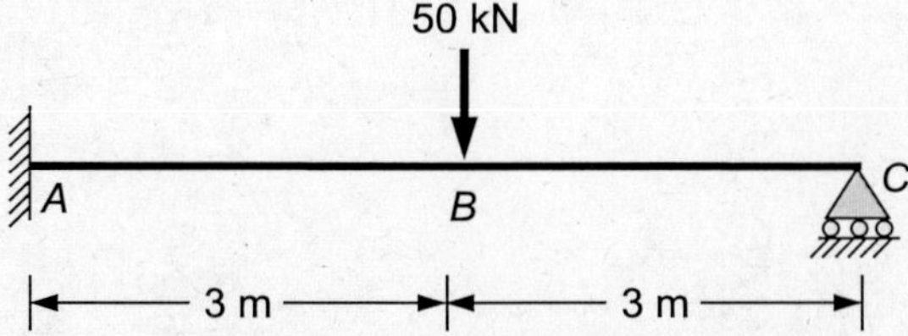

10.2 Determine the reactions and moments at the supports and the rotation at the supports B and C of the beam shown below using the Direct stiffness matrix method. Assume $EI = 5000$ kNm^2.

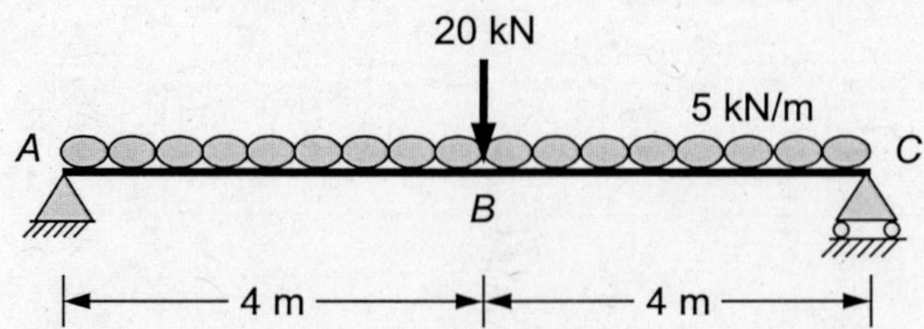

10.3 Analyze the continuous beam shown below using the Direct Stiffness matrix method. Determine the support reactions, the member end rotations, and the support moments.

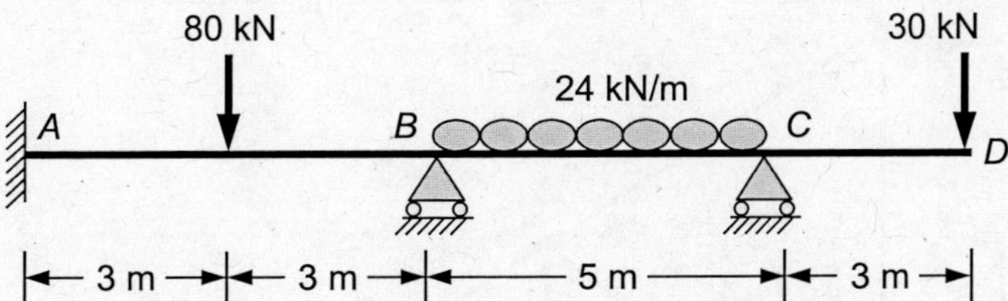

10.4 Determine the support reactions, the member forces and the joint displacements of the truss shown below using the Direct stiffness matrix method. All members are of same area of cross-section. Take AE = 8000 kN.

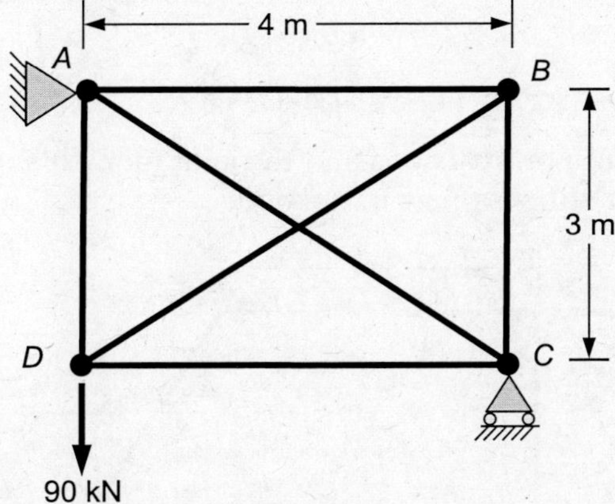

10.5 Determine the support reactions, the joint displacements, and the member forces of the truss shown below using the Direct Stiffness matrix method. Take E = 200 GPa.

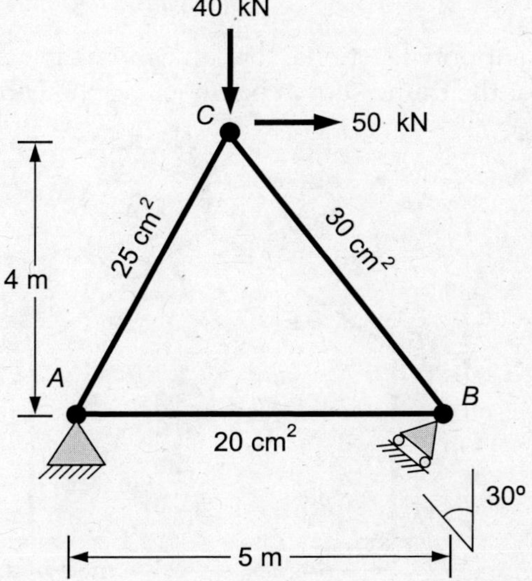

10.6 Determine the support reactions, the joint displacements, and the member forces of the truss shown below using the Direct stiffness matrix method. Take $E = 200$ GPa.

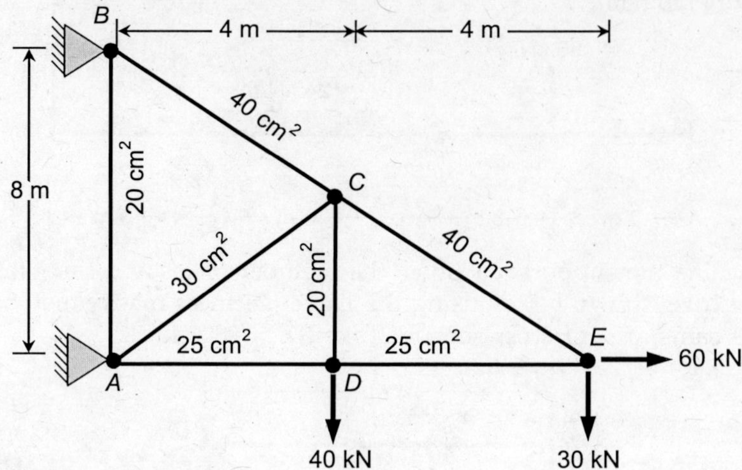

10.7 Determine the support reactions and the joint moments of the frame shown below using the Direct stiffness matrix method.

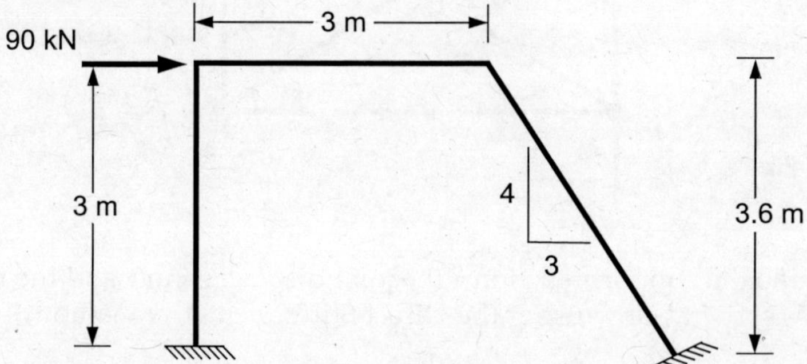

10.8 Determine the support reactions, the joint moments, the joint displacements, and joint rotations of the frame shown below using the Direct stiffness matrix method.

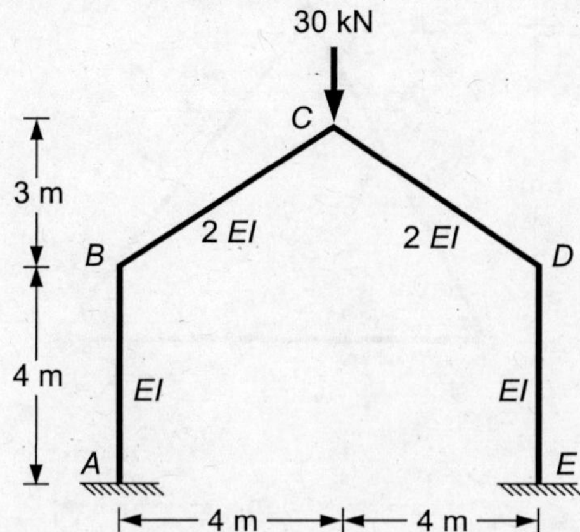

10.9 Determine the support reactions, the joint moments, the joint displacements, and joint rotations of the frame shown below using the Direct stiffness matrix method.

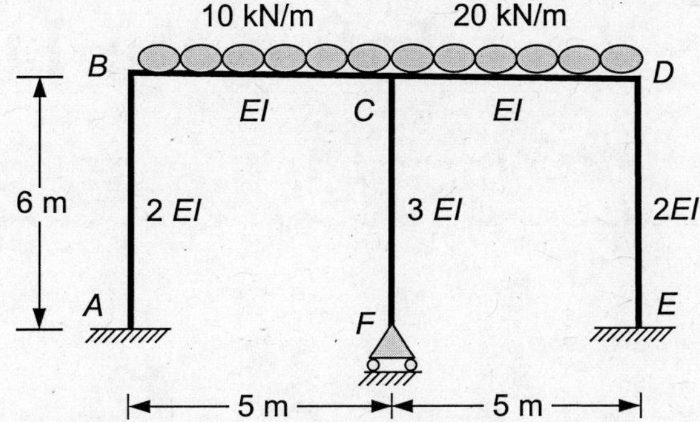

10.10 Determine the support reactions, the joint moments, and the joint displacements of the frame shown below using the Direct stiffness matrix method.

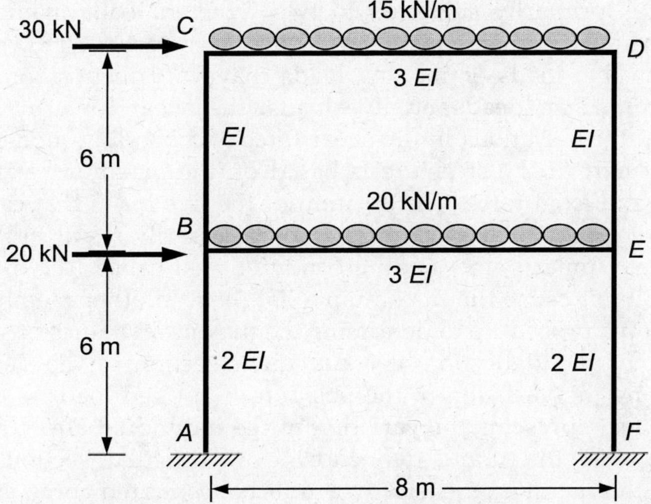

10.11 Determine the support reactions, the joint moments, and the joint displacements of the frame shown below using the Direct stiffness matrix method.

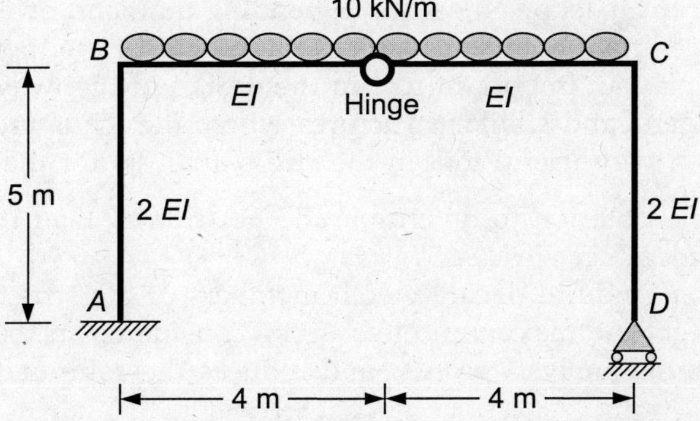

Influence Lines

11.1 GENERAL CONCEPT

A static structure is primarily subjected to two types of loads, namely, dead load and live load. The dead load remains stationary with the structure. However, the live loads in the form of movable loads or moving loads may vary in position and magnitude in the structure. For a given dead and live loads, the shear force and bending moment diagrams represent the variations in the shear force and bending moment in the structure. The design of any part of the structure is based on the maximum values of shear force, bending moment, and axial force. The position of the live loads that causes the maximum bending moment at a particular section may not necessarily result in the maximum shear at the same section. Similarly, the loading condition that causes the maximum axial force in one member will not cause the maximum axial force in other members. It is, therefore, necessary to develop a procedure to determine the maximum values of shear force, bending moment, axial force, and deflection at a particular section of the desired member of a structure considering the variation in the position of applied loads.

An *influence line* represents the variation in the individual effect, such as shear force, bending moment, axial force, and deflection at a particular section as a concentrated force moves over the member. Generally, a unit concentrated force is considered as the moving load to draw the influence line. The *influence line diagram* facilitates in determining the position of the moving load that would result in the maximum value of the desired parameters at a particular location of a member. The y-axis of the influence line diagram may present the support reaction, shear force, bending moment, or deflection, whereas the x-axis of the influence line diagram represents the length of the member.

Influence lines play an important role in the design of highway and railway bridges, conveyors, crane girders, and similar structures where the loads move along their spans. The influence line is a very useful tool in the stress analysis for the following reasons.

1. It serves as a criterion to determine the position of load to cause the maximum stress in a member.
2. It provides a guide to decide which members of the structure should be loaded in order to get the maximum effect to be considered in the design.
3. It simplifies the analysis process and reduces the computational effort.

11.2 INFLUENCE LINE DIAGRAM

Consider a simply supported beam of length 8 m as shown in Figure 11.1(a). Let us assume that it is required to draw the influence line diagram for the bending moment at the mid-span of the beam. The beam is divided into eight equal segments, i.e., *Aa, ab, bc, cd, ...,* etc. To draw the bending moment diagram at the mid-span, a unit load is moved from left to right along the length of the beam. First, assume that the unit load is applied at point A. For this load position, the bending moment at the mid-span will be zero. Next, when the unit load is placed at point *a* (i.e., *Aa* = 1 m), the bending moment at the mid-span will be 0.5. Similarly, the bending moment at the mid-span can be computed for different positions of the unit load. Figure 11.1(b) shows the ordinates of the bending moment at the mid-span of the beam for different positions of the unit load. The x-axis of the plot represents the length of the beam. Thus, Figure 11.1(b) is the influence line diagram for the bending moment at the midspan (i.e., point d) of the beam. A similar procedure can be adopted to draw the influence line diagrams for support reactions, shear force, and deflection at any section of the beam.

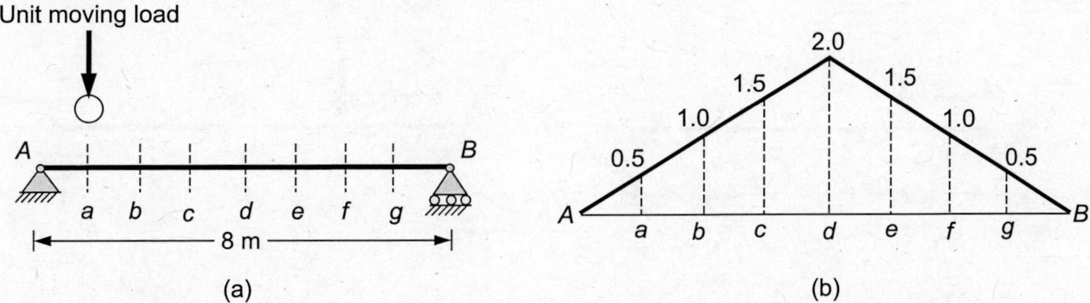

FIGURE 11.1 (a) Simply supported with moving load, (b) Influence line diagram.

The influence line diagram should not be confused with the shear force diagram or bending moment diagram of a member. In the above example, the shape of the *influence line diagram* is similar to the *bending moment diagram* for the beam. The basic difference between the *influence line diagram* and the *bending moment diagram* is the *influence line diagram* represents the bending moment at *a particular section* for a *moving load*, whereas the *bending moment diagram* shows the bending moment at *different sections* of the beam for a *particular load position*. This is true for any other parameters, namely, shear force, deflection, support reactions, etc.

11.3 INFLUENCE LINE EQUATION

It is not always possible to draw the inference line diagram for the desired parameter. It may also be time-consuming sometimes. Alternatively, the influence line can be expressed in the equation $y = f(x)$, where y is the desired parameter at a given section expressed in terms of the load position x. The curve representing the equation is the influence line for the desired parameter.

To illustrate this concept, consider a beam *AB* as shown in Figure 11.2(a) in which the unit load is applied at a distance x from the end *A*. Let us assume that it is required to determine the influence line for the bending moment at the mid-span. For this position of load, the reactions at A and B are computed as $(1 - x/8)$ and $(x/8)$, respectively.

Bending moment at the mid-span for the computed reactions is given by

$$M_C = \left(1 - \frac{x}{8}\right)(4) - (1)(4 - x) = \frac{x}{2} \tag{11.1}$$

The above equation is valid for $(0 \le x \le 4)$, i.e., if the unit load is placed in the segment *AC*. If the unit load is placed in the segment *CB* (i.e., $4 \le x \le 8$), the bending moment at the mid-span of the beam can be determined as follows:

$$M_C = 4\left(1 - \frac{x}{8}\right) = 4 - \frac{x}{2} \tag{11.2}$$

These equations representing the influence line for the bending moment at the mid-span is shown in Figure 11.2(c).

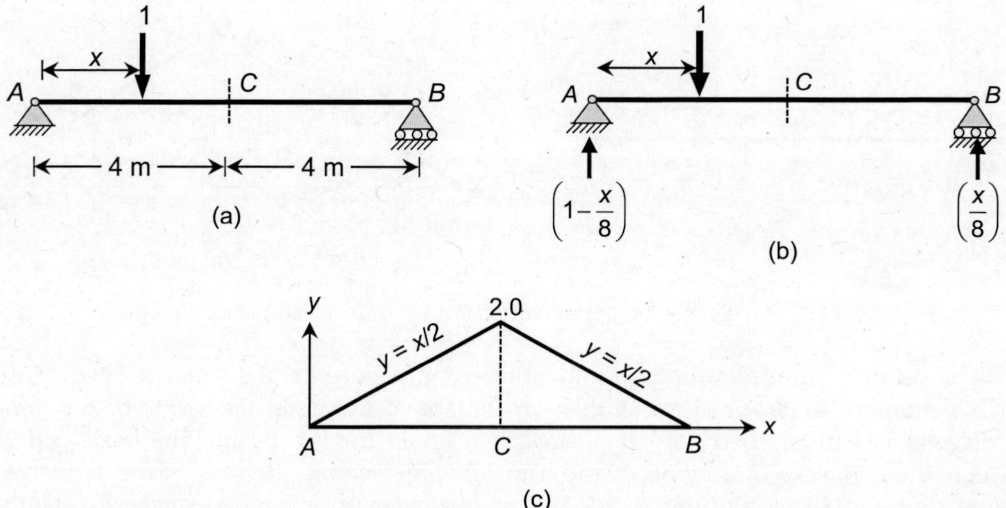

FIGURE 11.2 Beam with moving load and influence line diagram from moment at C.

11.4 INFLUENCE LINES FOR STATICALLY DETERMINATE BEAMS

The fundamental approach to draw the influence lines for a statically determinate beam is to apply the equilibrium condition for the free-body diagram of the beam as the unit load moves along the length of the beam. It is often convenient to draw the reacting influence diagram first and deduce the shear and moment influence line diagrams.

Two types of loads, namely, concentrated loads and distributed loads are considered for the discussion on the influence lines of beams as follows:

Concentrated load

The influence lines for a desired parameter of the beam are determined by applying a dimensionless unit load and moving the unit load along the length of the beam. The ordinates of the influence lines for a given load magnitude are computed by multiplying the ordinates of influence lines for the unit load with the load magnitude.

For example, consider the beam AB shown in Figure 11.3(a) subjected to a concentrated load, P at the mid-span. Figure 11.3(b) shows the influence line diagram for the reaction at A (R_A) for a unit load moving from one end to another. The value of R_A at the load position (i.e., $x = 0.5L$) is $0.5P$. Clearly, the absolute maximum value of R_A is P when the load is applied at point A.

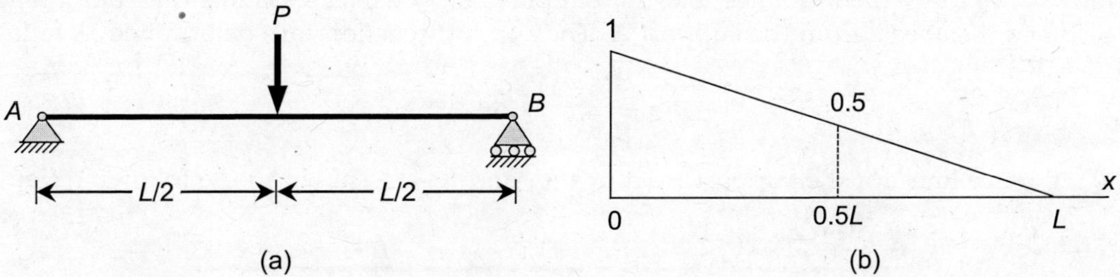

FIGURE 11.3   (a) Beam with concentrated loading, (b) Influence line for R_A.

Distributed load

The influence lines for beams with the distributed loading can be derived using the same concept used in those with the concentrated loads. The distributed loads can be considered as the sum of equivalent concentrated loads computed from the distributed loads on a series of small segments.

To illustrate this, consider a simply supported beam AB subjected to a uniformly distributed load w as shown in the Figure 11.4(a). Consider an infinitesimal segment dx at a distance x from the end A. The equivalent concentrated force on this segment is $dP = w.dx$. Figure 11.4(b) shows the influence line diagram for the support reaction R_A for a moving unit concentrated load. The ordinate of the influence line diagram at the point where the equivalent force dP is acting is assumed to be y. The value of reaction R_A for load magnitude dP would be $y.dP = y(w.dx)$. The effect of all the equivalent forces on the reaction R_A can be obtained by integrating over the entire length, i.e., $\int y(w.dx) = w\int y.dx$. This shows that the value of reaction (or any parameter) caused by a *uniformly distributed load* is the *area under the influence line diagram* for the reaction (or any parameter) multiplied by the intensity of the uniform load. For the given example, the reaction R_A is equal to $0.5\,wL$.

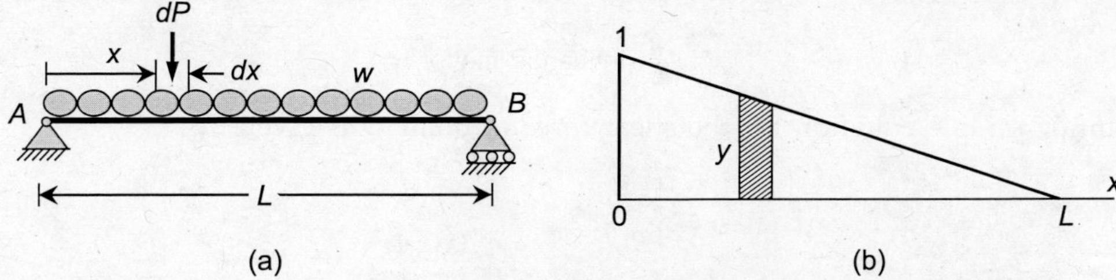

FIGURE 11.4   (a) Beam with distributed loading, and (b) Influence line for R_A.

EXAMPLE 11.1 Draw the influence lines for the reactions at supports A and B, the shear and moment at D and the bending moment at B of the overhanging beam shown in Figure E11.1.

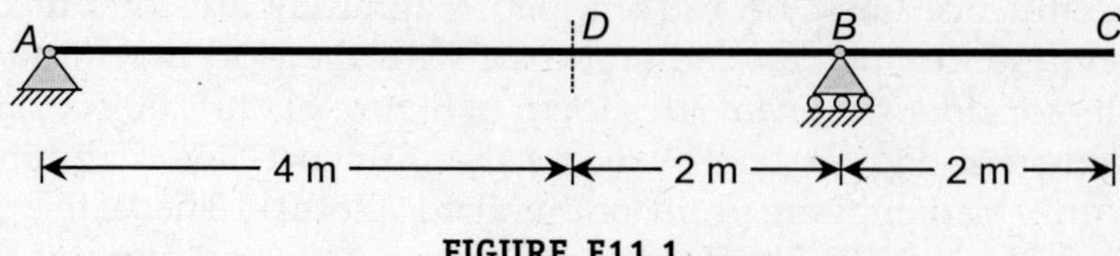

FIGURE E11.1

Solution: To draw the influence lines for support reactions, let's assume that a unit load is placed at a distance x from the support A. The support reactions are determined as follows:

$$R_A = \frac{(6-x)}{6}; \quad R_B = \frac{x}{6}$$

The influence lines for the support reactions R_A and R_B are shown in the following figures.

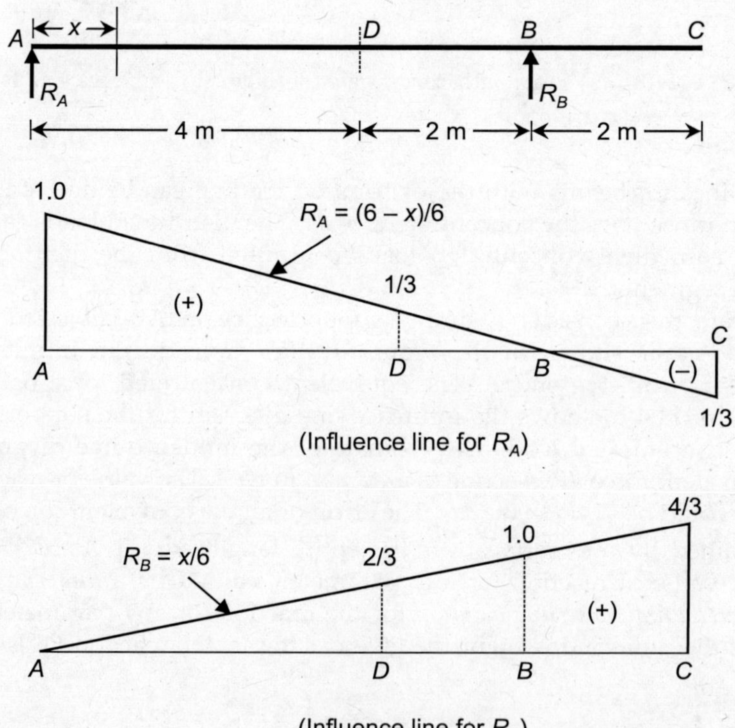

Influence line equation for the shear force at point D is given by

$$V_D = \begin{cases} -R_B & (0 \le x \le 4) \\ R_A & (x \ge 4) \end{cases}$$

$$V_D = \begin{cases} -\dfrac{x}{6} & (0 \le x \le 4) \\[2ex] \dfrac{(6-x)}{6} & (x \ge 4) \end{cases}$$

Influence line equation for the bending moment at point D is given by

$$M_D = \begin{cases} 2R_B & (0 \le x \le 4) \\ 4R_A & (x \ge 4) \end{cases}$$

$$M_D = \begin{cases} 2\left(\dfrac{x}{6}\right) & (0 \le x \le 4) \\[2ex] \dfrac{4(6-x)}{6} & (x \ge 4) \end{cases}$$

Influence lines diagrams for shear force and bending moment at D are shown below:

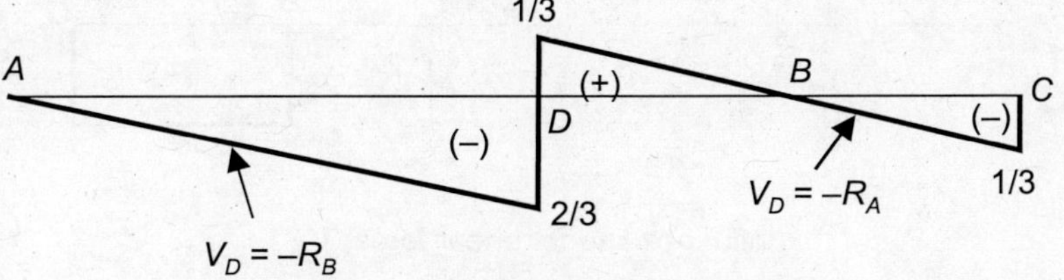

(Influence line for shear force at D)

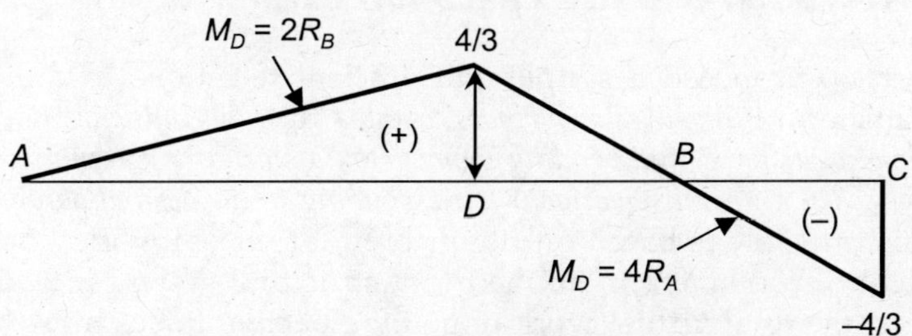

(Influence line for bending moment at D)

Influence line equation for the bending moment at B is given by

$$M_B = \begin{cases} 0 & (0 \le x \le 6) \\ (6-x) & (x \ge 6) \end{cases}$$

Influence lines diagram for the bending moment at B is shown below:

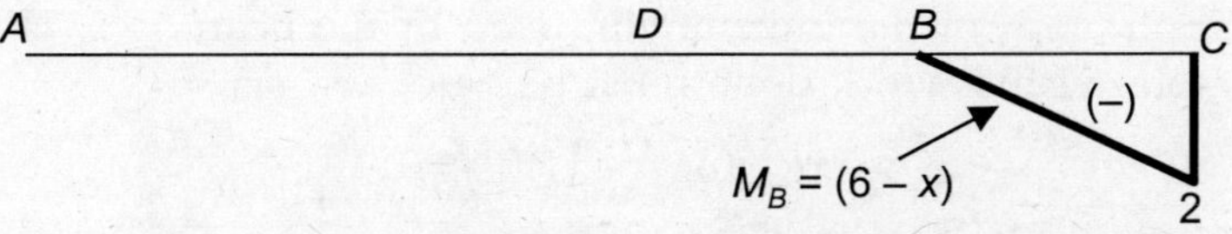

(Influence line for bending moment at B)

Similarly, the influence line diagrams for shear force at point B to the left (V_{BL}) and to the right (V_{BR}) of the support are shown below.

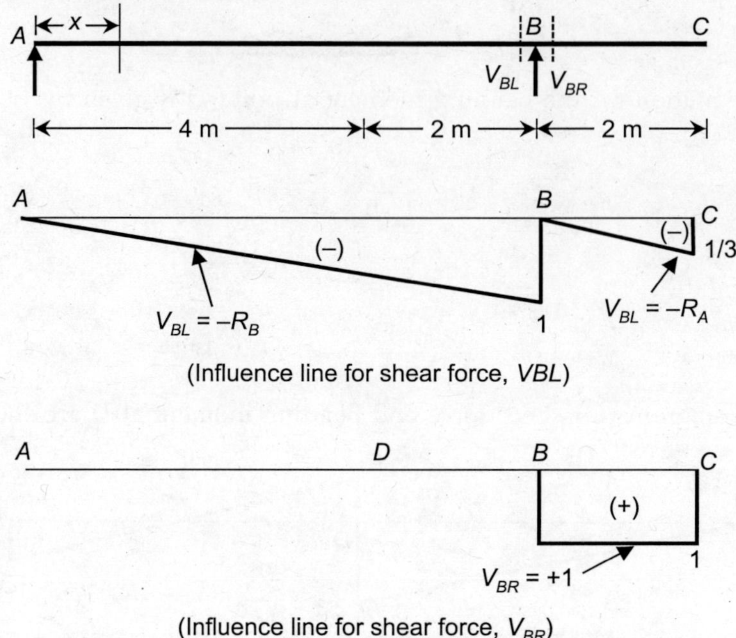

11.5   QUALITATIVE INFLUENCE LINES OF BEAMS

Heinrich Müller-Breslau proposed a simple and efficient technique to draw the influence lines for any parameter. *Müller-Breslau principle* states that *the influence line for a parameter (i.e., support reaction, shear force, and bending moment) of a statically determinate beam is similar to its deflected shape when a unit displacement corresponding to the desired parameter is applied at the same point.* This principle is based on the theorem of virtual work which states that if a compatible virtual displacement in applied to a structural system in equilibrium under the balanced forces, the total virtual work done by all active forces must be zero.

To illustrate this concept, consider a simply supported beam subjected to a single unit moving load as shown in Figure 11.5(a).

Influence line for support reaction at A

To determine the influence line for the support reaction R_A, let's apply a small virtual displacement at the support A along the direction of R_A by removing the support A. The deflected shape of the beam is $A'B$ as shown in Figure 11.5(b). The transverse displacement at the point of application of the unit moving load of the beam is y. Using the principle of virtual work, total virtual work done must be zero, i.e., $\delta W = 0$.

Thus, $$(R_A)(\delta\Delta_a) - (1)(y) = 0$$

Since, $(\delta\Delta_a) = 1$ $\therefore R_A = y$

Since the position of unit moving load is arbitrary and y represents the ordinate of the deflected shape of the beam, it can be concluded that *the influence line for the support reaction R_A is equal to the deflected shape of the beam in the absence of support A and a unit displacement is applied along the direction of the same support reaction at the support A.*

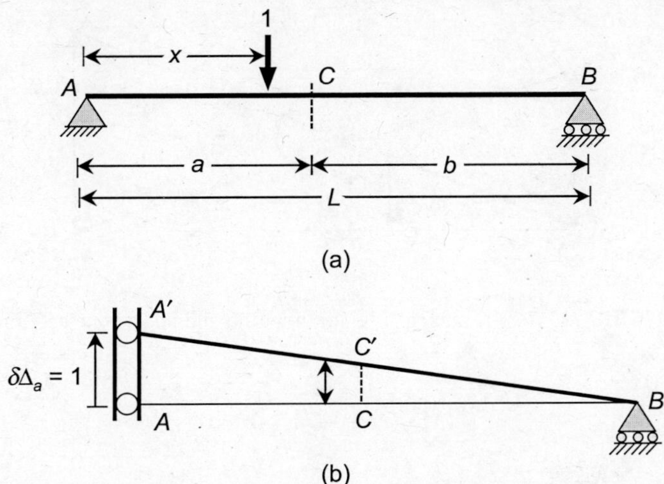

FIGURE 11.5 (a) Beam with a single unit moving load, (b) Unit displacement applied at the support A.

Influence line for shear force at a section

A similar procedure can be adopted to draw the influence line for shear force at any point. Consider a point C on the beam for this purpose. Cut the beam at point C and apply a virtual transverse displacement between two segment AC and BC such that there is no relative rotation between these segments. Figure 11.6 shows the virtual displacement applied at cut section C. Since both segments have the same angle of rotation at point C, $CC' = b/L$ and $CC'' = a/L$. Applying $\delta W = 0$, one can obtain

$$(V_C)(\delta\Delta_c) - (1)(y) = 0$$

Since $(\delta\Delta_c) = 1$ $\therefore$ $V_C = y$

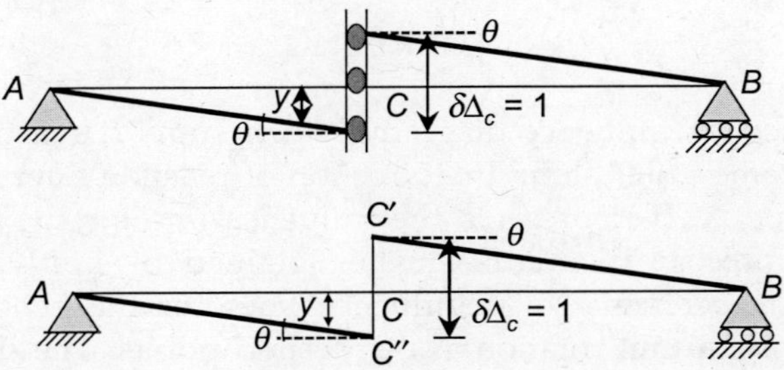

FIGURE 11.6 Virtual displacement applied at the cut section C of a beam.

Thus, it can be concluded that *the influence line for the shear force V_C is equal to the deflected shape of the beam obtained by applying a unit displacement in a direction normal to the beam at the cut section C while maintaining the same angle of rotation between two cut segments.*

Influence line for bending moment at a section

To construct the influence line for the bending moment at C, the beam has to be cut into two segments at the same point. Then, a virtual unit rotation is applied at point C without any relative transverse displacement between these segments as shown in Figure 11.7.

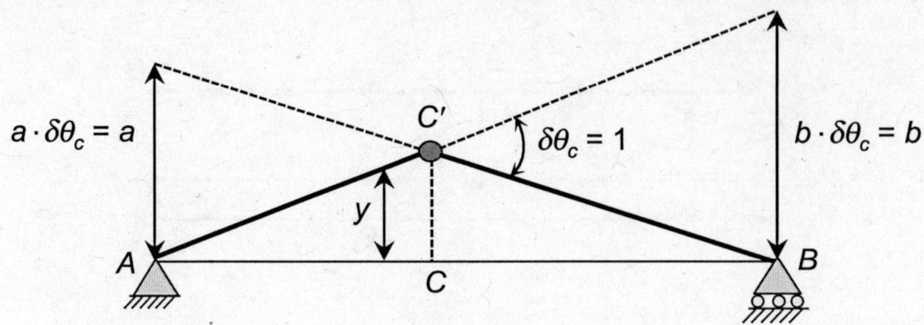

FIGURE 11.7   Influence line for bending moment at a section.

Since, $\delta W = 0$,

$$(M_C)(\delta\theta_c) - (1)(y) = 0$$

Since, $(\delta\theta_c) = 1$ $\therefore\ M_C = y$

Hence, *the influence line for the bending moment M_C is equal to the deflected shape of the beam obtained by applying a unit rotation at the cut section C while maintaining no relative transverse displacement between two cut segments.*

EXAMPLE 11.2   Draw the influence lines for the reaction at support B, the shear force and bending moments at points D and E of the beam AB shown in Figure E11.2. There is an internal hinge at the point C of the beam. Use Müller-Breslau principle.

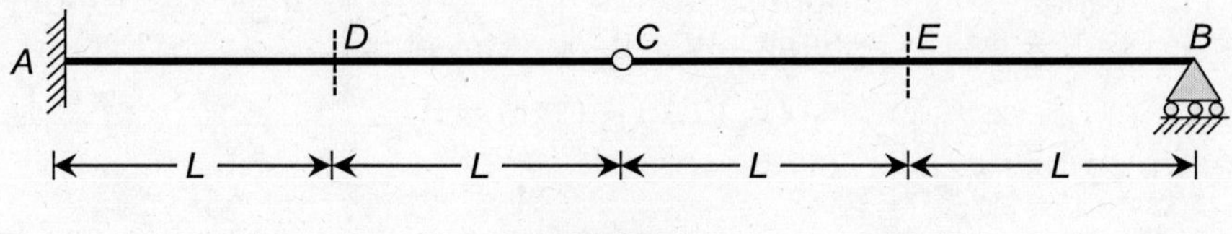

FIGURE E11.2

Solution: To construct the influence diagram for the support reaction at B, apply a unit displacement at the same point along the direction of reaction after removing the support B. The deflected shape ACB' represents the influence line for the reaction R_B. Influence lines for the shear force at D and E can be obtained by applying a unit transverse displacement at the respective points. Influence lines for the bending moments can be constructed by applying a unit rotation at the desired points. The deflected shapes of the beam for these conditions are shown in the following figures which are also the influence lines as per Müller-Breslau principle.

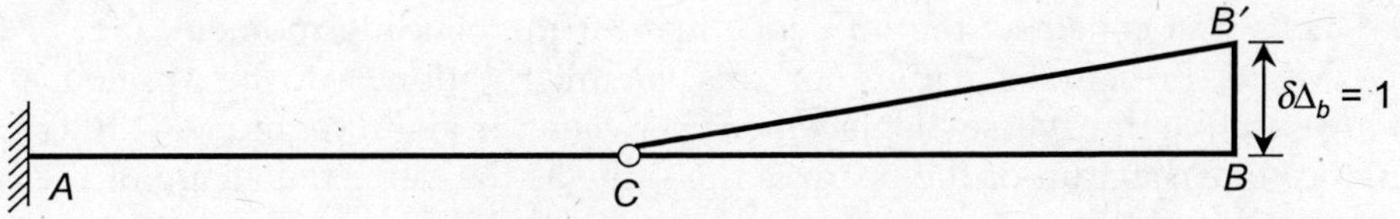

(Influence line for R_B)

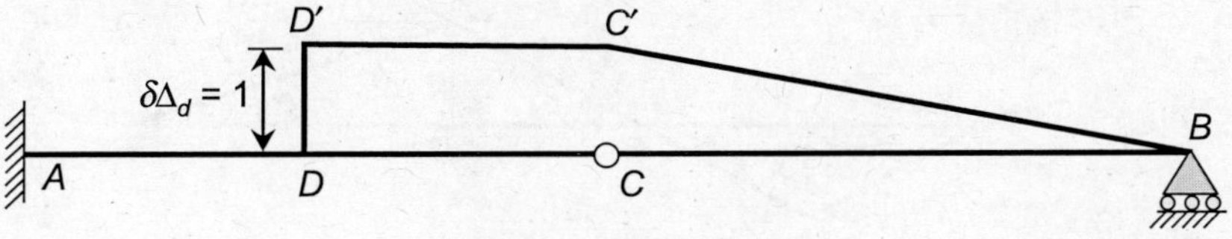

(Influence line for V_D)

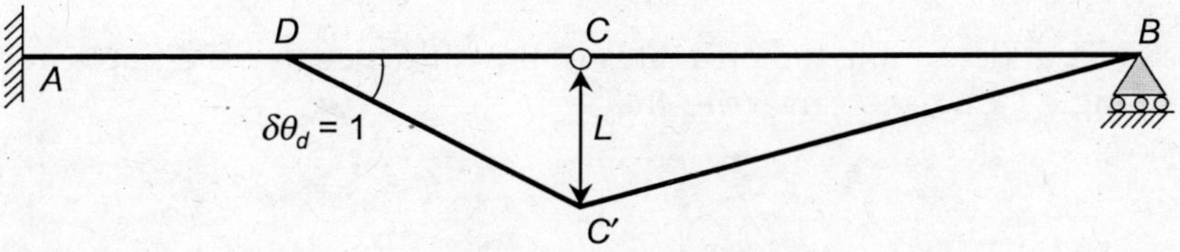

(Influence line for M_D)

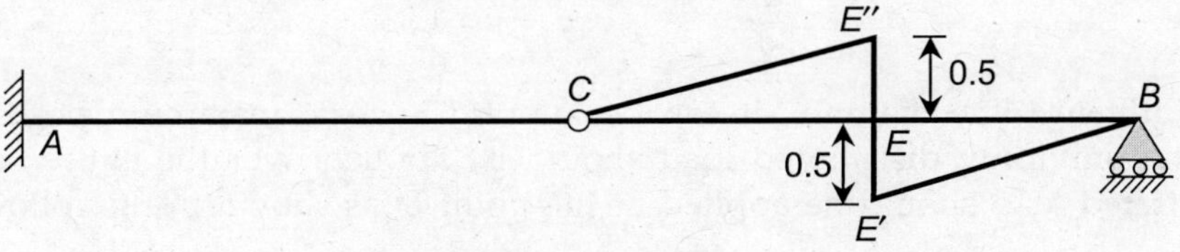

(Influence line for V_E)

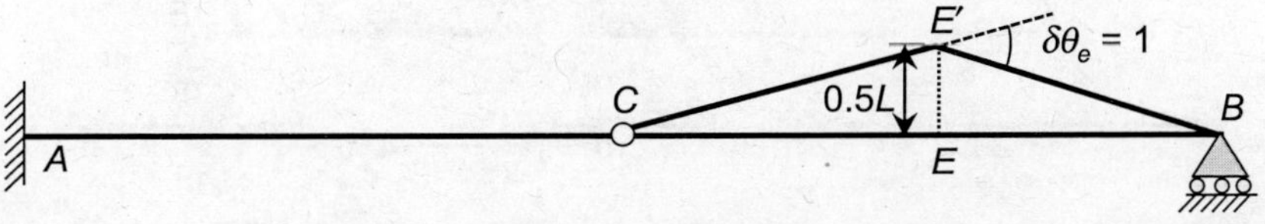

(Influence line for M_E)

EXAMPLE 11.3 Determine the maximum positive bending moment that would develop at point D due to a concentrated load of 30 kN and a uniformly distributed load of 5 kN/m of the beam shown in Figure E11.3. There is an internal hinge at point B of the beam. Use the Müller-Breslau principle.

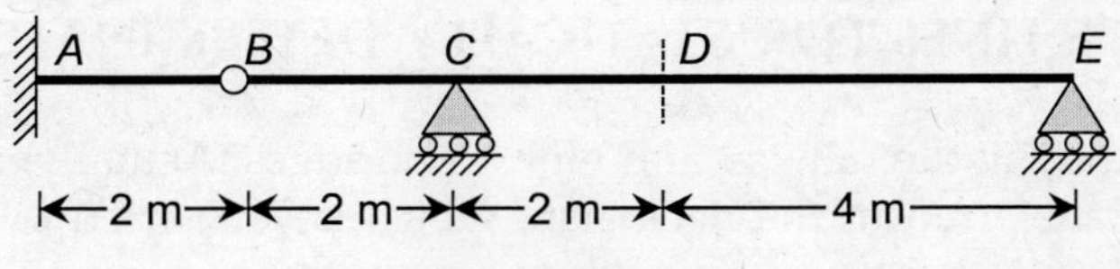

FIGURE E11.3

Solution: Let's first construct the influence line for the bending moment at D of the beam using the Müller-Breslau principle. For this, a unit rotation is to be applied at point D without any relative transverse displacement between the beam segments at the same point. The influence line diagram of the beam is the same as the deflected shape of the beam due to the application of this unit rotation as shown in the following figure.

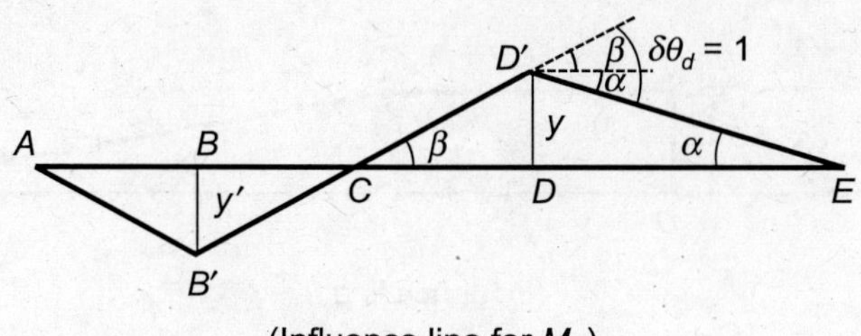

(Influence line for M_D)

Next step is to determine the ordinates of the influence line diagrams at points B and D. For the triangle CED', we can write that

$$\alpha + \beta = 1 \implies \frac{y}{2} + \frac{y}{4} = 1 \quad \therefore y = \frac{4}{3} = 1.33$$

Using the triangles CDD' and BCB'

$$\frac{y'}{2} = \frac{y}{2} \quad \therefore y' = y = 1.33$$

From the influence line diagram, it can be inferred, for the maximum positive bending moment, the uniformly distributed load should be applied over the entire span CE and the concentrated load should be applied at the point D as shown in the following figure.

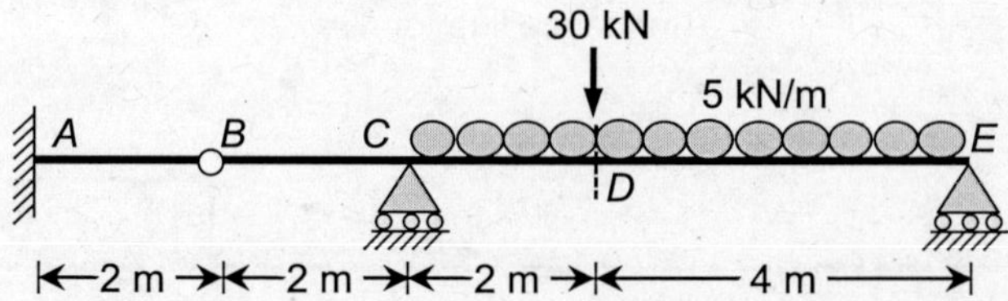

Accordingly, the maximum positive bending moment at D can be computed as follows:

$$M_D = (30)(1.33) + (5)\left(\frac{1}{2}\right)(6)(1.33) \text{ kNm}$$

$$\therefore \quad M_D = 60 \text{ kNm}$$

11.6   INFLUENCE LINES FOR STATICALLY DETERMINATE TRUSSES

Trusses are widely used in the railway and highway bridges. Actually, the load of a moving vehicle is directly transferred to the deck slab, which is supported by the stringers. Since these stringers run along the length of the bridges and are supported by the transverse floor beams, the vehicle load is transmitted to the floor beams. The floor beams and the bottom chords of the bridges are connected at the joints of the panels, where the vertical as well diagonal members are also connected. As a result, loads of moving vehicles are transmitted to the joints of the bridges. Therefore, it is assumed that all live loads are applied only at the joints of a truss in the analysis. All the joints of a truss are assumed to be pin-connected. Accordingly, all member forces are computed using the method of joints or method of sections as discussed earlier.

The influence lines of a truss are constructed by applying a unit force at each successive joint of the loaded chord and determining all the member forces. The computed bar forces are called as the *influence coefficients*. Figure 11.8 shows a segment of the bridge truss. The vehicle load is transmitted to the truss chord of the bridge through the floor beams and the stringers. Consider the loaded panel a–b of length L. When the unit load is placed just above a, the influence coefficient is y_a. When the unit load is placed at point b, the influence coefficient is y_b. If the unit load is placed at any intermediate point between a and b, say, at a distance x from end a , the corresponding influence coefficient y can be determined using the influence coefficients at both ends and corresponding the reactions due to the unit load. This can be computed as follows:

$$(1)(y) = \left(\frac{L-x}{L}\right)(y_a) + \left(\frac{x}{L}\right)(y_b)$$

The above expression represents a linear variation between the influence coefficients at the ends of the panel of the bridge truss. For the sign convention, the tension bar forces are considered as positive and vice-versa. Similar to the beams, the influence lines for the support reactions are constructed in the first step. This simplifies the process of drawing the influence lines for bar forces of a bridge truss.

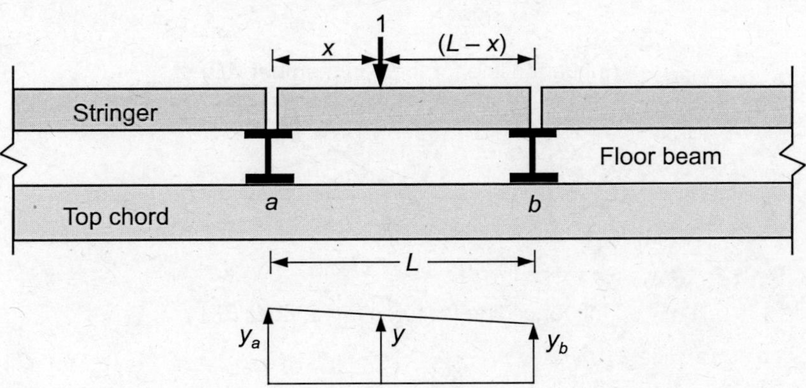

FIGURE 11.8 Bridge truss under a unit moving load.

EXAMPLE 11.4 For the bridge truss shown in the Figure E11.4, draw the influence lines for the support reaction at A and the forces in members AB, BE, BC, and EF.

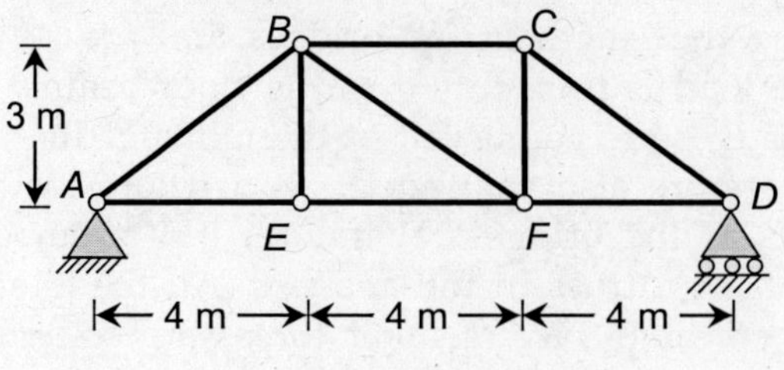

FIGURE E11.4

Solution: To draw the influence lines, a unit load is applied at the joints successively and the support reaction (R_A) and the member forces are computed accordingly. The influence lines of various parameters are shown in the following figure.

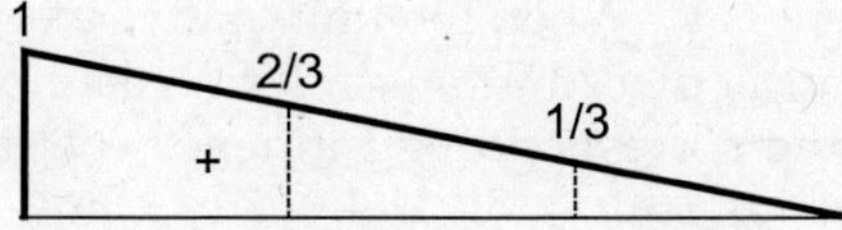

(Influence line for support reaction at *A*)

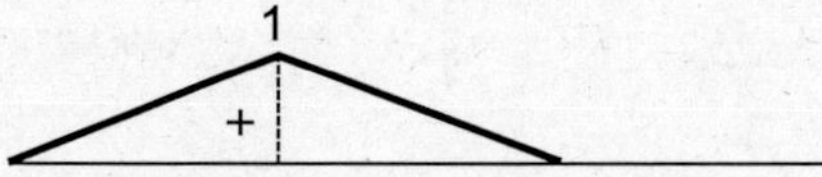

(Influence line for force in member *BE*)

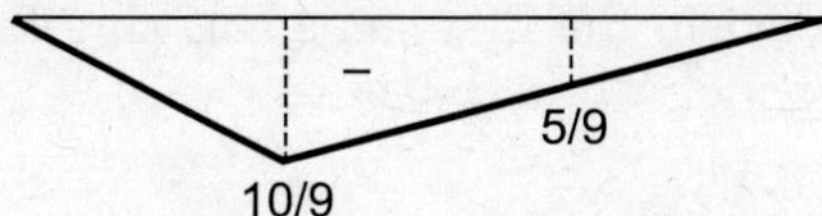

(Influence line for force in member *AB*)

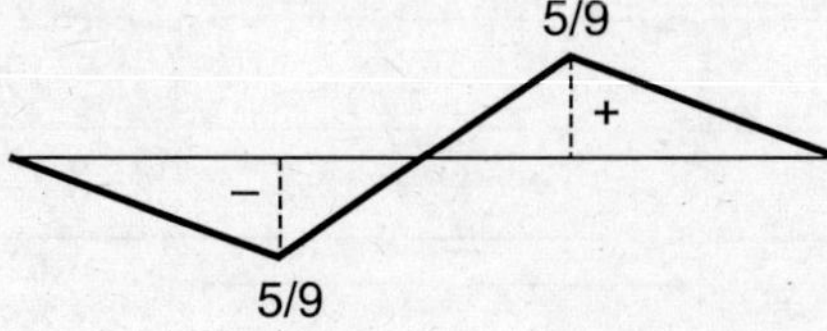

(Influence line for force in member *BF*)

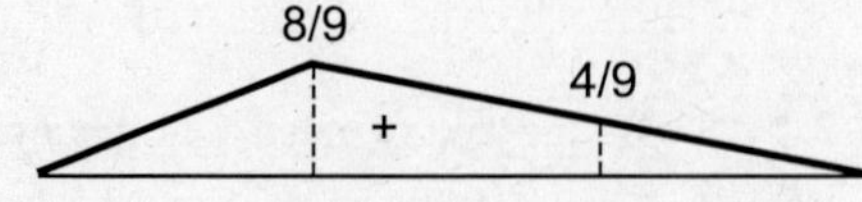

(Influence line for force in member *EF*)

EXAMPLE 11.5 Draw the influence lines for the members CD, *DH*, and *HJ* of the truss shown in Figure E11.5.

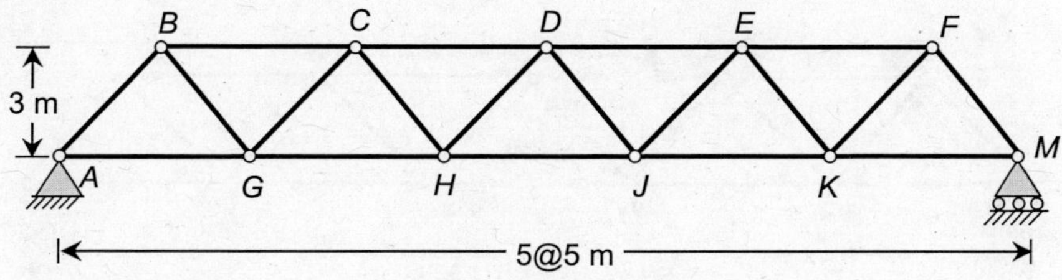

FIGURE E11.5

Solution: To draw the influence lines of member forces, it is convenient to draw the influence lines for the support reactions (i.e., R_A and R_M) by applying a unit load at the joints successively. The influence lines of the support reactions are as shown below.

To determine the influence lines for the member forces, let's take a cut section *x-x* as shown in the figure. The member forces can be determined for different positions of unit load in the truss considering the moment equilibrium about the joints.

When the unit load is applied at any joint of the truss in the portion left of the cut section, the force in member *CD* can be determined by taking moment about joint *H* for the right portion of the truss. When the unit is applied nay joint in the right portion, the force in member *CD* can be determined considering the moment equilibrium of the left portion about the same joint.

Accordingly,

$$F_{cd} = -\frac{10R_A}{3} \quad \text{(Unit load in the right portion)}$$

$$F_{cd} = -\frac{15R_M}{3} \quad \text{(Unit load in the left portion)}$$

The force in the member *HJ* can be computed taking moment equilibrium about the joint D. Thus,

$$F_{hj} = \frac{7.5R_A}{3} \quad \text{(Unit load in the right portion)}$$

$$F_{hj} = \frac{12.5R_M}{3} \quad \text{(Unit load in the left portion)}$$

Similarly, the force in the member *DH* can be computed taking vertical force equilibrium of the cut portions of the truss as a whole. Thus,

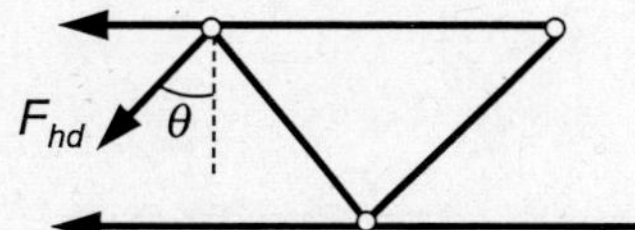

$$F_{dh} = \frac{R_A}{\cos \theta} = 1.30R_A \quad \text{(Unit load in the right portion)}$$

$$F_{dh} = \frac{R_M}{\cos \theta} = 1.30R_M \quad \text{(Unit load in the left portion)}$$

The influence lines for the support reactions *ad* the member forces of the truss are shown in the following figures.

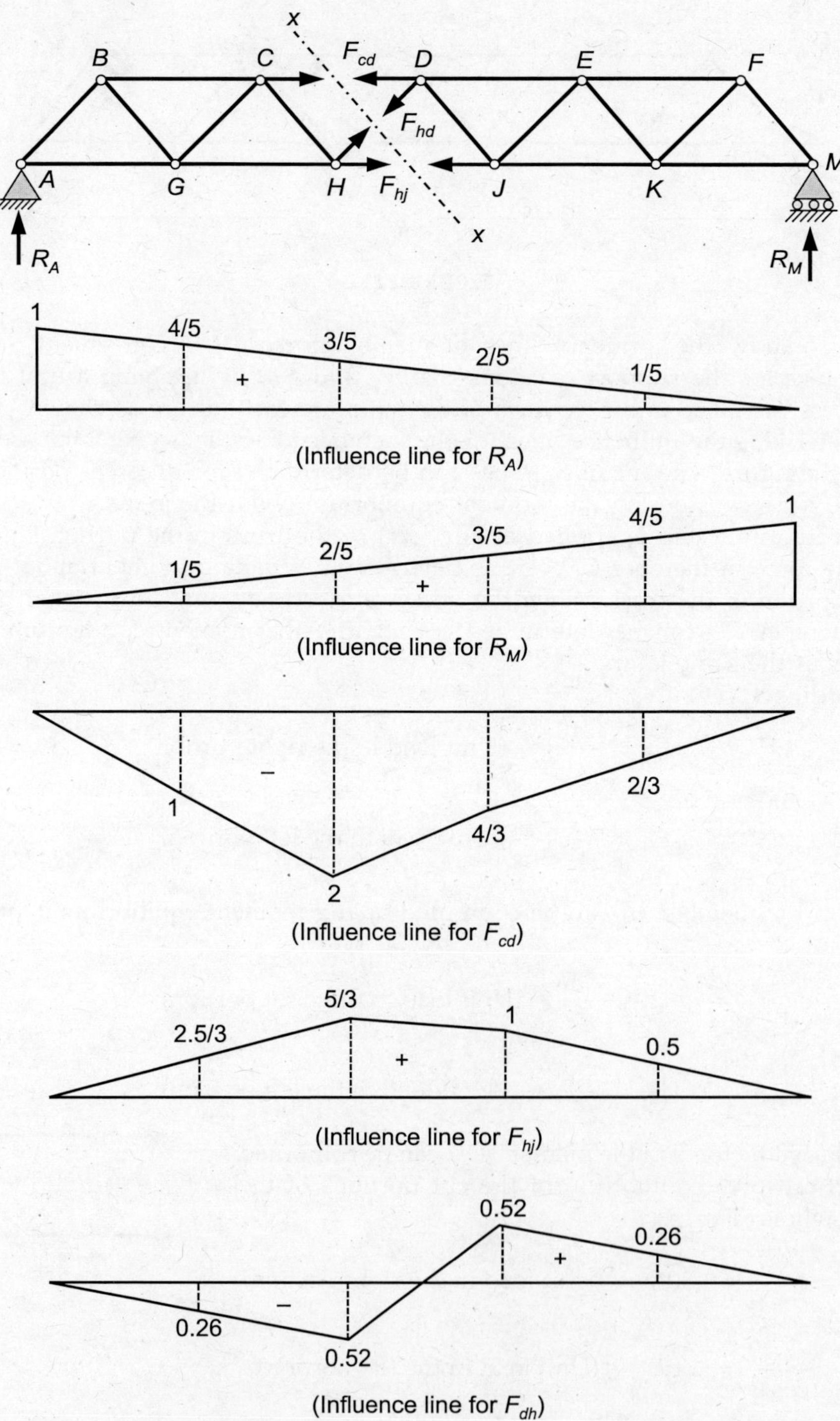

11.7　INFLUENCE LINES FOR A SYSTEM OF CONCENTRATED LOADS

The maximum influence caused by a concentrated load is determined by multiplying the peak ordinate of the influence line by the magnitude of the given load. Very often, the beams or trusses are subjected to a series of concentrated live loads. Some of the examples are the highway bridges that are subjected to the wheel loads of a moving truck or the railway bridges subjected to the wheel loads of a moving train. The maximum influence of a response parameter (i.e., shear force, bending moment, support reaction, deflection) due to a series of concentrated moving loads is determined either by the trial-and-error method as discussed in the following section.

Consider a simply supported beam subjected to a series of moving concentrated loads. Let's assume that it is required to determine the maximum shear at point C as shown in Figure 11.9(a). If a unit load is assumed to move over the beam, the influence line for the shear force at C is shown in Figure 11.9(b).

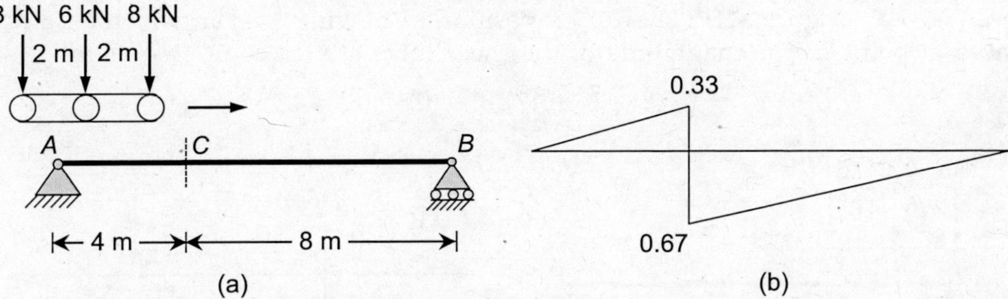

FIGURE 11.9　(a) Beam under multiple moving loads, (b) Influence line for shear force at C of the beam.

When a series of concentrated is allowed to move over the same beam, the maximum values of shear force at C can be determined by placing each load at the peak ordinate of the influence line diagram. Accordingly, three cases are possible for three concentrated loads.

CASE–I:　Figure 11.10 shows the position of moving load in which an 8 kN load is placed at point C. The magnitude of shear force at C is computed as follows:

$$V_{c1} = (0.67)(8) + (0.50)(6) + (0.33)(8) = 11.0 \text{ kN}$$

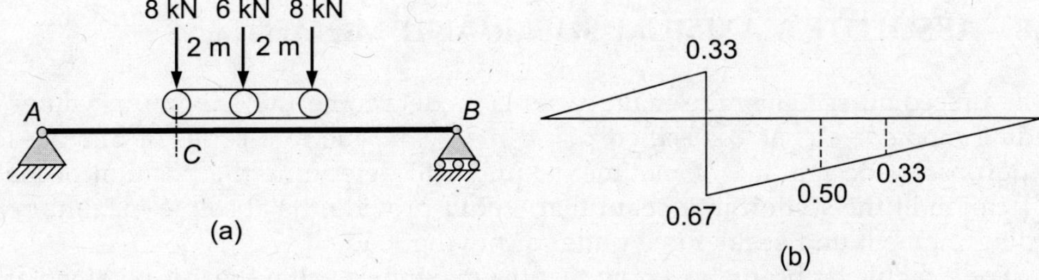

FIGURE 11.10　(a) Beam with 8 kN placed at point C, (b) Influence line for shear force at C of the beam.

CASE–II: Figure 11.11 shows the position of moving load in which a 6 kN load is placed at point C. The magnitude of the shear force at C is computed as follows:

$$V_{c2} = (-0.17)(8) + (0.67)(6) + (0.5)(8) = 6.7 \text{ kN}$$

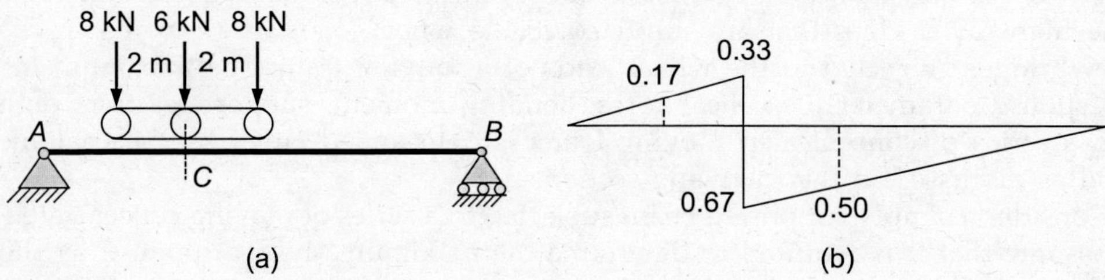

FIGURE 11.11 (a) Beam with 6 kN placed at point C, (b) Influence line for shear force at C of the beam.

CASE–III: Figure 11.12 shows the position of moving load in which an 8 kN load is placed at point C. The magnitude of the shear force at C is computed as follows:

$$V_{c3} = (0)(8) + (-0.17)(6) + (0.67)(8) = 4.3 \text{ kN}$$

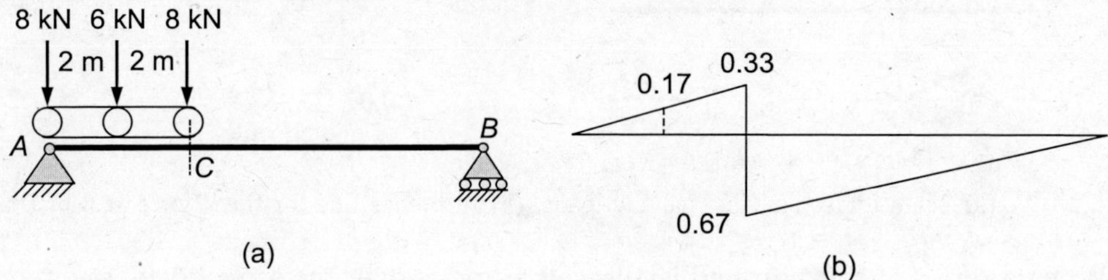

FIGURE 11.12 (a) Beam with 8 kN placed at point C, (b) Influence line for shear force at C of the beam.

Thus, it can be seen that the maximum value of shear force at C is computed as 6.7 kN corresponding to the critical position of loading. In this case, Case-I represents the critical position of loading for the maximum shear at point C. A similar procedure can be adopted to compute the maximum values of other parameters.

11.8 ABSOLUTE MAXIMUM SHEAR AND MOMENT

In the preceding sections, the main focus is to determine the maximum values of shear, bending moment, etc. at a specified point in a beam due to a series of moving loads. For the purpose of design, it is sometimes required to determine the position of loading and the location of the section in a beam that would provide the absolute maximum values of bending moment and shear force under a moving load.

For a cantilever beam, we know that the maximum value shear is obtained at the fixed support. Thus, the moving loads must be placed just next to support in order to get the absolute maximum value of shear. The same is also true of the simply supported beams.

If it is required to determine the absolute maximum bending moment of the cantilever beam, the position of the moving load should be far away from the fixed support as much as possible to increase the lever arm. Thus, the moving load should be placed near the free end of the cantilever beam. In the case of the simply supported beam, the moving loads must be close to the mid-span in order to achieve the absolute maximum bending moment. If the magnitude and spacing of the concentrated forces in the moving load are not exactly the same, it may not be possible to determine the critical position and the absolute maximum values just by inspection.

Consider a simply supported beam AB subjected to a moving load consisting of three concentrated forces, namely, F_1, F_2, F_3 and F_4 as shown in Figure 11.13. The maximum ordinate of bending moment may occur under any loading point depending on their magnitude and position. Assume that the absolute maximum bending moment occurs just below the force F_2. In order to determine the distance x of the force F_2 from the mid-span of the beam, we need to first determine the resultant force F_R and its position $\overline{x}$.

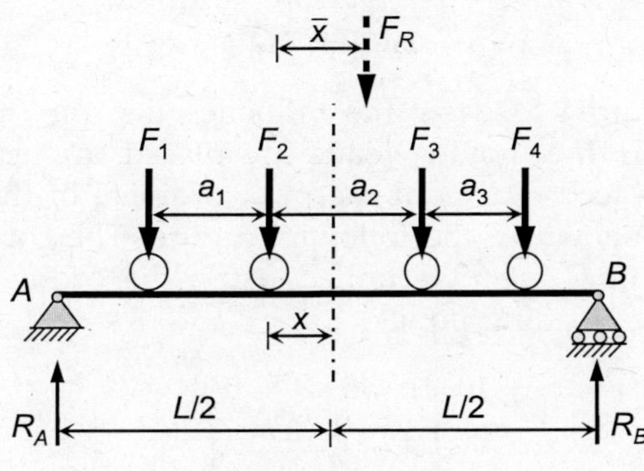

FIGURE 11.13

The support reaction at A, R_A can then be determined by taking the moment equilibrium of the entire beam about the support B as follows:

$$R_A = \frac{F_R}{L}\left(\frac{L}{2} - x + \overline{x}\right)$$

Bending moment at the point corresponding to the force F_2 is

$$M_2 = \frac{F_R}{L}\left(\frac{L}{2} - x + \overline{x}\right)\left(\frac{L}{2} - x\right) - F_1 a_1$$

For the maximum value of M_2,

$$\frac{dM_2}{dx} = \frac{F_R}{L}\left(\frac{L}{2} - x + \overline{x}\right)\left(\frac{L}{2} - x\right) - F_1 a_1 = 0 \quad \therefore \ x = \frac{\overline{x}}{2}$$

This shows that if *one concentrated load of a series of moving loads applied on a simply supported beam causes the absolute maximum bending moment at the point of load application, then the same load and the resultant of all loads must be placed at an equidistance from the mid-span of the beam.* In order to get the absolute maximum bending moment of a simply supported beam due to a series of moving loads, the values of the absolute maximum bending moment must be computed for each load. The largest value of the bending moment computed from all cases would represent the absolute maximum bending moment for the simply supported beam.

EXAMPLE 11.6 For the simply supported beam shown in Figure E11.6, determine the maximum bending moment at point C under the application of the moving load. Also, determine the absolute maximum bending moment for the beam.

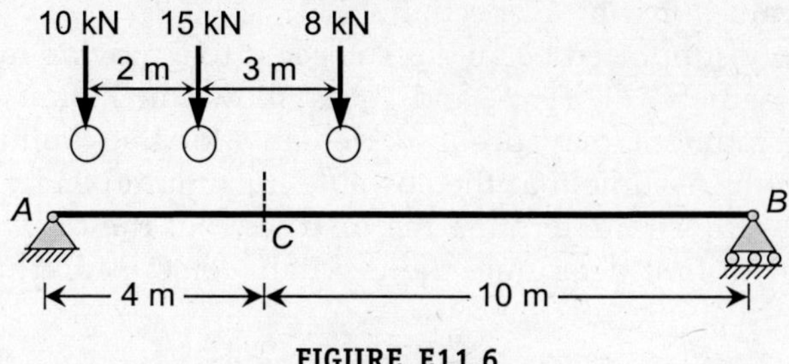

FIGURE E11.6

Solution: First, it is required to draw the influence line for the bending moment at C due to a unit load. Then, the moving loads are placed in such a way that one of the concentrated loads are placed exactly at the peak ordinate of the influence line diagram. These loading cases are shown in the following figure. The corresponding values of the bending moment at point C are determined using the ordinates of the influence line diagrams at the respective loading points.

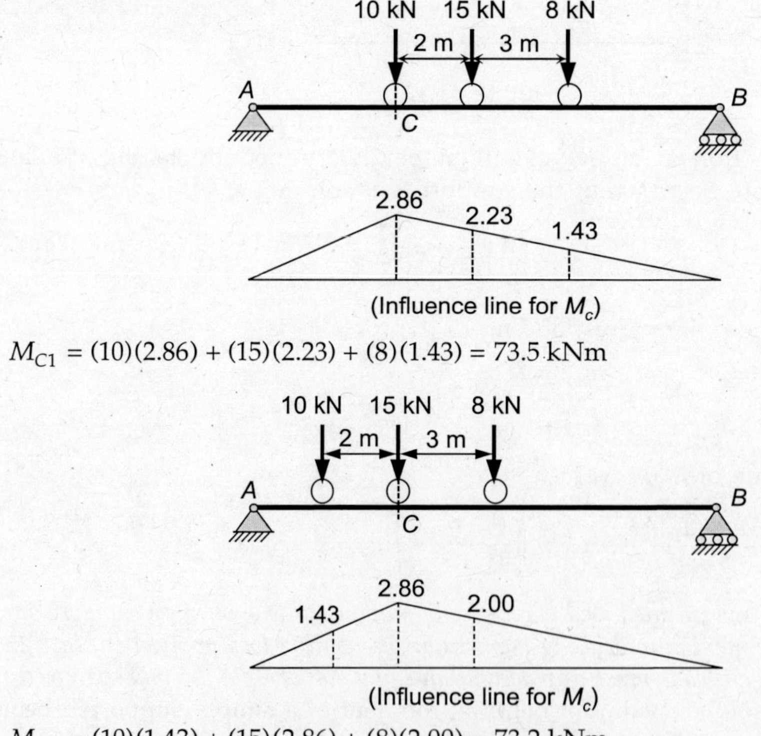

$$M_{C1} = (10)(2.86) + (15)(2.23) + (8)(1.43) = 73.5 \text{ kNm}$$

$$M_{C2} = (10)(1.43) + (15)(2.86) + (8)(2.00) = 73.2 \text{ kNm}$$

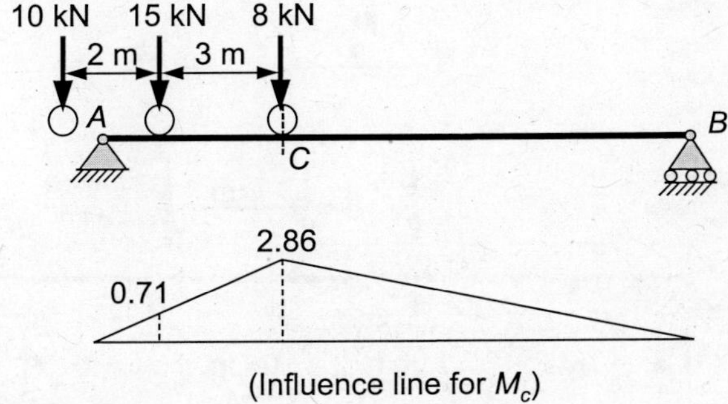

$$M_{C3} = (15)(0.71) + (8)(2.86) = 33.6 \text{ kNm}$$

In the third case, only two loads will act on the beam due to spacing between the consecutive loads, it would not lead to a higher bending moment than those computed in the above two cases. Hence, the bending moment corresponding to the third case is not required to be computed.

Comparing the computed values, the maximum bending moment at point C is $M_C = 73.5$ kNm.

Absolute maximum bending moment

It is required to compute the resultant force and its position.

The resultant force, $F_R = 10 + 15 + 8 = 33$ kN

The distance of the resultant force from the 10 kN force is computed as follows:

$$\bar{x} = \frac{(15)(2) + (8)(5)}{33} = 2.12 \text{ m}$$

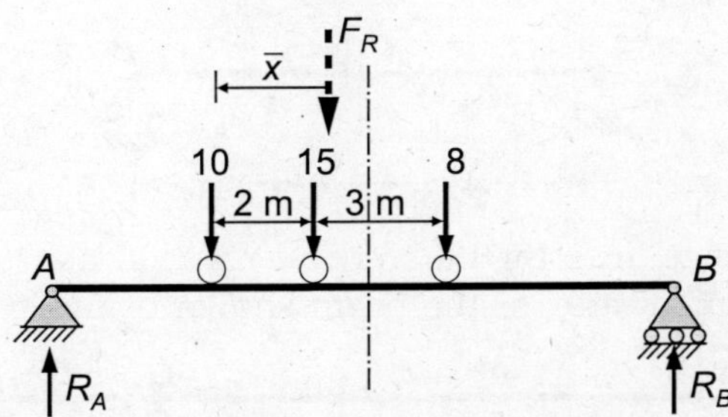

Assume that the absolute maximum bending moment occurs just below the 10 kN loading point. Then, this load should be placed at an equidistance from the resultant force about the mid-span of the beam. The load position and the computed reactions are shown in the following figure.

Bending moment at the 10 kN load point is

$$M_1 = (14)(5.94) = 83.16 \text{ kNm}$$

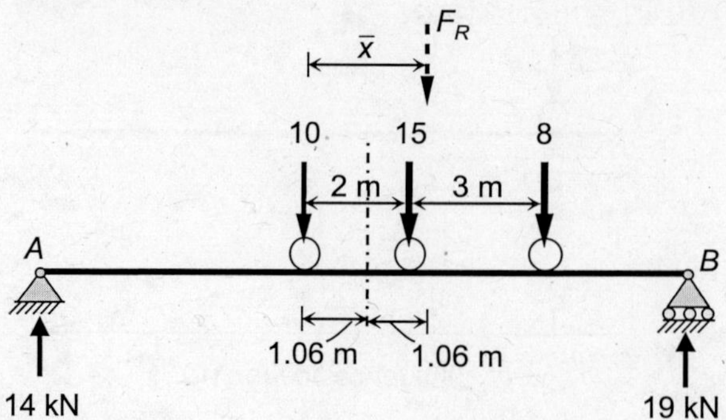

Next, assume that the absolute bending moment would occur just below the point where the 15 kN load is acting. In this case, the 15 kN load and the resultant load must be placed at equidistant about the mid-span of the beam as shown in the following figure. Thus, the distance between the mid-span and the 15 kN load would be $(2.12 - 2.0)/2 = 0.06$ m.

The corresponding value of the bending moment will be

$$M_2 = (16.36)(6.94) - (10)(2) = 93.54 \text{ kNm}$$

A similar procedure can be adopted for the third case in which it would be assumed that the absolute maximum bending moment would occur just below 8 kN load. However, one would infer that the bending moment computed for this case would be smaller than the earlier values obtained for two cases.

Hence, the absolute maximum bending will be 93.54 kN under the 15 kN load.

11.9   PROBLEMS

11.1   Draw the influence lines for the vertical reaction at B, the shear force at C and the bending moment at C of the beam shown below.

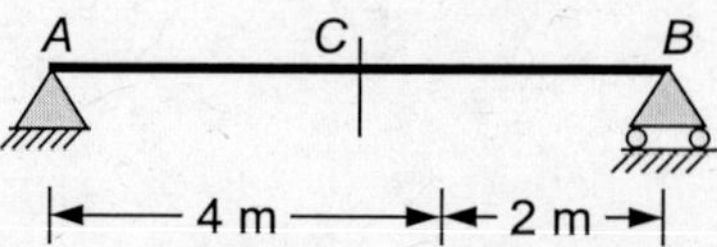

11.2   Draw the influence lines for the vertical reaction at D, the shear force at C and the bending moment at C of the beam shown below.

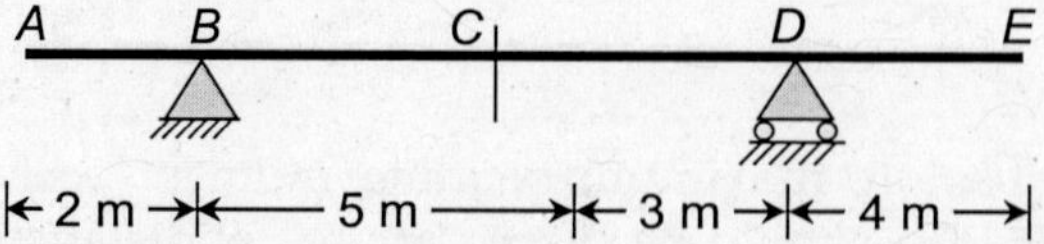

11.3   Draw the influence lines for the vertical reaction at A and D, the shear force and the bending moment at points B and E of the beam shown below.

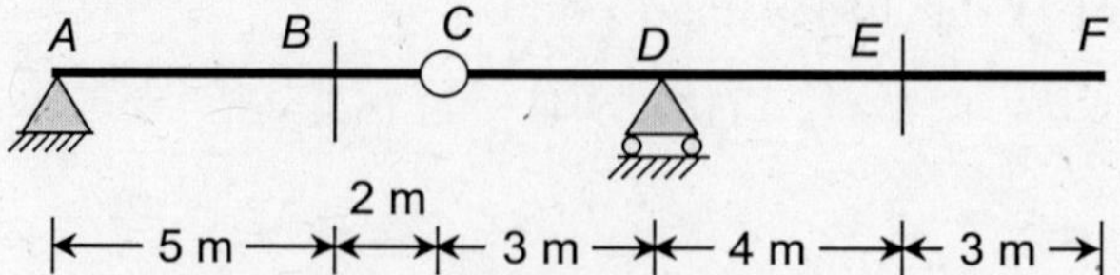

11.4 Draw the influence lines for the vertical reactions at A and D, the shear force and the bending moment at point B of the beam shown below.

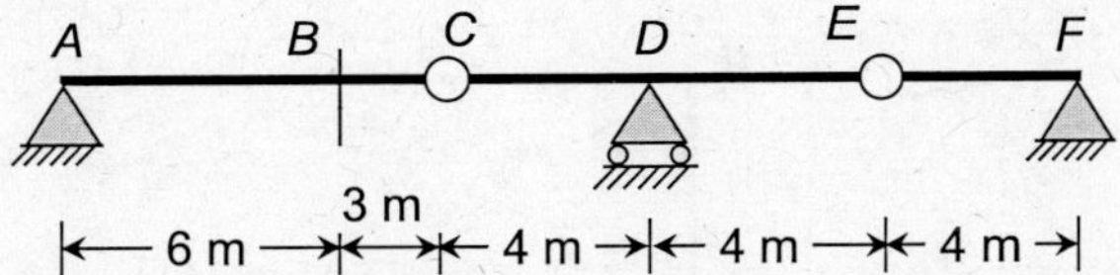

11.5 Draw the influence lines for the forces in members HC, DK, and GC of the truss shown below. Determine the maximum force that would be induced in the member HD due to a uniformly distributed load of 5 kN/m moving over the bottom chord.

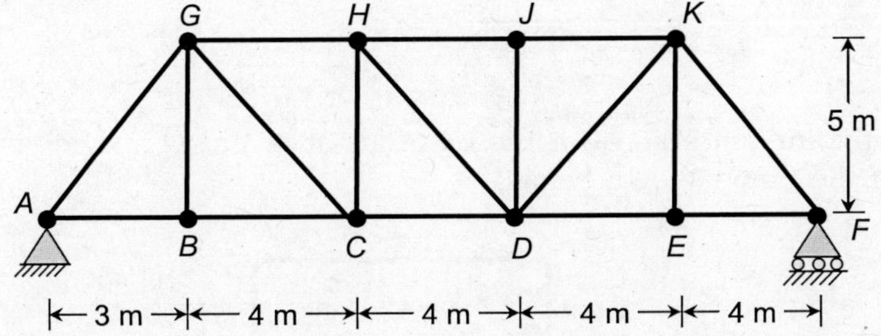

11.6 Draw the influence lines for the forces in members CD, and BG of the truss shown below. Determine the maximum force that would be induced in the member HD due to a point load of 25 kN moving over the bottom chord.

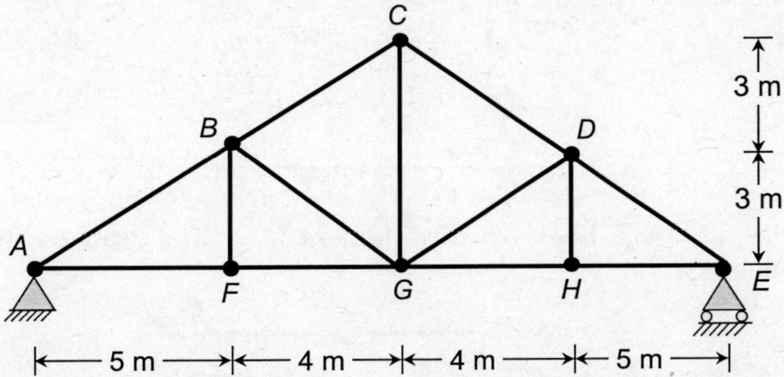

11.7 Draw the influence lines for the forces in members BF of the truss shown below. Determine the maximum force that would be induced in the member HD due to a point load of 100 kN moving over the bottom chord.

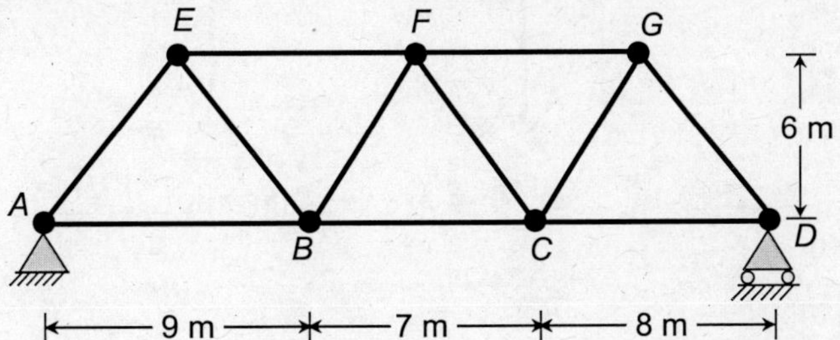

11.8 Draw the influence lines for the bending moments at *A*, *B*, and *C* and the reaction at *A* of the frame shown below.

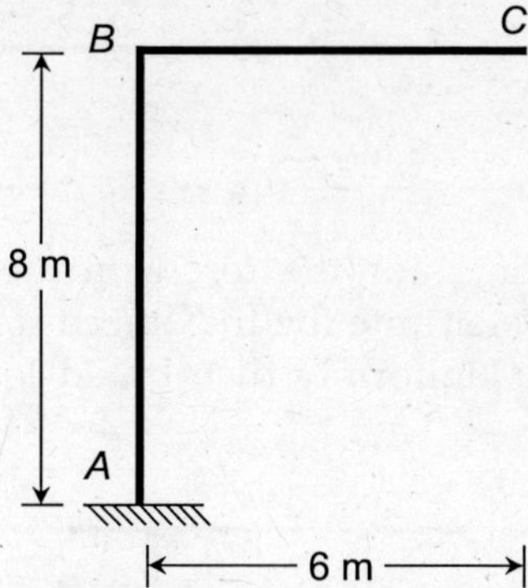

11.9 Draw the influence lines for the reactions at *A* and *D* and the bending moment at *B* of the frame shown below.

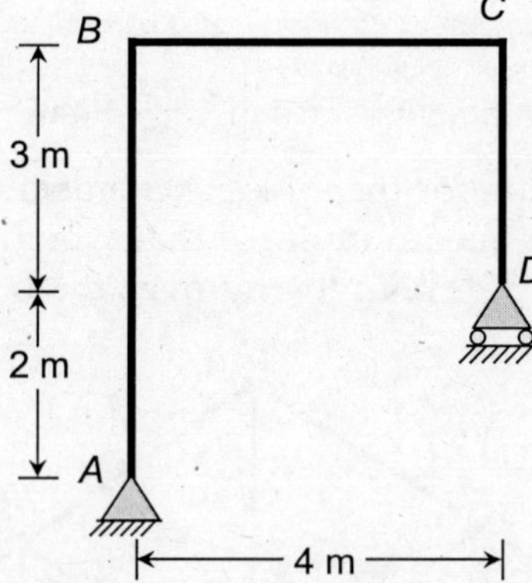

11.10 Draw the influence lines for the moment at *A* and reaction at *D* of the frame shown below.

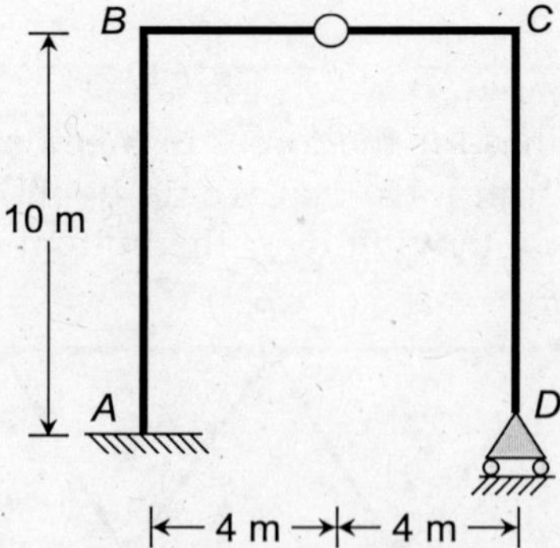

Approximate Methods of Analysis

12.1 GENERAL

The analysis of an indeterminate structure requires detailed information on the geometry and material properties of its members. These are defined by the area of cross-section, the moment of inertia, and the modulus of elasticity. Generally, the type of material (i.e., steel, concrete, etc.) to be used in a structure is known a priori. However, the actual member sizes are mostly unknown for a new structure as the design of this structure is dependent on the results of structural analysis. In such cases, it is required to assume the member properties based on the experience. The assumed section properties may not satisfy the design requirements in many cases, particularly, for a complex structure or a structure subjected to a variety of loads. Therefore, a trial-and-error procedure must be adopted in the design and analysis of a structure.

This chapter presents the approximate methods and an introduction to the energy methods used to analyze the statically indeterminate structures.

12.2 APPROXIMATE METHODS

In order to reduce this iterative process, *approximate methods of analysis* may be adopted in order to obtain a reasonable estimate of the initial member sizes. Since the axial force, shear force, and bending moments at any section of a statically determinate beam can be determined using the equilibrium equations, the analysis procedure is independent of the material and geometric properties. Accordingly, the main aim of *approximate methods* is to transform the indeterminate structures into a series of statically determinate structures and to determine the axial force, shear force, and bending moment at the desired sections.

The analyses of trusses and rigid frames using the *approximate methods* are presented in this chapter.

12.3 ANALYSIS OF TRUSSES

To illustrate the application of approximate methods in trusses, consider a statically indeterminate truss subjected to the generalized loadings at the joints as shown in

Figure 12.1(a). The degree of static indeterminacy of the given truss is three. This means that three additional conditions in excess of the available equilibrium equations are required to determine all member forces. In other words, three assumptions are required to convert the given indeterminate truss to a determinate one.

There are two methods available to establish these assumptions. Both the methods are dependent on the behaviour of cross-diagonals of trusses. In each panel under the applied loading, one diagonal is subjected to tensile force and the other diagonal of the same panel carries the compressive force. If the diagonal members are long and slender, the compressive force would cause the buckling of diagonal members and would not offer any resistance to the applied loading. Thus, each panel may be assumed to consist of only tension diagonals. This would establish three assumptions (one for each panel) for the truss shown in Figure 12.1(b). If the diagonal members are stocky and non-slender, both tension and compression diagonal would provide resistance to the applied loading. If a section x-x cut in a panel, the vertical shear V may be conservatively assumed to be resisted by two diagonals in equal proportion. This would also lead to three assumptions for the given truss.

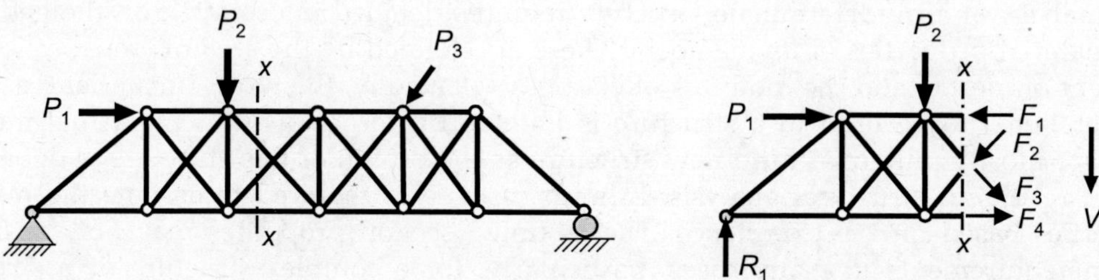

FIGURE 12.1   (a) Indeterminate truss with loading, (b) Free-body diagram at cut section.

Accordingly, two methods used to analyze an indeterminate truss are stated as follows:

1. If the diagonal members are *long and slender*, the vertical shear in each panel of the truss may be assumed to be resisted by the tensional diagonals only. Thus, the axial forces in the compression diagonals are assumed to be zero.
2. If both diagonal members are capable of carrying tension and compression forces, the vertical shear in each panel of the truss may be assumed to be resisted equally by both these diagonals.

The analysis of trusses using these approximate methods has been illustrated in the following examples.

EXAMPLE 12.1   Determine the forces in the members of the truss shown in Figure E12.1 using the approximate method. Assume that the diagonals are so slender that their resistance to the compressive forces can be neglected.

Solution: The first step is to determine the support reactions.
    Vertical reaction at *E*,

$$R_E = \frac{(15)(6) + (20)(3)}{9} \text{ kN} = 16.67 \text{ kN}(\uparrow)$$

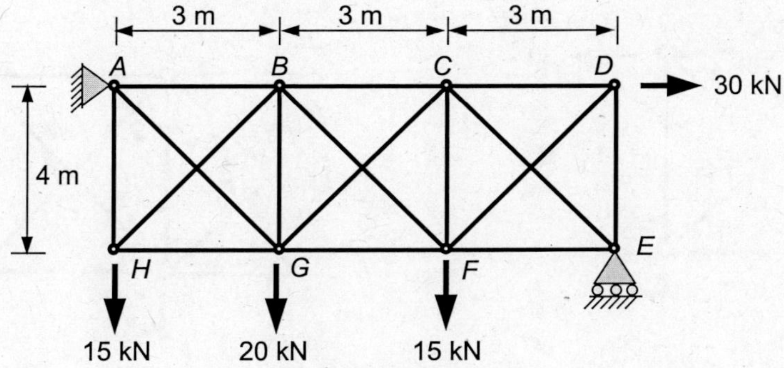

FIGURE E12.1

Vertical reaction at A, $R_A = (15 + 20 + 15 - 16.67)\,\text{kN} = 33.33\ \text{kN}\ (\uparrow)$

Horizontal reaction at A, $H_A = 30$ kN $(\leftarrow)$

Consider the free-body diagram of the first panel of the truss as shown below:

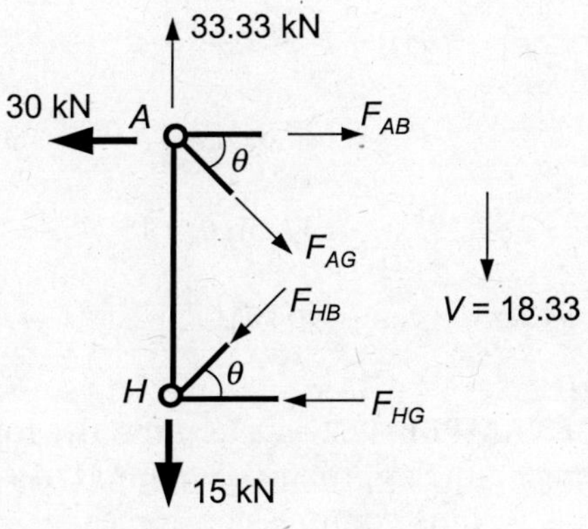

$F_{HB} = 0$; $F_{HG} = 0$; $F_{HA} = 15$ kN (T)

$F_{AG} = \sin\theta = 18.33$

$$F_{AG}\left(\frac{4}{5}\right) = 18.33 \qquad \therefore\ F_{AG} = 22.91\ \text{kN}\ (T)$$

$$F_{AB} + F_{AG}\cos\theta = 30$$

$$F_{AB} + (22.91)\left(\frac{3}{5}\right) = 30 \qquad \therefore\ F_{AB} = 16.25\ \text{kN}\ (T)$$

Next, consider the free-body diagram of third panel of the truss.

$F_{CE} = 0$; $F_{EF} = 0$; $F_{DE} = 16.67$ kN (C)

$F_{DF} = \sin\theta = 16.67$

$$F_{DF}\left(\frac{4}{5}\right) = 16.67 \qquad \therefore\ F_{DF} = 20.84\ \text{kN}\ (T)$$

$$F_{DC} + F_{DF}\cos\theta = 30$$

$$F_{DC} + (20.84)\left(\frac{3}{5}\right) = 30 \qquad \therefore\ F_{DC} = 17.5\ \text{kN}\ (T)$$

Finally consider the free-body diagram of the middle panel as shown below:

$$F_{FB} = 0$$

$$F_{CG}\sin\theta = 1.67$$

$$F_{CG}\left(\frac{4}{5}\right) = 1.67 \qquad \therefore\ F_{CG} = 2.08\ \text{kN}\ (T)$$

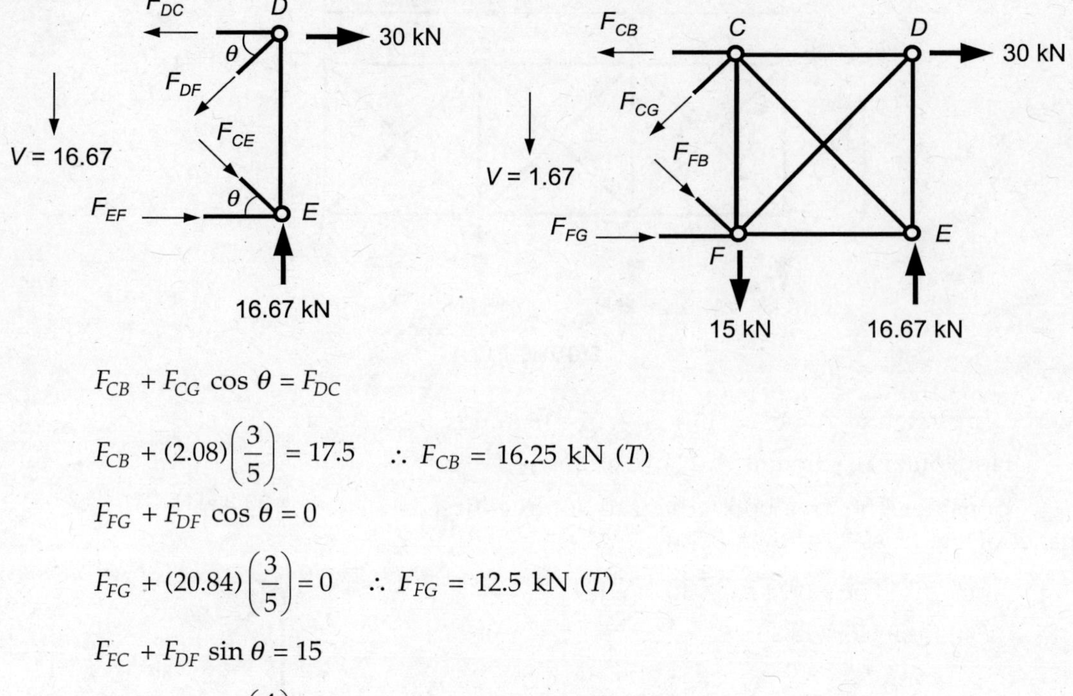

$$F_{CB} + F_{CG} \cos \theta = F_{DC}$$

$$F_{CB} + (2.08)\left(\frac{3}{5}\right) = 17.5 \quad \therefore F_{CB} = 16.25 \text{ kN } (T)$$

$$F_{FG} + F_{DF} \cos \theta = 0$$

$$F_{FG} + (20.84)\left(\frac{3}{5}\right) = 0 \quad \therefore F_{FG} = 12.5 \text{ kN } (T)$$

$$F_{FC} + F_{DF} \sin \theta = 15$$

$$F_{FC} + (20.84)\left(\frac{4}{5}\right) = 15 \quad \therefore F_{FC} = 1.67 \text{ kN } (C)$$

EXAMPLE 12.2 Determine the forces in the members of the truss shown in Figure E12.2 using the approximate method. Assume that all diagonals are designed to resist the tension as well as compression forces.

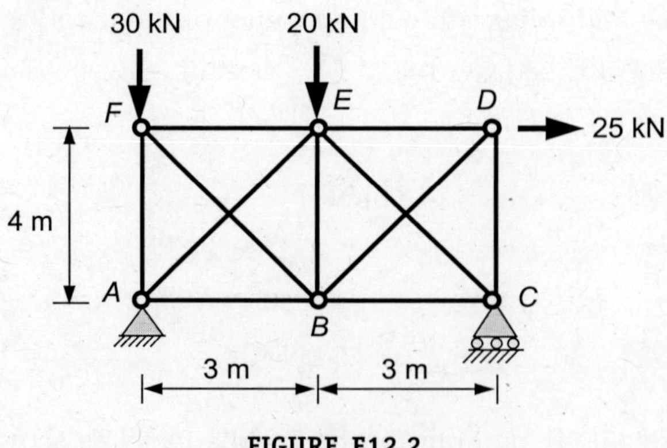

FIGURE E12.2

Solution: The support reactions are computed as follows.

$$R_C = \frac{(20)(3) + (25)(4)}{6} \text{ kN} = 26.67 \text{ kN}(\uparrow)$$

$$R_A = (30 + 20 - 26.67)\ \text{kN} = 23.33\ \text{kN}\ (\uparrow)$$

$$H_A = 25\ \text{kN}\ (\leftarrow)$$

Consider the free-body diagram of the first panel of the truss as shown below:

The axial forces in the cross-diagonals are considered to be the same and in the opposite directions. Thus, $F_{BF} = -F_{AE}$

$$(F_{BF} + F_{AE}) \sin \theta = 6.67$$

$$2F_{BF}\left(\frac{4}{5}\right) = 6.67 \quad \therefore F_{AE} = 4.17\ \text{kN}\ (T) \quad \therefore F_{BF} = 4.17\ \text{kN}\ (C)$$

$$F_{FE} + F_{BF} \cos \theta = 0$$

$$F_{FE} + (4.17)\left(\frac{3}{5}\right) = 0 \quad \therefore F_{FE} = 2.5\ \text{kN}\ (T)$$

$$F_{AF} + F_{BF} \sin \theta = 30$$

$$F_{AF} + (4.17)\left(\frac{4}{5}\right) = 30 \quad \therefore F_{AF} = 26.67\ \text{kN}\ (C)$$

$$F_{AB} + F_{AE} \cos \theta = 25$$

$$F_{AB} + (4.17)\left(\frac{3}{5}\right) = 25 \quad \therefore F_{AB} = 22.5\ \text{kN}\ (T)$$

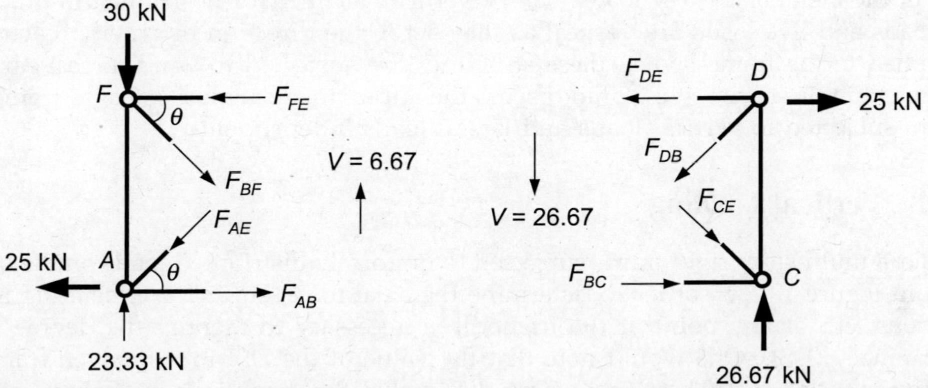

Next, consider the free-body diagram of second panel of the truss.

$$F_{CE} = -F_{DB}$$

$$(F_{DB} + F_{CE}) \sin \theta = 56.67$$

$$2F_{DB}\left(\frac{4}{5}\right) = 26.67 \quad \therefore F_{DB} = 16.67\ \text{kN}\ (T) \quad \therefore F_{CE} = 16.67\ \text{kN}\ (C)$$

$$F_{BC} + F_{CE}\cos\theta = 0$$

$$F_{BC} + (16.67)\left(\frac{3}{5}\right) = 0 \qquad \therefore\ F_{BC} = 10.0\ \text{kN}\ (T)$$

$$F_{DE} + F_{DB}\cos\theta = 25$$

$$F_{DE} + (16.67)\left(\frac{3}{5}\right) = 25 \qquad \therefore\ F_{DE} = 15.0\ \text{kN}\ (T)$$

$$F_{DC} + F_{CE}\sin\theta = 26.67$$

$$F_{DC} + (16.67)\left(\frac{4}{5}\right) = 26.67 \qquad \therefore\ F_{DC} = 13.33\ \text{kN}\ (C)$$

Consider the joint equilibrium at E

$$(F_{CE} - F_{AE})\sin\theta + F_{BE} = 20$$

$$(16.67 - 4.17)\left(\frac{4}{5}\right) + F_{BE} = 20$$

$$\therefore \qquad F_{BE} = 10\ \text{kN}\ (C)$$

12.4 ANALYSIS OF RIGID FRAMES

Rigid frames can be subjected to a variety of loads at joints as well as at any intermediate points of the members. These loads may be vertical, inclined, or horizontal. In practice, the dead loads and live loads are vertical as they act in the direction of gravity. Lateral loads are applied to the frames when these structures are subjected to wind forces, earthquake forces, etc. In this section, we would discuss the approximate analysis of rigid frames when they are subjected to vertical loads and lateral loads independently.

12.4.1 Vertical Loading

Consider a multi-story rigid frame subjected to uniformly distributed loads on the beams as shown in Figure 12.2. In order to determine the axial force, shear force, bending moment, deflections, etc. at any point of the frame, it is necessary to identify the degree of static indeterminacy (DSI). One would note that the value of the DSI for the given frame is 18, i.e., three in each panel. Therefore, we need to make 18 assumptions to analyze this frame.

This problem becomes simpler if one would consider one panel at a time. Let's consider the beam of an intermediate panel. If both ends of this span would have been pinned ends, one could quickly determine the response quantity (i.e., shear, bending moment, etc.) for this determinate beam. For example, the end bending moments at both ends of the simply supported beam would be zero. On the other hand, if both ends were assumed to be fixed, two assumptions are required to solve this fixed beam. Actually, the joints in a rigid frame are neither pinned nor fixed. They are allowed to rotate (like a simply supported beam) as well as can resist the bending moment (like a fixed ended beam). Thus, the behaviour of these members falls between the simply supported and fixed ended cases.

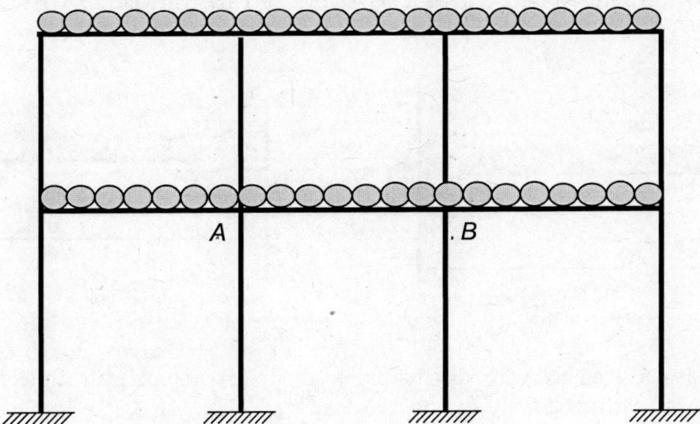

FIGURE 12.2   Multi-story frame subjected to gravity loading.

Consider a simply supported beam AB shown in Figure 12.3(a). The bending moments are zero at both ends of the beam. Thus, the points of zero moments (or contraflexures) are located at zero distance from both ends. Now consider a fixed-ended beam AB subjected to a uniformly distributed load as shown in Figure 12.3(b). In this case, the negative bending moment occurs at the ends and the positive bending moment is observed in the middle segments of the beam. Thus, there are two points of inflection in the deflected shape of the fixed ended beams, where the bending moments are zero. One would find that the exact distance of these points of contraflexure is $0.21L$ from each end, where L is the span of the fixed end beam.

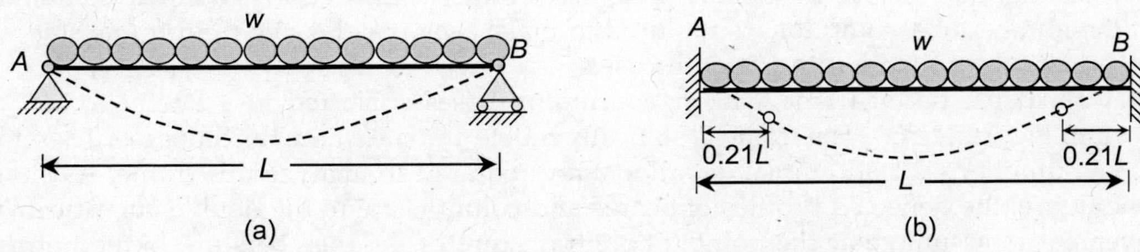

(a)   (b)

FIGURE 12.3   (a) Simply-supported beam with uniformly distributed loading, (b) Fixed beam with the position of points of contraflexure.

It can be inferred that if the points of zero bending moments in the beams of the rigid frames can be known approximately, then this can establish two assumptions in each panel. Figure 12.4(a) shows the beam AB of the rigid frame with loading. Since the ends A and B are neither pinned nor fixed, the point of contraflexure can be approximately considered as the average of the values corresponding to the hinged and the fixed cases, i.e., $(0 + 0.21)L/2 = 0.10L$. The locations of these points are shown in Figure 12.4. Since the bending moment at these points are zero, they can be considered as the hinge or pinned as shown in Figure 12.4(b).

One more assumption is required to make each panel of the frame determinate. This can be established by assuming the axial forces in the beams of the rigid frames to be zero. This is often true when a frame is subjected to uniformly distributed load in the vertical direction.

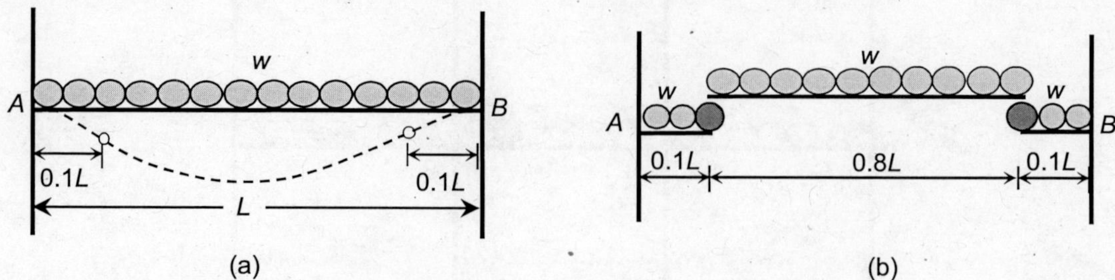

FIGURE 12.4   (a) Frame member with distributed loading, (b) Approximate inflection points of frame member with uniformly distributed loading.

Thus, the following assumptions are required to carry out approximate analysis of a rigid frame subjected to vertical loadings:

1. Bending moments at a distance of $0.1L$ from each joint are zero in all beams of the frame.
2. All beams do not carry axial forces.

12.4.2   Lateral Loading

If the frames are subjected to the lateral loads, the assumptions made for the analysis of frames with vertical (gravity) loadings are not applicable. This is primarily due to the different deflected shapes of members of these frames. However, it is worth mentioning that the number of assumptions is required to make each panel of the rigid frame statically determinate remains the same in both cases.

Consider a portal frame with fixed column bases subjected to a lateral force Q as shown in Figure 12.5(a). The frame is statically indeterminate to the third degree. Therefore, three assumptions (or additional equations) are required to analyze this frame. As shown in the figure, the deflected profiles of beams and columns are in the double curvature. It is convenient to assume that the points of contraflexure of the members are located at their mid-spans. Since there are three members in the rigid frame, we can get three additional equations for the given frame. Accordingly, the free-body diagrams of the members can be drawn considering these points of contraflexures as the internal hinges as shown in Figures 12.5(b). For each free-body diagram, the unknown forces and bending moments can be determined by using the equations of equilibrium. Accordingly, the bending moment, shear force, and axial force at any point in the frame can be obtained. Figure 12.5(c) shows the bending moment diagram of the given frame with the applied lateral loading.

A similar procedure can be adopted to analyze a portal frame with hinged bases and subjected to a lateral force, Q as shown in Figure 12.6(a). The degree of static indeterminacy of this frame is one. Hence, only one assumption is required to analyze the frame. The

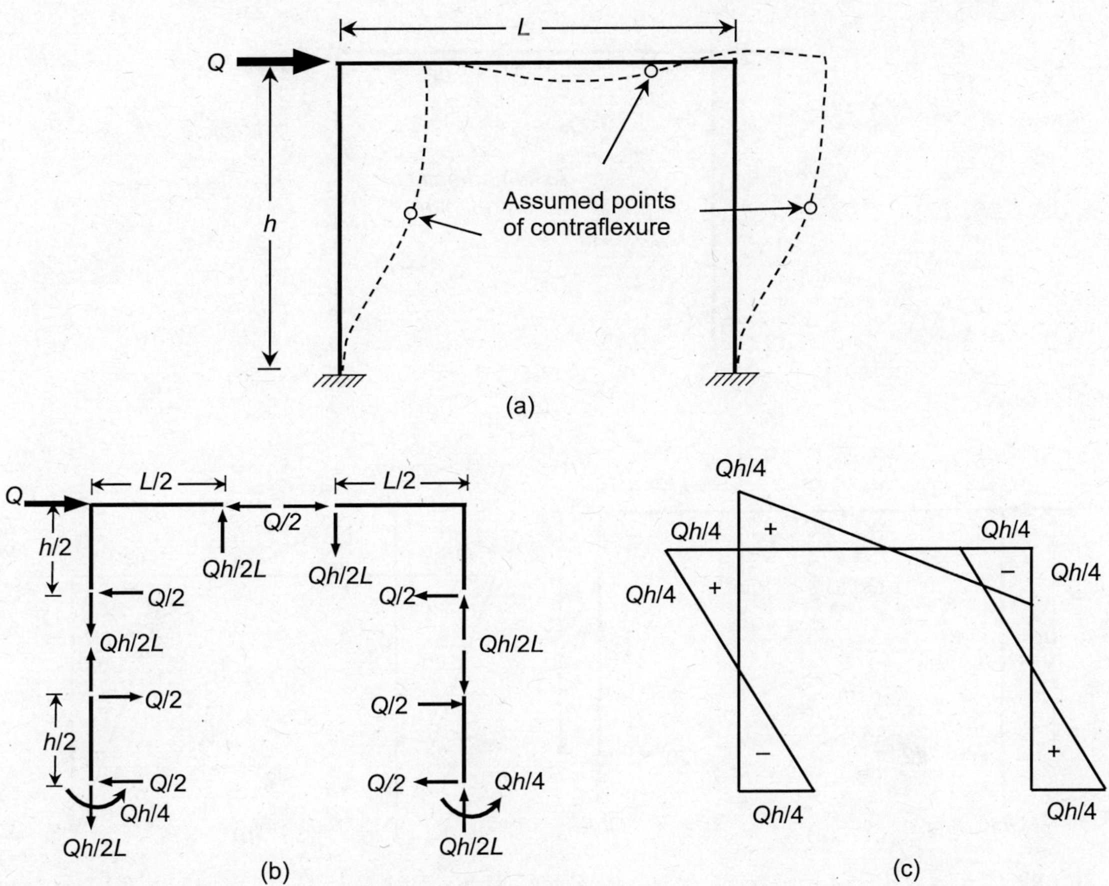

FIGURE 12.5   (a) Fixed base frame under lateral load with assumed points of inflections, (b) Free-body diagrams, (c) Bending moment diagram.

deflected shapes of the column are in single curvature, whereas the beam is deflected in the double curvature. Hence, the point of contraflexure of the beam is assumed to be located at its mid-span. The free-body diagrams of different parts of the frame considering the internal hinge at the beam are shown in Figure 12.6(b). Using the equations of equilibrium, the bending moment, shear force, and axial force at any section of all the members can be determined. Figure 12.6(c) shows the bending moment diagram of the frame under the lateral loading.

12.5   ANALYSIS OF MULTI-STORY FRAMES

The procedure discussed in the preceding section may not be adequate to analyze a multi-story multi-bay portal frame subjected to lateral loads. For example, consider a two-bay frame portal frame shown in Figure 12.7(a). The degree of static indeterminacy of the structure is six. There are three columns and two beams in the given frame. Assuming the

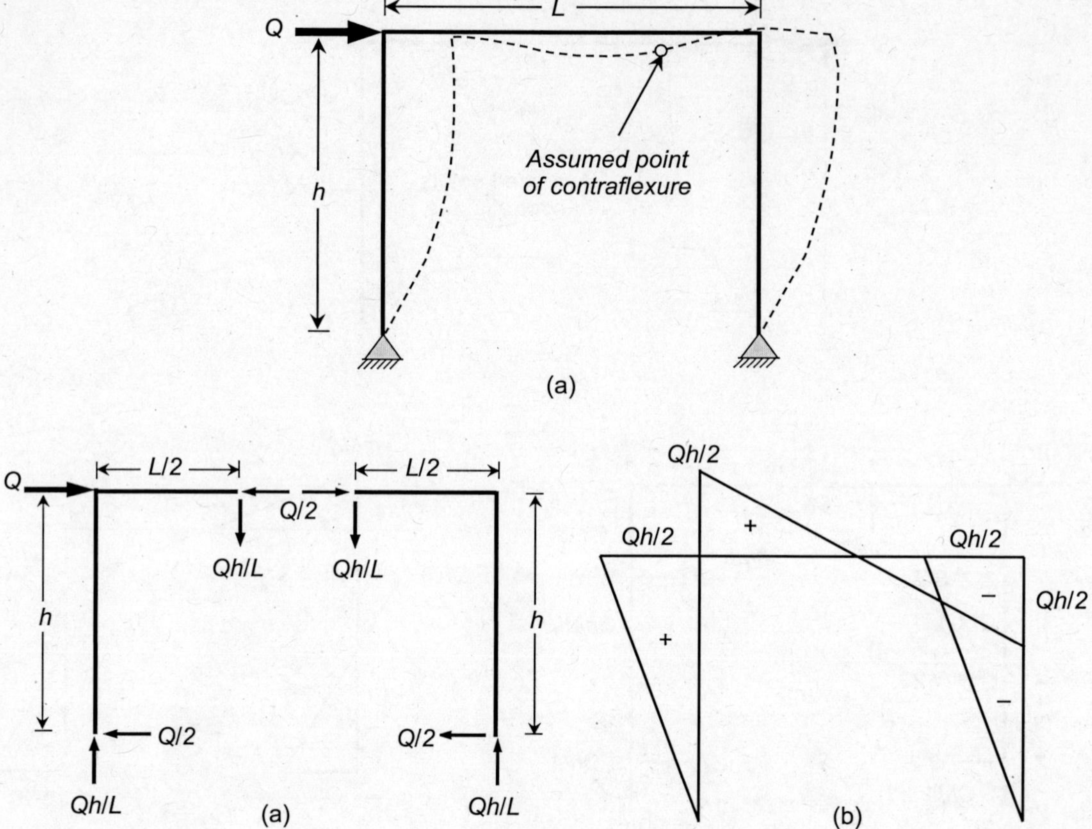

FIGURE 12.6   (a) Hinged-base frame under lateral load with assumed points of inflections, (b) Free-body diagrams, (c) Bending moment diagram.

point of contraflexure at the mid-span of each member as discussed earlier, five additional equations can be obtained. Therefore, one more additional assumption is required to analyze the frame using the approximate methods.

Two methods, namely, *Portal method* and *Cantilever method*, are widely used to approximately analyzing a multi-story frame structure subjected to lateral loads. These methods are discussed in the following sections.

12.5.1   Portal Method

The *portal method* is generally used to analyze the shorter and wider rigid frames. In addition to the assumptions of hinges at the mid-spans of all members (with rigid ends), one additional assumption is made in reference to the horizontal shear. In this method, the multi-bay frames are divided into a series of single-bay frames in which each column is assumed to resist the same magnitude of shear. For the given frame shown in Figure 12.7(a), two single-bay frames can be formed as shown in Figure 12.7(b). It can be seen that the interior column is subjected to twice of shear forces carried by the external columns. These assumptions are now sufficient to analyze the given frame.

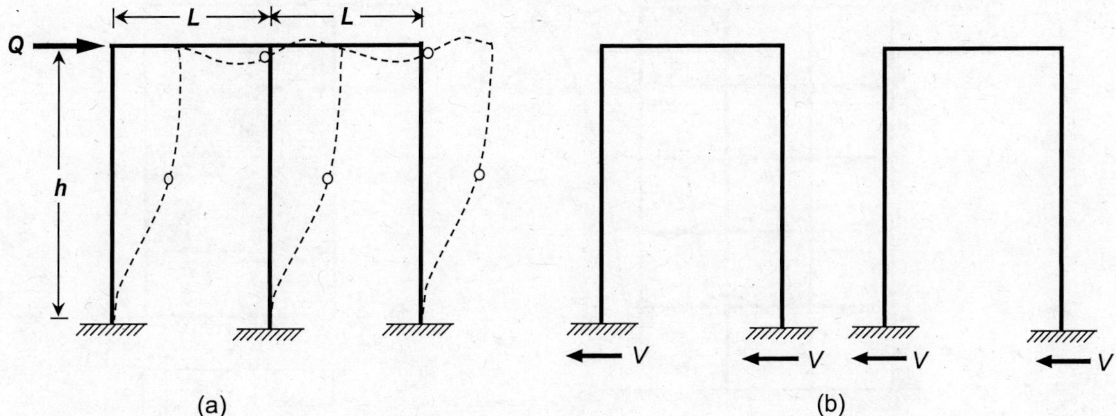

FIGURE 12.7 Portal method: (a) Frame with points of inflections, (b) Series of single-bay portal frames.

Accordingly, the following assumptions are made to analyze a frame (with fixed column bases) using the Portal method.

1. The points of zero bending moment are assumed to be located at the mid-spans of all beams.
2. The points of zero bending moment are assumed to be located at the mid-heights of all columns.
3. Each interior column is assumed to carry twice the shear resisted by the external columns.

12.5.2 Cantilever Method

The cantilever method is adopted to approximately analyze the tall and slender rigid frames. To illustrate this method, consider a rigid frame subjected to a lateral force shown in Figure 12.8(a). The deflected shape of the frame resembles that of a cantilever structure subjected to a lateral load as shown in the figure. Some of the columns will be subjected to tension forces, whereas others are subjected to compressive forces. Thus, the given frame can be replaced by an equivalent cantilever column. The resulting bending moment due to the applied lateral load would induce compressive and tensile flexural stresses, which varies linearly about the neutral axis (n.a.) of the column cross-section. The equivalent cantilever column with flexural stress distribution is shown in Figure 12.8(b). These flexural stresses are equal to the axial stresses in the columns of the actual frame at the corresponding locations.

Accordingly, the following assumptions are made to analyze a frame (with fixed column bases) using the Cantilever method.

1. The points of zero bending moment are assumed to be located at the mid-spans of all beams.
2. The points of zero bending moment are assumed to be located at the mid-heights of all columns.
3. The axial stress in a particular column is proportional to its distance from the centroid of cross-sectional areas of all columns at any floor level.

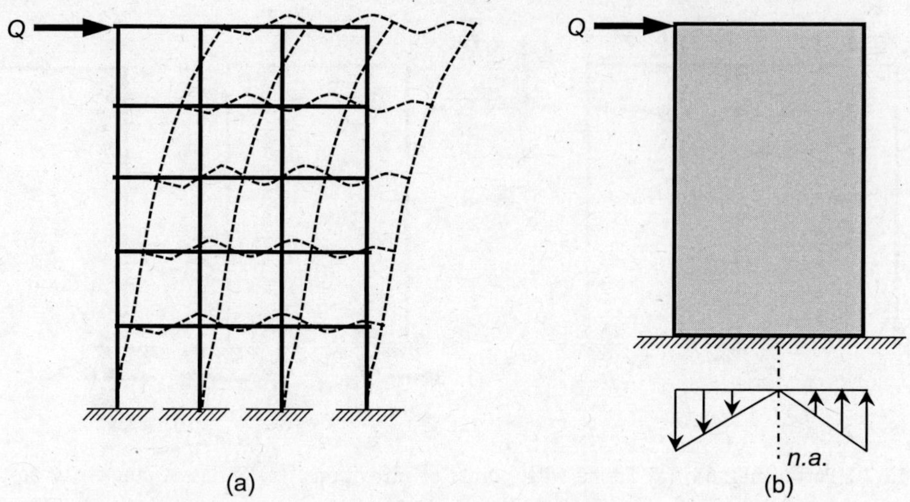

FIGURE 12.8   Cantilever method: (a) Deflected shape of frame, (b) Distribution of bending stress at the base of frame.

These approximate methods applied to the building frames are illustrated in the following examples.

EXAMPLE 12.3   Determine the support reactions and the bending moments at the joints of the frame shown in Figure E12.3 using the Portal method.

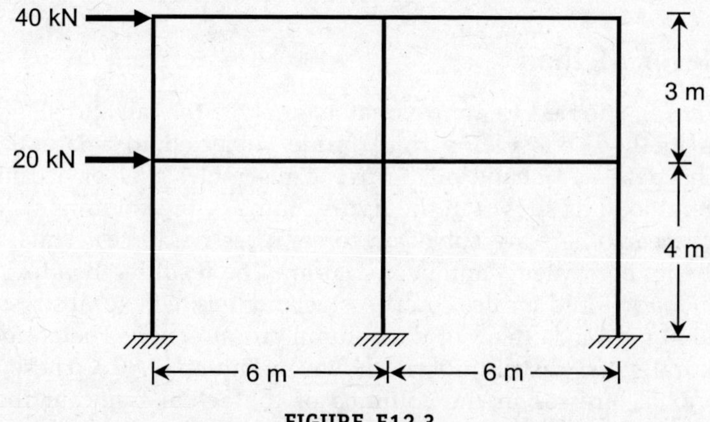

FIGURE E12.3

Solution: For the given frame, the points of zero bending moment (i.e., hinges) are considered at the mid-span of each member.

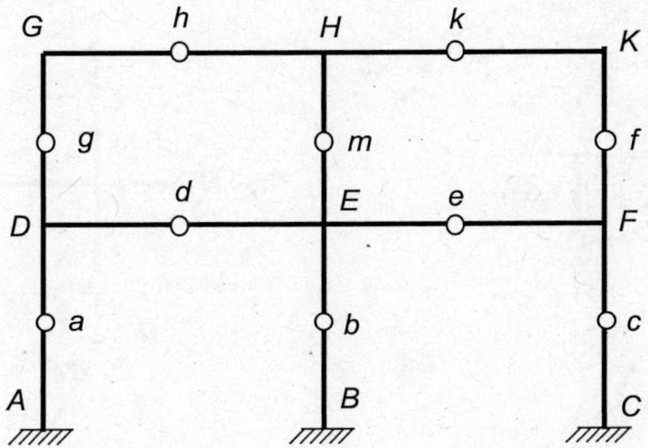

Next, the shear force resisted by the columns in each floor level can be determined considering the force equilibrium in the respectively free-body diagrams.

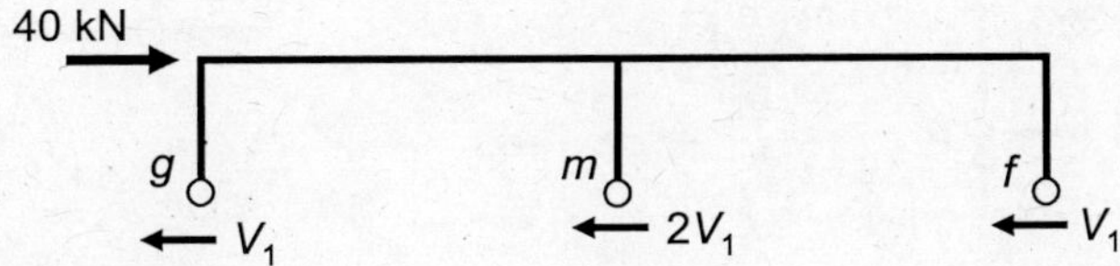

$$4V_1 = 40 \text{ kN} \quad \therefore V_1 = 10 \text{ kN}$$

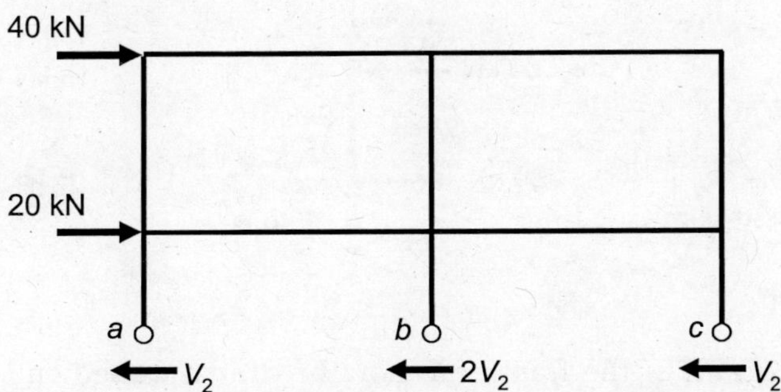

$$4V_2 = (40 + 20) \text{ kN} = 60 \text{ kN} \quad \therefore V_2 = 15 \text{ kN}$$

The forces and moments at the hinge points can be determined considering the free-body diagrams of the members between the consecutive hinge points.

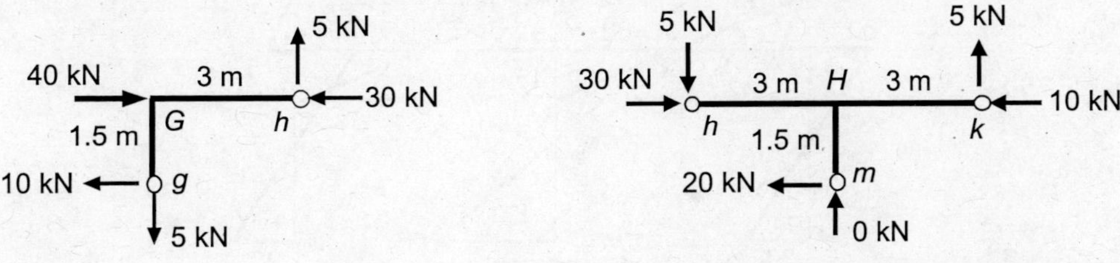

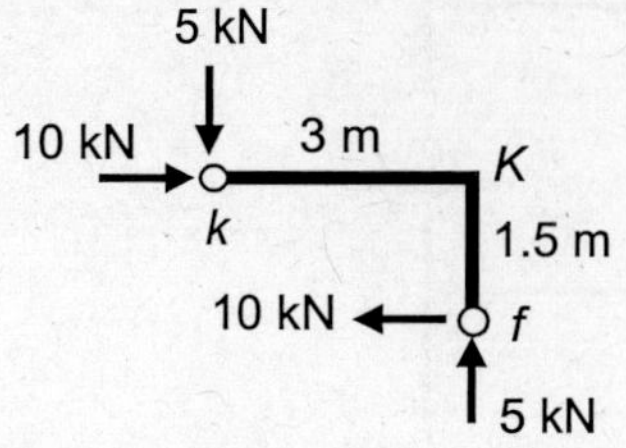

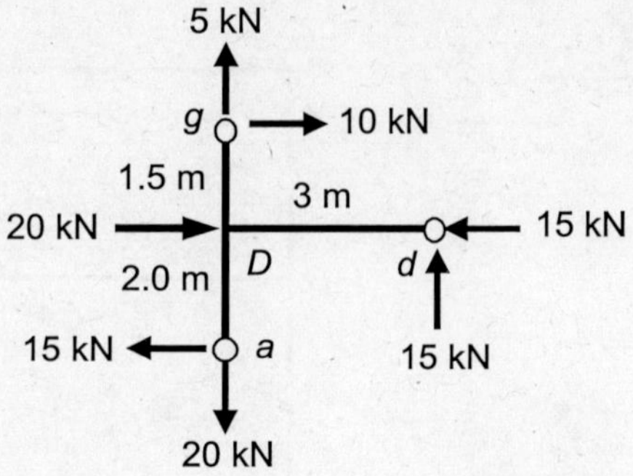

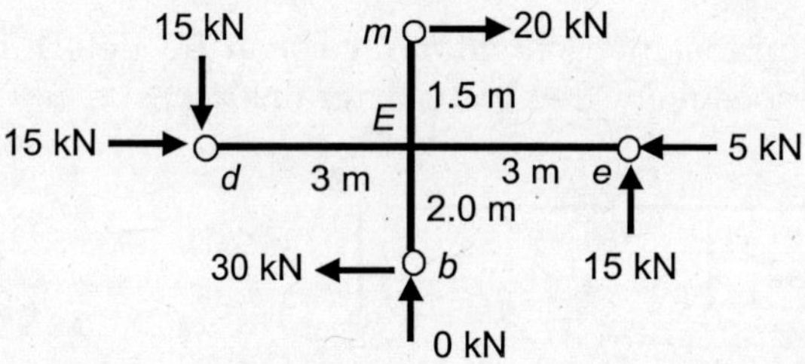

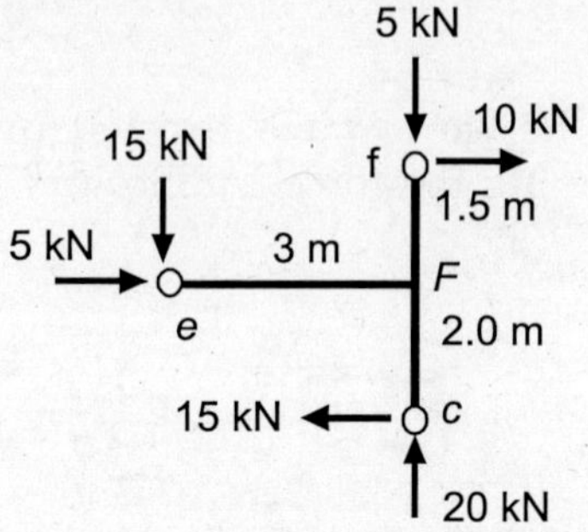

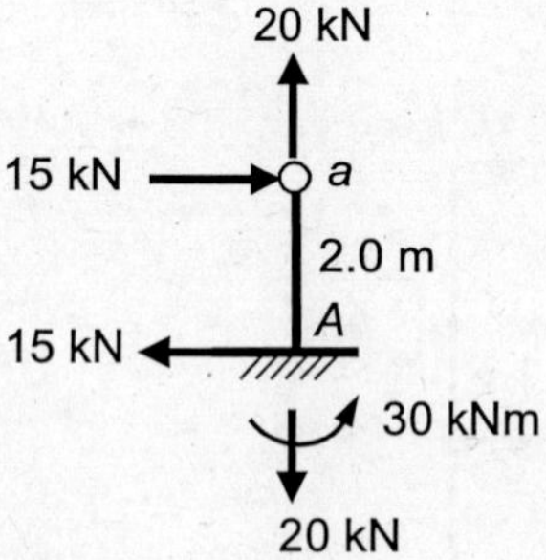

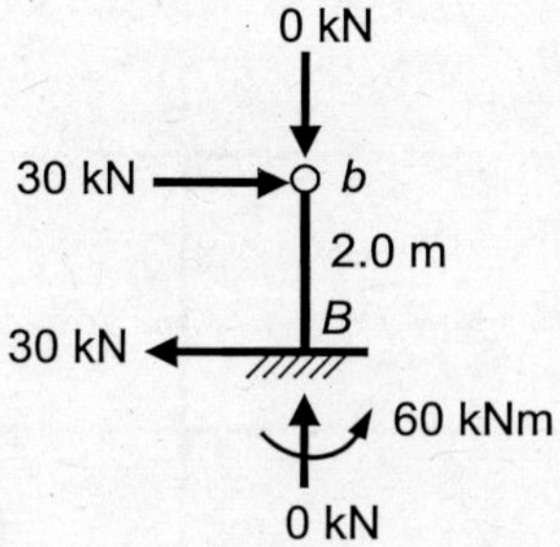

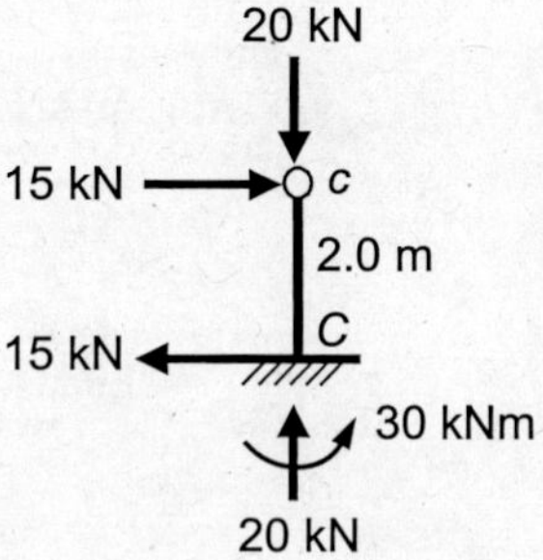

Bending moment diagram of the frame can now be drawn as shown below:

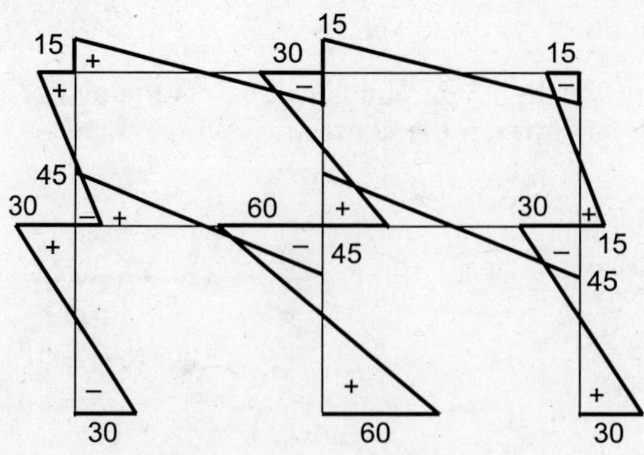

EXAMPLE 12.4 Determine the reactions and the bending moments at fixed supports of the frame shown in Figure E12.4 using the Cantilever method.

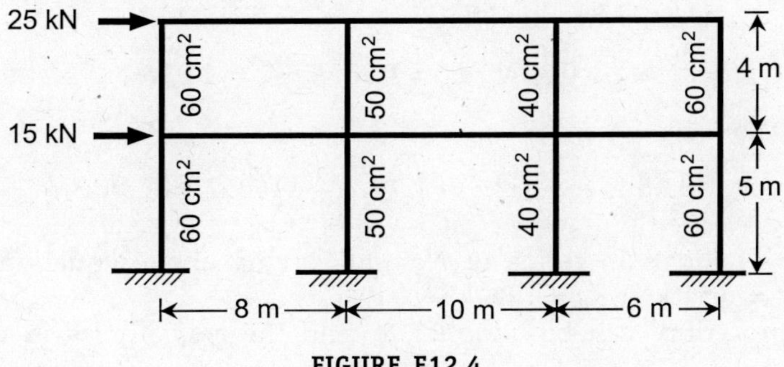

FIGURE E12.4

Solution: For the given frame, the points of zero bending moment (i.e., hinges) are considered at the mid-span of each member.

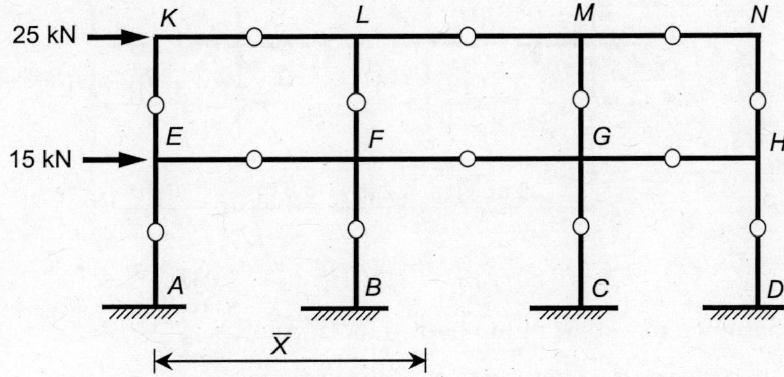

The distance of centroid of the frame from end A is determined as follows:

$$\bar{x} = \frac{(50 \times 8) + (40 \times 18) + (60 \times 24)}{(60 + 50 + 40 + 60)} = 12.2 \text{ m}$$

As per the cantilever method, the axial stresses in columns vary linearly from the centroid of the frame. Consider the free-body diagram of the story at the level of point of contraflexure. The shear forces and axial forces in columns of the top story are shown in the following figure.

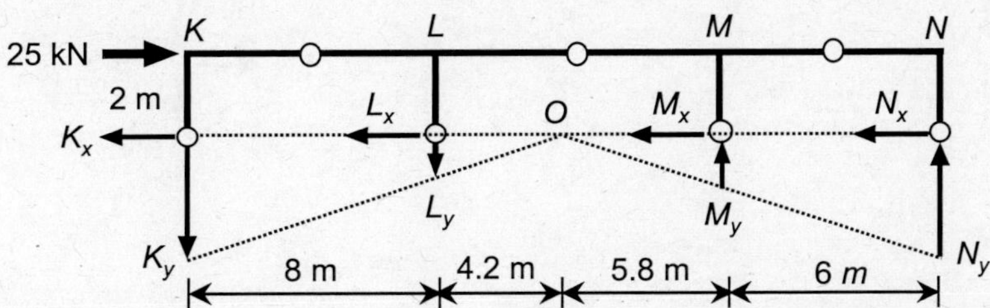

Using the linear variation in axial stresses in columns, one can get

$$\frac{(N_y/60)}{11.8} = \frac{(M_y/40)}{5.8} = \frac{(L_y/50)}{4.2} = \frac{(K_y/60)}{12.2}$$

$$\therefore \quad M_y = 0.33\,N_y;\; L_y = 0.30\,N_y;\; K_y = 1.03\,N_y$$

Taking moment about the centroid O,

$$M_y\,(5.8) + N_y\,(11.8) + L_y\,(4.2) + K_y\,(12.2) = (25)(2)$$

Expressing all forces in terms of N_y and solving above equation, $N_y = 1.81$ kN; $M_y = 0.59$ kN; $L_y = 0.54$ kN; $K_y = 1.86$ kN.

A similar procedure can be adopted to find the axial forces in the bottom story columns.

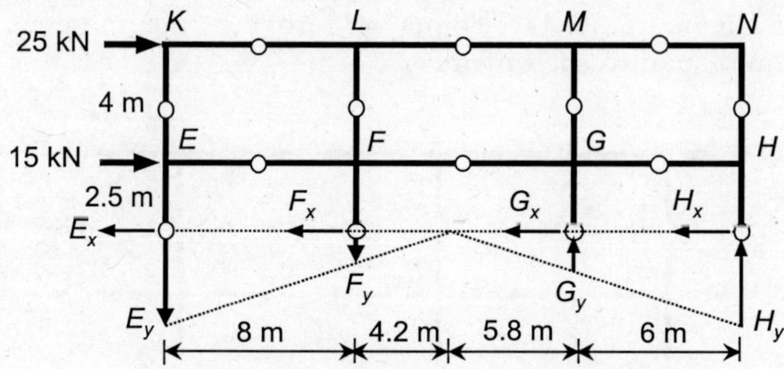

$$\therefore \quad G_y = 0.33\,H_y;\; F_y = 0.30\,H_y;\; E_y = 1.03\,H_y$$

Considering the moment equilibrium about the centroid,

$$G_y\,(5.8) + H_y\,(11.8) + F_y\,(4.2) + E_y\,(12.2) = (25)(6.5) + (15)(2.5)$$

$$H_y = 7.26 \text{ kN};\; G_y = 2.40 \text{ kN};\; F_y = 2.18 \text{ kN};\; E_y = 7.48 \text{ kN}$$

The forces and moments at the hinge points can be determined considering the free-body diagrams of the members between the consecutive hinge points.

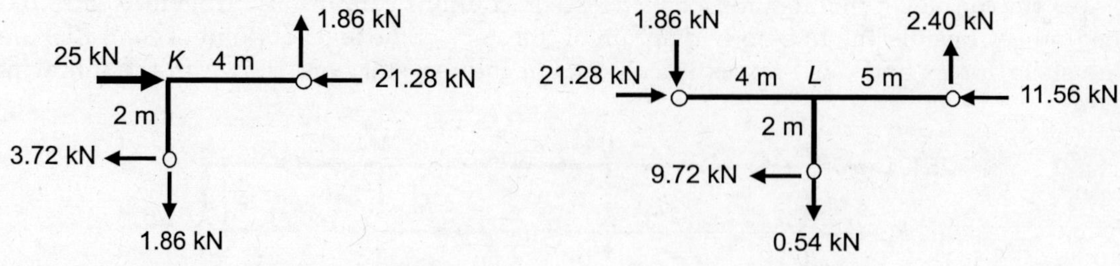

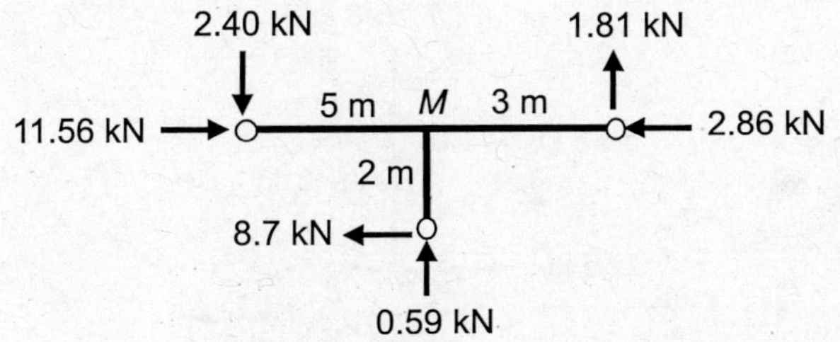

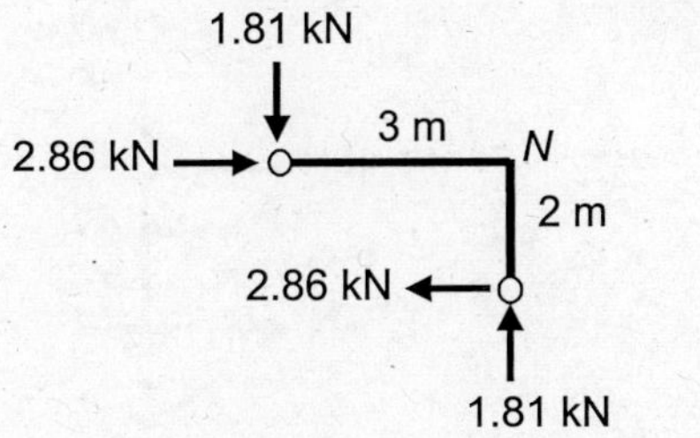

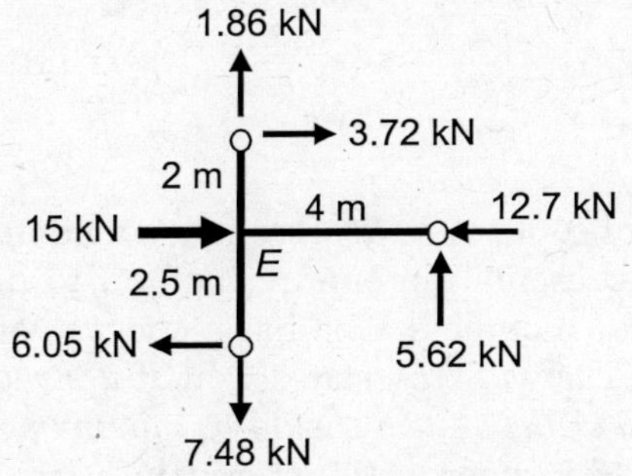

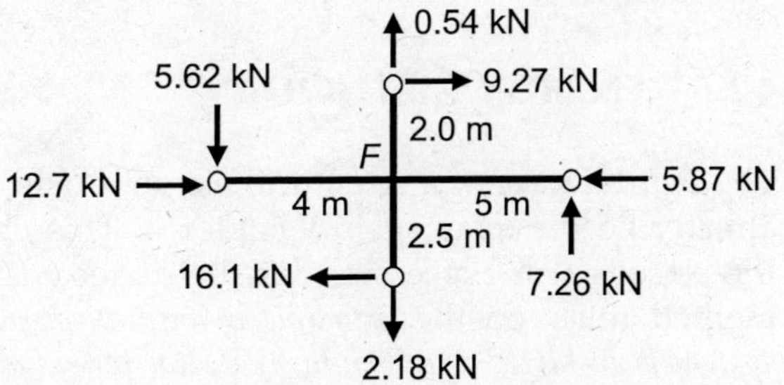

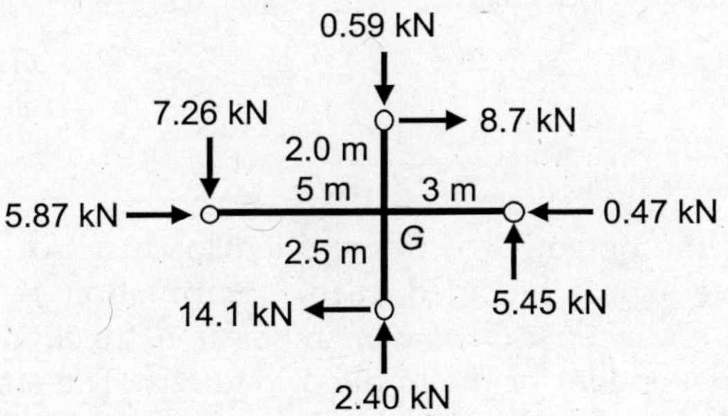

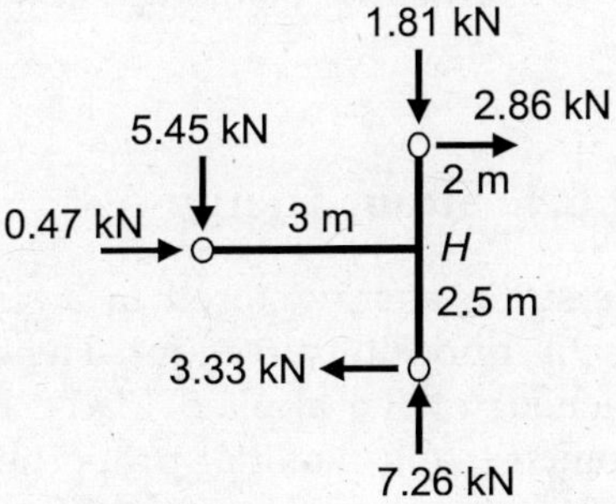

The value of reactions, shear forces, and bending moments at the fixed joints are determined as follows:

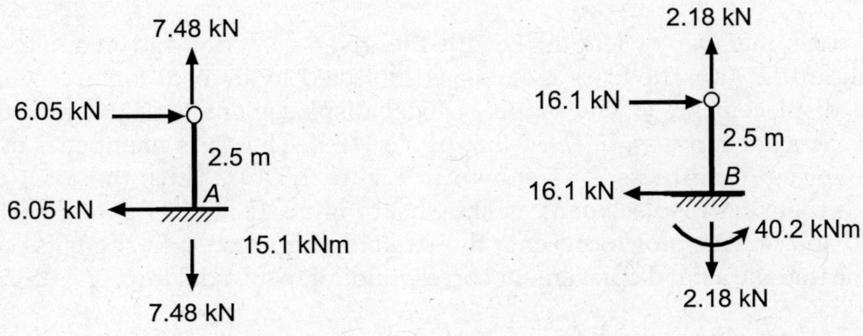

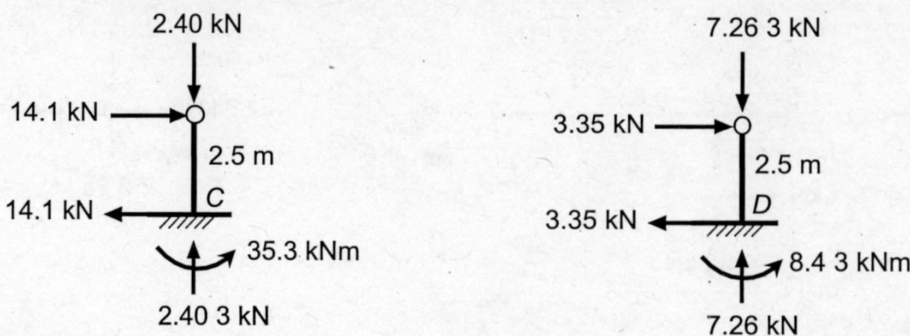

12.6   ENERGY METHOD

The deformation of a structural element using the concept of the work-energy method has already been explained in Chapter 4. This chapter presents the energy method to derive the closed-form expression for the deflected shape of members. The basic concept of this method relies on the *principle of minimum potential energy*. This states that *if a structural system is in static equilibrium, the total potential energy of the system should be minimum.*

Total potential energy (Π) is the sum of the strain energy (U) stored in a structure and the potential of the applied loads (V). Mathematically,

$$\Pi = U + V \tag{12.1}$$

12.6.1   Strain Energy

The strain energy stored in a structure and the potential of the applied loads act in the exactly opposite directions. Hence, a negative sign is included in the computation of the potential of the applied loads. Further, the strain energy of a member is a function of geometric and material properties and is independent of the applied loadings. The strain energy stored in a truss member and a beam member has been derived in the following sections.

Truss member

Consider a truss member of length, L with the given material and geometric properties shown in Figure 12.9(a). The truss member is subjected to an axial force, P which resulted in the axial displacement, y. The value of axial displacement, y is the ratio of the axial force P and the axial stiffness, k. The value of k is AE/L. The truss member is now replaced by an axial spring of stiffness, k as shown in Figure 12.9(b). Thus, the axial force varied linearly with the axial displacement as shown in Figure 12.9(c). The area enclosed in the axial force-displacement plot represents the strain energy stored in the truss member.

Assume that the axial displacement corresponding to an axial force, P is Δ. Accordingly,

strain energy $U = \dfrac{1}{2}(P\Delta) = \left(\dfrac{1}{2}\right)\left(\dfrac{P}{A}\right)\left(\dfrac{\Delta}{L}\right)(AL)$

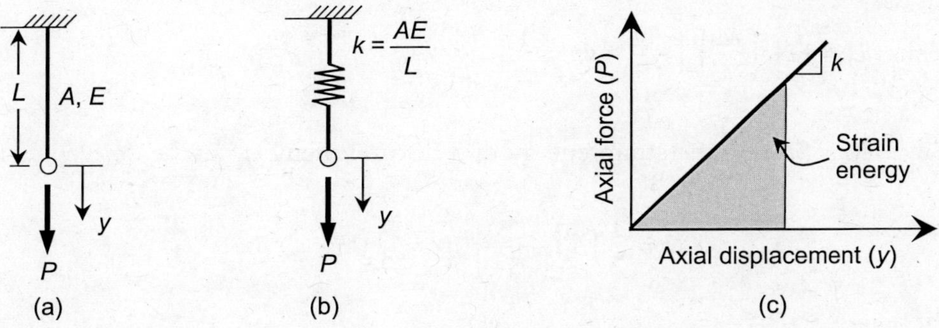

FIGURE 12.9 (a) Truss member under axial load, (b) Equivalent spring, (c) Axial force-displacement response.

Using $\sigma = P/A$; $\varepsilon = \Delta/L$ and $\sigma = E\varepsilon$

$$U = \left(\frac{1}{2}\right)(\sigma\varepsilon)\,(Vol.) = \left(\frac{1}{2}\right)(E\varepsilon^2)\,(Vol.) \tag{12.2}$$

Strain energy density is defined as the ratio of strain energy stored in a member per unit volume, which is equal to one-half of the product of the stress (σ) and strain (ε).

For a prismatic member of uniform cross-sectional area (A) and noting that $\varepsilon = dy/dx$, the strain energy for a truss member can be obtained from Eqn. (12.2) as follows:

$$U = \frac{1}{2}\int AE\left(\frac{dy}{dx}\right)^2 dx = \frac{1}{2}\int AE(y')^2 dx \tag{12.3}$$

Flexural member

Consider a beam of length, L and flexural rigidity, EI as shown in Figure 12.10(a). At any section a-a, one can find the bending moment (M) and rotation (θ). The strain energy for a beam can be related to the bending moment and rotation diagram as shown in Figure 12.10(b).

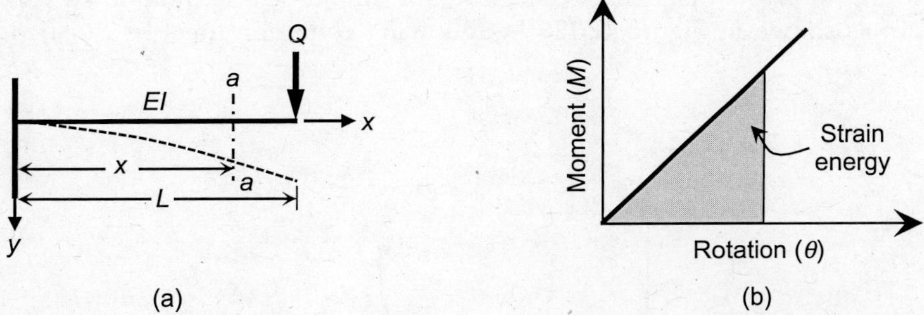

FIGURE 12.10 Beam under transverse loading, (b) Moment-rotation response.

Strain energy, $U = \dfrac{1}{2}\displaystyle\int_V \sigma\varepsilon(dV.) = \dfrac{1}{2}\displaystyle\int \sigma\varepsilon(A.dx) = \dfrac{1}{2}\displaystyle\int \dfrac{\sigma^2 A}{E}dx$

We know that $\dfrac{M}{I} = \dfrac{\sigma}{y}$ or $\sigma = \dfrac{My}{I}$

Thus, $U = \dfrac{1}{2} \int \left(\dfrac{M}{I}\right)^2 \left(\dfrac{Ay^2}{E}\right) dx = \dfrac{1}{2} \int \dfrac{M^2}{EI} dx$ (12.3)

Since $M = EI\dfrac{d^2y}{dx^2}$, the strain energy of a flexural member can be derived as follows:

$$U = \dfrac{1}{2} \int EI\left(\dfrac{d^2y}{dx^2}\right)^2 dx = \dfrac{1}{2} \int EI(y'')^2 dx$$ (12.4)

12.6.2   Potential of Applied Loads

The potential of applied loads is equal to the work done by the structure in the direction of the applied loads. Consider a beam subjected to generalized loadings as shown in Figure 12.11. The generalized expression for the potential of the applied loads is given by

$$V = -\sum_{i=1}^{n} F_i y_i - \sum_{i=1}^{m} M_i \left(\dfrac{dy}{dx}\right)_i - \int_0^L w(x).y.dx$$ (12.5)

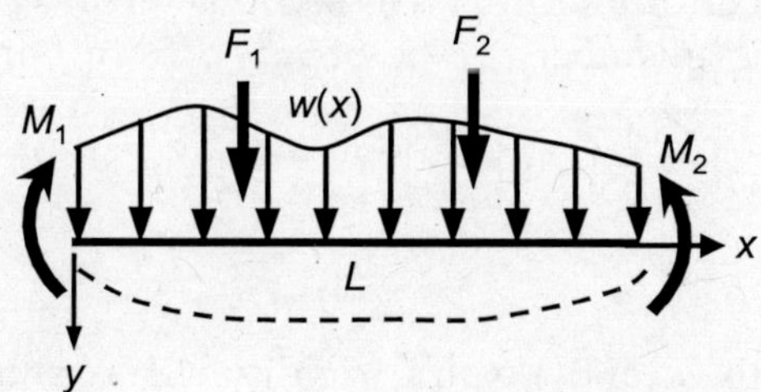

FIGURE 12.11   Beam under generalized loadings.

Since the total potential energy of a structure should be minimum to keep in the static equilibrium as shown in Figure 12.12, the following condition must be satisfied.

$$\dfrac{\partial \Pi}{\partial y} = 0$$ (12.6)

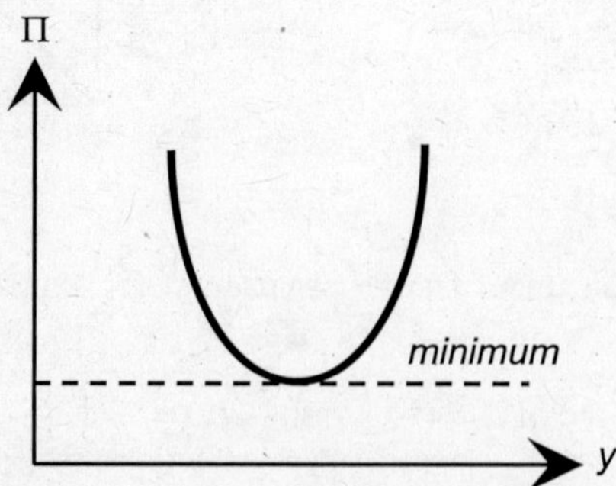

FIGURE 12.12   Minimum total potential energy.

This states that *the partial derivative of the total potential energy* (Π) *of a structure with respect to the displacement* (y) *is zero*. This forms the basis of energy methods to determine the deflection at the section of a structural member or structural system.

12.7   RAYLEIGH-RITZ METHOD

The energy method relying on the principle of minimum potential energy requires that the strain energy and the potential of applied loads must be expressed in terms of displacement. However, the expression for the deflected profile is not known. Rayleigh-Ritz method provides a unique solution to this problem. This method is based on an assumed shape of the deflected shape of the member.

For example, $y = C_1 f_1(x) + C_2 f_2(x) + C_3 f_3(x) + \ldots$

In the assumed expression for the deflected shape, the coefficients C_1, C_2, C_3, … are unknown and the functions $f_1(x)$, $f_2(x)$, $f_3(x)$, … are known (assumed). If the assumed expression represents the deflected shape, it must satisfy the given boundary conditions of the member.

The next step is to determine the total potential energy of the structure as follows:

$$\Pi(C_1,\ C_2,\ C_3,\ \ldots) = U + V$$

Note that the total potential energy expression is a function of unknown coefficients, C_1, C_2, C_3, …. These coefficients can be determined by applying the principle of minimum potential energy as defined in Eqn. (12.6). This, the following conditions must be satisfied.

$$\frac{\partial \Pi}{\partial C_1} = 0; \quad \frac{\partial \Pi}{\partial C_2} = 0; \quad \frac{\partial \Pi}{\partial C_3} = 0 \ldots \text{and so on}$$

The application of the Rayleigh-Ritz method has been illustrated in the following examples.

EXAMPLE 12.5   Determine the maximum deflection and the slope at both ends of the simply supported beam subjected to a uniformly distributed load w as shown in Figure E12.5 using the Rayleigh-Ritz Method. Also, compare with the exact values.

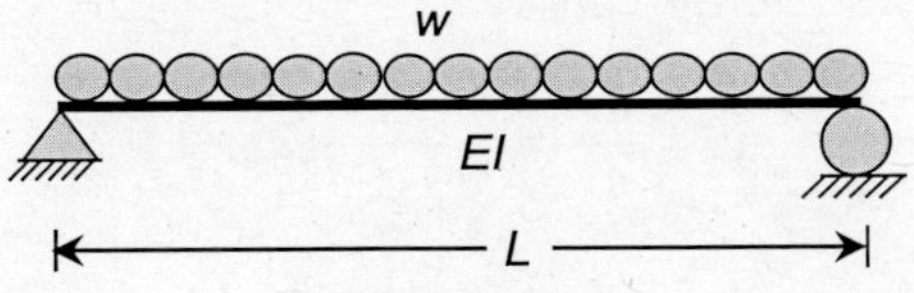

FIGURE E12.5

Solution: Let's assume the deflected shape of the beam as $y = Ax^2 + Bx + C$.

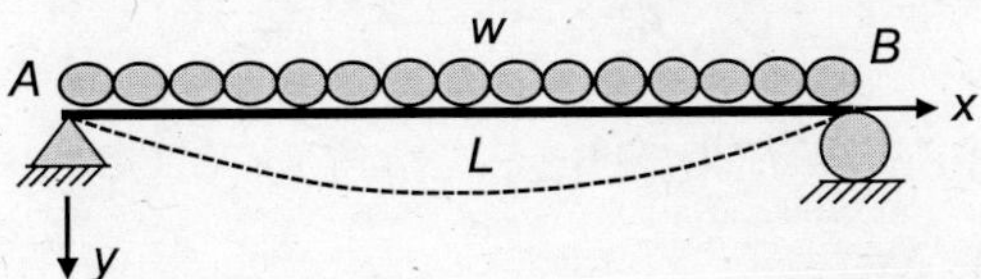

Boundary conditions: Deflection of the beam at the supports A and B are zero.

$$y\big|_{x=0} = 0 \quad \therefore C = 0$$

$$y\big|_{x=L} = 0 \quad \therefore B = -AL$$

Thus, $y = A(x^2 - Lx)$

$$y' = \frac{dy}{dx} = A(2x - L)$$

$$y'' = \frac{d^2y}{dx^2} = 2A$$

Strain energy, $U = \dfrac{1}{2} \int EI(y'')^2 dx = \dfrac{EI}{2} \int_0^L (2A)^2 dx = 2EIA^2L$

Potential of applied load, $V = -\int_0^L w.y.dx = -w\int_0^L A(x^2 - Lx)\,dx = \dfrac{wAL^3}{6}$

Total potential energy, $\Pi = U + V = 2EIA^2L + \dfrac{wAL^3}{6}$

Applying the principle of minimum potential energy, $\dfrac{\partial \Pi}{\partial A} = 0$

$$\frac{\partial\left(2EIA^2L + \dfrac{wAL^3}{6}\right)}{\partial A} = 0 \quad \therefore A = -\frac{wL^2}{24EI}$$

Putting the value of A, the final equation of deflected shape can be obtained as follows:

$$y = -\frac{wL^2}{24EI}(x^2 - Lx)$$

The equation for the slope is then given by

$$y' = \frac{dy}{dx} = -\frac{wL^2}{24EI}(2x - L)$$

Slope at support A (i.e., $x = 0$), $\theta_A = \dfrac{dy}{dx}\bigg|_{x=0} = \dfrac{wL^3}{24EI}$

Slope at support B (i.e., $x = L$), $\theta_B = \dfrac{dy}{dx}\bigg|_{x=L} = -\dfrac{wL^3}{24EI}$

For the maximum deflection, $\dfrac{dy}{dx} = 0$. Thus, $(2x - L) = 0$. $\quad \therefore x = \dfrac{L}{2}$

Hence, $y_{max} = -\dfrac{wL^2}{24EI}\left[\left(\dfrac{L}{2}\right)^2 - L\left(\dfrac{L}{2}\right)\right] = \dfrac{wL^4}{96EI}$

Exact value of maximum deflection, $y_{max} = \dfrac{5wL^4}{384EI}$

Thus, there is an error of 20% in the computation of the maximum deflection of the beam assuming the deflected shape. (*why!*)

Note that the value of curvature, $y'' = \dfrac{d^2y}{dx^2} = 2A$. This means that the curvature is constant throughout the beam span for the assumed deflected shape, which is not true in the actual beam under the applied loading. In spite of this, the values of slope computed using the assumed deflected shape are nearly equal to the exact values.

The accuracy in the maximum deflection can be improved by assuming a higher-order equation or any other function satisfying the boundary conditions.

EXAMPLE 12.6 Re-analyze the beam shown in Example 12.5 using a trigonometric function.

Solution: Assume, $y = A \sin\dfrac{\pi x}{L}$

One can see that the assumed expression satisfies the boundary conditions, i.e.,

$$y\big|_{x=0} = 0 \quad \text{and} \quad y\big|_{x=L} = 0$$

$$y' = \frac{dy}{dx} = \frac{A\pi}{L}\cos\frac{\pi x}{L}$$

$$y'' = \frac{d^2y}{dx^2} = -\frac{A\pi^2}{L^2}\sin\frac{\pi x}{L}$$

$$U = \frac{1}{2}\int EI(y'')^2 dx = \frac{A^2\pi^4 EI}{4L^3}$$

$$V = -\int_0^L w.y.dx = -w\int_0^L A\sin\frac{\pi x}{L}dx = -\frac{2wAL}{\pi}$$

$$\Pi = U + V = \frac{A^2\pi^4 EI}{4L^3} - \frac{2wAL}{\pi}$$

Since, $\dfrac{\partial \Pi}{\partial A} = 0$, $\dfrac{\partial\left(\dfrac{A^2\pi^4 EI}{4L^3} - \dfrac{2wAL}{\pi}\right)}{\partial A} = 0$ $\therefore A = \dfrac{4wL^4}{EI\pi^5}$

The final equation of deflected shape can be obtained as follows:

$$y = \frac{4wL^4}{EI\pi^5}\sin\frac{\pi x}{L}$$

The equation for the slope is then given by

$$y' = \frac{dy}{dx} = \frac{4wL^3}{EI\pi^4}\cos\frac{\pi x}{L}$$

Slope at support A (i.e., $x = 0$), $\theta_A = \left.\dfrac{dy}{dx}\right|_{x=0} = \dfrac{wL^3}{24EI}$

Slope at support B (i.e., $x = L$), $\theta_B = \left.\dfrac{dy}{dx}\right|_{x=L} = -\dfrac{wL^3}{24EI}$

For the maximum deflection, $\dfrac{dy}{dx} = 0$. $\therefore\; x = \dfrac{L}{2}$

Hence, $y_{\max} = \dfrac{4wL^4}{EI\pi^5}\sin\dfrac{\pi}{2} = \dfrac{5.02\,wL^4}{384\,EI}$

Exact value of maximum deflection, $y_{\max} = \dfrac{5\,wL^4}{384\,EI}$

The computed value is nearly equal to the exact value.

EXAMPLE 12.7 Establish a closed-form equation for the deflected shape of a cantilever beam subjected to a uniformly distributed load and a concentrated bending moment as shown in Figure E12.7 using the *Rayleigh-Ritz method*. Determine the slope and deflection at the free end. Compare with the exact solution.

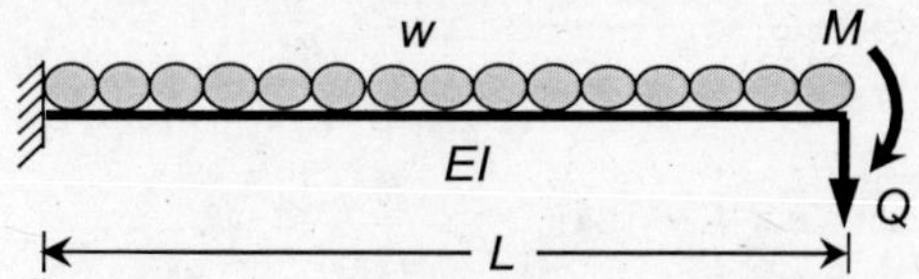

FIGURE E12.7

Solution: Assume a deflected shape $y = Ax^3 + Bx^2 + Cx + D$

$$y' = 3Ax^2 + 2Bx + C$$

The assumed curve must satisfy the boundary conditions.

$$\left.y\right|_{x=0} = 0 \;\Rightarrow D = 0$$

$$\left.y'\right|_{x=0} = 0 \;\Rightarrow C = 0$$

Thus, the deflected shape can be expressed as $y = Ax^3 + Bx^2$

$$y' = 3Ax^2 + 2Bx$$

$$y'' = 6Ax + 2B$$

$$U = \frac{1}{2}\int EI(y'')^2 dx = \frac{1}{2}\int_0^L EI(6Ax + 2B)^2\, dx$$

$\therefore$
$$U = EI(6A^2L^3 + 2B^2L + 6ABL^2)$$

$$V = -\int_0^L w.y.dx - Qy\big|_{x=L} - My'\big|_{x=L}$$

$\therefore$
$$V = -w\left(\frac{AL^4}{4} + \frac{BL^3}{3}\right) - Q(AL^3 + BL^2) - M(3AL^2 + 2BL)$$

$$\Pi = U + V$$

$$\frac{\partial \Pi}{\partial A} = 0,\ EI(12AL^3 + 6BL^2) - \frac{wL^4}{4} - QL^3 - 3ML^2 = 0$$

$$\frac{\partial \Pi}{\partial B} = 0,\ EI(4BL + 6AL^2) - \frac{wL^3}{3} - QL^2 - 2ML = 0$$

Solving for A and B, one would get

$$A = -\left(\frac{wL}{12EI} + \frac{Q}{6EI}\right)\ \text{and}\ B = \left(\frac{5wL^2}{24EI} + \frac{QL}{6EI} + \frac{M}{2EI}\right)$$

The final equation of deflected shape can be obtained as follows:

$$y = -\left(\frac{wL}{12EI} + \frac{Q}{6EI}\right)x^3 + \left(\frac{5wL^2}{24EI} + \frac{QL}{6EI} + \frac{M}{2EI}\right)x^2$$

The equation for the slope is then given by

$$y' = -3\left(\frac{wL}{12EI} + \frac{Q}{6EI}\right)x^2 + 2\left(\frac{5wL^2}{24EI} + \frac{QL}{6EI} + \frac{M}{2EI}\right)x$$

Maximum deflection at free end (i.e., $x = L$), $y\big|_{x=L} = \dfrac{wL^4}{8EI} + \dfrac{QL^3}{3EI} + \dfrac{ML^2}{2EI}$

Slope at the free end (i.e., $x = L$), $y'\big|_{x=L} = \dfrac{wL^3}{6EI} + \dfrac{QL^2}{2EI} + \dfrac{ML}{EI}$

The computed values are the same as the exact value.

12.8 PROBLEMS

12.1 Determine the force in each member of the truss shown below using the approximate method. Assume the diagonals cannot support compressive forces.

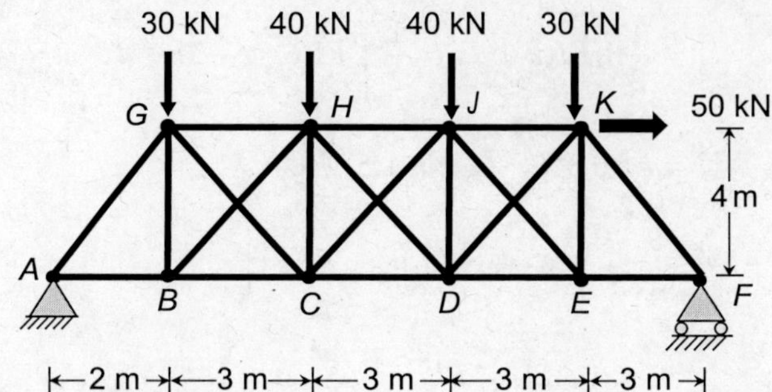

12.2 Re-analyze the truss shown in Problem 12.1 if the diagonals can support both tension and compressive forces.

12.3 Determine the force in each member of the truss shown below using the approximate method. Assume the diagonals can support both tension and compressive forces.

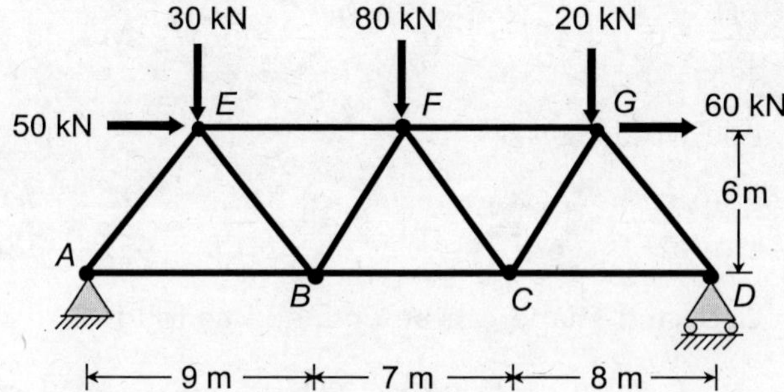

12.4 Determine the force in each member of the truss shown below using the approximate method. Assume the diagonals can support both tension and compressive forces.

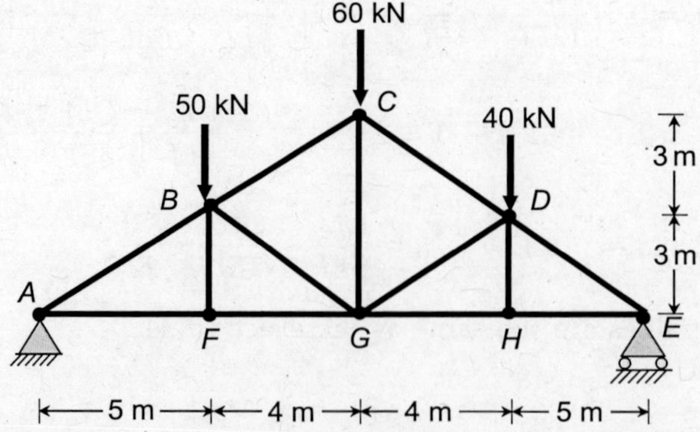

12.5 Analyze the frame shown below using the approximate method.

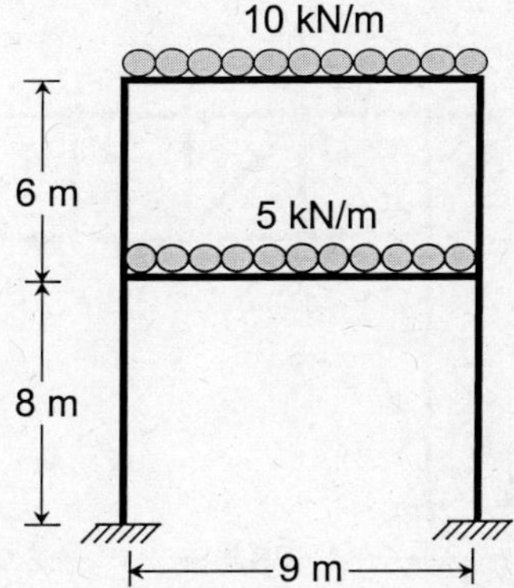

12.6 Analyze the frame shown below using the portal method.

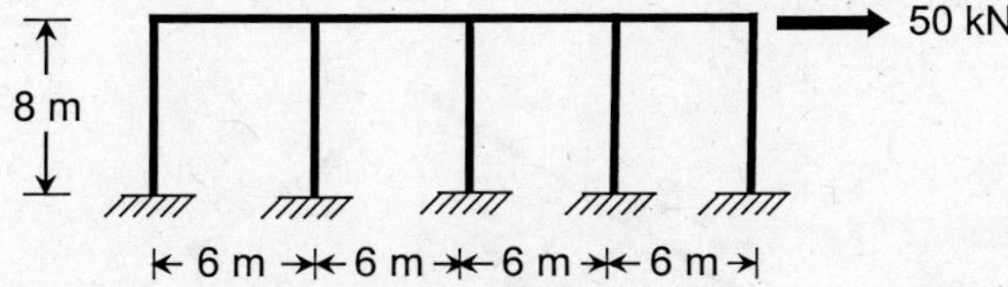

12.7 Analyze the frame shown below using the cantilever method.

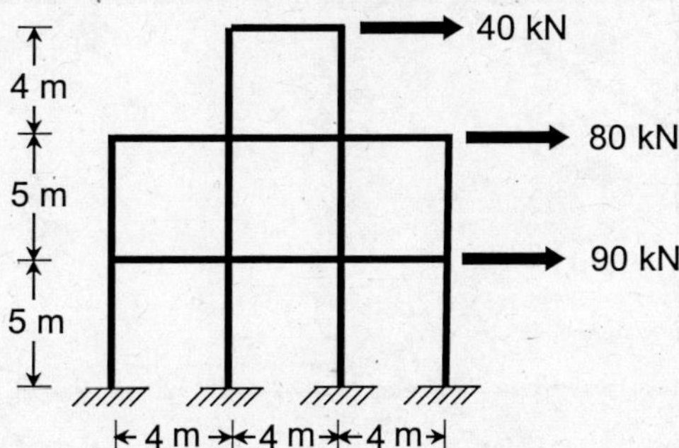

12.8 Analyze the frame shown below using the cantilever method.

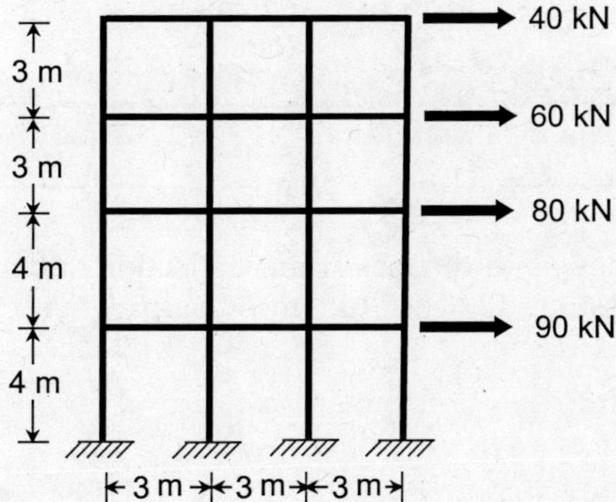

12.9   Determine the member forces in the truss of the structure shown below using the approximate method.

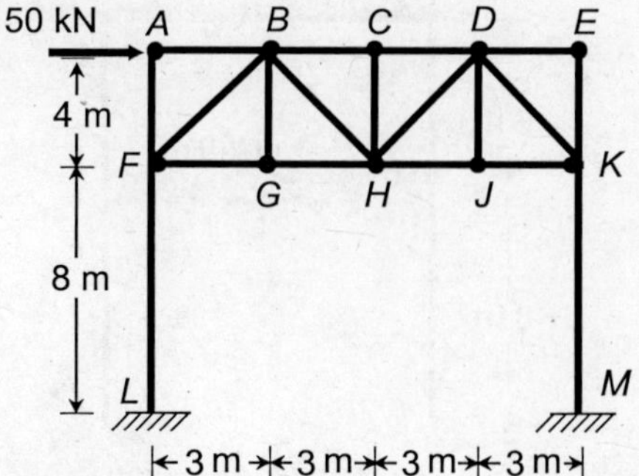

12.10   Determine the member forces in the truss of the structure shown below using the approximate method.

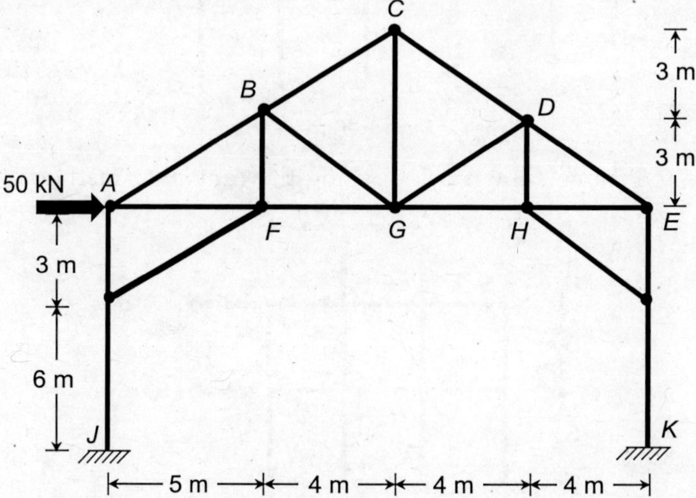

12.11   Determine the slope and the maximum deflection of the beam using the Rayleigh-Ritz method. Assume $EI = 10000 \text{ kNm}^2$.

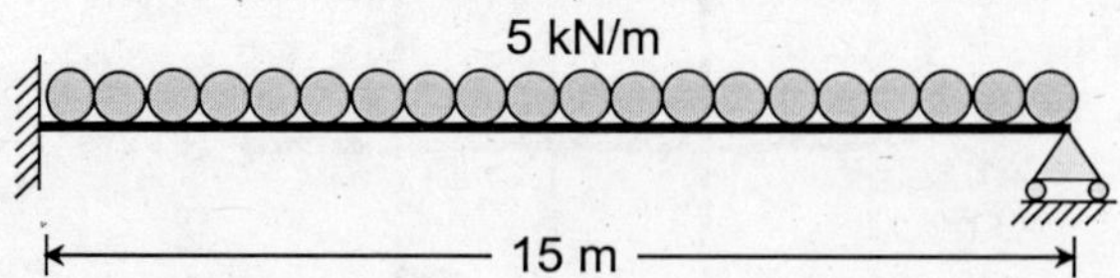

12.12   Determine the slope and the maximum deflection of the beam using the Rayleigh-Ritz method. Assume $EI = 8000 \text{ kNm}^2$. Compare with the exact solutions.

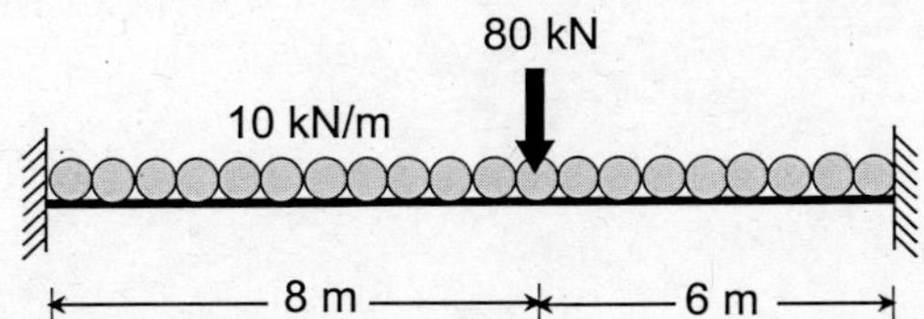

12.13 Determine the slope and the maximum deflection of the beam using the Rayleigh-Ritz method. Assume $EI = 9000$ kNm^2.

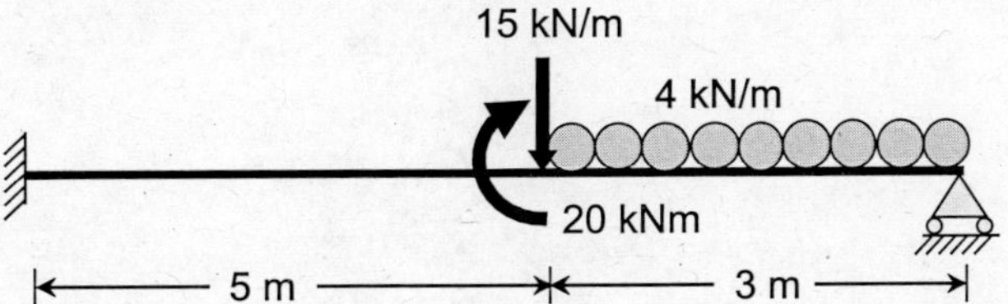

12.14 Determine the slope and the maximum deflection of the beam using the Rayleigh-Ritz method. Assume $EI = 6000$ kNm^2.

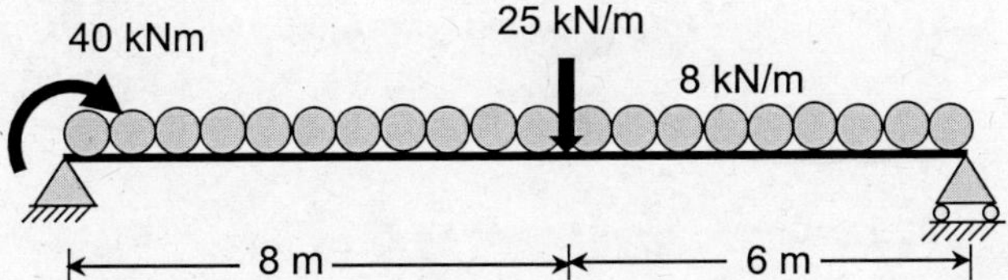

12.15 Determine the slope and the maximum deflection of the beam using the Rayleigh-Ritz method. Assume $EI = 12000$ kNm^2.

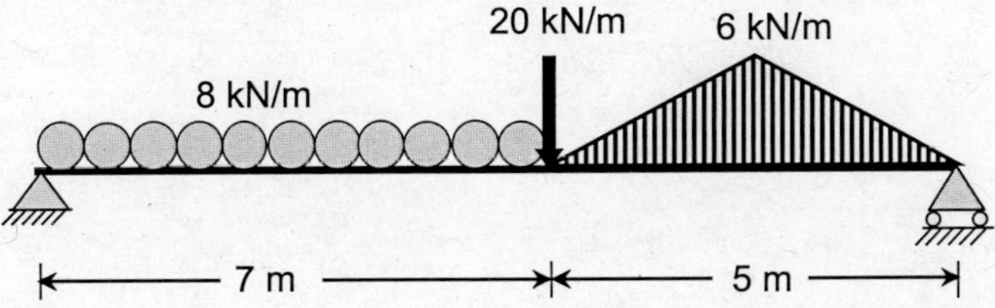

Answers to Exercise Problems

Chapter 2

(i) 1, 8; (ii) 0, 2; (iii) 1, 11; (iv) 0, 9; (v) 0, 9; (vi) 1 ,7; (vii) 1, 2; (viii) 3, 5; (ix) 0, 8; (x) 1, 9; (xi) 0, 8; (xii) 2, 6; (xiii) 4, 11; (xiv) 2, 11; (xv) 3, 7; (xvi) 4, 8; (xvii) 1, 7; (xviii) 1, 18; (xix) 1, 13; (xx) 2, 12; (xxi) 1, 10; (xxii) 1, 4; (xxiii) 0, 6; (xxiv) 0, 8; (xxv) 0, 43; (xxvi) 8, 8; (xxvii) 11, 9; (xxviii) 2, 32; (xxix) 192, 80.

Chapter 3

Please check Chapter 4.

Chapter 4

4.1 $\quad \theta_A = \dfrac{PL^2}{12EI} - \dfrac{wL^3}{48EI}; \theta_B = -\dfrac{PL^2}{6EI}; \Delta_C = \dfrac{PL^3}{8EI} - \dfrac{wL^4}{128EI}$

4.2 $\quad \theta = \dfrac{ML}{EI} - \dfrac{wL^3}{48EI}; \Delta = -\dfrac{ML^2}{2EI} + \dfrac{7wL^4}{384EI};$

4.3 $\quad \theta_A = -\left(\dfrac{3PL^2}{4EI} + \dfrac{wL^3}{24EI}\right); \Delta_A = \dfrac{19PL^3}{48EI} + \dfrac{wL^4}{48EI}; \Delta_{mid} = \left(\dfrac{PL^3}{8EI} - \dfrac{wL^4}{128EI}\right)$

4.4 $\quad \Delta_{mid} = \dfrac{5wL^4}{768EI}$

4.5 $\quad (\Delta_V)_O = 29.8$ mm

4.6 $\quad (\Delta_H)_B = 2.61$ mm

4.7 $\quad (\Delta_V)_A = 287.7$ mm; $\theta_A = 0.024$ rad; $(\Delta_V)_B = 223.4$ mm; $\theta_B = -0.061$ rad

4.8 $\quad (\Delta_H)_A = 0.30$ mm; $(\Delta_V)_A = 15.5$ mm

4.9 $(\Delta_H)_E = 3.33$ mm; $(\Delta_V)_E = 6.34$ mm

4.10 $(\Delta_H)_C = 43.78$ mm

4.11 $(\Delta_H)_A = 810$ mm; $(\Delta_V)_A = 1356$ mm

4.12 $(\Delta_H)_A = 1049.3$ mm; $(\Delta_V)_A = 0.034$ mm

Chapter 5

5.1 $R_A = \dfrac{wL}{10}; R_B = \dfrac{2wL}{5}; M_B = -\dfrac{wL^2}{15}$

5.2 $R_A = R_B = \dfrac{wL}{4}; M_A = -M_B = -\dfrac{5wL^2}{96}$

5.3 $R_{\text{Left}} = 40.6$ kN; $R_{\text{Right}} = 59.4$ kN; $M_{\text{Fixed}} = 109.4$ kNm

5.4 $R_A = 5.8$ kN; $R_B = 31.6$ kN; $R_C = 65.8$ kN; $M_B = 75$ kNm; $M_C = 200$ kNm

5.5 $R_A = 43$ kN; $R_C = 29$ kN; $H_A = 2$ kN; $M_A = 75$ kNm; $M_{BA} = -M_{BC} = 12$ kNm

5.6 $R_A = 63.8$ kN; $R_D = 36.2$ kN; $H_A = 21$ kN; $H_D = 54$ kN; $M_{BA} = -M_{BC} = 82.7$ kNm

5.7 $R_A = -11.7$ kN; $R_D = 71.7$ kN; $H_A = 20.7$ kN; $H_D = 29.3$ kN; $M_{BA} = -M_{BC} = 103.7$ kNm; $M_{CB} = -M_{CD} = -146.3$ kNm

5.8 $F_{AB} = F_{CD} = 0.98$ kN; $F_{BC} = F_{AD} = 1.46$ kN; $F_{AC} = F_{BD} = -1.76$ kN

5.9 $F_{AB} = F_{CD} = -0.8$ kN; $F_{BC} = F_{AD} = -1.07$ kN; $F_{AC} = F_{BD} = 1.33$ kN

5.10 $R_A = -23.8$ kN; $R_B = 223.8$ kN; $M_{AB} = 277.5$ kNm; $M_{BA} = -M_{BC} = 960$ kNm

5.11 $R_A = 0.53wL$; $R_B = 0.47wL$; $M_A = 0.104wL^2$; $M_B = 0.074wL^2$ ($L_1 = L/3; L_2 = 2L/3$)

5.12 $F_{AB} = 106.7$ kN; $F_{BC} = 76.3$ kN; $F_{CD} = 53.3$ kN; $F_{EF} = -43.7$ kN; $F_{EB} = -20.3$ kN; $F_{FC} = -15.2$ kN; $F_{BF} = 36.5$ kN; $F_{EC} = -27.8$ kN; $F_{AE} = -80.1$ kN; $F_{FD} = -64.1$ kN

5.13 $R_A = 74.5$ kN; $R_B = 223.3$ kN; $R_C = 74.2$ kN; $M_{AB} = 81.5$ kNm; $M_{BA} = 174.5$ kNm

5.14 $F_{AB} = F_{CE} = 35.3$ kN; $F_{BC} = 49.9$ kN; $F_{BD} = F_{CD} = -35.3$ kN; $F_{AD} = F_{DE} = -25$ kN; $M_D = 38.9$ kNm

5.15   Axial force in steel member = 50 kN; Axial force in RC member = 35.3 kN, Bending moment at joints = 0 kNm

Chapter 6

6.1 $R_A = 305.9$ kN; $R_B = 435.7$ kN; $R_C = -154.3$ kN; $M_{AB} = -1295$ kNm; $M_{BA} = -M_{BC} = 867$ kNm; $M_{CB} = -416$ kNm

6.2　$R_A = 63.4 \text{ kN}$; $M_A = M_B = \dfrac{5wL^2}{96}$; $R_C = 79.7 \text{ kN}$; $M_{AB} = -66.7 \text{ kNm}$;

　　$M_{BA} = -M_{BC} = 62.0 \text{ kNm}$; $M_{CB} = 60 \text{ kNm}$

6.3　$R_A = 1.0 \text{ kN}$; $R_B = 21.7 \text{ kN}$; $R_C = 17.3 \text{ kN}$; $\Delta = \dfrac{232}{10k - 4EI/9}$

6.4　$R_A = 68.7 \text{ kN}$; $R_B = 103.0 \text{ kN}$; $R_C = 141.1 \text{ kN}$; $M_B = 75 \text{ kNm}$;. $M_{AB} = -55.7 \text{ kNm}$;

　　$M_{BA} = -M_{BA} = 21.0 \text{ kNm}$; $M_{CB} = -M_{CD} = 54.4 \text{ kNm}$; $M_C = 3.4 \text{ kNm}$

6.5　$R_A = 166.5 \text{ kN}$; $R_C = 195.9 \text{ kN}$; $H_A = 2 \text{ kN}$; $M_A = 279.0 \text{ kNm}$; $M_C = 227.0 \text{ kNm}$;

　　$M_D = 112.5 \text{ kNm}$

6.6　$R_A = 63.8 \text{ kN}$; $R_D = 36.2 \text{ kN}$; $H_A = 21 \text{ kN}$; $H_D = 54 \text{ kN}$; $M_B = 82.7 \text{ kNm}$;

　　$M_{DC} = 169.8 \text{ kNm}$

6.7　$R_A = 18.7 \text{ kN}$; $R_D = 71.7 \text{ kN}$; $H_A = 20.7 \text{ kN}$; $H_D = 29.3 \text{ kN}$; $M_{AB} = 38.3 \text{ kNm}$;

　　$M_{BA} = 38.9 \text{ kNm}$; $M_{CD} = 39.8 \text{ kNm}$; $M_{DC} = 35.7 \text{ kNm}$

6.8　$F_{AB} = F_{CD} = 0.98 \text{ kN}$; $F_{BC} = F_{AD} = 1.46 \text{ kN}$; $F_{AC} = F_{BD} = -1.76 \text{ kN}$; $H_D = 17.2 \text{ kN}$;

　　$M_{AB} = 29 \text{ kNm}$; $M_{BA} = 27 \text{ kNm}$; $M_{CB} = 21.1 \text{ kNm}$; $M_{DC} = 16.4 \text{ kNm}$

6.9　$\Delta = 89 \text{ mm}$; $M_{AB} = -119.2 \text{ kNm}$; $M_{CB} = 272.6 \text{ kNm}$

6.10　$R_A = 67.9 \text{ kN}$; $H_C = 30 \text{ kN}$; $R_C = 52.1 \text{ kN}$; $M_{AB} = 0 \text{ kNm}$;

　　$M_{BA} = -M_{BC} = -47.5 \text{ kNm}$; $M_{CB} = 102.5 \text{ kNm}$

6.11　$R_A = 0.53 \, wL$; $R_B = 0.47 \, wL$; $M_A = 0.104 \, wL^2$; $M_B = 0.074 \, wL^2$; $L_1 = L/3 ; L_2 = 2L/3$;

　　$M_{CB} = 3.3 \text{ kNm}$; $M_{DC} = 34.5 \text{ kNm}$

6.12　$F_{AB} = 106.7 \text{ kN}$; $F_{BC} = 76.3 \text{ kN}$; $F_{CD} = 53.3 \text{ kN}$; $F_{EF} = -43.7 \text{ kN}$; $F_{EB} = -20.3 \text{ kN}$

6.13　$R_A = 74.5 \text{ kN}$; $R_B = 223.3 \text{ kN}$; $R_C = 74.2 \text{ kN}$; $M_A = 81.5 \text{ kNm}$; $M_B = 174.5 \text{ kNm}$;

　　$M_{BA} = 11.3 \text{ kNm}$; $M_{DE} = 7.6 \text{ kNm}$

6.14　$R_A = -3.1 \text{ kN}$; $H_A = 13.9 \text{ kN}$; $R_B = 40.9 \text{ kN}$; $H_B = 13.9 \text{ kN}$; $M_{AC} = 31 \text{ kNm}$;

　　$M_{CA} = 62 \text{ kNm}$; $M_{BC} = 46 \text{ kNm}$; $M_{CB} = 90 \text{ kNm}$

6.15　$R_A = R_D = \dfrac{11wL}{28} ; R_B = R_C = \dfrac{31wL}{28} ; M_{BA} = M_{CD} = \dfrac{3wL^2}{28}$

Chapter 7

7.1　$R_A = 132.8 \text{ kN}$; $R_B = 143.1 \text{ kN}$; $R_C = 28.5 \text{ kN}$; $R_D = 20.9 \text{ kN}$; $M_{AB} = -233.6 \text{ kNm}$;

　　$M_{BA} = -M_{BC} = 132.9 \text{ kNm}$; $M_{CB} = -M_{CD} = 14.9 \text{ kNm}$

7.2　$R_A = 68.25 \text{ kN}$; $R_B = -39 \text{ kN}$; $R_C = 120.75 \text{ kN}$; $M_{AB} = 0 \text{ kNm}$; $M_{BA} = -33 \text{ kNm}$;

　　$M_{BC} = 43 \text{ kNm}$; $M_{CB} = -M_{CD} = 200 \text{ kNm}$

7.3 $R_A = 20.2$ kN ; $R_B = -1.2$ kN ; $R_C = 23.4$ kN ; $R_D = 22.6$ kN ; $M_{AB} = -27.9$ kNm ;
$M_{BA} = -M_{BC} = -10.7$ kNm ; $M_{CB} = -M_{CD} = 19.5$ kNm ; $M_{DC} = 35.2$ kNm

7.4 $R_A = 74.5$ kN ; $R_B = 223.3$ kN ; $R_C = 74.2$ kN ; $M_{AB} = -81.5$ kNm ;
$M_{BA} = -M_{BC} = 174.5$ kNm ; $M_{CB} = 0$ kNm

7.5 $R_A = R_D = 5.5$ kN ; $R_B = R_C = 6.5$ kN ; $M_{AB} = 0$ kNm ; $M_{BA} = -M_{BC} = 1.86$ kNm ;
$M_{CB} = -M_{CD} = -1.86$ kNm ; $M_{DC} = 0$ kNm

7.6 $R_A = 32.8$ kN ; $R_B = 3.4$ kN ; $R_C = 23.8$ kN ; $M_{AB} = -30.6$ kNm

7.7 $R_A = 22.2$ kN; $H_A = 18.8$ kN; $R_C = 17.8$ kN; $H_C = 1.2$ kN; $M_{AB} = -20.2$ kNm;
$M_{BA} = -M_{BC} = 9.6$ kNm ; $M_{CB} = 0.82$ kNm

7.8 $M_{AB} = M_{DE} = -15.6$ kNm ; $M_{BA} = M_{ED} = 12.2$ kNm ; $M_{BE} = M_{AB} = 22.5$ kNm ;
$M_{CB} = M_{CF} = -18.8$ kNm ; $M_{FC} = M_{FE} = 18.8$ kNm ; $R_A = -23.8$ kN ;
$H_A = 12.5$ kN ; $R_D = 23.8$ kN ; $H_D = 12.5$ kN

7.9 $M_{AB} = -M_{BA} = 2.7$ kNm ; $M_{BC} = M_{CB} = -M_{CF} = -6.85$ kNm ; $M_{DE} = 7.42$ kNm ;
$M_{ED} = 7.72$ kNm ; $M_{EF} = 7.32$ kNm ; $M_{EB} = 15.0$ kNm ; $M_{FC} = -M_{FE} = 6.4$ kNm ;
$R_A = 33.7$ kN ; $H_A = 0$ kN ; $R_D = 26.3$ kN ; $H_D = 5$ kN

7.10 $M_{AB} = 0$ kNm ; $M_{BA} = 75.7$ kNm ; $M_{BC} = 3.7$ kNm ; $M_{BE} = 79.4$ kNm ;
$M_{CD} = -M_{CB} = -37.2$ kNm ; $M_{FC} = -M_{FE} = 43.7$ kNm ; $M_{DE} = 0$ kNm ;
$M_{ED} = -74.3$ kNm ; $M_{EF} = 5.4$ kNm ; $M_{EB} = 79.7$ kNm ; $R_A = -54.0$ kN ;
$H_A = 25.2$ kN ; $R_D = 66.0$ kN ; $H_D = 24.8$ kN

7.11 $M_{AB} = M_{ED} = 0$ kNm ; $M_{BA} = -M_{BC} = 25.5$ kNm ; $M_{CB} = -M_{CD} = -3.3$ kNm ;
$M_{DC} = -M_{DE} = -34.5$ kNm ; $R_A = 6.4$ kN ; $H_A = 8.5$ kN ; $R_E = 21.4$ kN ;
$H_E = 11.5$ kN

7.12 $M_{AB} = -6.1$ kNm ; $M_{BA} = -11.4$ kNm ; $M_{BC} = -5.6$ kNm ; $M_{BD} = 17$ kNm ;
$M_{CB} = -2.5$ kNm ; $M_{DB} = -M_{DE} = 9.2$ kNm ; $M_{ED} = 3.5$ kNm ; $R_A = 10.4$ kN ;
$H_A = 5.8$ kN ; $R_E = 8.4$ kN ; $H_E = -4.2$ kN ; $R_C = 6.2$ kN ; $H_C = -1.6$ kN (At top)

7.13 $M_{AB} = M_{DC} = 0$ kNm ; $M_{BA} = -M_{BC} = 19.2$ kNm ; $M_{CD} = -M_{CB} = -5.6$ kNm

7.14 $M_{AB} = 7$ kNm; $M_{BA} = -M_{BC} = 31.8$ kNm ; $M_{CD} = -M_{CB} = -10.6$ kNm ; $M_{DC} = 0$ kNm

7.15 $M_{AB} = M_{DB} = M_{EC} = 0$ kNm ; $M_{BA} = -33.9$ kNm ; $M_{BD} = 10.4$ kNm ;
$M_{CB} = -M_{CE} = 36.8$ kNm ; $M_{BC} = 23.5$ kNm

Chapter 8

8.1 $\theta_{Left} = 192.6/EI$; $\theta_{Right} = -3180/EI$; $\Delta_{40} = 1657.3/EI$; $\Delta_{60} = 1478.2/EI$

8.2 Fixed end, $R_A = \dfrac{\dfrac{wL^4}{8EI}}{f_s + \dfrac{L^3}{3EI}} - \dfrac{3wL}{2}$; $M_A = \dfrac{\dfrac{wL^5}{8EI}}{f_s + \dfrac{L^3}{3EI}} - \dfrac{9wL^2}{8}$

8.3 $R_{Left} = 13.75$ kN ; $R_{Right} = 21.25$ kN ; $\Delta = 134.4/EI$; $\theta_{Left} = 62.7/EI$; $\theta_{Right} = -71.7/EI$

8.4 $R_A = 74.6$ kN ; $R_B = 223.2$ kN ; $R_C = 74.2$ kN ; $M_A = -82$ kNm ; $M_B = -174.1$ kNm

8.5 $R_A = 32.3$ kN ; $R_B = 37.7$ kN ; $M_A = -28.7$ kNm ; $M_B = -17.4$ kNm ; $\theta_B = -8.96/EI$;
$\Delta = 8.23/EI$

8.6 $H_A = 66.7$ kN ; $H_B = -6.7$ kN ; $V_A = 50$ kN ; $V_B = 20$ kN ; $\Delta_{Hor} = 0.03$ m $(\rightarrow)$;
$\Delta_{Ver} = 0.17$ m $(\downarrow)$; $F_{BC} = -33.3$ kN ; $F_{CD} = 46$ kN ; $F_{DE} = 20$ kN ; $\Delta_{Eh} = 0.03$ m ;
$\Delta_{Ev} = 0.17$ m ; $\Delta_{Dh} = 0.015$ m ; $\Delta_{Dv} = 0.114$ m

8.7 $H_{Top} = 66.7$ kN $(\leftarrow)$; $H_{Bot} = 6.7$ kN $(\rightarrow)$; $V_{Top} = 50$ kN $(\uparrow)$; $V_{Bot} = 20$ kN $(\uparrow)$;
$\Delta_{Hor} = 0.03$ m $(\rightarrow)$; $\Delta_{Ver} = 0.17$ m $(\downarrow)$; $F_{BD} = -14.6$ kN ; $F_{DC} = -6.7$ kN

8.8 $R_A = 2.7$ kN ; $R_C = -2.7$ kN ; $H_A = -H_C = -1.3$ kN ; $M_{BA} = 13.3$ kNm ;
$M_{BC} = 6.7$ kNm

8.9 $R_A = 38.1$ kN ; $R_D = 81.9$ kN ; $H_A = 35.9$ kN ; $H_D = 74.1$ kN ; $M_{AB} = -198.8$ kNm ;
$M_{BA} = -M_{BC} = -182.1$ kNm ; $M_{CD} = -M_{CB} = -7.7$ kNm ; $M_{DC} = -359.5$ kNm

8.10 $R_A = 26.9$ kN ; $R_B = 23.1$ kN ; $\Delta_{Bh} = 0.02$ m ; $\Delta_{Bv} = -0.16$ m ; $\Delta_{Ch} = -0.015$ m ;
$\Delta_{Cv} = -0.143$ m

8.11 $R_A = 7.2$ kN ; $R_D = 22.8$ kN ; $H_A = -20$ kN ; $M_{AB} = 63.2$ kNm ;
$M_{BA} = -M_{BC} = 36.8$ kNm ; $M_{CB} = 10$ kNm

8.12 $F_1 = 29.3$ kN ; $F_2 = 28.2$ kN ; $F_3 = 6.2$ kN ; $F_4 = -6.7$ kN ; $\Delta_H = 0.044$ m ;
$\Delta_V = 0.023$ m

8.13 $F_{AB} = -53.9$ kN ; $F_{BC} = -34.6$ kN ; $F_{CD} = -62.8$ kN ; $F_{AE} = 63.1$ kN ; $F_{EF} = 58.7$ kN ;
$F_{FD} = 50.3$ kN ; $F_{BE} = -2.6$ kN ; $F_{BF} = -9.9$ kN ; $F_{CE} = 5.1$ kN

8.14 $F_{AB} = 45.6$ kN ; $F_{BC} = 15.9$ kN ; $F_{CD} = -8.8$ kN ; $F_{BE} = -18$ kN ; $F_{AF} = 30.8$ kN ;
$F_{AE} = 49$ kN ; $F_{BF} = 38.5$ kN ; $F_{BD} = 11$ kN ; $F_{CE} = -26.5$ kN ; $F_{DE} = -6.6$ kN ;
$F_{EF} = -51.9$ kN

8.15 $F_{AB} = 83.9$ kN ; $F_{BC} = 6.4$ kN ; $F_{CD} = -21.9$ kN ; $F_{BE} = -13.4$ kN ; $F_{AE} = 89.4$ kN ;
$F_{BF} = -123.1$ kN ; $F_{BD} = 27.4$ kN ; $F_{CE} = -72.6$ kN ; $F_{DE} = -56.4$ kN ;
$F_{EF} = -153.6$ kN

Chapter 9

9.1 $\theta_A = -\theta_B = 45/EI$; $\Delta_{mid} = 90/EI$

9.2 $R_A = 10.8 \text{ kN}$; $R_B = 22.2 \text{ kN}$; $R_C = 7.0 \text{ kN}$; $M_A = -16.6 \text{ kNm}$; $\theta_B = -2.4/EI$;
$\theta_C = -5.6/EI$

9.3 $R_A = 16.2 \text{ kN}$; $R_B = 39 \text{ kN}$; $R_C = 43 \text{ kN}$; $R_D = 21.8 \text{ kN}$; $M_{AB} = -33.2 \text{ kNm}$;
$M_{DC} = 33.6 \text{ kNm}$; $\theta_B = -6.3/EI$; $\theta_C = 5.5/EI$

9.4 $F_{AB} = 10.7 \text{ kN}$; $F_{BC} = 0 \text{ kN}$; $F_{BD} = 44.8 \text{ kN}$; $F_{AC} = F_{CD} = -31.7 \text{ kN}$; $R_A = 8.3 \text{ kN}$;
$R_D = 31.7 \text{ kN}$; $H_A = -25 \text{ kN}$; $\Delta_{Bh} = 0.73 \text{ mm}$; $\Delta_{Bv} = 0.73 \text{ mm}$

9.5 $F_{AB} = 0 \text{ kN}$; $F_{BC} = 70.7 \text{ kN}$; $F_{AC} = -28.3 \text{ kN}$; $F_{CE} = 42.4 \text{ kN}$; $F_{CD} = 40 \text{ kN}$;
$F_{AD} = 30 \text{ kN}$; $F_{DE} = 30 \text{ kN}$; $\Delta_{Eh} = 0.48 \text{ mm}$; $\Delta_{Ev} = -0.65 \text{ mm}$

9.6 $R_A = 20.4 \text{ kN}$; $R_B = 19.3 \text{ kN}$; $R_C = 1.3 \text{ kN}$; $M_{AB} = -22.8 \text{ kNm}$;
$M_{BC} = -M_{BA} = -8.5 \text{ kNm}$; $\theta_B = -7.14/EI$; $\theta_C = -0.93/EI$

9.7 $k = \dfrac{M}{\theta} = \dfrac{3EIl_1^2}{l_1^3 + l_2^3}$

9.8 $F_1 = 14.9 \text{ kN}$; $F_2 = 17.8 \text{ kN}$; $F_3 = 13 \text{ kN}$; $F_4 = 2.75 \text{ kN}$; $\Delta_H = 23 \text{ mm}$; $\Delta_V = 15.6 \text{ mm}$

9.9 $H_A = H_D = -75 \text{ kN}$; $R_A = -R_D = -126.6 \text{ kN}$; $M_{AB} = M_{DE} = -212 \text{ kNm}$;
$\theta_B = \theta_E = 122/EI$; $\theta_C = \theta_F = 46.7/EI$; $\Delta_B = 1086.3/EI$; $\Delta_C = 1406.9/EI$

9.10 $\Delta = 253.2/EI$; $\theta = 84.6/EI$

Chapter 10

10.1 $\theta_C = 0.007 \text{ rad}$; $\Delta_{mid} = 14.8 \text{ mm}$; $R_A = 34.15 \text{ kN}$; $R_C = 15.85 \text{ kN}$; $M_{AB} = -54.9 \text{ kNm}$

10.2 $\theta_A = -\theta_C = 0.037 \text{ rad}$; $\Delta_{mid} = 96 \text{ mm}$; $R_A = R_C = 30 \text{ kN}$

10.3 $M_{AB} = -67.9 \text{ kNm}$; $M_{BA} = -M_{BC} = 44.2 \text{ kNm}$; $M_{CB} = -M_{CD} = 90 \text{ kNm}$;
$R_B = 86.9 \text{ kN}$; $R_C = 99.1 \text{ kN}$

10.4 $F_{AB} = F_{CD} = -7.5 \text{ kN}$; $F_{AD} = 84.4 \text{ kN}$; $F_{BC} = -5.6 \text{ kN}$; $F_{AC} = F_{BD} = 9.4 \text{ kN}$;
$\Delta_{Dh} = 0.011 \text{ m}$; $\Delta_{Dv} = -0.032 \text{ m}$

10.5 $F_{AB} = 141.4 \text{ kN}$; $F_{BC} = -70.7 \text{ kN}$; $F_{AC} = 23.6 \text{ kN}$; $\Delta_{Ch} = 4.07 \text{ mm}$; $\Delta_{Cv} = -2.83 \text{ mm}$

10.6 $F_{AB} = 0 \text{ kN}$; $F_{BC} = 70.7 \text{ kN}$; $F_{AC} = -28.3 \text{ kN}$; $F_{CE} = 42.4 \text{ kN}$; $F_{CD} = -40 \text{ kN}$;
$F_{AD} = 30 \text{ kN}$; $F_{DE} = 30 \text{ kN}$; $\Delta_{Eh} = 0.48 \text{ mm}$; $\Delta_{Ev} = -0.65 \text{ mm}$; $\Delta_{Dh} = 0.24 \text{ mm}$;
$\Delta_{Bv} = -0.95 \text{ mm}$; $\Delta_{Ch} = 0.16 \text{ mm}$; $\Delta_{Cv} = -0.55 \text{ mm}$

10.7 $M_{AB} = 69.0$ kNm ; $M_{BA} = -M_{BC} = 56.3$ kNm ; $M_{CB} = -M_{CD} = -42.6$ kNm ;
$M_{DC} = 42.0$ kNm

10.8 $M_{AB} = -13.1$ kNm ; $M_{BA} = -M_{BC} = -16.4$ kNm ; $M_{CB} = -M_{CD} = 21.5$ kNm ;
$M_{DC} = -M_{DE} = -16.4$ kNm ; $M_{ED} = 13.1$ kNm ; $H_A = -H_E = -7.4$ kN

10.9 $M_{AB} = -15.2$ kNm ; $M_{BA} = -M_{BC} = -14.3$ kNm ; $M_{CB} = -M_{CD} = -36.6$ kNm ;
$M_{DC} = -M_{DE} = -25.3$ kNm ; $M_{CF} = M_{FC} = 0$ kNm ; $M_{ED} = 4.1$ kNm ;
$H_A = -H_E = -5.0$ kN ; $R_A = 21.1$ kN ; $R_F = 80.7$ kN ; $R_E = 48.2$ kN

10.10 $M_{AB} = 70.8$ kNm ; $M_{CB} = -M_{CD} = 6.8$ kNm ; $M_{DC} = -M_{DE} = -90.7$ kNm ;
$M_{EB} = -180.5$ kNm ; $M_{FE} = 111.7$ kNm ; $H_A = -14.3$ kN ; $H_F = -35.7$ kN ;
$R_A = 102.8$ kN ; $R_F = 177.1$ kN

10.11 $M_{AB} = -M_{BA} = 160$ kNm ; $M_{BC} = 160$ kNm ; $R_A = 60$ kN ; $R_D = 20$ kN

Index